The ARRL UHF/Microwave Projects Manual

Published By The American Radio Relay League
225 Main Street
Newington, CT 06111 USA

Printed in USA

Library of Congress Catalog Card Number:

ISBN: 0-87259-449-1

Second Printing, 1996

Acknowledgments

This book would not have been possible without the hard work and dedication of a number of people. First, thanks to the many authors who shared their work.

Jim Kearman, KR1S, Rus Healy, NJ2L; Zack Lau, KH6CP; Bob Schetgen, KU7G; and Joel Kleinman, N1BKE, assisted with editing. Special words of thanks are also due the following: Charles L. Hutchinson, K8CH; Dennis Lusis, W1LJ; Mark Wilson, AA2Z; and Kirk Kleinschmidt, NTØZ.

Shelly Bloom, WB1ENT; Deborah Strzeszkowski; Dave Pingree, N1NAS; Sue Fagan; Jodi Morin, KA1JPA; and Dan Wolfgang did yeoman service in designing, typesetting, illustrating and laying out this book. Sue Fagan also designed the cover.

Photos: Meyers Studio, Kirk Kleinschmidt, NTØZ, and Brad Thomas, KC1EX.

Foreword

Radio amateurs have been experimenting with UHF and microwave frequencies almost as long as the amateur service has existed. That enough projects exist to fill a book (in fact, several books) of this size is testimony to the frontier spirit that still exists in Amateur Radio, leading adventurous experimenters to explore the upper end of the radio spectrum.

In preparing this book of projects, the editors tried to select designs that would remain usable for years to come. Some of the designs are already seasoned, having appeared in conference proceedings and *QST* several years ago. UHF/microwave experimenters are always seeking a little better performance. The fact that no one has improved on these designs in the ensuing years is proof enough of their ability to get the job done.

At present, it may appear that the state-of-the-art transverter design is stagnant, but you don't have to look back very many years to see how far we've come. Anyone who, not that many years ago, fiddled with frequency multiplier chains to generate a few milliwatts at 2304 MHz, or listened to the roaring IF stage of a Gunn-diode transceiver, will agree. It may be that we have all the technology we need, at least for the bands through 10 GHz. As HF technology stabilized some 60 years ago, the ranks swelled with more and more operators, and the experimenters moved to higher frequencies. As access to easily constructed and operated microwave equipment fills the bands with more operators, the authors represented here will surely look for new frontiers. Perhaps the next collection of microwave projects produced by the ARRL will *start* where this one leaves off: at 24 GHz.

David Sumner, K1ZZ
Executive Vice President
Newington, Connecticut
December 1994

Contents

1

Introduction

Contents

Do We Really Want Our Microwave Bands?

By David Sumner, K1ZZ

(Adapted from *QST*, March 1996)

Do we really want our microwave bands?!

What a dumb question. What a no-brainer. Of course we do! Why even ask the question?

Why? Because, frankly, sometimes we seem to act as if we don't.

Consider this. Microwave equipment has been available to amateurs ever since the first World War II surplus gear hit the market. The basic framework of our microwave allocations has been in place for about 50 years. During that time, technology has advanced to the point where many consumer-level products use microwave components.

Home reception of satellite television has been fairly common for about 15 years. This means that sensitive 4-GHz receivers have been in people's hands for that long. "Wireless cable" has been operating in the vicinity of 2 GHz for about the same length of time. In the past two years the 12-GHz satellite receiver has become the hottest-ever consumer electronics product—a complete digital video and audio satellite system retails for less than $600 and most of the cost is in the digital processing, not the RF section. Wireless LANs operating at 2.4 GHz are springing up in stores and offices everywhere. Some radar detectors (remember, that's 10.5 and 24-GHz technology) rival in sophistication what you'd have found in military aircraft not so long ago. That's not to mention the proliferation of more mundane products such as microwave ovens, intrusion alarms, and the like.

Amateurs have been doing some interesting and fun things with microwaves. On page 99 of March 1996 *QST* you can read about one of the brighter spots on our radio landscape, the 10-GHz band, where pockets of critical masses of enthusiasts have sprung up in most corners of the country. We're bridging distances just for fun that the professionals wouldn't think of trying for profit. But let's be blunt: Our application of microwave technology to Amateur Radio has not kept pace with the possibilities. And that's a problem for us, because we face enormous pressure for commercial development of our microwave bands. Defending our access to these valuable bands won't be easy, and might well be impossible if we had to do it solely on the basis of existing amateur activity.

Some readers will protest, "A lot of hams are involved in all of those commercial microwave ventures that you've mentioned, and many more besides." That's true enough. Defense contractors, the telephone companies, broadcasters, and other employers with a heavy investment in microwave technology have depended on Amateur Radio's brain trust ever since the dawn of the technology itself. It's also true that the rest of us can't expect that very many of the microwave professionals in our midst will take their work home with them, and make it their hobby. But if we're not using our allocations *as amateurs*, it's a little difficult to explain to policy-makers at the national or international level why they ought to continue to preserve those allocations in their present form, free of the restrictions that would be required to accommodate new services.

There is no accurate census of the hams who are active on the microwave bands, but any reasonable estimate of the percentage would have to be in the single digits. Every amateur licensee, including Novices, can use the 1270 to 1295-MHz band. Every Technician, meaning practically everyone who has entered Amateur Radio in the past five years, has full privileges in *all* of the microwave bands—yet, few newcomers (or old timers, for that matter) are using them. Techs don't need to wait until they are proficient in Morse code, or for the requirement to disappear, to enjoy worldwide communication; they can do it now, by satellite. So can anyone else whose HF capabilities are limited by neighborhood or family considerations. Why aren't more of us doing it? Why, with such an exciting frontier waiting to be exploited, do we persist in doing the same old thing, year after year? Why is it that we'd rather complain about crowding in one part of the spectrum than to make the effort to use a different, underutilized part?

It isn't the expense; a Mode S satellite station can be put together for less than the cost of a full-featured HF station, but relatively few of us have done so. It isn't the absence of information; microwave construction is one of the best-documented facets of Amateur Radio. It isn't the complexity; modern microwave circuits actually are quite simple.

What it is, we suspect, is that not enough of us appreciate the rewards that come from doing something a little different. There's the joy of building something that works better than anything you can buy. There's the satisfaction of solving problems with ingenuity instead of dollars (some microwavers' uses of common kitchen utensils come to mind). There's the gratification of being recognized as an expert, which is a lot easier to come by in an uncrowded field. There's the thrill of doing something for the first time, ever. There's the camaraderie, which exists in great heaps in the microwave community.

Enough words. Actions speak louder. Make this the year you take the road less traveled, and explore the microwaves. And tell us what you find!

Getting Started on the Microwave Bands

By Rick Campbell, KK7B

(from *QST*, February 1992)

Several pockets around the country contain much of the US amateur microwave activity. These pockets come into being when one amateur in a geographic location builds a station for the "next higher band," and then serves as a mentor to a small group of interested enthusiasts—often contesters who want to improve their scores. Recent advances in microwave technology have made access to the microwave bands easier than ever before. Building a complete microwave station in the 1990s is about as challenging as assembling a mobile rig in the 1950s—and just as much fun!

I intend this tutorial to help you get your feet wet in engineering, building and operating Amateur Radio stations on the microwave bands, with a focus on the 5.7-GHz band (which I'll sometimes refer to as "5760"). Although the focus of this article is weak-signal contest and "grid-expedition" style communication, the information provided here is useful to anyone interested in getting on any of the microwave bands. After a little introductory theory, I'll describe a basic 5760 station that you can use as-is for line-of-sight hilltopping, or that you can enhance in stages all the way to moonbounce (also known as earth-moon-earth, or EME) capability.

Introduction

Commercial equipment is now available for narrowband, weak-signal work on all amateur bands from 1.8 MHz through 10 GHz. Until recently, the only gap in that coverage was the 5760-MHz band, a void that Down East Microwave filled with a no tune 5760-MHz transverter.[1] Amateurs who have operated on 2.3, 3.4 or 10 GHz will find no surprises at 5.7 GHz; those who haven't done so will find 5760 to be an interesting and surprisingly far-reaching band.

It's easy to find other hams interested in getting on 5760—their calls are published in the VHF and UHF contest scores in *QST*. Look for a nearby call with lots of letters after it, but no *h*.

A few of the developments responsible for the dramatic increase in amateur microwave activity are:

• Unconditionally stable MMIC amplifiers are now available at far less than the cost of the surplus amplifiers used by amateurs a few years ago. Destroying a component is now a three-dollar mistake instead of a total catastrophe!

• Printed-circuit no-tune bandpass filters, widely used in industry, have been designed for the amateur bands. These filters include the manufacturing and circuit-board tolerances in the design, and use many low-Q resonators in a flat-response bandpass filter, instead of a few high-Q, critically tuned resonators. It's now possible to obtain better performance with *no tuning at all* than was previously available with multiple interacting microwave tuning adjustments made by carefully tweaking a device while watching its output with a spectrum analyzer!

• You can build a complete microwave transverter system, using all new components, at a cost competitive with radio equipment for the lower frequencies. This makes the use of surplus components optional and removes the greatest barrier to amateur microwave operation: the lack of reasonable priced equipment.

• A pair of basic transverters can be used to test receive and transmit amplifiers, filters and antennas, *with essentially no other test equipment.*

Propagation

Amateurs commonly use four propagation modes at 5.7 GHz: line-of-sight, obstructed paths, over-the-horizon tropospheric, and EME. Line-of-sight paths provide the strongest signals, and an excellent way for microwave experimenters to get their feet wet. The other modes are discussed in more detail in my article in the *Proceedings of the 1991 Central States VHF Conference.*[2]

Free-Space Path Loss

The Friis path-loss formula, often quoted but seldom understood by communications-systems engineers (amateur or professional), states that: *The loss between a pair of isotropic antennas increases as the square of the distance and the square of the frequency.* The statement, "Path loss increases with frequency," taken out of the context of the Friis path-loss formula, has been repeated so many times in the literature that many amateurs and professionals believe it is a physical law. *It's not!*

Friis actually worked with *three* "path loss" formulas, the most basic of which states that the loss between an isotropic transmit antenna and a one-square-meter receiving antenna goes up as the square of the distance—independent of the frequency. This is called "geometric spreading," and it *is* a physical law.[3] The second formula is for the loss between a pair of isotropic antennas, and includes a correction factor for the effective size of an "isotropic receiver." The third formula states that the loss between two antennas of a given size goes up as the square of the distance, *and down as the square of the frequency.*[4,5,6]

Because antenna gain can be converted to effective area and vice versa, any of the three formulas can be used to obtain the correct path loss for any transmit and receive antennas. The only difference between the three formulas is in the correction factors needed for the different antenna combinations.

Most of us would rather remember a few simple rules of thumb than a bunch of formulas in a textbook. Let's convert the various "path loss" formulas into English:

1) For systems with antennas like dipoles and Yagis, where we specify antenna *gain*, the "path loss" goes up as the frequency goes up. A dipole is a fine receive antenna for 80 meters, but a 5760 dipole isn't very useful—it doesn't intercept much signal, and it's too small for a tie clip!

2) For systems with dishes and lenses, where we specify antenna *size*, "path loss" goes *down* as frequency goes up. If you use the same dish for 2304, 3456 and 5760, *and run the same power and noise figure on all three bands*, signals will be 8 dB stronger on 5760 than 2304.

Reference Data for Radio Engineers concludes its section on free-space path loss with this statement: "As frequency is increased, the transmit power or the antenna sizes can be reduced; it is clearly better to use the highest frequency for which generators and receivers are available."[7]

Remember: *Loss on a free-space path is independent of frequency*; the antenna gain-size correction factors are frequency dependent.

On line-of-sight paths with no ground reflections, path loss increases as 20 times the log of the path distance. In very useful terms, that works out to 6 dB every time you double the distance, or 20 dB if you multiply the distance by 10.

Let's put all this into practice. Suppose you set up two 5760-MHz systems 100 meters apart. Using a calibrated variable attenuator in the transmit feed line, you determine that the signal received on the 100-meter path is 46 dB above the noise. How far apart can these two stations be and still have a signal above the noise level? First, double the distance to 200 meters—this reduces the signal by 6 dB, to 40 dB above the noise. Next, multiply the distance by 10, to 2 km—this reduces the signal by 20 dB, to 20 dB above the noise. Multiplying the distance by 10 again, to 20 km, reduces the signal another 20 dB, down to the noise level.

Setting up portable (rover) systems on a known path and measuring the signal-to-noise ratio is an excellent way to both determine the maximum range of the system on line-of-sight paths and to verify system performance.

Obstructed Paths, Trees and Tropospheric Scatter

When you're attempting to work over non-line-of-sight paths, many other variables become important. A few of these are frequency dependent, such as antenna height above ground. Most loss mechanisms are not very frequency dependent. In my experience, all of the microwave bands have very similar propagation on obstructed paths, *after all the bugs are worked out at both ends of the system.*

Over much of the country, even the best microwave paths have a few trees. How much attenuation do a few small trees have? If they are close to one end of an otherwise line-of-sight path, it's about 10 dB. If a few small trees occlude each end of an otherwise line-of-sight path, the tree attenuation will be about 20 dB. In other words, if the microwave system is capable of operating over a 100 mile line-of-sight path with no trees, it should be able to work over a 10-mile path with a few trees at each end.[8,9]

What about dense or big trees? Once the direct signal is attenuated by 20 dB or so, the signals bounced (or "scattered") from the treetops will be stronger than the signals coming directly through the trees. These scattered signals can be used to communicate when line-of-sight signals are attenuated by more than 20 dB. If one end of the path is in the clear and the other is inside a dense forest, expect 20 or 30 dB of additional path loss. If both ends of the path are forested, but with a clear line-of-sight path between the treetops near the receiver and transmitter, 40 to 60 dB of additional path loss may occur. The path from the forest floor to the nearby treetops near the receiver, and then down to the receiver is called a "scattered lateral wave."[10] A microwave system that can communicate 100 miles in free space may be able to communicate 1 mile in dense forest.

These numbers are speculative, but they represent a cur-

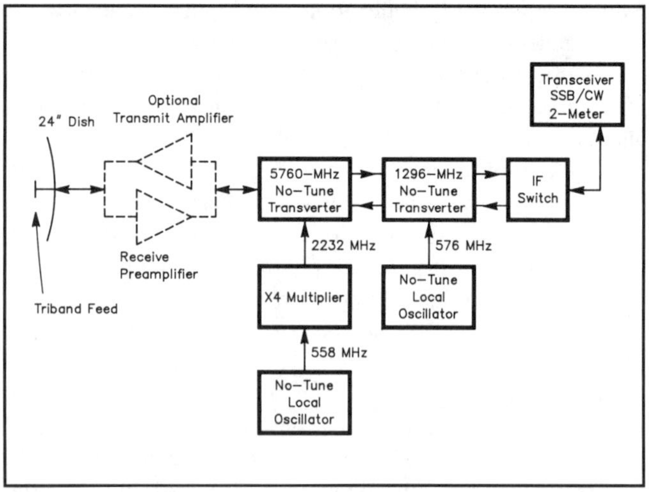

Fig 1—Block diagram of a basic 5760-MHz transmit/receive system.

Approximate Gain, Size and Beamwidth of Dish Antennas

$$\text{gain} = \eta \frac{4\pi A}{\lambda^2} \cong 5D^2$$

$$\text{Gain (dB)} \cong 7 + 20 \log (D)$$

where

η = efficiency, typically about 0.5

A = area of dish = πr^2

D = dish diameter in wavelengths

$$\text{3-dB beamwidth} \cong \frac{90}{D} \text{ degrees}$$

A_e (effective receive aperture) = η (area of dish) $\cong$ 0.4 (diam)^2

These formulas are generally accurate to within 10%, or less than 1 dB.—*KK7B*

rent consensus of experimental and anecdotal data by professional and Amateur Radio scientists. In any case, they are the best information currently available. This is an area in which amateurs can make a meaningful contribution to radio science.

Tropospheric scatter (also known as *troposcatter*) is a practical propagation mode on all of the amateur microwave bands. The US Army has operated long-range portable troposcatter links above 3 GHz for many years. Amateur experience indicates that a 5.7-GHz station equipped with a small TVRO dish, a GaAsFET preamp and a 10-watt amplifier is approximately equal to a 150-watt, single-Yagi 2-meter station—reliable for SSB communications out to about 250 miles and occasionally supporting CW work out to about 400 miles. During enhancements, much greater distances are possible.

Antennas

Selecting an antenna for 5760 MHz is easy: Nothing matches the performance of a small dish. Approximate formulas relating the gain, diameter and beamwidth for dish antennas are given in the sidebar "Approximate Gain, Size and Beamwidth of Dish Antennas." For example, a 24-inch dish at 5.76 GHz has the following gain, beamwidth and effective aperture, according to the formulas given in the sidebar:

Dish diameter: 11.7 wavelengths
Gain: 28 dB
3-dB beamwidth: 7.7 degrees
Effective aperture = 0.15 m^2

The gain can be improved by up to about 2 dB by using an optimized dish feed. Several excellent articles have been published on such feeds. The recent one by Barry Malowanchuk, VE4MA, is particularly useful.[11]

Unlike the lower-frequency bands, it's fairly easy to develop *too much* antenna gain at 5.7 GHz. Recall that gain and beamwidth are inversely related. My 19-inch dish on a small camera tripod has a nice 10-degree beamwidth. It is ideal for loaning to high-school students running a grid expedition to a windy hilltop for the first taste of microwave

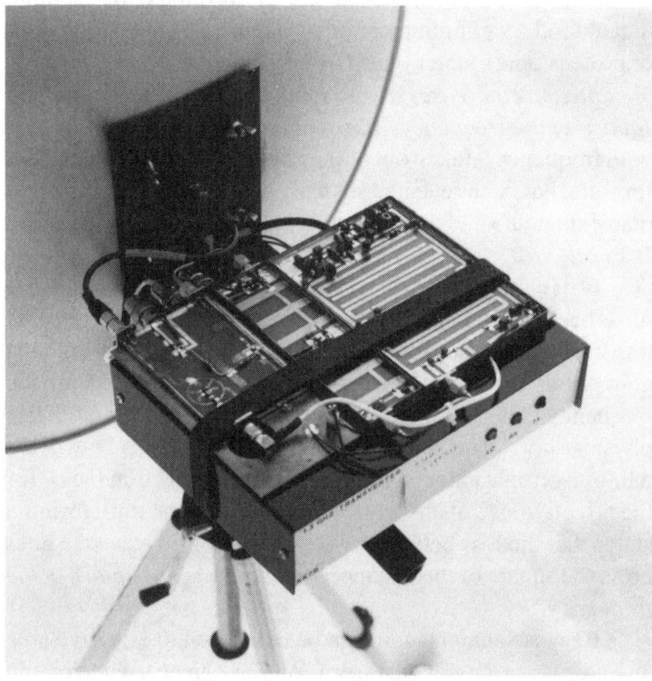

Fig 2—A complete 5760-MHz station and 24-inch dish antenna, less IF transceiver and 12-V power supply, mounts on an inexpensive camera tripod and is ready to go anywhere!

Amateur Radio. Even a 10-degree beamwidth is awful narrow when you can't see the other station and aren't sure where to point the antenna. In addition to antenna aiming uncertainty, there's also frequency uncertainty. For the contact to occur, the antenna aiming and radio tuning must be correct at the same time. For beginning operators, or experienced operators on cold, wet hilltops with state troopers slowing down to ask questions, aiming a narrow-beamwidth antenna can be frustrating.

Building a Basic Station for 5760 MHz

This section presents a basic 5.7-GHz station. You can improve this station by adding amplifiers, TR switching and a bigger dish. This is in contrast to another common microwave system concept, in which an operator builds a "simple" duplex station to make a few contacts with one other station, and then has to build a whole new station if he or she becomes interested in working someone else on the weak-signal calling frequency (5760.1 MHz).

The entry-level station I'll describe here is simple, almost foolproof, can be built at low cost using only new components, and can serve as the core of a moonbounce or troposcatter system. Of equal importance is that the entry-level station serves as test equipment for all of the antennas, feed lines and amplifiers as they're added to the system.

There are two attractive approaches to entry-level 5760 stations that can be easily enhanced up through the EME level (and beyond). The first approach, used by many hams already on the band, is to convert a Frequency West-type surplus phase-locked oscillator to 5760 plus or minus 144 MHz, use a surplus microwave mixer, and retune a surplus filter to pass 5760 MHz. The surplus route has some advantages—if you have access to the parts, expertise and test equipment needed to get it up and running:

- It can be very inexpensive
- Converting surplus is a long and honorable Amateur Radio tradition.
- There may be a psychological advantage—it seems easier to start with something that already works and modify it rather than to make something new from a little Teflon board and a teaspoonful of parts.
- Converting surplus is fun—and is the perfect excuse for buying and using a spectrum analyzer—which is even more fun, but negates the first advantage.

The second approach to an expandable entry-level 5760 station involves assembling a station from no-tune boards, as discussed in October 1990 *QST*.[12] This approach has numerous advantages, mostly because microwave technology has come a long way since those Frequency West local oscillators were built a quarter century ago. Some of the advantages of the no-tune approach are:

- You don't need microwave engineering talent to get the basic station on the air.
- It uses all new, inexpensive parts, and you can duplicate or repair it at low cost.
- You don't need any RF test equipment.
- It is very easy to assemble and very reliable.
- Most importantly, a station can be put on the air by anyone with basic soldering and construction skills—not just mystical guru spectrum-analyzer engineers.

System Block Diagram

The block diagram of a 5760-MHz transverter system using no-tune boards is shown in Fig 1. (The block diagram of the entry-level system using surplus components is not shown—it depends on the available components, and if you're capable of assembling a microwave station from surplus components, you don't need my block diagram!) Note that a 1296-MHz IF is needed with the no-tune 5760 transverter board. Three options for a 1296-MHz IF are (1) a 1296-MHz multimode transceiver (expensive but increasingly popular); (2) an existing 1296 transverter with a multimode IF transceiver; and (3) a no-tune 1296 board with a multimode 2-meter IF radio.

Where to Get the Pieces

In addition to flea markets and VHF conferences, here are a few places where you can get the parts you'll need to get on any microwave ham band:

Down East Microwave, 954 Rte 519, Frenchtown, NJ 08825, tel 908-996-3584, fax 908-946-3072: no-tune transverters, antennas, triband dish feeds, components, etc; catalog available.

Microwave Components of Michigan, PO Box 1697, Taylor, MI 48180, evening tel 313-753-4581: components; price list available.

The Antenna Center, 505 Oak St, Calumet, MI 49913, tel 906-337-5062: spun-aluminum dishes; price list available.

2-Meter IF Transceiver

Any 2-meter multimode radio can be used as a microwave IF. Older rigs are simple, inexpensive, and have poor sensitivity, which is fine, because the system noise figure is set by the 1296-MHz IF or 5760-MHz preamp. My favorite is the ICOM IC-202, a nearly ideal microwave IF rig in the \$100 to \$150 price range. These radios, unfortunately, are becoming quite hard to find. The original Yaesu FT-290R is my second choice, as it is small, has a few useful bells and whistles and low battery drain. The typical price range for these rigs is about \$200 to \$300, depending on condition and accessories. For those who prefer a "real" radio, the analog-tuned Kenwood and Yaesu radios from the TS-700 era are widely available in the \$250 to \$350 price class. I personally prefer digging for weak CW signals with an analog knob, rather than a 100-Hz click-step synthesizer knob.

Antennas

As mentioned earlier, for an entry-level antenna, it's hard to beat a 24-inch dish (UPS shippable and currently about \$70 from The Antenna Center [see the sidebar "Where to Get the Pieces"]). Use the triband feed designed by Tom Hill, WA3RMX[13] (available for \$15 when this book was being prepared from Down East Microwave), mounted on an inexpensive tripod from K-Mart or another variety store. Feed efficiency can be improved later, but the 24-inch dish is just about optimum for both hilltopping and long-haul tropo work. For what other band can you buy the same antenna the big guns are using, brand new, for \$70?

Communication with Entry-Level Stations

As Table 1 shows, a pair of entry-level stations are capable of surprising DX from hilltop to hilltop—roughly the same as a pair of 10-GHz Gunnplexer stations with small dishes. A pair of operators willing to locate the appropriate set of grid squares and hilltops can easily work VUCC. Entry-

Table 1

Basic 5760-MHz Station Characteristics

Power output: –10 dBm (100 μW, or 0.1 mW)
Noise figure (NF): 10 dB
Bandwidth: 100 Hz or 2.5 kHz*
Antenna gain (G_a): 25 dB†
Effective aperture (A_e): 0.1 m^2†
Maximum communication distance: 8000 km (5000 miles)**
Routine SSB communication distance: 160 km (100 miles)††
SN100:*** 84 dB

*100 Hz is an approximation for the ear-brain bandwidth when copying weak CW through an SSB filter.

†G_a and A_e are for a 24-inch dish with 25% efficiency (includes illumination and feed-line losses).

**Represents a 0-dB signal-to-noise CW signal at the human hearing threshold.

††Represents "armchair copy" SSB (20 dB above the noise in a 2.5-kHz bandwidth).

***Expected signal-to-noise ratio in a 2.5-kHz bandwidth of two identical 5760-MHz systems aimed at each other over an unobstructed 100-mile path.

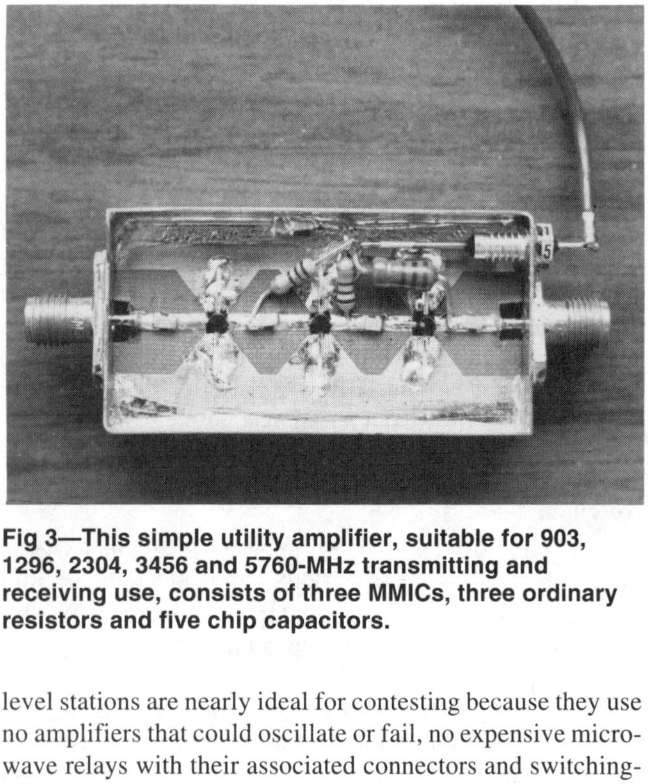

Fig 3—This simple utility amplifier, suitable for 903, 1296, 2304, 3456 and 5760-MHz transmitting and receiving use, consists of three MMICs, three ordinary resistors and five chip capacitors.

level stations are nearly ideal for contesting because they use no amplifiers that could oscillate or fail, no expensive microwave relays with their associated connectors and switching-voltage complications, and they can be mounted right at the dish to eliminate feed-line loss and all the other complications of using feed lines. The entry-level station is all many operators will ever need, and smart contesters experiment with entry-level stations before adding the complexity of amplifiers and relays. Fig 2 shows a complete entry-level system built entirely from Down East Microwave no-tune boards.

Entry-Level Stations as Test Equipment

The nature of microwave hamming is such that you either own two complete systems for each microwave band or are within driving distance of another station. The ideal piece of test equipment for a 5760-MHz station is a second 5760 station. If each is an entry-level station made from no-tune boards, then the approximate power output, noise figure, antenna gain, and so forth, are known and, more importantly, stable.

One rule needs to be followed: The local oscillators (LOs) in the two transverters need to be on slightly different frequencies. Otherwise, it's impossible to separate the up- and down-converted microwave signals from IF-signal leakage (sometimes called *breakthrough*). Normally the natural offsets of fifth-overtone crystals are more than enough to provide the needed LO frequency offset, but if you're not sure whether the signal you're hearing is at 5760 MHz or IF breakthrough, wave your hand near the antenna and listen for Doppler fluctuations, or breathe on the LO crystal to change its temperature, which should provide enough frequency shift for you to hear easily.

A number of useful tests are possible by setting up two stations as described in Table 1 facing each other over a path of about 100 m.

- Place a 40-dB pad in the 1296-MHz IF line of each system. SSB signals should be just barely readable both ways. With pads in the IF lines and signals at the "just-barely readable" level, you can compare the effects of different dish feeds, feed lines and filters. By using CW, disabling receiver AGC and using an audio-output meter calibrated in decibels, you can make quantitative comparisons. If a standard-gain antenna is available, you can also make antenna-gain measurements.
- You can determine frequency offsets and study the effects of time and temperature on stability. I write the frequency offset (ie, "5760.100 MHz = 144.126 MHz") on each transverter with an indelible marker.
- On the bench, an entry-level system can serve as a low-level source for aligning filters and measuring amplifier gain, filter loss and transmission-line loss. In receive, the entry-level system provides a way to listen and compare signal levels in transmitters.

Enhancing the Entry-Level Station

The entry-level transverter is the heart of a more elaborate station for working over obstructed paths, and via troposcatter and moonbounce. Moving up to the next level requires adding 5760-MHz gain to the transmit and receive signal paths. You need three things to do this: (1) a way to separate the transmit and receive signal paths; (2) a receive preamplifier; and (3) a transmit power amplifier.

The easiest way to split the transmit and receive paths is with a pair of microwave relays. These are expensive, but are often available surplus at hamfests and flea markets. Once the transmit and receive paths are separated, you can add amplifiers. A particularly useful amplifier for low-level transmit and receive use is shown in Figs 3 and 4. At 5760 MHz, the three-MMIC circuit has a gain of about 15 dB, a noise figure (NF) of about 7 dB and a few milliwatts output. Station enhancements are discussed in more detail in my article in the *Proceedings of the 1991 Central States VHF Conference.*[14]

Grid-Hopping Operating Hints

Here are four suggestions that have been useful in the last

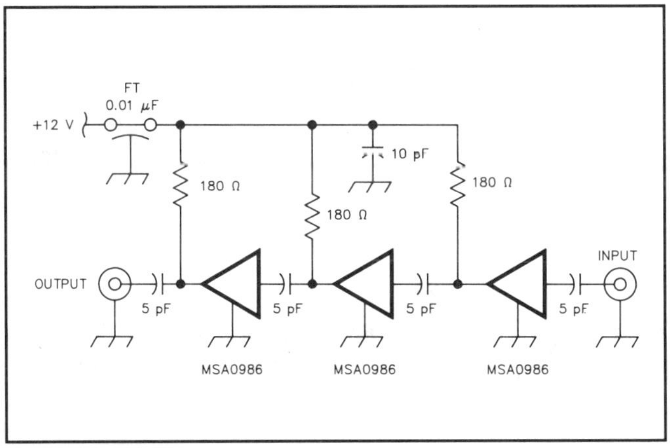

Fig 4—Schematic of the utility amplifier. Chip capacitors and MMICs are available in small quantities from Down East Microwave, Microwave Components of Michigan, and other sources.

ten years of microwave grid hopping at KK7B:

• Set up and test everything (both stations) the week before the contest or grid expedition. Fix any loose connectors, charge batteries and make sure everything is ready to go.

• Pick a reasonable liaison frequency. Two-meter FM handhelds are useless for all but the shortest microwave paths. Two-meter SSB radios with 5/8-wavelength whips are okay for medium-length paths, but remember that even the simplest pair of 5760 stations with 24-inch dishes is capable of operating SSB over a 100-mile path. With amplifiers, the 5760-MHz system will outperform 2-meter SSB stations with 150-watt "bricks" and Yagis! During the last few KK7B/8 multiband grid expeditions, I've used 5760 MHz as the liaison frequency for the lower bands!

• Work your own grid square first by setting up both stations a couple of hundred feet apart. This confirms that everything still works, permits a final check on operating frequencies, and gives you at least one contact on the band, in case the weather turns foul or something fails.

• Most "line-of-sight" microwave paths have a ground-reflected component. Ground-reflected signals may add or subtract with the direct signal. Sometimes aiming a dish a few degrees above the horizon improves signal strength by reducing the ground-reflected component.

Conclusions

The 5760 stations described in this article are portable, lightweight, reliable and can be battery powered. They can be operated from home stations, but the *real* fun begins when you take the station to a hilltop or beach. You can earn a 5760-MHz VHF-UHF Century Club (VUCC) award by working just five grid squares on 5760. Obtaining VUCC from a portable location is not only practical, but is generally *easier* than working five grids from a home station. This is one of the few areas in ham radio where you can be a "big gun" while living in an apartment, dormitory, barracks or "controlled community"! Two-foot dishes fit nicely in the trunk or back seat of a car, and a backpack station is entirely feasible.

An operator who concentrates on 5760 and coordinates with other 5760 operators can stay happily busy for an entire contest weekend. And this is without even considering adding a couple more bands to your setup, which you can do easily using no-tune transverters and a multiband dish feed.[15] You don't even have to get permission in advance to set up a camera tripod and small dish at a scenic overlook.

Perhaps best of all, you get to drive around and enjoy the hilltops and beaches between contacts, instead of gritting your teeth in front of an overloaded 2-meter rig while your neighbor works the rare grids with his legal-limit amplifier!

Notes

[1] R. Campbell, "A Single-Board Bilateral 5760-MHz Transverter," *QST*, Oct 1990, pp 27-31 (reprinted in Chapter 3 of this book).

[2] R. Campbell, "A 5760 MHz Primer," *Proceedings of the 1991 Central States VHF Conference* (Newington: ARRL, 1991), pp 7-24.

[3] A. Ishimaru, *Electromagnetic Wave Propagation, Radiation and Scattering* (Englewood Cliffs, NJ: Prentice-Hall, 1991), pp 123-127.

[4] *Reference Data for Radio Engineers*, 6th Ed. (Indianapolis: Howard W. Sams, 1975), pp 33-3, 33-7.

[5] R. Campbell, "Does Path Loss Increase with Frequency?" Technical Correspondence, *QST*, Jan 1991, p 38.

[6] A. Hambley, *An Introduction to Communications Systems* (New York: W. H. Freeman and Co, 1990) pp 168-169.

[7] See note 4.

[8] D. Hilliard and B. McCaa, "Preliminary Results of Foliage Attenuation Measurements at 902 MHz and 3456 MHz in the Boulder, Colorado Area," *Proceedings of Microwave Update '89* (Newington: ARRL, 1989), pp 133-147.

[9] W. Vogel and J. Goldhirsch, "Roadside Tree Attenuation Measurements at UHF for Land Mobile Satellite Systems," *IEEE Transactions on Antennas and Propagation.*

[10] R. Campbell and H. Wang, "A Theoretical and Experimental Study of Scattered Lateral Wave Propagation in Forest," *Digest of URSI North American Radio Science Meeting*, London, Ontario, Jun 1991.

[11] B. Malowanchuck, "Use of Small TVRO Dishes for EME," *Proceedings of the 21st Conference of the Central States VHF Society* (Newington: ARRL, 1987), pp 66-77.

[12] See note 1.

[13] T. Hill, "A Triband Microwave Dish Feed," *QST*, Aug 1990, pp 22-27 (reprinted in the Antennas chapter of this book).

[14] See note 2.

[15] See note 13.

2

Construction Practices and Tips

Contents

RF Radiation and Electromagnetic Field Safety

By The ARRL RF Safety Committee

(From *The ARRL Handbook*)

Although Amateur Radio is basically a safe activity, in recent years there has been considerable discussion and concern about the possible hazards of electromagnetic radiation (EMR), including both RF energy and power frequency (50-60 Hz) electromagnetic fields. Extensive research on this topic is underway in many countries. This section was prepared by members of the ARRL RF Safety Committee and coordinated by Dr Robert E. Gold, WBØKIZ. It summarizes what is now known and offers safety precautions based on the research to date.

All life on Earth has adapted to survive in an environment of weak, natural low-frequency electromagnetic fields (in addition to the Earth's static geomagnetic field). Natural low-frequency EM fields come from two main sources: the sun, and thunderstorm activity. But in the last 100 years, man-made fields at much higher intensities and with a very different spectral distribution have altered this natural EM background in ways that are not yet fully understood. Much more research is needed to assess the biological effects of EMR.

Both RF and 60-Hz fields are classified as *nonionizing radiation* because the frequency is too low for there to be enough photon energy to ionize atoms. Still, at sufficiently high power densities, EMR poses certain health hazards. It has been known since the early days of radio that RF energy can cause injuries by heating body tissue. In extreme cases, RF-induced heating can cause blindness, sterility and other serious health problems. These heat-related health hazards are called *thermal effects*. In addition, there is evidence that magnetic fields may produce biologic effects at energy levels too low to cause body heating. The proposition that these athermal effects may produce harmful health consequences has produced a great deal of research.

In addition to the ongoing research, much else has been done to address this issue. For example, the American National Standards Institute, among others, has recommended voluntary guidelines to limit human exposure to RF energy. And the ARRL has established the RF Safety Committee, a committee of concerned medical doctors and scientists, serving voluntarily to monitor scientific research in the fields and to recommend safe practices for radio amateurs.

THERMAL EFFECTS OF RF ENERGY

Body tissues that are subjected to very high levels of RF energy may suffer serious heat damage. These effects depend upon the frequency of the energy, the power density of the RF field that strikes the body, and even on factors such as the polarization of the wave.

At frequencies near the body's natural resonant frequency, RF energy is absorbed more efficiently, and maximum heating occurs. In adults, this frequency usually is about 35 MHz if the person is grounded, and about 70 MHz if the person's body is insulated from the ground. Also, body parts may be resonant; the adult head, for example is resonant around 400 MHz, while a baby's smaller head resonates near 700 MHz. Body size thus determines the frequency at which most RF energy is absorbed. As the frequency is increased above resonance, less RF heating generally occurs. However, additional longitudinal resonances occur at about 1 GHz near the body surface.

Nevertheless, thermal effects of RF energy should not be a major concern for most radio amateurs because of the relatively low RF power we normally use and intermittent nature of most amateur transmissions. Amateurs spend more time listening than transmitting, and many amateur transmissions such as CW and SSB use low-duty-cycle modes. (With FM or RTTY, though, the RF is present continuously at its maximum level during each transmission.) In any event, it is rare for radio amateurs to be subjected to RF fields strong enough to produce thermal effects unless they are fairly close to an energized antenna or unshielded power amplifier. Specific suggestions for avoiding excessive exposure are offered later.

ATHERMAL EFFECTS OF EMR

Nonthermal effects of EMR may be of greater concern to most amateurs because they involve lower level energy fields. Research about possible health effects resulting from exposure to the lower level energy fields, the athermal effects, has been of two basic types: epidemiological research and laboratory research.

Scientists conduct laboratory research into biological mechanisms by which EMR may affect animals including humans. Epidemiologists look at the health patterns of large groups of people using statistical methods. These epidemiological studies have been inconclusive. By their basic design, these studies do not demonstrate cause and effect, nor do they postulate mechanisms of disease. Instead, epidemiologists look for associations between an environmental factor and an observed pattern of illness. For example, in the earliest research on malaria, epidemiologists observed the association between populations with high prevalence of the disease and the proximity of mosquito infested swamplands. It was left to the biological and medical scientists to isolate the organism causing malaria in the blood of those with the disease and identify the same organisms in the mosquito population.

In the case of athermal effects, some studies have identified a weak association between exposure to EMF at home or at work and various malignant conditions including leukemia and brain cancer. However, a larger number of equally well designed and performed studies have found no association. A risk ratio of between 1.5 and 2.0 has been observed in positive studies (the number of observed cases of malignancy being 1.5 to 2.0 times the "expected" number in the population). Epidemiologists generally regard a risk ratio of 4.0 or greater to be indicative of a strong association between the cause and effect under study. For example, men who smoke one pack of cigarettes per day increase their risk for lung cancer tenfold compared to nonsmokers, and two packs per day increase the risk to more than 25 times the nonsmokers' risk.

However, epidemiological research by itself is rarely conclusive. Epidemiology only identifies health patterns in groups—it does not ordinarily determine their cause. And there are often confounding factors: Most of us are exposed to many different environmental hazards that may affect our health in various ways. Moreover, not all studies of persons likely to be exposed to high levels of EMR have yielded the same results.

There has also been considerable laboratory research about the biological effects of EMR in recent years. For example, it has been shown that even fairly low levels of EMR can alter the human body's circadian rhythms, affect the manner in which cancer-fighting T lymphocytes function in the immune system, and alter the nature of the electrical and chemical signals communicated through the cell membrane and between cells, among other things.

Much of this research has focused on low-frequency magnetic fields, or on RF fields that are keyed, pulsed or modulated at a low audio frequency (often below 100 Hz). Several studies suggested that humans and animals can adapt to the presence of a steady RF carrier more readily than to an intermittent, keyed or modulated energy source. There is some evidence that while EMR may not directly cause cancer, it may sometimes combine with chemical agents to promote its growth or inhibit the work of the body's immune system.

None of the research to date conclusively proves that low-level EMR causes adverse health effects. Given the fact that there is a great deal of research ongoing to examine the health consequences of exposure to EMF, the American Physical Society (a national group of highly respected scientists) issued a statement in May 1995 based on its review of available data pertaining to the possible connections of cancer to 60-Hz EMF exposure. This report is exhaustive and should be reviewed by anyone with a serious interest in the field. Among its general conclusions were the following:

1. "The scientific literature and the reports of reviews by other panels show no consistent, significant link between cancer and powerline fields."
2. "No plausible biophysical mechanisms for the systematic initiation or promotion of cancer by these extremely weak 60-Hz fields has been identified."
3. "While it is impossible to prove that no deleterious health effects occur from exposure to any environmental factor, it is necessary to demonstrate a consistent, significant, and causal relationship before one can conclude that such effects do occur."

The APS study is limited to exposure to 60-Hz EMF. Amateurs will also be interested in exposure to EMF in the RF range. A 1995 publication entitled *Radio Frequency and ELF Electromagnetic Energies, A Handbook for Health Professionals* includes a chapter called "Biologic Effects of RF Fields." In it the authors state: "In conclusion, the data do not support the finding that exposure to RF fields is a causal agent for any type of cancer" (page 176). Later in the same chapter they write: "Although the data base has grown substantially over the past decades, much of the information concerning nonthermal effects is generally inconclusive, incomplete, and sometimes contradictory. Studies of human populations have not demonstrated any reliably effected end point." (page 186).

Readers may want to follow this topic as further studies are reported. Amateurs should be aware that exposure to RF and ELF (60 Hz) electromagnetic fields at all power levels and frequencies may not be completely safe. Prudent avoidance of any avoidable EMR is always a good idea. However, an Amateur Radio operator should not be fearful of using his equipment. If any risk does exist, it will almost surely fall well down on the list of causes that may be harmful to your health (on the other end of the list from your automobile).

Safe Exposure Levels

How much EM energy is safe? Scientists have devoted a great deal of effort to deciding upon safe RF-exposure limits. This is a very complex problem, involving difficult public health and economic considerations. The recommended safe levels have been revised downward several times in recent years—and not all scientific bodies agree on this question even today. A new Institute of Electrical and Electronics Engineers (IEEE) guideline for recommended EM exposure limits went into effect in 1991 (see Bibliography). It replaced a 1982 American National Standards Institute guideline that permitted somewhat higher exposure levels. ANSI-recommended exposure limits before 1982 were higher still.

This new IEEE guideline recommends frequency-dependent and time-dependent maximum permissible exposure levels. Unlike earlier versions of the standard, the 1991 standard recommends different RF exposure limits in *controlled environments* (that is, where energy levels can be accurately determined and everyone on the premises is aware of the presence of EM fields) and in *uncontrolled environments* (where energy levels are not known or where some persons present may not be aware of the EM fields).

The graph in **Fig 1** depicts the new IEEE standard. It is necessarily a complex graph because the standards differ not only for controlled and uncontrolled environments but also for electric fields (E fields) and magnetic fields (H fields). Basically, the lowest E-field exposure limits occur at frequencies between 30 and 300 MHz. The lowest H-field exposure levels occur at 100-300 MHz. The ANSI standard sets the maximum E-field limits between 30 and 300 MHz at a power density of 1 mW/cm² (61.4 V/m) in controlled environments—but at one-fifth that level (0.2 mW/cm² or 27.5 V/m) in uncontrolled environments. The H-field limit drops to 1 mW/cm² (0.163 A/m) at 100-300 MHz in controlled environments and 0.2 mW/cm² (0.0728 A/m) in uncontrolled environments. Higher power densities are permitted at frequencies below 30 MHz (below 100 MHz for H fields) and above 300 MHz, based on the concept that the body will not be resonant at those frequencies and will therefore absorb less energy.

In general, the IEEE guideline requires averaging the power level over time periods ranging from 6 to 30 minutes for power-density calculations, depending on the frequency and other variables. The ANSI exposure limits for uncontrolled environments are lower than those for controlled environments, but to compensate for that the guideline allows exposure levels in those environments to be averaged over much longer time periods (generally 30 minutes). This long averaging time means that an intermittently operating RF source (such as an Amateur Radio transmitter) will show a much lower power density than a continuous-duty station for a given power level and antenna configuration.

Time averaging is based on the concept that the human body can withstand a greater rate of body heating (and thus, a higher level of RF energy) for a short time than for a longer period. However, time averaging may not be appropriate in considerations of nonthermal effects of RF energy.

The IEEE guideline excludes any transmitter with an output below 7 W because such low-power transmitters would not be able to produce significant whole-body heating. (However, recent studies show that hand-held transceivers often produce power densities in excess of the IEEE standard within the head.)

There is disagreement within the scientific community about these RF exposure guidelines. The IEEE guideline is still intended primarily to deal with thermal effects, not exposure to energy at lower levels. A small but significant number of researchers now believe athermal effects should also be taken into consideration. Several European countries and localities in the United States have adopted stricter standards than the recently updated IEEE standard.

Another national body in the United States, the National Council for Radiation Protection and Measurement (NCRP), has also adopted recommended exposure guidelines. NCRP urges a limit of 0.2 mW/cm² for nonoccupational exposure in the 30-300 MHz range. The NCRP guideline differs from IEEE in two notable ways: It takes into account the effects of modulation on an RF carrier, and it does not exempt transmitters with outputs below 7 W.

Cardiac Pacemakers and RF Safety

It is a widely held belief that cardiac pacemakers may be adversely affected in their function by exposure to electromagnetic fields. Amateurs with pacemakers may ask whether their operating might endanger themselves or visitors to their shacks who have a pacemaker. Because of this and similar concerns regarding other sources of electromagnetic fields, pacemaker manufacturers apply design methods that for the most part shield the pacemaker circuitry from even relatively high EM field strengths.

It is recommended that any amateur who has a pace-

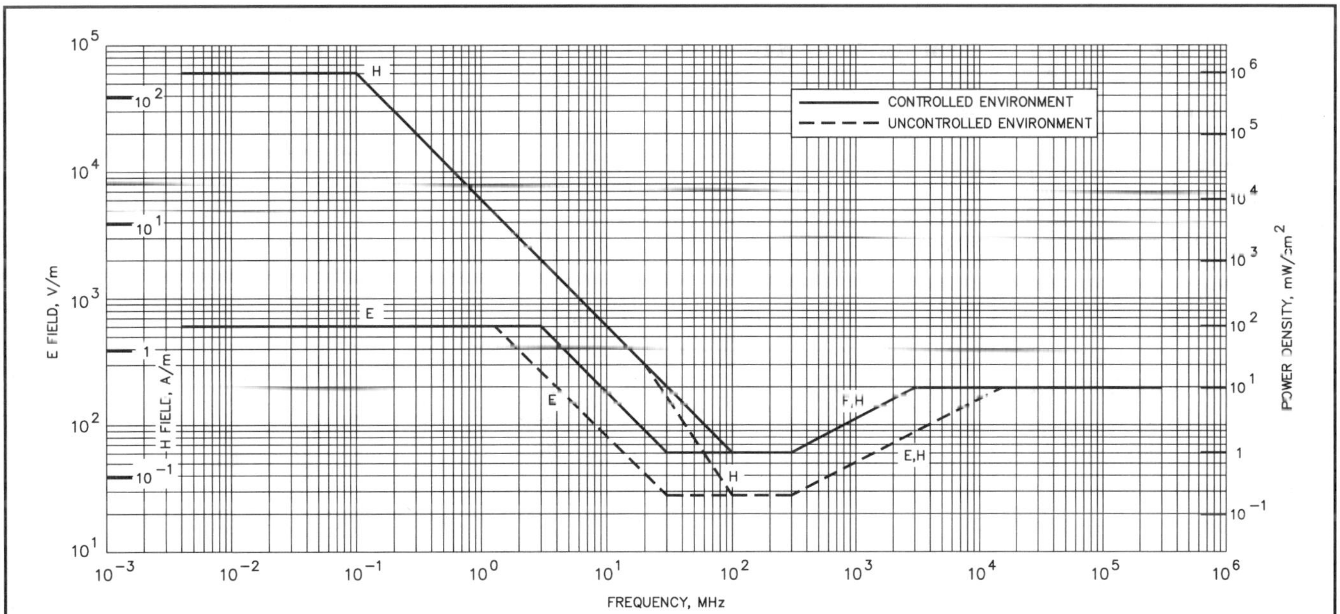

Fig 1—1991 RF protection guidelines for body exposure of humans. It is known officially as the "IEEE Standard for Safety Levels with Respect to Human Exposure to Radio Frequency Electromagnetic Fields, 3 kHz to 300 GHz."

maker or is being considered for one discuss this matter with his or her physician. The physician will probably put the amateur into contact with the technical representative of the pacemaker manufacturer. These representatives are generally excellent resources and may have data from laboratory or "in the field" studies with pacemaker units of the type the amateur needs to know about.

One study examined the function of a modern (dual chamber) pacemaker in and around an Amateur Radio station. The pacemaker generator has circuits that receive and process electrical signals produced by the heart and also generate electrical signals that stimulate (pace) the heart. In one series of experiments the pacemaker was connected to a heart simulator. The system was placed on top of the cabinet of a 1-kW HF linear amplifier during SSB and CW operation. In addition, the system was placed in close proximity to several 1 to 5-W 2-meter hand-held transceivers. The test pacemaker connected to the heart simulator was also placed on the ground 9 meters below and 5 meters in front of a three-element Yagi HF antenna. No interference with pacemaker function was observed in this experimental system.

Although the possibility of interference cannot be entirely ruled out by these few observations, these tests represent more severe exposure to EM fields than would ordinarily be encountered by an amateur with an average amount of common sense. Of course prudence dictates that amateurs with pacemakers using hand-held VHF transceivers keep the antenna as far from the site of the implanted pacemaker generator as possible and use the lowest transmitter output required for adequate communication. For high power HF transmission, the antenna should be as far from the operating position as possible and all equipment should be properly grounded.

Table 1

Typical 60-Hz Magnetic Fields Near Amateur Radio Equipment and AC-Powered Household Appliances

Values are in milligauss.

Item	*Field*	*Distance*
Electric blanket	30-90	Surface
Microwave oven	10-100 1-10	Surface 12"
IBM personal computer	5-10 0-1	Atop monitor 15" from screen
Electric drill	500-2000	At handle
Hair dryer	200-2000	At handle
HF transceiver	10-100 1-5	Atop cabinet 15" from front
1-kW RF amplifier	80-1000 1-25	Atop cabinet 15" from front

(Source: measurements made by members of the ARRL RF Safety Committee)

Low-Frequency Fields

Recently, much concern about EMR has focused on low-frequency energy rather than RF. Amateur Radio equipment can be a significant source of low-frequency magnetic fields, although there are many other sources of this kind of energy in the typical home. Magnetic fields can be measured relatively accurately with inexpensive 60-Hz dosimeters that are made by several manufacturers.

Table 1 shows typical magnetic field intensities of Amateur Radio equipment and various household items. Because these fields dissipate rapidly with distance, "prudent avoidance" would mean staying perhaps 12 to 18 inches away from most Amateur Radio equipment (and 24 inches from power supplies with 1-kW RF amplifiers) whenever the ac power is turned on. The old custom of leaning over a linear amplifier on a cold winter night to keep warm may not be the best idea!

There are currently no non-occupational US standards for exposure to low-frequency fields. However, some epidemiological evidence suggests that when the general level of 60-Hz fields exceeds 2 milligauss, there is an increased cancer risk in both domestic environments and industrial environments. Typical home environments (not close to appliances or power lines) are in the range of 0.1-0.5 milligauss.

Table 2

Typical RF Field Strengths Near Amateur Radio Antennas

A sampling of values as measured by the Federal Communications Commission and Environmental Protection Agency, 1990

Antenna Type	*Freq (MHz)*	*Power (W)*	*E Field (V/m)*	*Location*
Dipole in attic	14.15	100	7-100	In home
Discone in attic	146.5	250	10-27	In home
Half sloper	21.5	1000	50	1 m from base
Dipole at 7-13 ft	7.14	120	8-150	1-2 m from earth
Vertical	3.8	800	180	0.5 m from base
5-element Yagi at 60 ft	21.2	1000	10-20 14	In shack 12 m from base
3-element Yagi at 25 ft	28.5	425	8-12	12 m from base
Inverted V at 22-46 ft	7.23	1400	5-27	Below antenna
Vertical on roof	14.11	140	6-9 35-100	In house At antenna tuner
Whip on auto roof	146.5	100	22-75 15-30 90	2 m from antenna In vehicle Rear seat
5-element Yagi at 20 ft	50.1	500	37-50	10 m from antenna

Determining RF Power Density

Unfortunately, determining the power density of the RF fields generated by an amateur station is not as simple as measuring low-frequency magnetic fields. Although sophisticated instruments can be used to measure RF power densities quite accurately, they are costly and require frequent recalibration. Most amateurs don't have access to such equipment, and the inexpensive field-strength meters that we do have are not suitable for measuring RF power density. The best we can usually do is to estimate our own RF power density based on measurements made by others or, given sufficient computer programming skills, use computer modeling techniques.

Table 2 shows a sampling of measurements made at Amateur Radio stations by the Federal Communications Commission and the Environmental Protection Agency in 1990. As this table indicates, a good antenna well removed from inhabited areas poses no hazard under any of the various exposure guidelines. However, the FCC/EPA survey also indicates that amateurs must be careful about using indoor or attic-mounted antennas, mobile antennas, low directional arrays or any other antenna that is close to inhabited areas, especially when moderate to high power is used.

Ideally, before using any antenna that is in close proximity to an inhabited area, you should measure the RF power density. If that is not feasible, the next best option is make the installation as safe as possible by observing the safety suggestions listed in **Table 3**.

It is also possible, of course, to calculate the probable power density near an antenna using simple equations. However, such calculations have many pitfalls. For one, most of the situations in which the power density would be high enough to be of concern are in the near field—an area roughly bounded by several wavelengths of the antenna. In the near field, ground interactions and other variables produce power densities that cannot be determined by simple arithmetic.

Computer antenna-modeling programs such as MININEC or other codes derived from NEC (Numerical Electromagnetics Code) are suitable for estimating RF magnetic and electric fields around amateur antenna systems. (See the **Propagation** chapter of *The ARRL Handbook* for more information about MININEC.) And yet, these too have limitations. Ground interactions must be considered in estimating near-field power densities. Also, computer modeling is not sophisticated enough to predict "hot spots" in the near field—places where the field intensity may be far higher than would be expected.

Intensely elevated but localized fields often can be detected by professional measuring instruments. These "hot spots" are often found near wiring in the shack and metal objects such as antenna masts or equipment cabinets. But even with the best instrumentation, these measurements may also be misleading in the near field.

One need not make precise measurements or model the exact antenna system, however, to develop some idea of the relative fields around an antenna. Computer modeling using close approximations of the geometry and power input of the antenna will generally suffice. Those who are familiar with MININEC can estimate their power densities by computer modeling, and those who have access to professional power-density meters can make useful measurements.

While our primary concern is ordinarily the intensity of the signal radiated by an antenna, we should also remember that there are other potential energy sources to be considered. You can also be exposed to RF radiation directly from a power amplifier if it is operated without proper shielding. Transmission lines may also radiate a significant amount of energy under some conditions.

Further RF Exposure Suggestions

Potential exposure situations should be taken seriously. Based on the FCC/EPA measurements and other data, the "RF awareness" guidelines of Table 3 were developed by the ARRL RF Safety Committee. A longer version of these guidelines, along with a complete list of references, appeared in a *QST* article by Ivan Shulman, MD, WC2S (see Bibliography).

In addition, *QST* carries information regarding the latest developments for RF safety precautions and regulations at the local and federal levels.

Table 3

RF Awareness Guidelines

These guidelines were developed by the ARRL RF Safety Committee, based on the FCC/EPA measurements of Table 2 and other data.

- Although antennas on towers (well away from people) pose no exposure problem, make certain that the RF radiation is confined to the antennas' radiating elements themselves. Provide a single, good station ground (earth), and eliminate radiation from transmission lines. Use good coaxial cable, not open-wire lines or end-fed antennas that come directly into the transmitter area.
- No person should ever be near any transmitting antenna while it is in use. This is especially true for mobile or ground-mounted vertical antennas. Avoid transmitting with more than 25 W in a VHF mobile installation unless it is possible to first measure the RF fields inside the vehicle. At the 1-kW level, both HF and VHF directional antennas should be at least 35 ft above inhabited areas. Avoid using indoor and attic-mounted antennas if at all possible.
- Don't operate high-power amplifiers with the covers removed, especially at VHF/UHF.
- In the UHF/SHF region, never look into the open end of an activated length of waveguide or point it toward anyone. Never point a high-gain, narrow-bandwidth antenna (a paraboloid, for instance) toward people. Use caution in aiming an EME (moonbounce) array toward the horizon; EME arrays may deliver an effective radiated power of 250,000 W or more.
- With hand-held transceivers, keep the antenna away from your head and use the lowest power possible to maintain communications. Use a separate microphone and hold the rig as far away from you as possible.
- Don't work on antennas that have RF power applied.
- Don't stand or sit close to a power supply or linear amplifier when the ac power is turned on. Stay at least 24 inches away from power transformers, electrical fans and other sources of high-level 60-Hz magnetic fields.

Bibliography

IEEE Standard for Safety Levels with Respect to Human Exposure to Radio Frequency Electromagnetic Fields, 3 KHz to 300 GHz, IEEE Standard C95.1-1991, Institute of Electrical and Electronics Engineers, New York, 1992.

For an unbiased assessment of ELF hazards, read the series in *Science*, Vol 249 beginning 9/7/90 (p 1096), continuing 9/21/90 (p 1378), and ending 10/5/90 (p 23). Also see *Science*, Vol 258, p 1724 (1992). You can find *Science* in any large library.

An excellent and timely document is available on the Internet by an anonymous FTP from: rtfm.mit.edu, /pub/usenet-by-group/news.answers powerlines-cancer-faq/part1 and part2.

The Environmental Protection Agency publishes a free consumer-level booklet entitled, "EMF in Your Environment," document 402-R-92-008, dated December 1992. Look for the nearest office of the EPA in your phone book.

W. R. Adey, "Tissue Interactions with Nonionizing Electromagnetic Fields," *Physiology Review,* 1981; 61:435-514.

W. R. Adey, "Cell Membranes: The Electromagnetic Environment and Cancer Promotion," *Neurochemical Research,* 1988; 13:671-677.

W. R. Adey, "Electromagnetic Fields, Cell Membrane Amplification, and Cancer Promotion," in B. W. Wilson, R. G. Stevens, and L. E. Anderson, *Extremely Low Frequency Electromagnetic Fields: The Question of Cancer* (Columbus, OH: Batelle Press, 1989), pp 211-249.

W. R. Adey, "Electromagnetic Fields and the Essence of Living Systems," Plenary Lecture, 23rd General Assembly, International Union of Radio Sciences (URSI), Prague, 1990; in J. Bach Andersen, Ed., *Modern Radio Science* (Oxford: Oxford Univ Press), pp 1-36.

Q. Balzano, O. Garay and K. Siwiak, "The Near Field of Dipole Antennas, Part I: Theory," *IEEE Transactions on Vehicular Technology (VT) 30*, p 161, Nov 1981. Also "Part II; Experimental Results," same issue, p 175.

R. F. Cleveland and T. W. Athey, "Specific Absorption Rate (SAR) in Models of the Human Head Exposed to Hand-Held UHF Portable Radios," *Bio-electromagnetics,* 1989; 10:173-186.

R. F. Cleveland, E. D. Mantiply and T. L. West, "Measurements of Environmental Electromagnetic Fields Created by Amateur Radio Stations," presented at the 13th annual meeting of the Bioelectromagnetics Society, Salt Lake City, Utah, Jun 1991.

R. L. Davis and S. Milham, "Altered Immune Status in Aluminum Reduction Plant Workers," *American J Industrial Medicine,* 1990; 131:763-769.

F. C. Garland, et al, "Incidence of Leukemia in Occupations with Potential Electromagnetic Field Exposure in United States Navy Personnel," *American J Epidemiology,* 1990; 132:293-303.

A. W. Guy and C. K. Chou, "Thermograph Determination of SAR in Human Models Exposed to UHF Mobile Antenna Fields," Paper F-6, Third Annual Conference, Bioelectromagnetics Society, Washington, DC, Aug 9-12, 1981.

C. C. Johnson and M. R. Spitz, "Childhood Nervous System Tumours: An Assessment of Risk Associated with Paternal Occupations Involving Use, Repair or Manufacture of Electrical and Electronic Equipment," *International J Epidemiology,* 1989; 18:756-762.

D. L. Lambdin, "An Investigation of Energy Densities in the Vicinity of Vehicles with Mobile Communications Equipment and Near a Hand-Held Walkie Talkie," *EPA Report ORP/EAD 79-2,* Mar, 1979.

D. B. Lyle, P. Schechter, W. R. Adey and R. L. Lundak, "Suppression of T-Lymphocyte Cytotoxicity Following Exposure to Sinusoidally Amplitude Modulated Fields," *Bioelectromagnetics,* 1983; 4:281 -292.

G. M. Matanoski et al, "Cancer Incidence in New York Telephone Workers," *Proc Annual Review, Research on Biological Effects of 50/60 Hz Fields,* U.S. Dept of Energy, Office of Energy Storage and Distribution, Portland, OR, 1989.

D. I. McRee, *A Technical Review of the Biological Effects of Non-Ionizing Radiation,* Office of Science and Technology Policy, Washington, DC, 1978.

G. E. Myers, "ELF Hazard Facts" *Amateur Radio News Service Bulletin*, Alliance, OH, Apr 1994.

S. Milham, "Mortality from Leukemia in Workers Exposed to Electromagnetic Fields," *New England J Medicine,* 1982; 307:249.

S. Milham, "Increased Mortality in Amateur Radio Operators due to Lymphatic and Hematopoietic Malignancies," *American J Epidemiology,* 1988; 127:50-54.

W. W. Mumford, "Heat Stress Due to RF Radiation," *Proc IEEE,* 57, 1969, pp 171-178.

W. Overbeck, "Electromagnetic Fields and Your Health," *QST*, Apr 1994, pp 56-59.

S. Preston-Martin et al, "Risk Factors for Gliomas and Meningiomas in Males in Los Angeles County," *Cancer Research,* 1989; 49:6137-6143.

D. A. Savitz et al, "Case-Control Study of Childhood Cancer and Exposure to 60-Hz Magnetic Fields," *American J Epidemiology,* 1988; 128:21-38.

D. A. Savitz et al, "Magnetic Field Exposure from Electric Appliances and Childhood Cancer," *American J Epidemiology, 1990;* 131:763-773.

I. Shulman, "Is Amateur Radio Hazardous to Our Health?" *QST,* Oct 1989, pp 31-34.

R. J. Spiegel, "The Thermal Response of a Human in the Near-Zone of a Resonant Thin-Wire Antenna," *IEEE Transactions on Microwave Theory and Technology (MTT)* 30(2), pp 177-185, Feb 1982.

B. Springfield and R. Ely, "The Tower Shield," *QST*, Sep 1976, p 26.

T. L. Thomas et al, "Brain Tumor Mortality Risk among Men with Electrical and Electronic Jobs: A Case-Controlled Study," *J National Cancer Inst,* 1987; 79:223-237.

N. Wertheimer and E. Leeper, "Electrical Wiring Configurations and Childhood Cancer," *American J Epidemiology,* 1979; 109:273-284.

N. Wertheimer and E. Leeper, "Adult Cancer Related to Electrical Wires Near the Home," *Internat'l J Epidemiology,* 1982; 11:345-355.

"Safety Levels with Respect to Human Exposure to Radio Frequency Electromagnetic Fields (300 kHz to 100 GHz)," ANSI C95.1-1991 (New York: IEEE-American National Standards Institute).

"Biological Effects and Exposure Criteria for Radiofrequency Electromagnetic Fields," NCRP Report No. 86 (Bethesda, MD: National Council on Radiation Protection and Measurements, 1986).

US Congress, Office of Technology Assessment, "Biological Effects of Power Frequency Electric and Magnetic Fields—Background Paper," OTA-BP-E53 (Washington, DC: US Government Printing Office), 1989.

Make Etching Patterns with a Hobby Knife and Transparent Tape

By Dave Mascaro, WA3JUF

Anyone can reproduce a microstrip printed-circuit board without using photosensitive boards and negatives. The copper is etched off, so this method is neater than simply using a hobby (X-Acto) knife to cut away the copper. Transparent tape is used as the resist material. I use 4-inch-wide tape, available at stationery stores. Here's how I do it:

1. Cut the board to size and clean the foil so it's shiny.
2. Draw the artwork on the board with a pencil.
3. Cover the board with tape, making sure there are no bubbles.
4. Using a hobby knife, cut the tape along the pencil lines, with the aid of a straight edge if necessary.
5. Remove the tape from the board where you want the copper to be removed.
6. Press the remaining tape firmly against the board.
7. Etch the board in ferric-chloride solution. I put the board and ferric chloride in a plastic bottle with a lid. Then I agitate the bottle in a bucket of hot water.
8. After rinsing the board, remove the tape and clean the board with steel wool.
9. Plate the board with solder and a flat-bladed soldering tip. Use liquid flux so the solder flow is even and thin.

A Milled Brass Amplifier Case

By Dave Mascaro, WA3JUF

Refer to Figs 1, 2 and 3. The amplifier housing is milled from brass stock 1.5" wide by 0.5" thick. Cut the piece 0.30" longer than the printed-circuit board to be installed. Square off the stock on the milling machine. Then hog out inside material to a depth of 0.30". Take out the corners to clear the board. Then mill out the slot for the transistor flange. Drill and tap necessary holes.

The SMA connectors can be secured with machine screws, or sweat soldered in place after the board is installed. Cut a hole in the PC board the size of the transistor flange. Sweat solder the board (and connectors, if desired) by heating the brass on a hot plate or stove burner. The feedthrough capacitor is a Spectrum Control Inc. (SCI) 729-303 or equivalent.

Spread a thin coating of silicone grease on the housing surface that mates with the heat sink.

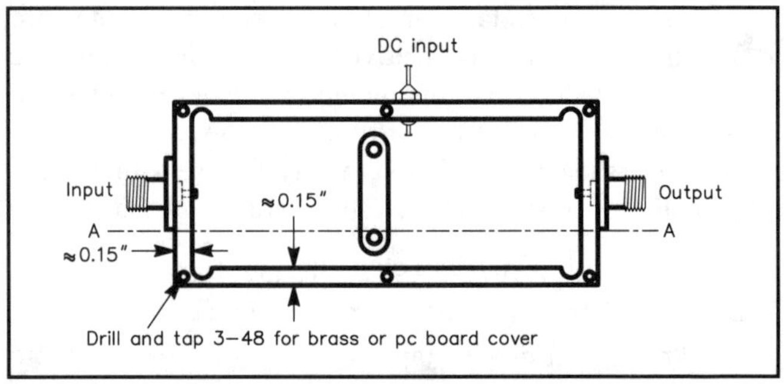

Fig 1—Top view of the brass amplifier case used by WA3JUF, before the PC board is installed. The bottom of the case is milled out to clear the device mounting flange.

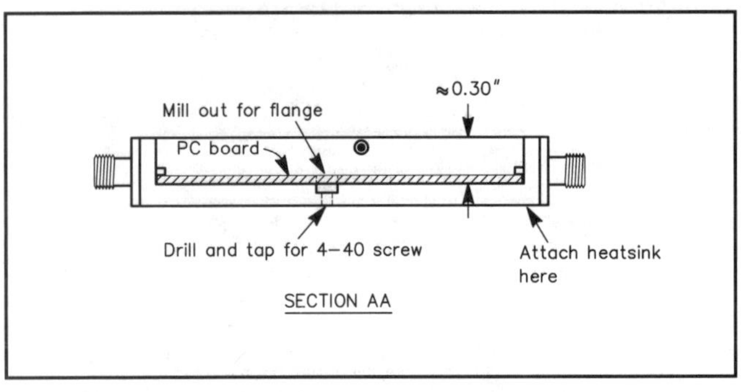

Fig 2—Side view of the brass case. The SMA connectors can be secured with machine screws or sweat soldered (see text). A heat sink is secured to the bottom of the case.

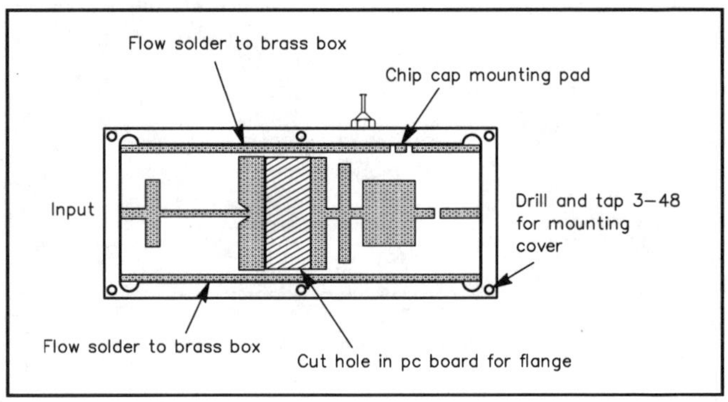

Fig 3—Top view of the case with a board installed. Run a continuous bead of solder around the perimeter of the board to provide a good RF ground. A hole cut in the center of the board provides clearance for the device mounting flange. A piece of copper-clad board or thin, sheet copper can be used as a cover.

Microwave Layout Tips

By Dave Mascaro, WA3JUF

Here's an example of a typical microwave amplifier layout using the end-launcher method. The PC board is soldered to the connectors on the groundplane side of the board, to provide a continuous RF ground. Rivets are used at all RF and dc grounds. A cover or box can be fabricated of scrap pieces of PC board. A U-shaped piece of copper foil is soldered to the bottom of the PC board beneath the device flange. When the device is secured to the heat sink, the foil provides a good RF ground.

Power-supply bypassing is important when several microwave stages are powered from the same source. I recommend series-resonant feedthrough capacitors, like the Spectrum Control Inc. 729-303.

Fig 4 shows a typical layout from the top. Ferrite beads, lumped-constant RF chokes and chip bypass capacitors are used liberally to decouple the power supply. Fig 5 shows the same layout from the side. The heat sink is milled out to just clear the device flanges. You can use a milling machine, or a milling bit in a Dremel drill. (Caution: Wear safety glasses.) The PC board should be secured to the heat sink in several places. The heat sink can be drilled and tapped to receive machine screws, or you can simply rivet the board to the heat sink.

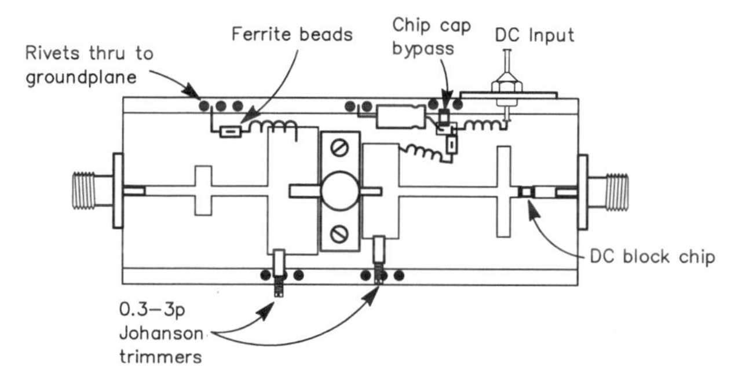

Fig 4—Typical microwave layout. Liberal use of ferrite beads, RF chokes and chip capacitors is recommended to prevent RF coupling through power leads.

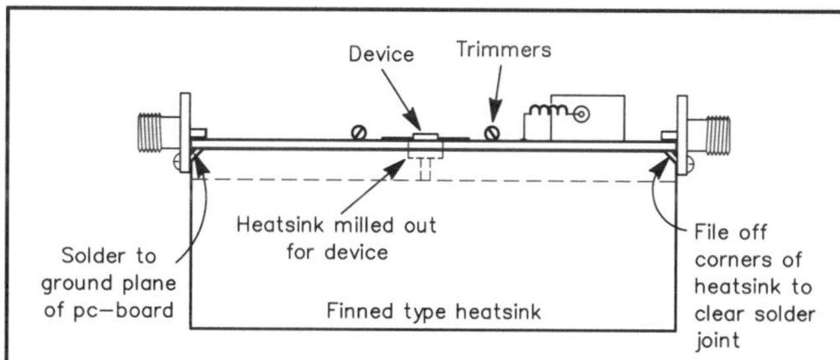

Fig 5—Side view, showing attention to RF grounding. Good layout practice makes for more-reliable operation—especially important for portable operation.

How to Safely Handle FETs

By Dick Jansson, WD4FAB

The safe handling of very-high-impedance FET devices requires great preparation to avoid destroying the device with stray static charges. When I was a practicing engineer, I had the use of an anti-static bench and floor mats (the operator was also grounded to this system). I don't have this type of equipment at home, yet I've been successfully working with GaAsFET devices using the simple grounding system described here. Since these devices are important to the quality and performance of low-noise UHF and VHF receivers, you may want to practice some of these safety techniques.

My grounding system starts with the use of a *grounded* (and transformer isolated) soldering iron (a Weller model W-TCP, in my case). Since this iron has a three-prong plug, I use the ground of an adjacent duplex outlet as an additional common ground connection. A banana plug which has been "fattened" a bit to make contact in the ground socket is connected to several limber wires of suitable length. Each wire has an alligator clip on the free end.

A length of ball chain is formed into a loop and slipped over one of the operator's wrists and clipped to one of the ground wires. Another wire is clipped to the chassis ground of the device being assembled. I usually ground the vice that the chassis is clamped in. Finally, the package containing the FET should be made of conductive material (black foam or foil-lined envelopes). This material should be connected to ground before you extract the device for installation. I tear back one corner of the envelope and connect the ground wire. Granted, this system is not up to commercial standards, but it seems to work well, and it certainly protects the FETs better than doing nothing.

Surface-Mount Soldering

By Paul D. Husby, WØUC

(From *QST*, June 1991)

Surface-mount devices and boards are a great invention and a joy to work with once you get comfortable handling and soldering the tiny devices. My soldering routine is slightly different from that suggested by Bryan Bergeron, NU1N[1]. Tinning both pads may lead to an installation in which the device is not flat and close to the board. Or, worse, a fragile chip device may be left physically stressed. I prefer this routine:

- Tin only one of the pads, and let it cool.
- Set the device in place. While pressing the device very slightly with a toothpick, reheat the tinned pad until the device sinks flat to the board. Allow the connection to cool. Solder the second terminal, touching the iron only to the pad.
- After the device has again cooled, touch up the first solder joint as necessary.
- Use silver solder.

With this method, I get a 100% success rate of devices that are flat, straight and well-soldered.

Finding silver solder can be a problem in some locations. (Radio Shack carries 62/36/2 solder [RS 64-013].)

[1]B. Bergeron, "A Surface-Mount Technology Primer—*Part 2*," *QST*, Jan 1991, pp 27-30; see p 29, Fig 12.

Caveats For Choosing Microwave Capacitors

By Bob Atkins, KA1GT

(From *QST*, August 1989)

Physically, chip capacitors are simply small, leadless capacitors. But all physically small, leadless capacitors are not necessarily microwave-quality components. Very small chip capacitors have become much more common as a result of their use in miniaturized circuits. Circuits that operate at only a few megahertz (or tens of megahertz) can get by using inexpensive chip capacitors; the problem is that the dielectric materials used in inexpensive chip capacitors show very low loss at VHF/ UHF, but are entirely unusable at 10 GHz.

One necessary characteristic of microwave-rated components is low dielectric loss at microwave frequencies. A second characteristic of all capacitors—including chip versions—that comes into play is the presence of undesired series and parallel inductances that result from the device packaging. These inductances not only have reactances; they also result in series and parallel device resonances. Where capacitors are used as bypass or dc-blocking devices, series resonances aren't a problem because impedance is minimized. Parallel resonances, however, can be highly detrimental to circuit performance. A third consideration when chip capacitors are used in microstripline circuits is that they create physical discontinuity, and cause some reflection of incident power as a result. This, too, can give rise to losses. Well-designed circuits use capacitors that cause minimal impedance discontinuities.

When a circuit calls for a particular type and value of capacitor, use of the specified component may be critical to circuit performance. Thorough designers take into account the factors discussed above, and select capacitors with low loss when designing equipment.

Loss data on a capacitor can be obtained by testing its effect when used as a dc block in a microstripline circuit. To do this, etch a board with a microstripline of the desired impedance, and leave a small gap in the line. The gap can be bridged by either the capacitor under test or a length of copper foil. First measure the circuit loss when the gap is bridged by the copper foil; then with the capacitor in place of the foil. The difference in attenuation is the additional loss caused by the capacitor.

Loss data for one line of commercial microwave chip capacitors rated for use at frequencies up to 4.2 GHz shows that one particular 120-pF capacitor has a loss of almost 0.4 dB at 3 GHz. A 100-pF capacitor from the same series shows less than 0.05 dB loss at that frequency. In this case, substitution of the 120-pF capacitor for the 100-pF capacitor would result in considerable performance degradation in a 3-GHz circuit—not from the change in capacitance, but from associated packaging effects. This also applies to nominally equivalent capacitors from different manufacturers. If you make substitutions without knowledge of these factors, you may find yourself in unexpected trouble.

Further references to microwave components and microstrip circuitry can be found in the "New Frontier" columns in January 1981, December 1981, April 1982 and June 1988 *QST*.

3

Transmitting and Receiving Equipment

Contents

A Simple Beacon System

By David G. Meier, N4MW

(From *Proceedings of Microwave Update '91*)

Introduction

Beacon transmitters provide several primary benefits. They expose enhanced propagation, indicating conditions we might otherwise not observe. They produce a constant signal source, helping assure that receiving systems are functioning properly. On some bands, beacons mark the narrowband/weak-signal portion of the band, discouraging encroachment of wide-band activities. Band plans and a beacon list are published in the ARRL *Repeater Directory*. The beacon system described in this paper is proving beneficial on all counts.

Perceiving operational and technical obstacles, few are willing to equip stations for relatively unoccupied bands. Many amateurs are unfamiliar with weak-signal VHF/UHF/microwave propagation characteristics, believing the "line-of-sight" myth. Demonstrations of beacon reception to potential VHF/UHF/microwave enthusiasts can help dispel the myth.

This paper documents progress on the N4MW/B beacon system (Fig 1). Since the system design depends on availability of surplus components, a general description emphasizing integration of diverse equipment is presented, rather than a particular construction plan. I hope that the information provided in this paper assists and encourages others to implement needed beacons on multiple bands and in diverse areas.

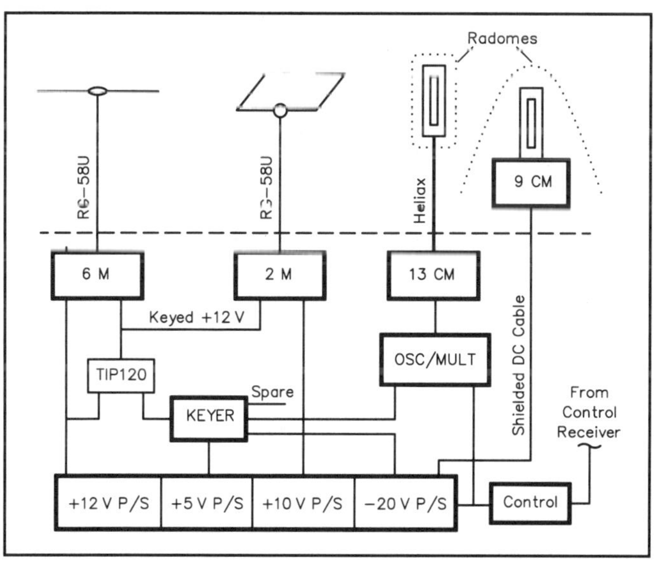

Fig 1—Block diagram of the N4MW Beacon System.

Background

FCC rules permitting beacon operation are found in Part 97 Sections 97.84 and 97.87. Beacon transmissions are defined as one-way transmissions conducted to facilitate measurement of radio-equipment characteristics, observation of propagation or transmission phenomena, or other related experimental activities. Rules for unattended beacons are stricter than for attended beacons. Here is a summary of the rules pertaining to unattended (automatically controlled) beacons:

- Operation is limited to one concurrently operating transmitter per band.
- Appropriate control provisions/procedures are required.
- If the FCC orders, operation must cease until problems are resolved.
- Operation is permitted on the following frequencies:
  - 28.2-28.3 MHz
  - 50.06-50.08 MHz
  - 144.275-144.3 MHz
  - 222.05-222.06 MHz
  - 432.3-432.4 MHz

Operation is permitted on any authorized (non-specific) frequency above 450 MHz.

- Below 450 MHz, permitted emissions are N0N A1A, F1B or J2A. Above 450 MHz, any authorized emission type is permitted. For F1B and J2A emissions, maximum radio- or audio-frequency shift is 1000 Hz.
- A Technician-class or higher license is required.

Power output is limited to the minimum necessary, but may not exceed 100 watts.

- For Morse identification, use the /BCN or /B call-sign suffix.
- The maximum interval between identifications is one minute.

ARRL has adopted beacon subband recommendations in

addition to those specified in the FCC Rules. These subbands are:

902.2-902.4 MHz
1296.3-1296.4 MHz
2304.3-2304.4 MHz
3456.3-3456.4 MHz
5760.3-5760.4 MHz
10368.3-10368.4 MHz

Design

All transmitters are commercial or military surplus, keyed simultaneously using a single keyer. Optoisolators are used instead of mechanical keying relays, providing a flexible and reliable means of keying different types of transmitters with different keying requirements from a common keyer. Common power supplies provide power to multiple transmitters where needed. Equipment is located inside a controlled access room, mounted in a locked EIA cabinet rack. On-off control of primary power is provided by a 23-cm FM link and a DTMF (Touch-Tone) control-code sequence.

Keyer

The beacon code keyer is the standard W4RFR EPROM design with the relay eliminated (Fig 2). Four 4N30 Darlington-output optoisolators provide outputs to key the various transmitters. Ideally, the transmitters would all be keyed by switching a logic-level input to ground. More often, current must be switched at the power supply input to the transmitter amplifier stages. The 4N30 will not handle enough current, so a TIP20 Darlington transistor is added as a current amplifier.

The keyer is constructed on a Radio Shack project board. The four optoisolators are mounted on the keyer board. The board is mounted in an aluminum box. A barrier terminal strip allows for connection to a +5-volt supply and to the various keyed devices. An externally visible LED flashes in sync with the keyer.

Programming the EPROM is simple, but requires an EPROM "burner." The 2716 EPROM is a 2k × 8-bit device. The beacon keyer uses only 2 of the 8 bits: one for the code sequence and the other for resetting at the end of the message. Each bit represents one element of a Morse character. Dots are 1 bit; dashes are 3 bits. One bit separates dots/dashes within characters. Character spaces are 3 bits and word spaces are 7 or 9 bits. Long dashes are 50 or more bits long. The EPROM is programmed to contain byte values of 1 or 0 to represent Morse elements and element spaces. A string of 2s forces reset to the beginning of the message. The rest of the EPROM con-

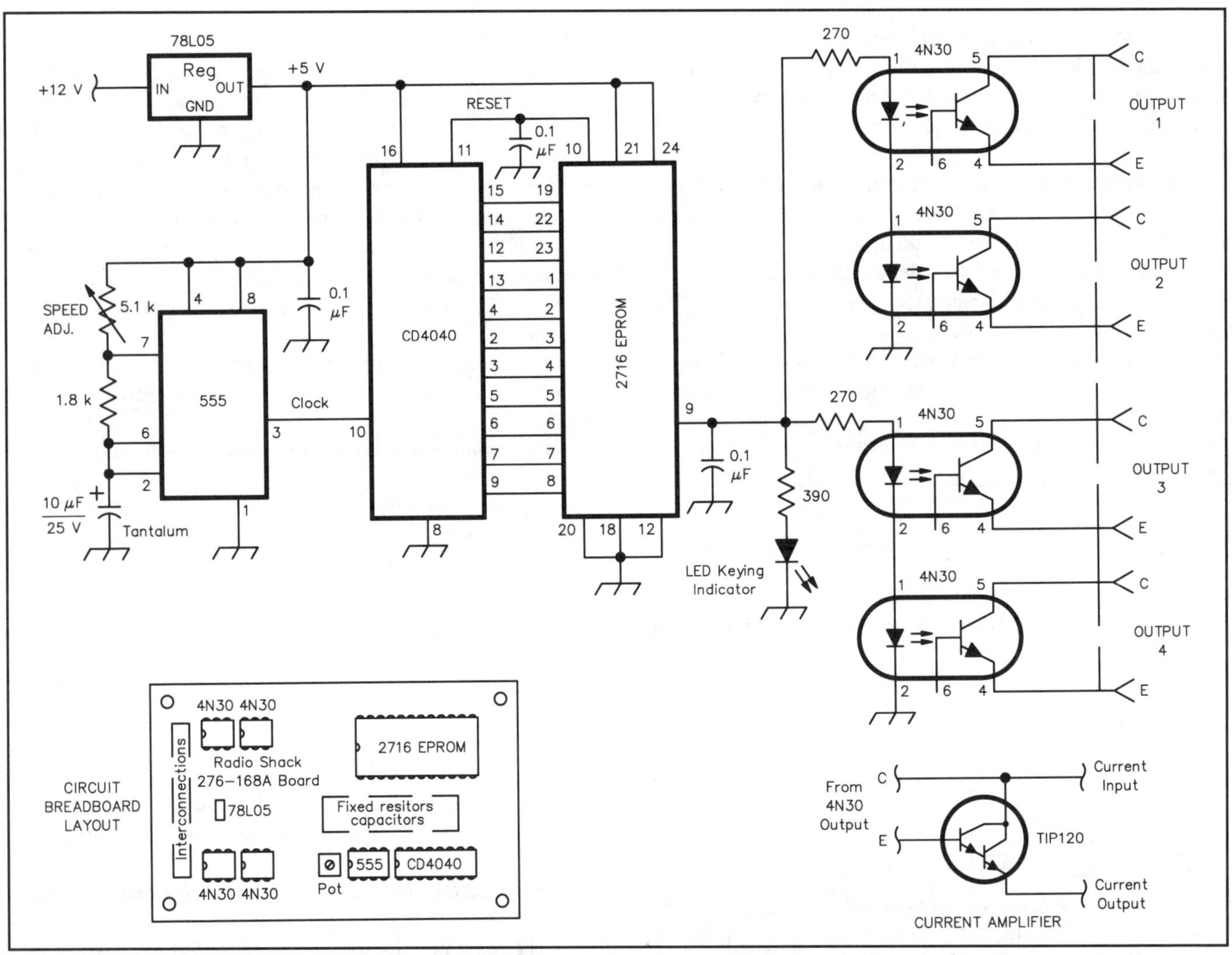

Fig 2—Schematic and layout of the modified W4RFR Beacon Keyer.

tents do not matter. The EPROM message can be seen in the following example, which uses 215 of 2048 bytes, or about 11% of the potential message capacity. For the code sequence:

DE*(WORD SPACE)*N4MW/B*(WORD SPACE)*EM55*(LONG DASH)(SHORT PAUSE)(RESET)*

Program the EPROM as follows:

11101010001000000011101000101010101110001110111000101110111
0001110101011101000111010101000000010001110111000101010101000
10101010100011111111111111111111111111111111111111111111111
11111000000000000000000000000022222

The message content must be arranged to provide at least one identification each minute. This can be done by keeping the message short, or by interspersing two or more identification sequences within the message string. Keying speeds between 10 and 15 WPM are reasonable. A long dash is desirable for equipment adjustment and strength measurement.

Equipment

With 70- and 23-cm beacons already operating locally, 2 meters seemed both the easiest and most immediately useful band to implement. I mounted an exciter board from an RF Communications Corp. RF-403 FM transmitter in an aluminum box. A BNC connector and 2 feedthrough/bypass capacitors provide all necessary connections. A constant +10 V supplies the oscillator and low-level transmitter stages. A keyed +12-V line to the high-level stages produces a chirp-free 4-W signal on 144.280 MHz. A simple gamma-matched "squalo" built from copper pipe produces reasonably omnidirectional, horizontally polarized radiation.

The next band I added was 13 cm. I retuned a Motorola transmit strip to provide 72-MHz output. A homebrew multiplier produces 288-MHz drive and is keyed by one of the keyer optoisolators. A surplus AN/GRC-144 telemetry transmitter provided a multiplier/amplifier assembly and diode doubler capable of transforming the 288-MHz input to 4 watts at 2304 MHz. I constructed an Alford slot from copper pipe, and fed it through 10 meters of ½-inch Heliax. I made a weatherproof cover for the antenna from a length of PVC pipe, capped at the top. The cable end and antenna slips inside the PVC cover, which is clamped to the top of a mast, so that the antenna clears the mast.

The 6 meter transmitter, a General Electric exciter board, requires a constant +12 V and keyed +12 V. To minimize current switched by the primary-keyed +12-V source, I installed another optoisolator and Darlington transistor inside the transmitter box, to switch the constant +12 V. The antenna is a dipole oriented east/west.

The 9-cm beacon uses a Frequency West PLO "brick" source producing 50 mW to an Alford slot. Source and antenna are mounted on a mast pipe under a weatherproof cover made from a TVRO feed cover. dc voltage is supplied through a shielded cable. Power supply current must be continuous to maintain frequency lock. An expedient keying method is to frequency-shift key the source, by slightly varying the negative 20-V supply. One of the keyer optoisolators modulates the dc supply by about 0.1 V, producing a distinctive frequency shift. The supply voltage is modulated by one of the keyer optoisolators, which switches a resistor in the power-supply feedback circuit.

TABLE 1

Alford Slot Antenna Scaling Dimensions

Dimensions (Inches)	*FREQUENCY (MHz)* 902.3	1296.3	2304	3456	5760	10368
Tube Diameter	1.86	1.29	0.73	0.49	0.29	0.16
Tube Length	24.23	16.86	9.49	6.33	3.8	2.11
Slot Width	0.26	0.18	0.1	0.07	0.04	0.02
Slot Length	23.02	16.02	9.02	6.01	3.61	2
Balun Length	2.61	1.82	1.02	0.68	0.41	0.23

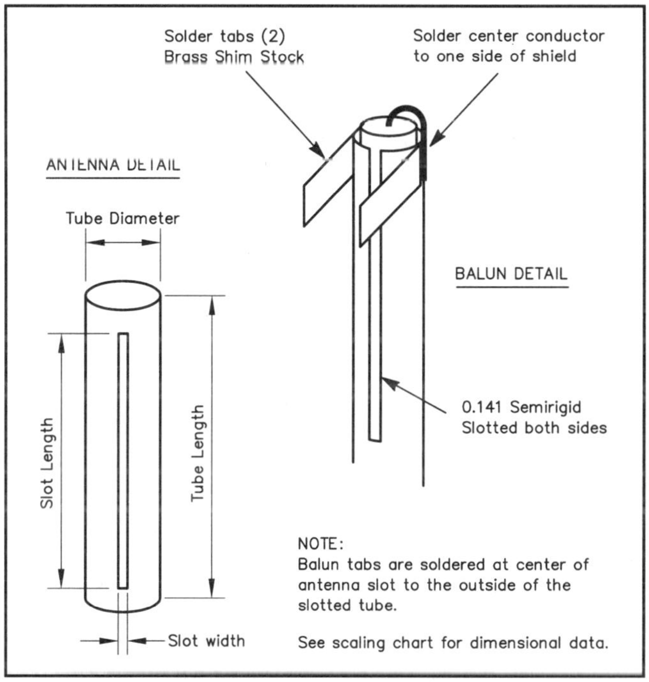

Fig 3—Alford Slot Antenna Construction.

The 13- and 9-cm Alford slot antennas are based on the G3JVL 2304-MHz Alford slot (Table 1 and Fig 3).[1] Tube wall thickness and length beyond the slot should not be critical. Coaxial 4:1 balun length is for PTFE dielectric line (0.141-in. diameter semirigid coax). Where feasible, make the balun from the feed line, thus eliminating connectors.

[1]M. Walters, G3JVL, "An Alford slot antenna for 2.3 GHz," *Radio Communication*, June 1983, p 527.

Results

Reception reports have been received from several distant stations. Paul Wilson, W4HHK has used the 13-cm beacon to help diagnose a failed preamp. Rex Turner, W5RCI, receives the 13-cm and lower-frequency beacons consistently over a 100-km (60-mile) path. Al Ward, WB5LUA, has received the 13-cm beacon 20 dB above the noise over a 672-km path. At my home, 14-km from the beacon site, the 13-cm beacon is received with no antenna connected to the converter! It is interesting to monitor the 13- and 9-cm beacons while mobile in motion—the Doppler and multipath effects resulting from vehicle motion are extremely pronounced, sometimes sounding like aurora. I receive the 9-cm beacon using a TVRO LNA as both antenna and preamplifier inside my shack. W4HHK receives it at ground level, over a 28-km obstructed path.

Conclusion and Acknowledgments

I hope this paper encourages additional beacon operation. Beacon operators everywhere appreciate reception reports, which help to establish normal performance expectations, as well as record unusual propagation trends.

I acknowledge and thank Ray Escue, K4RDK, for securing the beacon site, Ken Schildt, N4VSD, for locating the site, James Butler, KB4LJV, for programming the keyer EPROM, Bill Dearing, N4HKS for the logistics, Paul Wilson, W4HHK for inspiration and assistance, and my wife Cissy, N4ZRW for her understanding and support.

A 432-MHz Transverter

By Greg McIntire, AA5C

I recently have become interested in VHF/UHF and microwave activities. Not having built any equipment that operates on the higher frequencies I decided to pursue the construction of a 432-MHz transverter as a means to learn about the design and construction techniques for UHF. A survey of Amateur Radio publications did not yield any "modern" transverter articles. The few that I did find used tubes.

A block diagram of the station is shown in Fig 1. The IF for the project was to be a Kenwood TS-820S HF transceiver. This radio has transverter input and output ports. The transmit port provides 0 dBm and the receive port is switched between the transverter input and the primary SO-239 connector. Eventually, I plan to add a high-power amplifier to the 432 station. The TR relay to connect the transverter input and output to the antenna is external to the transverter so that it can be moved to the amplifier at a later time.

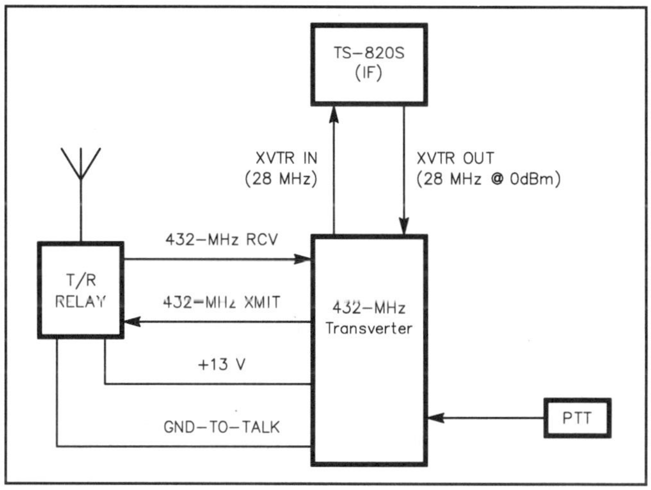

Fig 1—Block diagram of the AA5C 432-MHz station.

Transverter Block Diagram

The block diagram of the transverter is shown in Fig 2. Other than the LO and IF frequencies, there is nothing specific to the 432-MHz band in the block diagram, and this approach is applicable to other frequencies. The basic transverter is a four-port system with the TR relay external to the main unit. This permits additional transmitter power amplifiers to be added to the transmit chain with the TR relay downstream.

The local oscillator multiplies the 101-MHz oscillator by four to 404 MHz. Two times-two multiplier stages are used. Separate mixers are used for transmit and receive converters. A power splitter provides independent outputs for the transmit and receive mixers.

The receive converter chain consists of a low-noise RF amplifier followed by a 432-MHz filter. (The tuned circuits of the RF amplifier serve to limit some of the out-of-band energy.) The two-pole filter provides attenuation of the image frequency ahead of the mixer. The mixer is followed by a two-pole IF filter and an IF amplifier. The IF-amplifier gain can be adjusted to control signal level passed to the HF rig used as the 28-MHz IF.

The transmit mixer inputs are the 28.1 MHz IF (at about 1 mW) and the 404 MHz LO. The transmit mixer is followed by a two-pole filter to attenuate the 376 MHz signal from the output and pass the desired 432-MHz signal. (This should be done right after the mixer before any amplification, else the undesired products will be amplified by the untuned MMIC chain.) The MMIC chain amplifies the 432-MHz signal to the maximum MMIC power levels: Transistors then take over. Impedance-matching circuits in the transistor stages serve to limit out-of-band signals. The final stages bring the transmit signal up to about 7 W.

Construction Techniques

Double sided G10 circuit board material was chosen because the second side of the board makes a good ground plane. This ground plane is a necessity for microstrip circuits, which are used in the power amplifier of this transverter.

I started with a 5- × 18- in. board (Fig 2). The 5-in. dimension was determined by a heat sink I had on hand. The maximum width of the transverter would match that of the TS-820S because I wanted to place the unit on top of the TS-820S.

General construction guidelines that I followed in the unit include:

1) Placing all active devices and circuits on the top side of the board

2) Using 1000-pF ceramic feed-through capacitors to get to the bottom side of the board. I used the solder-in type and soldered around the entire circumference of the feed-through on both sides of the board. This connects the two ground planes together.

3) I bypassed each feedthrough with a 0.01-µF ceramic ca-

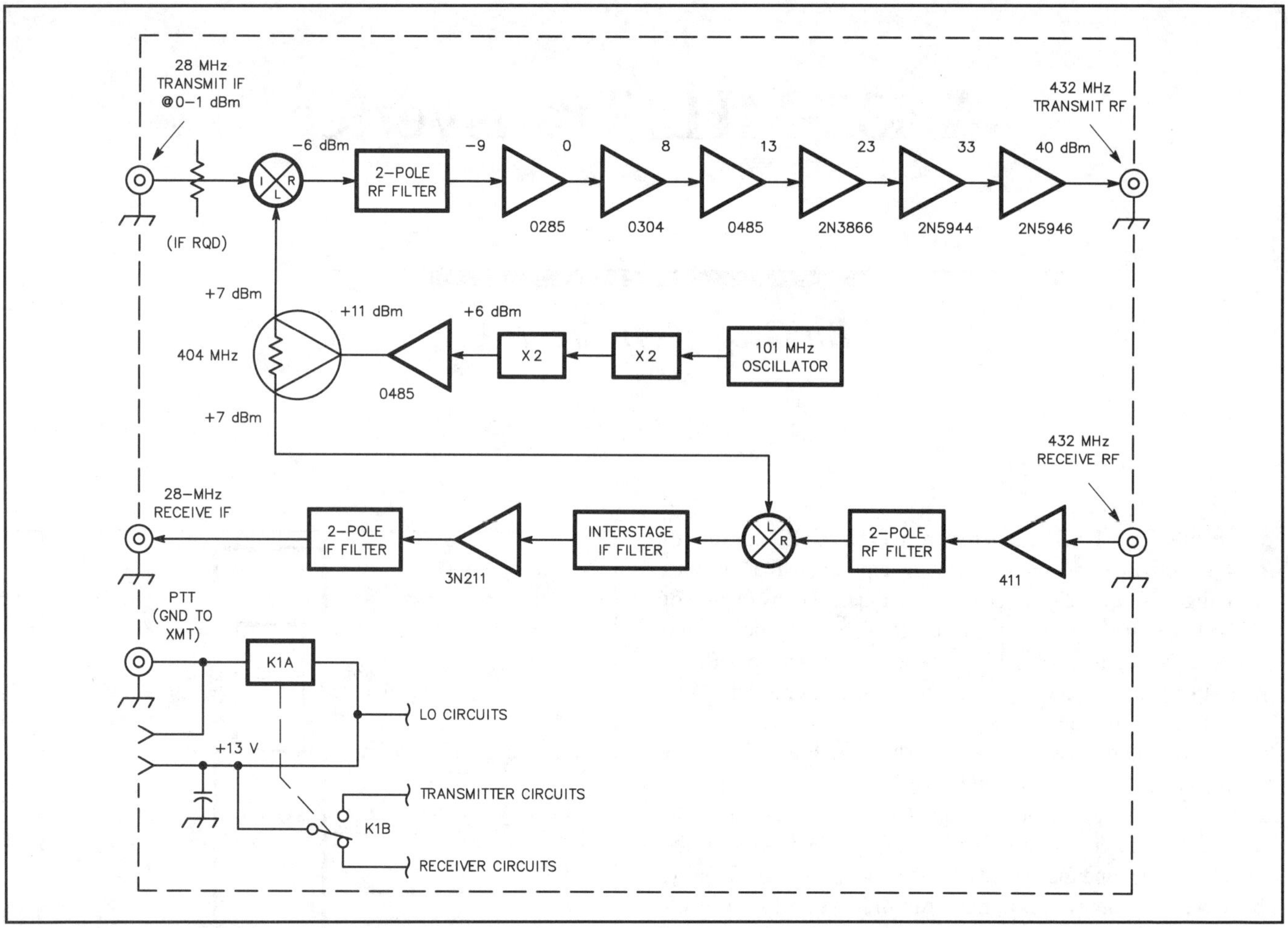

Fig 2—Block diagram of the transverter.

pacitor on the bottom side of the board as additional decoupling to prevent any HF oscillations through the supply leads.

4) All supply leads and biasing circuitry are located on the bottom side of the board.

5) Shielding was included as shown in Fig 2. The shields are placed to divide either major functions or circuits that could couple easily: high-gain, high-power or high-impedance stages, for example. The receiver RF amplifier MMIC chain and the power amplifier are thus appropriately shielded. Most of the system is at the 50-Ω impedance level.

I didn't have a metal chassis, so I tried building the chassis and shields out of PC board material. One-sixteenth-inch thick material is easily worked with a good pair of tin snips and a 60-W soldering iron. The chassis covers are secured by means of nuts soldered into the chassis corners and screws through the covers. Small pieces of scrap PC board are also handy: I used ⅜-inch squares for terminals on the bias side of the transverter. To attach a square to the board, I tinned one side of the square and the area of the board where I planned to locate it. With a soldering iron in contact with both pieces, I lowered each square into position. Be sure not to slap the squares down, as solder splashes can result.

The transverter grew around the local LO. Placing the LO between the receiver and transmitter sections seemed ideal to minimize the need for coaxial interconnects. The LO was fully tested first. A shield running the depth of the board separates the LO from the receiver section, which was built next. Once complete, the board was cut and one edge of the "chassis" was installed. A shield was added to the other side of the LO and the transmitter front end was added. A partition was added to separate the power amplifier from the transmitter low-level stages. After completing the PA the board was cutoff at the other end and that end of the chassis was installed. Small partitions of PC material were cut to match the height of the heat sink. These permit a cover to be installed over the bias side of the board, which then became the top of the chassis. The heat sink was exposed and placed on top for ventilation. Fig 3 shows the layout I used for the transverter.

Local Oscillator

A good, stable LO is critical to any UHF or microwave transverter. I adapted the 384-MHz stage of an 1152-MHz LO designed by Al Ward, WB5LUA, for my 404-MHz LO. The original circuit had the crystal in the emitter-collector circuit of the oscillator with no buffer amplifier following. This circuit drives the crystal pretty hard, and drift and "FMing" on current peaks were problems—even with the circuit supply Zener regulated to 9 V. Al was having drift problems with a 2160 LO that used the same circuit. He redesigned the oscillator using a Butler configuration and a buffer stage. That is the circuit presented in Fig 4. The parts list is in Table 1.

The oscillator, Q1, uses a 101-MHz, fifth-overtone, 0.001%-tolerance crystal. The 1-10 pF trimmer in series with the crystal can be used to adjust the LO frequency. An emitter follower, Q2,

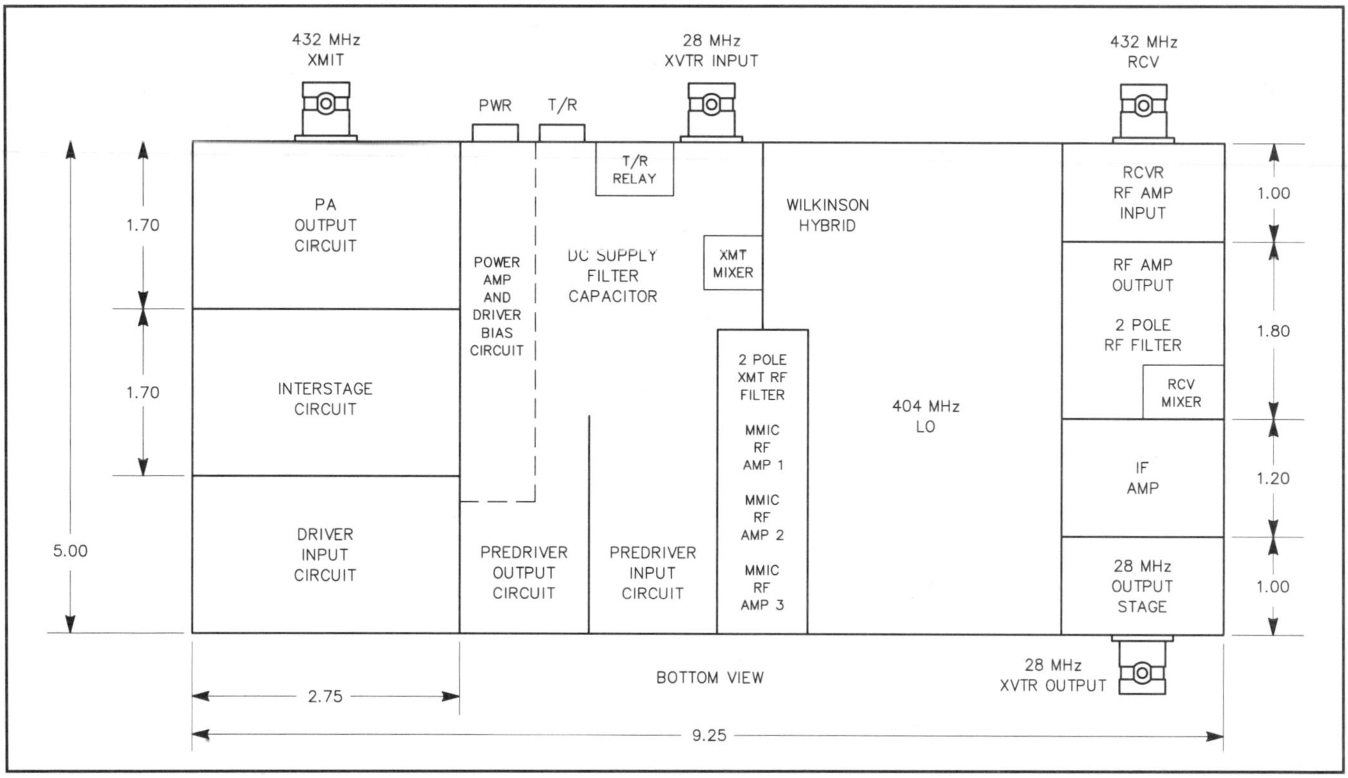

Fig 3—Layout of the transverter.

buffers the oscillator from the first doubler, Q3. The second doubler, Q4, produces the desired 404-MHz signal at +3 dBm. L6 is tapped at the 50-Ω point (a half turn) and the signal is fed into a '0485 MMIC. The '0485 amplifies the signal to about +10 dBm. A Wilkinson hybrid power splitter then produces two 50-Ω ports with +7 dBm signal levels. (This LO signal level is ideal for the SRA-1 transmit and receive mixers.) These signals are routed to the receive and transmit mixers.

The LO is best tuned with the aid of a spectrum analyzer. I was able to tune up the LO using a diode RF detector, and the results were verified on a spectrum analyzer. Access to a spectrum analyzer obviously makes the job easier.

In order to tune the LO with an RF detector, first connect the detector to the emitter of Q and tune C4 and L1 for maximum signal. The turns of L1 can be compressed or spread for adjustment. The positions of L1 and C4 affect the startup characteristics of the oscillator, and they should be positioned for consistent start up when power is applied to the circuit.

It is advisable to use a frequency counter to verify that the oscillator circuit is indeed tuned to 101 MHz before proceeding. A wide variation in circuit parameters could result in the crystal resonating on an incorrect overtone.

Move the RF detector to the collector of Q3 and tune the coupling circuit between Q2 and Q3 for maximum signal. The first doubler should be tuned for maximum signal at the emitter of Q4. Initial indications can be obtained across the 330-Ω base resistor of Q4. The 404-MHz output should be peaked by examining the signal level at each port of the power splitter, with the other port terminated in 50 Ω.

Since the final doubler circuit tuning is somewhat touchy, it may be easier to initially leave the MMIC input disconnected and terminate the output side of the 100-pF silver-mica capacitor that taps L6 in 50 Ω. Then tune for maximum signal across the resistor. Finally, tune the trimmer on the input side of the power divider for maximum signal at both output ports.

Receive Converter

The receive section of the transverter converts incoming 432-MHz RF to a 28-MHz IF. This circuit is shown in Fig 5, and the parts list is in Table 2. The front-end circuit uses an NEC 41137 GaAsFET. This circuit effectively sets the noise figure for the receiver. The 41137 is a high-gain, low-noise device, and is ideally suited for weak-signal receiver front ends. The specified noise figure of the 41137 is less than 0.8 dB at 500 MHz. I adapted a preamplifier circuit designed by Kent Britain, WA5VJB, for this application. This circuit provides a high output impedance for driving the two pole RF filter instead of a 50-Ω output for driving coax. A 1-kΩ resistor is placed inside inductor L2 to "de-Q" the drain circuit. The 41137 has about 20 dB gain and the circuit will easily break into oscillation without the resistor. A two-pole 432-MHz filter attenuates the image frequency prior to the mixer. Inductor L4 is tapped at the 50-Ω point for connection to the receive mixer. All three ports of the double balance SRA-1 have 50-Ω input or output impedances. This signal is fed to the RF port of the mixer. The LO port connects to one leg of the LO power divider.

An excellent discussion of double balanced mixers starts on page 304 of the 1976 *Radio Amateur's Handbook*. This article, by K1AGB, discusses the merits of proper termination of a DBM. One instinctively designs a mixer circuit to provide a proper termination for the desired signal. This article discusses the merits of properly terminating the DBM so that reflections from the undesired mixer products do not interfere with the desired signal. I applied the circuit included in this article to my transverter. The IF amplifier should be a good low noise device such as a 3N204 or

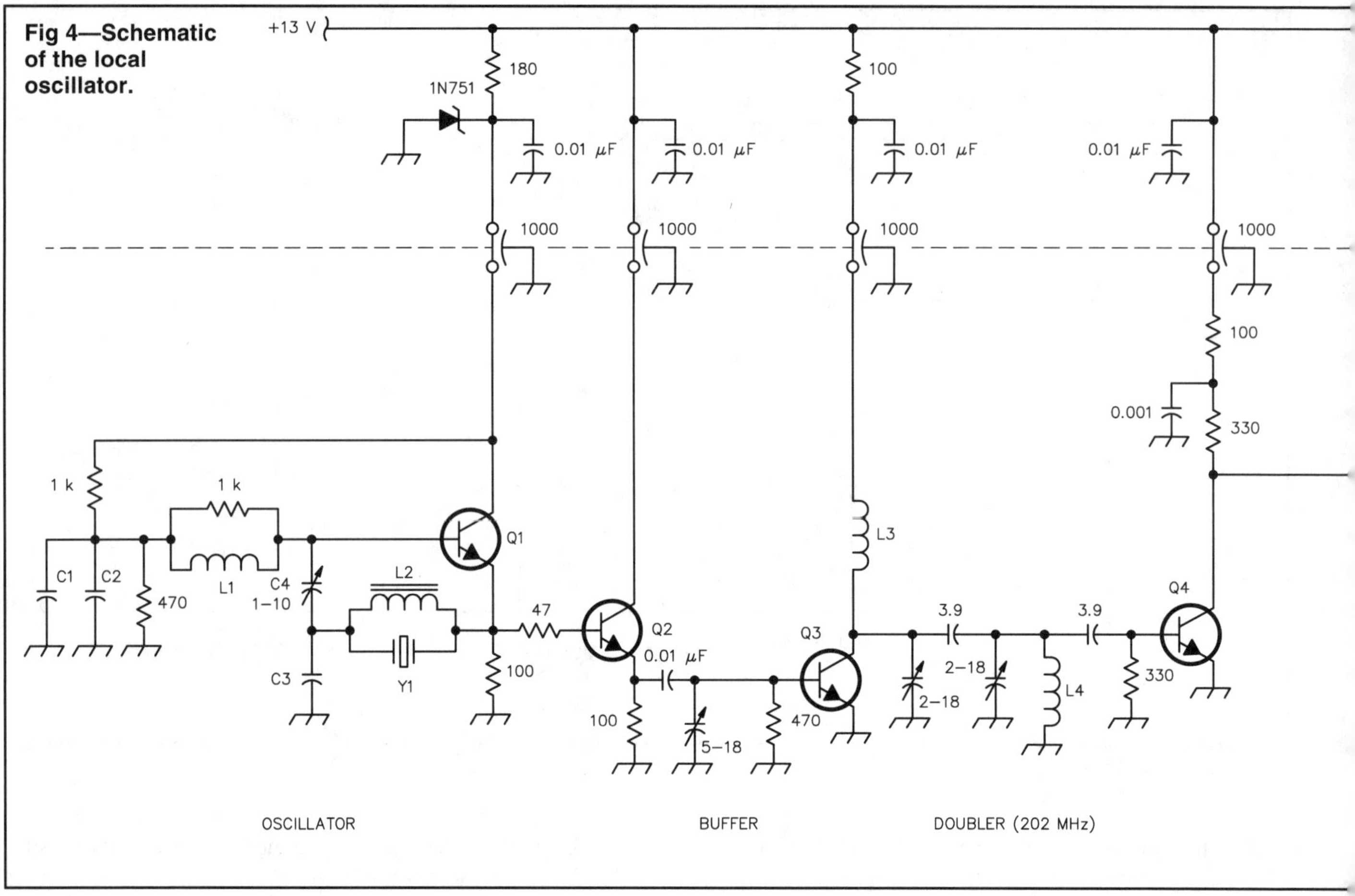

Fig 4—Schematic of the local oscillator.

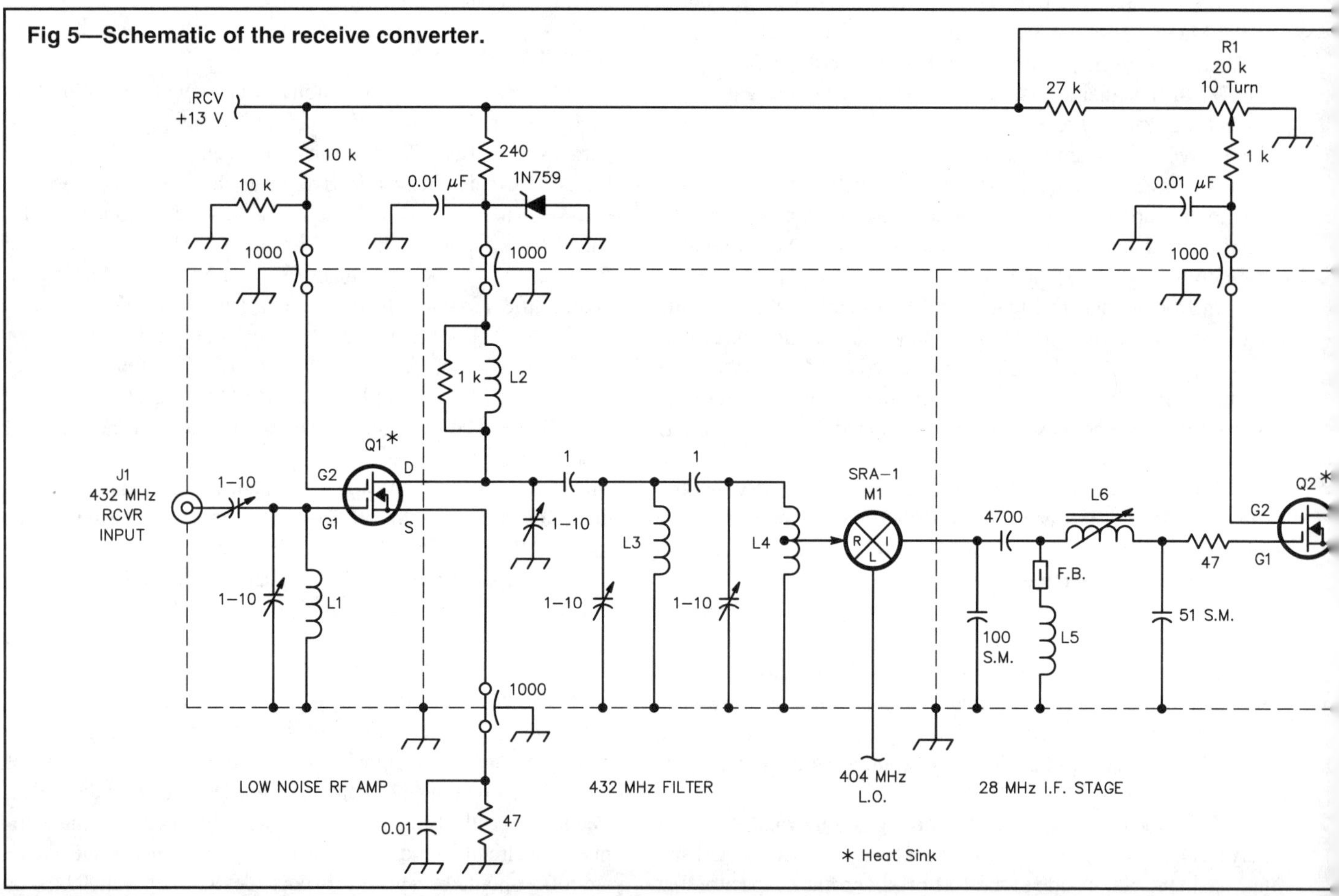

Fig 5—Schematic of the receive converter.

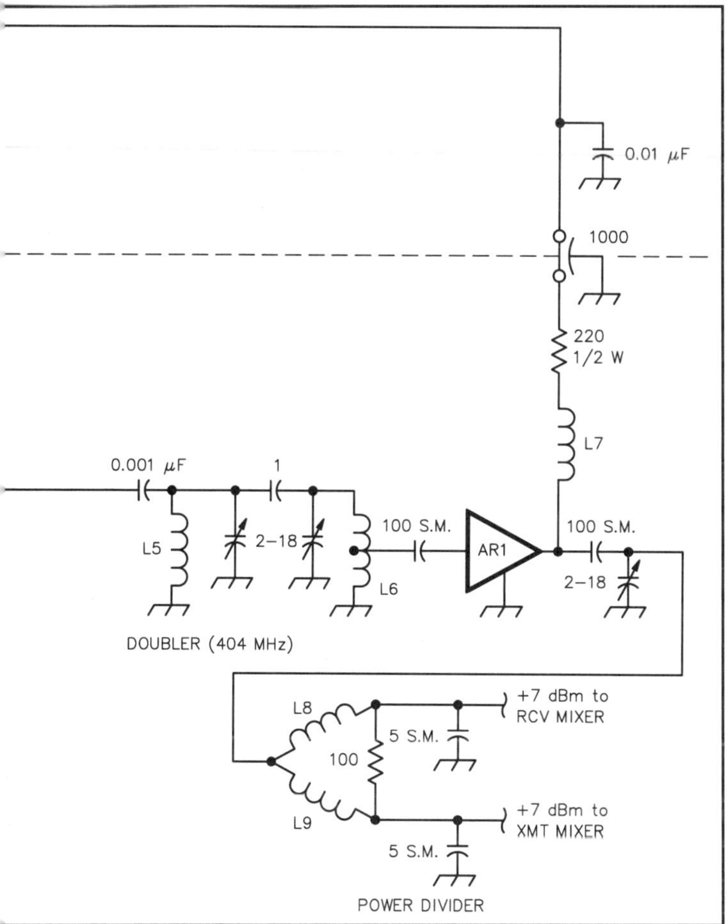

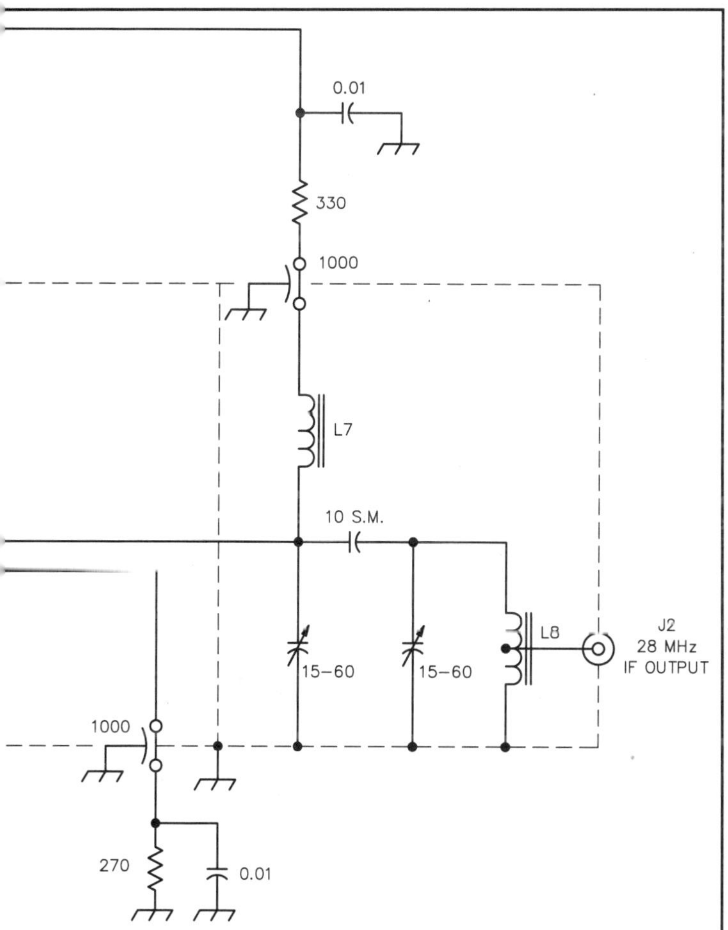

Table 1

Local Oscillator Parts

C1—0.001-μF ceramic.
C2—0.1-μF ceramic.
C3—33-pF NP0.
C4—1-10 pF trimmer.
L1—7 t no. 26, 0.25 ID, 1-kΩ resistor inside.
L2—39-μH molded choke.
L3, L4—3 t no. 16 0.25 ID.
L5—1 t no. 16, 0.25 ID.
L6—1 t no. 16, 0.25 ID, tapped ½ t.
L7—6 t no. 20 on ½-W 1 MW resistor as a form.
L8, L9—1 t no. 20, 0.25 ID.
Y1—101-MHz 5th overtone crystal, HC-18/U case.

Notes

All resistor values ¼-W unless otherwise noted.
All capacitor values in pF unless otherwise noted.

Table 2

Receive Converter Parts

L1—½ t no. 20 0.25 ID, 0.5 in. long.
L2—7 t no. 20, 0.25 ID, 1-kΩ resistor inside.
L3—1 t no. 20 20, 0.25 ID.
L4—1 t no. 20, 0.25 ID, tapped ½ t.
L5—56-μH molded choke.
L6—19 t no. 26 on 0.25 ID coil form, per slug.
L7—18 t no. 28 on T-30-6 core.
L8—18 t no. 28 on T-30-6 core tapped 4 t from cold end.
M1—SRA-1 Minicircuits DBM.
Q1—NEC 41137 GaAsFET.
Q2—3N204, 3N211.

Notes

All resistors ¼ W unless otherwise noted.
All capacitor values in pF unless otherwise noted.

3N211. A 40673 could be used but is not recommended. The drain circuit of the IF amplifier is connected to a doubly tuned output circuit. A dual-gate FET is used to provide stage gain adjustment. The adjustment is effected by using potentiometer R1 to control the voltage on G2 of the FET. A preamplifier or transverter will inject additional noise into the front end of the rig being used either as a baseband rig or an IF. The result is an S-meter reading higher than if the gain stage preceding the rig was turned off. R1 permits setting the S meter reading with the transverter turned on to the same value as if the transverter were turned off. S-meter readings can thus be exchanged. Good weak-signal reporting practice should still include reports of signal strength over the noise floor.

A two-pole 28-MHz filter follows the IF amplifier. I used the same inductors that were used in the "Rochester converters" shown in the 1976 *Radio Amateurs Handbook*. In that application, the circuit was stagger tuned for broad bandwidth and no tuning adjustments, other than taking turns off the toroids, were provided for. I replaced the fixed capacitors with trimmers. Since most activity is at 432.100 MHz, I tuned both sections of the filter for the corresponding IF frequency of 28.100 MHz. Tune up of the

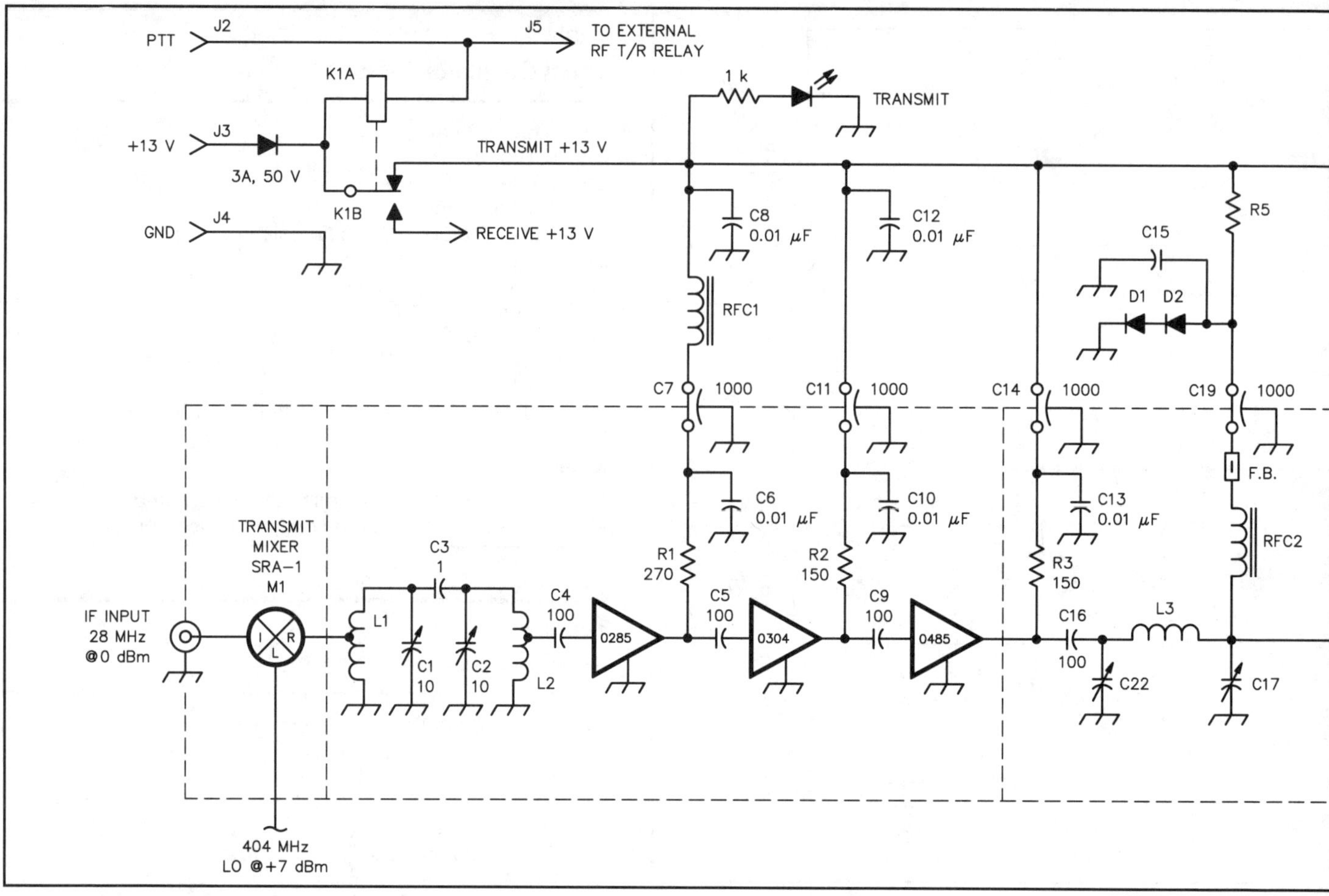

Fig 6—Schematic of the transmitter front end.

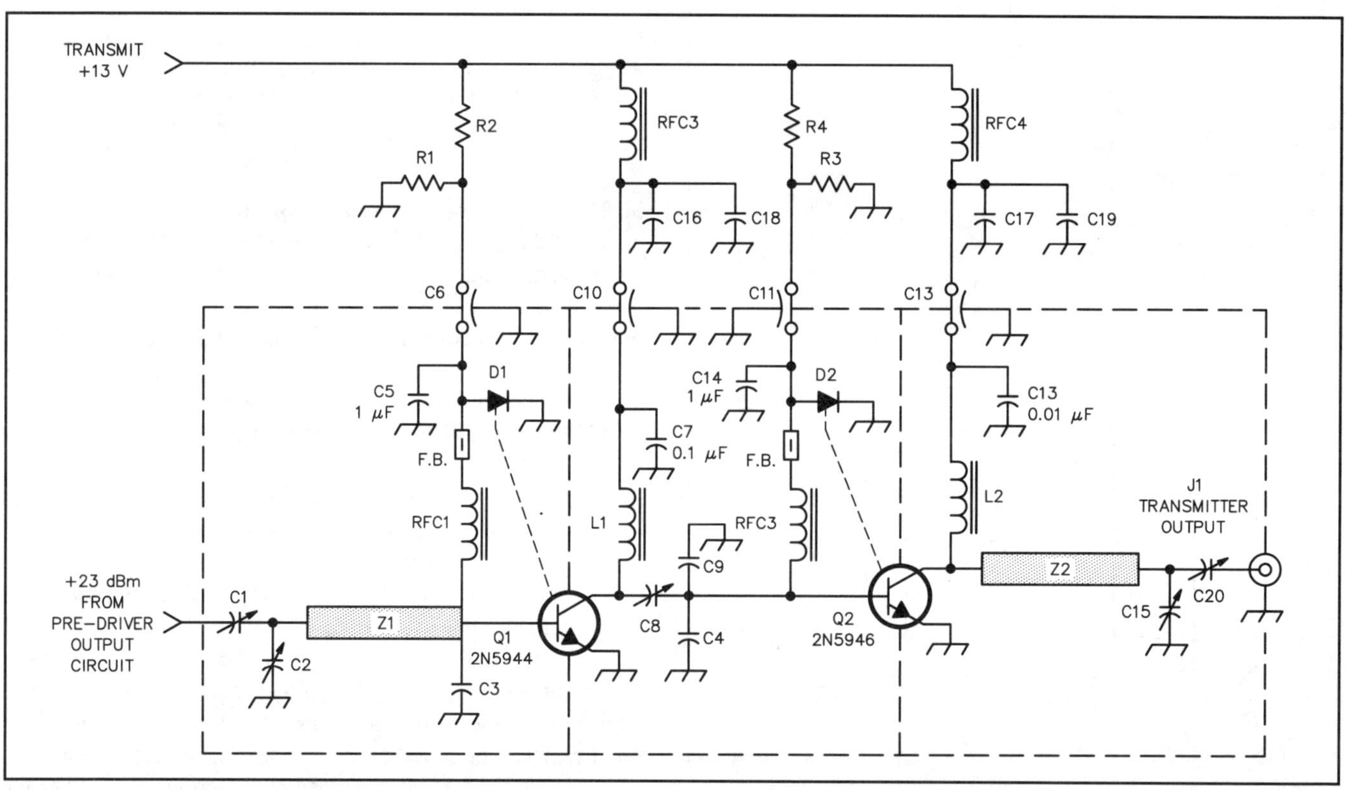

Fig 7—Schematic of the transmitter power amplifier.

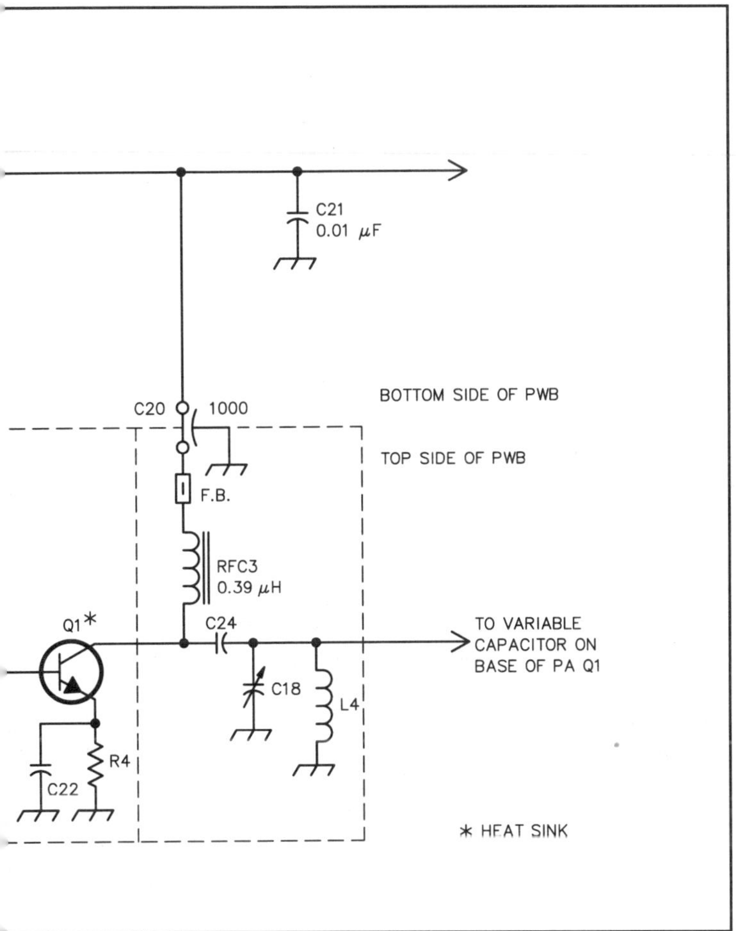

Table 3

Transmitter Front End Parts

Q1—MMIC, 0285 or equiv.
Q2—MMIC, 0304 or equiv.
Q3—MMIC, 0485 or equiv.
C1, C2, C22—miniature trimmer, 1-10 pF.
C3—Ceramic, 1 pF.
C4, C5, C9, C16—silver mica, 100 pF.
C6, C8, C10, C12, C13, C15, C21, C23—ceramic, 0.01 μF.
C7, C11, C14, C19, C20—feedthrough, 1000 pF.
C17, C18—miniature trimmer, 5-25 pF.
C24—0.01-μF ceramic.
D1, D2—silicon diode, 50 PIV, 1N4001 or equiv.
K1—DPDT relay, 5-A contacts, 12-V dc coil.
L1, L2—1 t no. 20 enam, ¼-in. diam, tapped at ½ t.
L3—2t no. 16 enam, ¼-in. diam.
L4—1 t no. 20 enam, ¼-in. diam.
Q1—2N3866 or equiv.
R1—270-Ω, ¼-W carbon composition.
R2, R3—150-Ω, ½-W carbon composition.
R4—15-Ω, ½-W carbon composition.
R5—270-Ω, ½-W carbon composition.
RFC1—molded choke, 1 μH.
RFC2, RFC3—molded choke, 0.39 μH.

circuit can be done several ways. A 432-MHz signal source with variable attenuation is necessary. The rig being used as the IF serves well as the detector. With the LO connected to the mixer, apply a strong 432-MHz signal (–30 dBm) to the 432-MHz input. Make several passes though the stages, tuning for maximum indication on the S-meter. Decrease the signal level as the circuits come into alignment, so that the S-meter reading stays below S9. You should be able to reduce the signal level to less than –120 dBm and still hear a beat note. The circuit could be split if tuning the full receiver proves difficult.

In this case, break the circuit at the input to the mixer and terminate the connection from L4 to the mixer with a 50-Ω resistor. Then, tune the 41137 circuit and the filter for maximum signal strength. Keep the input signal level low enough that the 41137 is not running into compression; for example, below –30 dBm. On the output side of the mixer, drive the circuit connected to the IF port of the mixer with a 28-MHz signal and tune L6 and the two 15-60 pF trimmers for maximum signal. The radio can still be used as the detector for this section of the circuit. A good initial value for R1 should result in about 1.5 volts on G2 of the IF amp. Finally, connect everything back up and make final adjustments for best overall signal conversion gain. Kent Britain discusses tune up of preamplifiers in the 1986 *Handbook for Radio Amateurs*. As noted there, adjustment of the two trimmers on the input of the 41137 are not necessarily the same for best gain or best noise figure. After the receiver section is tuned up enough to be connected to an antenna, find a friend who can transmit a weak signal and tune up on it for best intelligible signal. I had the opportunity to put the transverter on a Hewlett-Packard 8970 noise-figure meter. The overall receiver noise figure for the transverter was measured to be 1.60 dB. Not perfect, but very acceptable. I have assumed 13 dB gain in the 41137 stage. Although more than 20 dB of gain is available with the 41137, the damping to stabilize the stage probably results in about 13 dB of stage gain.

Transmit Converter

This portion of the transverter converts a 1-mW 28-MHz IF input to a +40-dBm (10-watt) 432-MHz output. Refer to the transverter block diagram for signal level and flow. These are described below. The transmitter front end schematic is shown in Fig 6 and the transmitter power amplifier in Fig 7. Associated parts are listed in Tables 3 and 4 respectively. The 28-MHz IF signal is input directly into the IF port of the SRA-1 mixer. This is the specified input level. Attenuators or amplifiers may be necessary if the 28-MHz signal level available is more than a few dB different from 0 dBm. 404-MHz LO energy from the power splitter is input to the LO port of the SRA-1.

The mixer output is taken from the RF port. The desired 432-MHz signal level is about –6 dBm at this point. A two-pole 432-MHz filter follows the mixer. This passes the desired 432-MHz signal and attenuates the 376 MHz lower mixer product. A DBM is useful here, as the 404-MHz and 28-MHz signals are greatly attenuated, due to the inherent nature of a DBM. Filter loss was measured at 3 dB and the resulting signal at the output of the filter is –9 dBm. The –9 dBm signal is input to a series of three MMICs for amplification. The upper limit of signal level that commonly available silicon MMICs can work with is about +13 dBm, for the 04 series of parts.

The data sheet specifications are 12 dB for the 0285 and 0304 MMICs, and 8 dB for the 0485 device. Overall three stage gain would be 32 dB. I got 22 dB overall. I don't find this disturbing, as the specification sheet values are based on tests where the devices

Table 4

Power Amplifier Parts

C1, C2—compression trimmer, 4-20 pF; ARCO 402 or equiv.
C3—UNELCO, 10 pF.
C4—UNELCO, 25 pF.
C5, C14—ceramic, 1 μF.
C6, C10, C11, C12—feedthrough, 1000 pF.
C7, C13—ceramic, 0.1 μF.
C8—compression trimmer, 55-250 pF; ARCO 426 or equiv.
C15—Johanson trimmer, 1-10 pF.
C16, C17—dipped tantalum, 1 μF.
C18, C19—aluminum electrolytic, 47 μF, 25 V.
C20—compression trimmer, 8-45 pF, ARCO 403 or equiv.
D1, D2—silicon diode, 50 PIV, 1N4001 or equiv.
J1—BNC connector, chassis mount.
L1—6 t no. 20 enam, ¼-in. diam.
L2—5 t no. 20, ¼-in. diam.
R1, R3—10-Ω, ½-W carbon composition.
R2, R4—150-Ω, 1-W carbon composition.
RFC1—molded choke, 0.1 μH.
RFC2—molded choke, 0.39 μH.
RFC3, RFC4— 2½ t no. 20 ferrite core.
Z1—0.95 × 0.25 microstrip; see text.
Z2—see text.

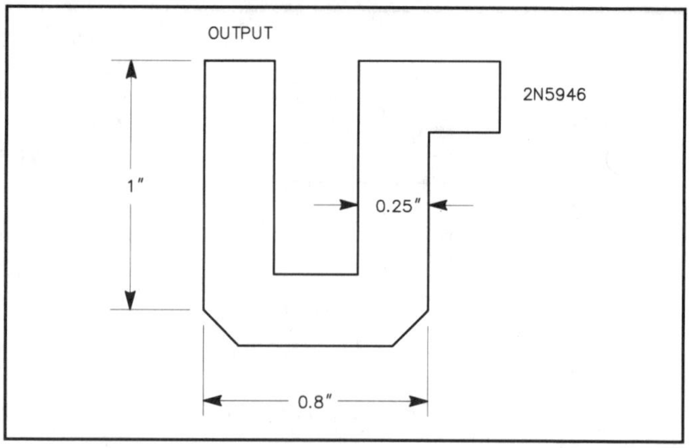

Fig 8—This copper microstrip is used in the 2N5946 output circuit.

are well matched. Mismatch losses between stages will decrease the available gain. The +13 dBm output of the 0485 is input to a 2N3866 transistor. This stage produces 10-dB gain and outputs the signal at +23 dBm to the next stage. L networks are used for input and output impedance matching. Both input and output circuits are impedance matched to 50 Ω. This permits testing of the stage independently, if desired. The quiescent collector current is set up by controlling the current into D1. The quiescent collector current is set to 6 mA, about 5% of the expected maximum collector current. This puts the device into the linear AB mode of operation. Next, a 2N5944 amplifies the +23 dBm level to +33 dBm (2 watts). This is followed by a 2N5946, which provides up to 10 watts output. Several circuits in the Motorola *Power Devices* catalog served as the model for the circuits shown here. A capacitive L and a 0.25-inch wide by 0.9-inch long section of microstrip is used to match the 50-Ω input impedance to the base impedance of the 2N5944. A "U" shaped section of microstrip shown in Fig 8, and a mirrored capacitive L transform the low collector impedance of the 2N5946 back to 50 Ω. This section of microstrip was shaped this way to conserve board space. There was enough inductance in the ARCO trimmer capacitor in the interstage circuit that a microstrip section was not necessary. The biasing technique for the driver and final is similar to that used for the 64020. The maximum collector current for the 2N5944 is 400 mA and the quiescent collector current was set to 20 mA. Maximum collector current for the 2N5946 is 2 A and the quiescent current was adjusted to 100 mA.

This current adjustment is again made by varying the current into a forward-biased diode across the base. It is important to thermally couple the bias diode to the transistor for these higher power devices. I put the bias diode directly across the lid of the power transistors and applied heat-sink grease, to effect a close thermal coupling. Experimentation may be required in the decoupling circuits to achieve stable operation. I initially started out with the decoupling circuits shown in the Motorola catalog. The end result was somewhat different. Overall, this portion of the transverter has required the most trial and error. Stabilization was achieved by varying the placement of the chokes and ferrite beads and by changing choke values. The impedance should be low. I used values in the 200- to 500-Ω range. Larger values may permit VHF oscillations in the stage. Chokes of 0.1 and 0.18 μH worked well.

I have achieved 7 watts output. Tests indicate that the driver is loaded up to the full 400 mA collector current. Key-down collector current of the 2N5946 measured 1.2 A. This indicates that the match to the input of the 2N5946 isn't perfect. This is an area for improvement. Testing the transmitter section requires some means of measuring 432-MHz energy from about −10 dBm to +10 watts. A calibrated diode detector gave me satisfactory results for the lower-level stages. A microwave power meter, such as an HP 430, 431 or 432 is better. If you do use one of these instruments, always put enough attenuation in front of the thermistor to prevent burning out the thermocouple if the stage breaks into oscillation. (I learned the hard way.)

A Bird 43 or other suitable wattmeter is needed to measure the output power of the final stage. Since the interstage impedances are mostly 50 Ω, the transmitter can be brought up as it is constructed. That is, you can measure power at the output of the first second and third MMICs, the 2N3866 and the 2N5946. The interstage circuit between the 2N5944 and 2N5946 is not at 50 Ω so these two stages must be tuned at the same time. A quick check to see if the MMIC is working is to measure the voltage on the output lead of the device. It should be about 5 volts for the 0285 and 0304, and about 6 volts for the 0485. Bias current is a better check. The bias current for the 0285 should be about 30 mA, with 40 mA for the 0304 and 50 mA for the 0485. Proper operation of the 2N5944 and 2N5946 can be determined by measuring the collector current in a key-down condition. It should be close to 400 mA for the 2N5944 and 2 A for the 2N5946.

A Note On The MMICs

The last two digits of the MMIC part numbers denote the package type. A 0404 will perform essentially the same as a 0485. These part numbers are Avantek numbers. Mini-Circuits equivalent parts are MAR-1, MAR-2, MAR-3 and MAR-4 (corresponding to the Avantek 0185, 0285, 0385 and 0485 devices.)

A Single-Board, No-Tune 902-MHz Transverter

By Rick Campbell, KK7B

(This article, which originally appeared in *Microwave Update '89*, is reprinted from *QST*, July 1991.)

The sole obstacle many amateurs face to getting on the UHF and microwave bands is a lack of equipment. It's not that the equipment and antennas for these bands are expensive or unobtainable; it's more that many would-be microwavers feel intimidated by the prospect of building their own gear for these bands. If you have a 2-meter multimode transceiver, you're already well along in getting on the 902-MHz and higher bands. The complete 902-MHz transverter I'll describe here is printed on a single 5- × 7-inch G10 circuit board and costs less than $150 to build. You'll soon see that you don't need lots of money or much building skill to put a 33-cm station on the air!

This transverter, unlike similar designs for higher-frequency bands,[1-4] has its local oscillator (LO) on the main PC board; only a 12-V supply, 144- and 902-MHz signals enter and exit the board. All the gain stages employ low-cost plastic MMIC (monolithic microwave integrated circuit) amplifiers. All the band-pass filters, traditionally the trickiest part of UHF-equipment design, are hairpin, third-order Chebyshev types printed on the PC board. The filters require no adjustments, like those of the higher-frequency transverters. Kits of parts and assembled transverters are available from Down East Microwave.[5]

Introduction

This transverter, shown in block-diagram form in Fig 1,

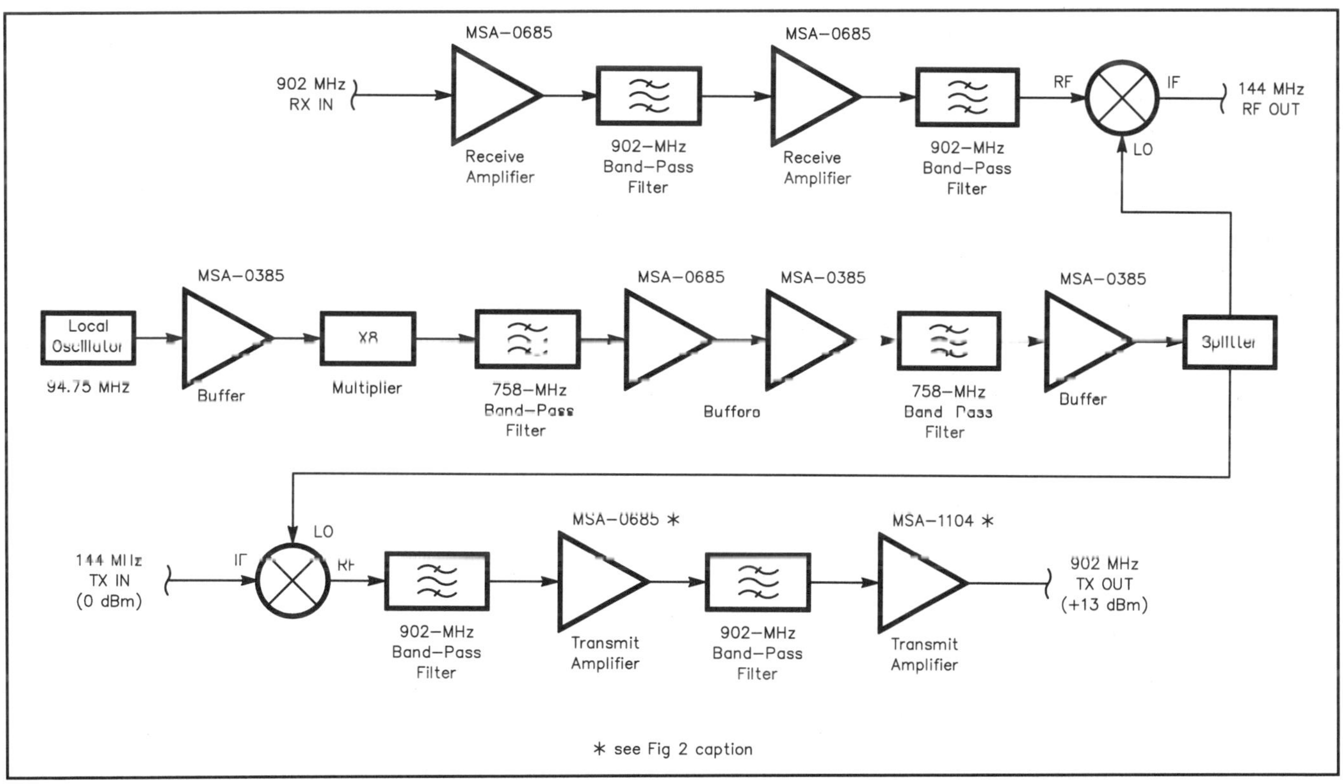

Fig 1—Block diagram of the single-board 902-MHz transverter. Unlike similar designs for the higher bands, this unit has an on-board local oscillator and discrete mixers. The only external connections are for the IF transceiver, antenna and power supply.

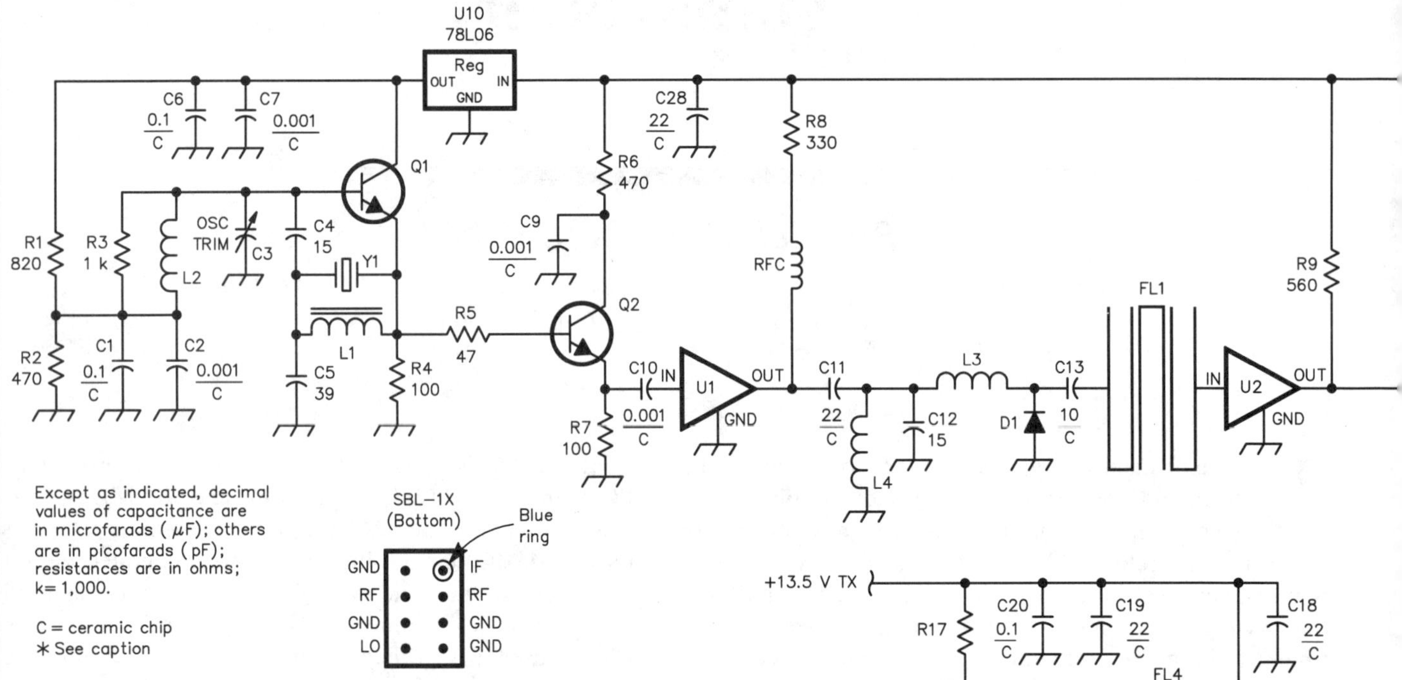

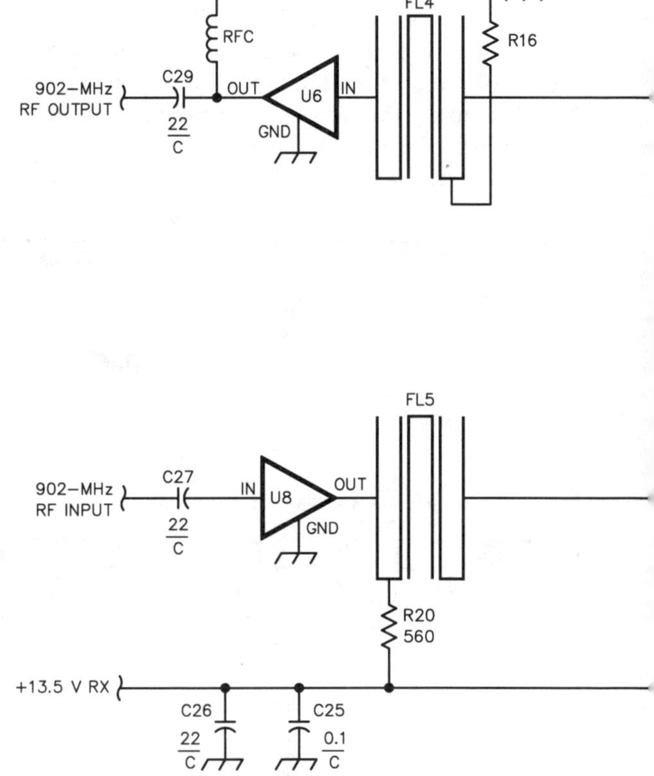

Fig 2—Schematic of the 902-MHz transverter. Heed the manufacturer's specifications for MMIC connections. The boxed part of the circuit isn't on the PC board, but may be needed to drive your 144-MHz IF receiver. A pi-network attenuator can be used here, if appropriate (see text and Fig 3). Filters FL1-FL6 are etched band-pass units. All resistors are ¼-W carbon-composition units unless specified otherwise. See the parts list for differences between 0-dBm-output and 13-dBm-output versions.

C3—3- to 10-pF ceramic or piston trimmer.
C4, C12—15 pF, miniature ceramic, NPO.
C5—39 pF, miniature ceramic, NPO.
C11, C14-19, C21, C24, C26-29—22 pF, ceramic chip, NPO.
C13—10 pF, ceramic chip, NPO.
D1—Multiplier diode; HP2800, HP2835 or 1N5711.
L1—10 turns no. 28 enam wire on T-25-6 toroid core.
L2—7 turns no. 24 enam wire, 0.1 inch diam, closewound.
L3, L4—8 turns no. 24 enam wire, 0.1 inch diam, closewound.
Q1, Q2—AT42085, NE021, BFX89 or equiv.
RFC—8 turns no. 24 enam wire, 0.1-in. diam, closewound.
R13, R14—150 Ω, ¼ W.
R15—33 Ω, ¼ W.
R16—360 Ω for low-power version; 560 Ω for high-power version.
R17—½ W; 330 Ω for low-power version; 130 Ω for high-power version.
U1—Avantek MSA-0385 or Mini-Circuits MAR-3.
U2, U7, U8—Avantek MSA-0685 or Mini-Circuits MAR-6.
U3, U4—Avantek MSA-0385 or Mini-Circuits MAR-3.
U5—Avantek MSA-0685 or Mini-Circuits MAR-6 for high-power version; MSA-0285 or MAR-2 for low-power version.
U6—Avantek MSA-0385 or Mini-Circuits MAR-3 for low-power version; MSA-1104 for high-power version.
Y1—94.75-MHz crystal. See text.

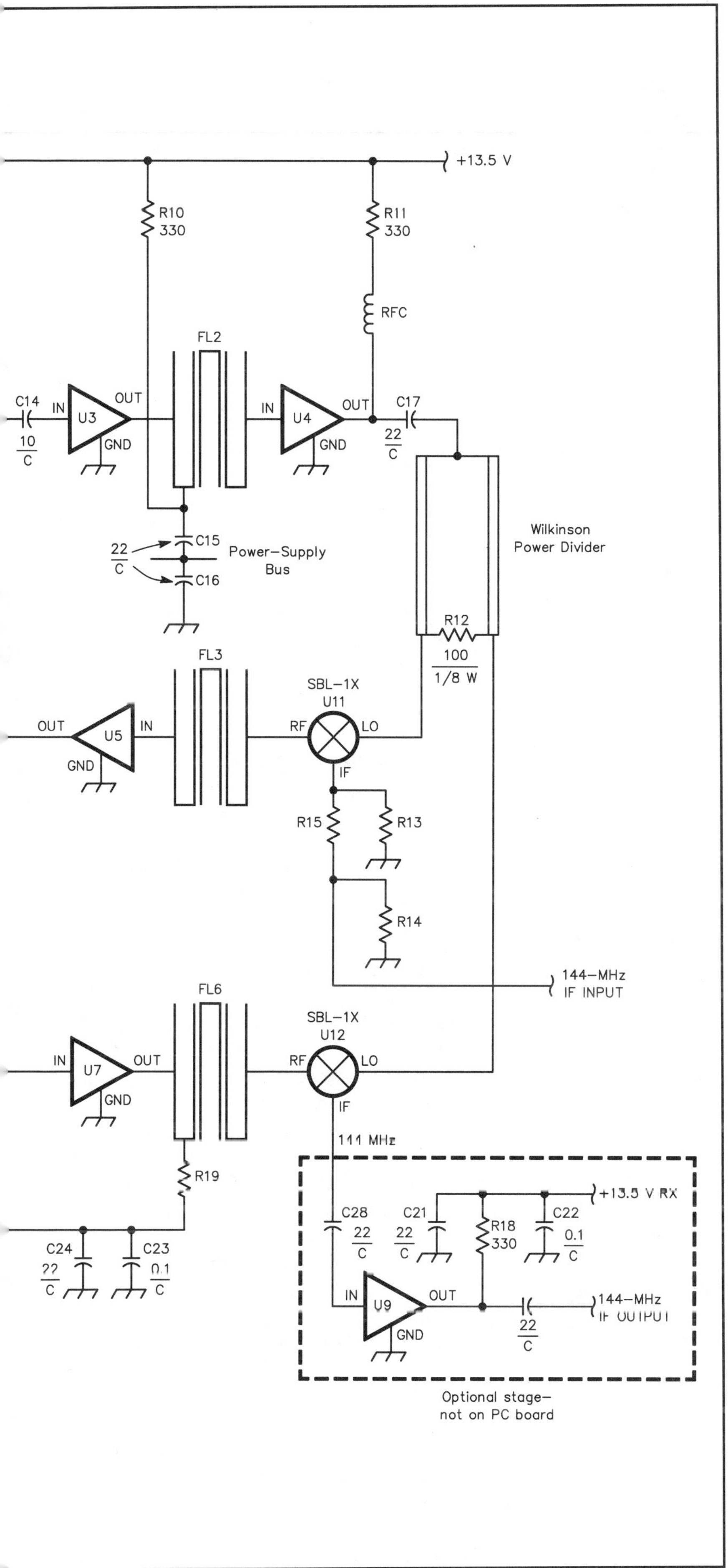

was a natural development following my 1.3-GHz transverter[6] and 576-MHz local-oscillator (LO) board.[7] The signal filters were scaled to pass 900-940 MHz and the LO filters were scaled to pass 740-780 MHz. The LO is a Butler type operating in the 100-MHz range, followed by a broad-tuned Schottky-diode multiplier and printed bandpass filter to reject all but the desired 758-MHz harmonic.

The 1.3-GHz transverter uses etched quadrature hybrid mixers. Scaled for the 758-MHz LO needed to convert a 144-MHz IF signal to 902 MHz, these mixers are too large to fit on a reasonably small PC board. A number of reasonably priced, discrete double-balanced mixers (DBMs) are available for frequencies below 1 GHz, so a pair of such mixers are used in this transverter. Prepackaged discrete DBMs need far less board space than etched mixers.

The transverter prototypes were built using Mini-Circuits SRA-11H and SRA-2CM mixers. The second-generation design shown here uses less-expensive Mini-Circuits SBL-1X mixers. Fig 2 shows the schematic diagram.

The Local Oscillator

Using prepackaged mixers freed up enough circuit-board space for an on-board local oscillator. The LO circuit is nearly identical to the 540- to 580-MHz version described in July 1989 *QST*.[8] For 758-MHz output, the Butler oscillator's frequency can be 94.75 MHz (x8), 108.286 MHz (x7) or 126.333 MHz (x6). As it turns out, 94.75 MHz is the best choice, as the oscillator is easiest to adjust and the undesired oscillator harmonics at 853 and 948 MHz are farther (and therefore easier to filter) from the RF-amplifier and transmit-amplifier passbands. The prototypes used 126.333-MHz seventh-overtone crystals. The first prototype built at KK7B uses SRA-11H mixers, which are specified for +17 dBm LO injection. A few decibels more output from the harmonic generator is obtained by biasing the diode as described in "A Clean Microwave Local Oscillator."[9] An 820-Ω bias resistor was used for this.

With an MSA-1104 LO amplifier and careful tweaking of the inductors in the harmonic generator, +18.5 dBm was available before the Wilkinson power divider that splits the LO for the receive and transmit mixers. This is sufficient for good performance with SRA-11H mixers. For the low-level SRA-2CM and SBL-1X mixers, +10 dBm output from the LO is optimum, which is easily obtained with the parts indicated in Fig 2.

All spurious LO outputs below 1.3 GHz are more than 40 dB below the desired output, but the LO filters have some spurious responses in the 1.4-GHz range. The spurs in this range may be stronger, depending on drive level to the MMICs

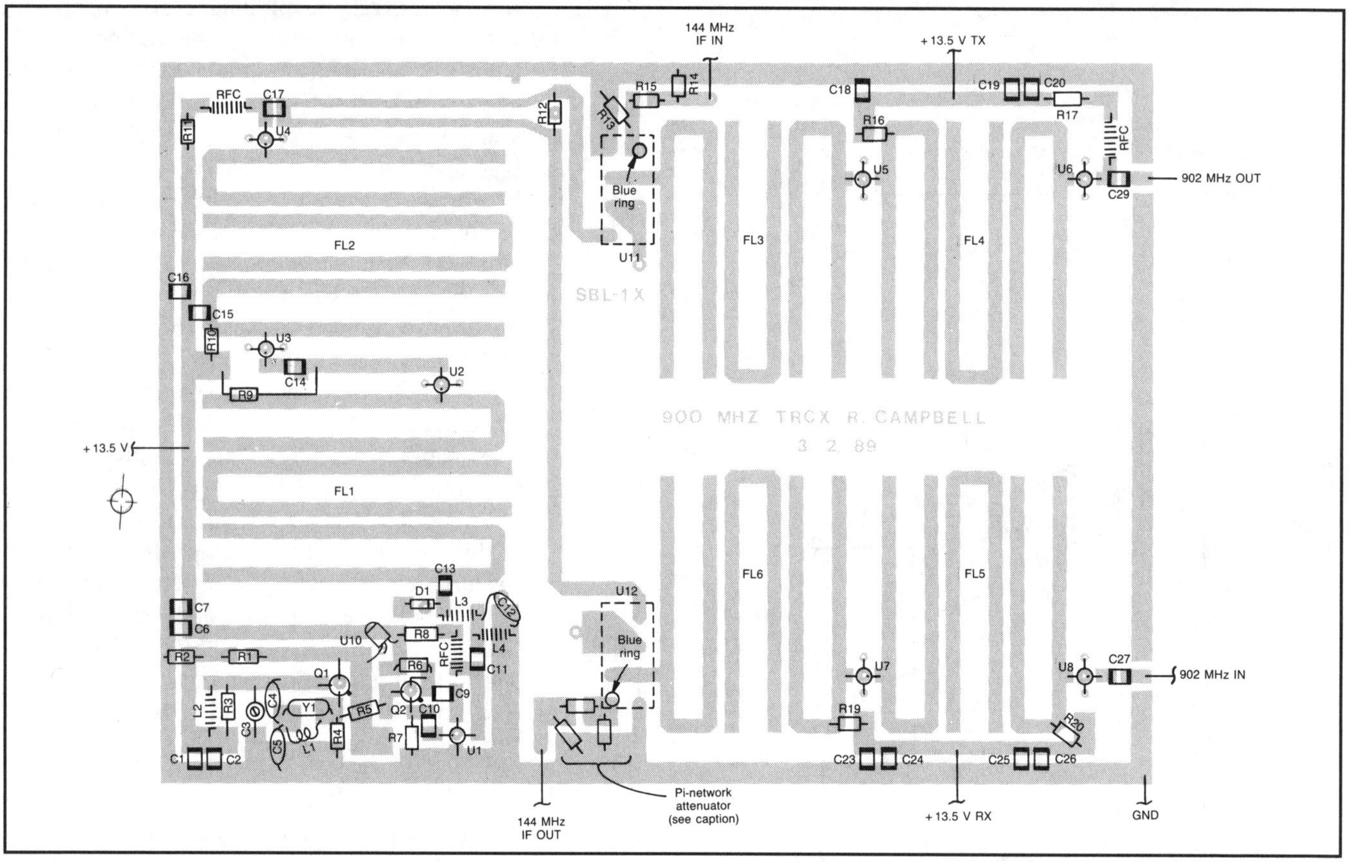

Fig 3—Part-placement diagram for the transverter. Although either MCL or Avantek MMICs are acceptable for use in this project, the MMICs shown here are marked like MCL parts (the colored dot signifies the input lead).

in the LO chain. The worst-case spurs in any of the prototype LOs were 20 dB below the desired output. Adding a low-pass filter after the LO would reduce the high-frequency LO spurs; I didn't do this, because these low-level, high-frequency spurs aren't visible in the RF-output spectrum and don't degrade receiver performance.

RF Amplifiers

The transmit and receive amplifiers are similar to those in the 1.3-GHz transverter, except that the filters are somewhat closer to being ideal at 902 MHz, so only two MMICs per chain are needed to obtain the required gain. MSA-0685s are used for both receive stages, providing a noise figure (NF) under 4 dB, more than 20 dB of conversion gain (including filter and mixer loss), and a stable 50-Ω termination for an external GaAsFET preamplifier. The first prototypes used an MSA-0685 IF amplifier after the receiver mixer, but total gain was excessive.

The SBL-1X version uses a 3-dB resistive pad at the receiver IF port to keep the receive-converter output to an appropriate level. The transmit amplifier provides 13 dBm (20 mW) output at the 1-dB compression point. This is appropriate for driving a discrete amplifier chain, or for direct connection to the antenna for local, line-of-sight or hilltop operation. Most hybrid linear-amplifier modules suitable for 33-cm use, such as the MC5874 and PF0011,[10] require a lower drive level. Fig 2 and the parts list specify components for versions with 13 dBm (20 mW) output and with 0 dBm (1 mW) output to best suit your requirements.

Construction, Tuning and Operation

Because of the tight dimensional tolerances required for etched microwave filters that require no adjustments, and because there are many variables in the printing process, a PC-board etching pattern is not included with this article. If you want to make a board for your own use, send an SASE to the ARRL Technical Department Secretary for a dimensioned copy of the artwork.[11] PC boards, parts and kits are available, however, as mentioned earlier.[12] Follow the construction guidelines discussed in Jim Davey's 1989 article,[13] and use only high-quality, microwave-rated porcelain chip capacitors in building the circuit. Avantek and Mini-Circuits (MCL) make MMICs suitable for use in this project. See the parts list in the Fig 2 caption for equivalent parts. There is some variation in the packaging of MMICs. Some use a colored dot at the bevel-cut input lead; others use a raised dot at the output lead. Be sure to install the MMICs in the correct orientation. Fig 3 shows placement of parts on the transverter PC board. The 144-MHz IF output comes off the board via a pad adjacent to U12. If necessary, you can use the traces in that area for a pi-network attenuator, or as a take-off point for the receiver signal or for a subsequent amplifier stage, such as the boxed section in the lower right corner of Fig 2.

Once the PC board is populated, the only adjustment this

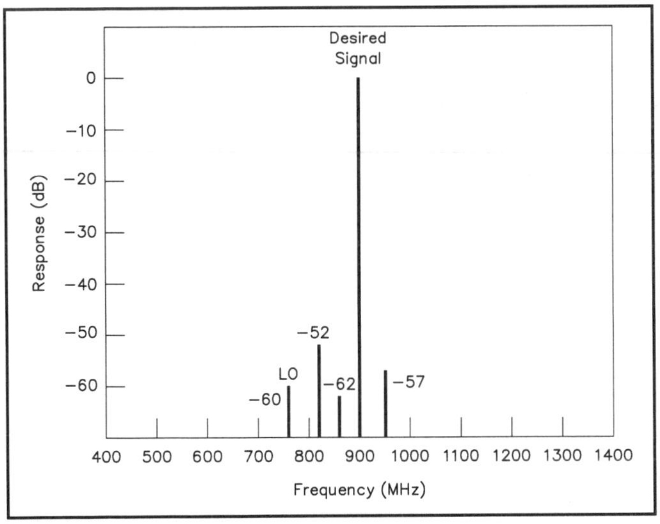

Fig 4—The transverter's spectral output. Vertical divisions are 10 dB; horizontal divisions are 100 MHz. All spurious outputs are at least 50 dB down from the carrier (–50 dBc). This plot shows that it doesn't take expensive or complex circuitry to generate clean signals—even above 900 MHz.

transverter requires is tweaking of its LO trimmer (C3) to ensure reliable oscillator starting. To do this, apply 13.5 V to the LO. If you can, observe the LO signal at 758 MHz and adjust the trimmer until the oscillator restarts every time power is removed and reapplied. If you like, you can also use either the 902-MHz transmit or receive section to verify this. The 94.75-MHz LO is also audible on standard FM-broadcast receivers. In operation, apply 13.5 V to the LO and the transmit or receive section, depending on which is in use. (It's good practice to remove power from the unused stage.)

Performance

When driven with 1 mW of 144-MHz RF, this transverter provides a clean, low-power 902-MHz signal. All spurious outputs are more than 50 dB below the desired output, as shown in Fig 4. The transmit-converter output is suitable for direct connection to a linear amplifier without additional filtering.

On receive, the transverter's under-4-dB noise figure and unconditionally stable 50-Ω input termination are hard to beat. Image rejection is more than 70 dB. No input filtering is done before the first amplifier stage, so the input stage is susceptible to overload in high RF environments. For use in such environments, there are several options:

- Replace the receive amplifiers with MSA-1104s. This increases the noise figure by about 1 dB and increases dynamic range by about 10 dB. This is usually not enough of an improvement to cure overload, though.
- Omit the first amplifier stage. This increases the noise figure to 7 or 8 dB, but will probably cure the problem.
- The best alternative: Use an external, low-loss filter. If this or any other transverter is to be used around other transmitters, it is good practice to use a low-loss cavity filter before the first receive-amplifier stage.

Conclusions

When used with a suitable outboard linear amplifier and a GaAsFET preamp, this transverter easily outperforms older designs—at a fraction of the cost. The performance advantages gained by the use of printed band-pass filters, combined with the elimination of all microwave adjustments and the need for a spectrum analyzer, are significant advances in the state of the art that are common to this family of microwave transverters. When this transverter is coupled with the right antenna system, you can work tremendous DX on the 902-MHz band. Dave (WA3JUF) Mascaro's *QST* article, "A High-Performance UHF and Microwave System Primer," which begins on page 30 of May 1991 *QST*, has antenna ideas and lots of other useful information on how to get the most out of your UHF station.

Acknowledgments

Thanks to Don Hilliard, WØPW, for building and testing a prototype, and for preparing the schematic.

Notes

[1]R. Campbell, "A Single Board No-Tuning 23 cm Transverter," *Proceedings of the 23rd Conference of the Central States VHF Society* (Newington: ARRL, 1989), pp 44-52. (Reprinted in this book.)

[2]J. Davey, "No-Tune Transverter for 2304 MHz," Proceedings of Microwave Update '89 (Newington: ARRL, 1989), pp 30-34. (Reprinted in this book.)

[3]J. Davey, "A No-Tune Transverter for 3456 MHz," *QST*, Jun 1989, pp 21-26. Also see Feedback, Oct 1990 *QST*, p 31. (Reprinted in this book.)

[4]R. Campbell, "A Single-Board Bilateral 5760-MHz Transverter," *QST*, Oct 1990, pp 27-31. This article lists, in its end notes, sources for lots of information on microwave antennas and other subjects of general interest. (Reprinted in this book.)

[5]Etched PC boards, crystals, kits of parts, assembled boards and complete transverters are available for this rig and the projects referenced in notes 1-4 and 7, from Down East Microwave, 954 Rte 519, Frenchtown, NJ 08825, tel 908-996-3584, fax 908-946-3072. Catalog available.

[6]See note 1.

[7]R. Campbell, "A Clean, Low Cost Microwave Local Oscillator," *QST*, Jul 1989, pp 18-23.

[8]See note 7.

[9]R. Campbell, "A Clean Microwave Local Oscillator," *Proceedings of the 1296 and 2304 Conference*, reprinted in *Proceedings of the 21st Conference of the Central States VHF Society* (Newington: ARRL, 1987), pp 51-57.

[10]Down East Microwave (see note 5) carries these hybrid amplifier modules.

[11]Send a no. 10 SAE with one unit of First-Class postage to the ARRL Technical Department Secretary; request the July 1991 *QST* 902-MHz transverter template package. The PC-board artwork is copyrighted by Down East Microwave. Feel free to use it to make boards for your personal, noncommercial use, but not for commercial purposes.

[12]See note 5.

[13]See note 3.

Reflections on the KK7B 903- and 1296-MHz No-Tune Transverters

By Ron Neyens, NØCIH

Introduction

I have built both the 903- and 1296-MHz versions of the KK7B No-Tune Transverters designed by Rick Campbell. Both are complete and ready to operate even if only at the QRP-output level. These transverters are about the fastest, easiest way of getting active on 903- and 1296-MHz. They're easy to assemble and, fortunately, don't require any of those hard-to-find, one-of-kind components or parts.

A few hours with the soldering iron is all that's required to put these transverters on the air. All the frequency-sensitive inductors and bandpass filters are etched directly on the circuit board. This eliminates the need for fancy test equipment and the tedious tuning process usually required with home-brew transverters. I found that I spent more time putting the assembled board into a chassis than soldering the parts on the circuit board and testing it.

As with other projects, someone always looks for ways to improve the design or layout. Hopefully this effort results in an improvement on the original design. What I am presenting here are some of the components I've used and the layout changes I've made. These changes came about during the process of building and testing the transverters. Hopefully this information will help others that are using these transverters now or are thinking of building equipment for 903 and/or 1296 MHz. This paper refers to the 903-MHz transverter but the information presented here applies just as well to the 1296-MHz version.

Background

To begin with, I want to supply a little background information on the 903-MHz transverter. I purchased the etched circuit board from Down East Microwave while I was attending the 1989 Central States VHF Conference in Chicago. The quality of the G-10 circuit board and the etched pattern on it was excellent. A *902K Preliminary Instruction Manual* was included with the circuit board and it contained a description of operation, a parts list and several component layout diagrams. The only error with the preliminary manual was that the component layout diagrams were for an earlier version of the transverter layout. These were only minor errors and I had several articles on the transverter to refer to. The most recent release of the *902K Assembly And Operational Guide* has corrected some of these errors. The ones that have not been corrected are pretty obvious.

The recommended parts list in the preliminary manual provided excellent information about the parts needed in the transverter. All parts are easily obtained through a variety of mailorder suppliers and local vendors. If you don't feel like collecting the individual parts, a packaged kit for the transverter is available from Down East Microwave. Going a bit further in cost, Down East Microwave also sells a complete ready-to-run transverter with an optional aluminum housing and IF switching board.

In my case, I decided to build the transverter from scratch, in the process saving a few dollars and at the same time learning something. The local oscillator, receive downconverter, and transmit upconverter are contained on one 5" × 7" circuit board (Note: The 1296-MHz version requires two circuit boards for a complete transverter.)

The local oscillator (LO) circuit is simple and straightforward. It contains a crystal controlled Butler oscillator, X8 multiplier, two triple stage stripline bandpass filters, and a stripline power divider. A 94.75000-MHz, fifth-overtone crystal is multiplied by 8 to provide a final output frequency of 758-MHz at approximately +13 dBm. The crystal (Y1) was purchased from International Crystal (ICM) but is also available directly from Down East Microwave. I used a Motorola 78L05 for the +5-volt regulator and a Motorola 2N5179 for transistors Q1 and Q2. For the ×8 multiplier diode I used a 1N5711 Schottky diode in place of the recommended HP 2835 diode. I installed the diode so that the anode lead extended through the circuit-board pad and was soldered on both the top and bottom of the circuit board.

A combination of one Avantek MSA-0685 monolithic microwave integrated circuits (MMIC) and four MSA-1104s are used in the ×8 multiplier/bandpass filter circuit. These provide the required LO level for two double-balanced mixers. I purchased all of the MMICs from Microwave Components of Michigan. The double-balanced mixers I used are Mini-Circuits SBL-1Xs. These also can be purchased from Microwave Components of Michigan.

The receive downconverter circuit uses a pair of Avantek

MSA-0685s MMICs. This combination is presently the best to use as far as gain, low noise figure, and device cost. A Mini-Circuits MAR-6 can be used as a replacement for the Avantek MSA-0685s without any changes to the bias resistor values.

In the final receiver configuration of the transverter, I have a plate-line preamp using an MGF-1302 in front of the receiver. This was added to improve the overall noise figure of the transverter. The design of the preamp is similar to one described by Al Ward, WB5LUA, for 1296 MHz. Without the preamp the noise figure of the transverter is between 3.8 to 4.0 dB with a gain of approximately 35 dB.

The transmit upconverter circuit uses a single Avantek MSA-0685 driving an Avantek MSA-1104. This combination is also the best set up with what MMICs are presently available. One-mW drive at 145 MHz provides approximately +13 dBm (20-mW) output at 903 MHz. A Mini-Circuits MAV-11 can be used as a replacement for the Avantek MSA-1104 without any changes to the bias resistor value. If a 0-dBm (1-mW) power output is desired, it can be achieved by using two MSA-0685s.

Several RF chokes are used on the circuit board. These were hand wound on a 3⁄32" drill bit using #24 AWG enameled copper wire. All are 8 turns, close wound, with the exception of a 7-turn, close-wound choke across the base/collector of transistor Q1 and a 0.33-µH molded RF choke across crystal Y1. If the oscillator fails to start properly or the 903-MHz signal is distorted, spread or compress the 7-turn choke slightly.

I followed the recommended parts list in the preliminary manual, but made a change to the capacitors used. Instead of a mixture of ceramic disk caps and chip caps, I decided to use 0.110" × 0.110" ceramic chip caps throughout (including the local oscillator). I also made a change in the values of bypass capacitors by using three (22 pF, 0.01 µF and 0.001 µF) instead of the recommended two (22 pF and 0.01 µF).

That about sums it up. The ceramic chip caps are the only real deviation I made from the original Down East Microwave parts list. Overall, the placement of components was per the assembly instructions supplied by Down East Microwave. I recommend that you refer to the notes at the end of this article for more information.

A Housing to Put It In

I cut the circuit board down to 5" × 7" to expose the copper-clad ground plane along all the edges. After installing the parts, I formed a box around the circuit board using 0.32" thick × 1.00" wide brass strips. I had access to a sheet metal brake and this allowed me to bend the strips to fit the circuit board, and form the corners of the box. The component side of the circuit board was placed ½-inch up from the bottom edge of the strip. The walls were then soldered to the ground plane on both the top and the bottom sides of the board to provide a continuous RF ground.

To complete the RF-tight assembly, I needed top and bottom covers. Refer to Fig 1. The covers were made by using two pieces of 0.060" sheet aluminum cut to 5⅜" × 7⅛". I "sandwiched" the transverter between the two covers and used 1" long, 4-40 threaded, hex standoffs on each corner to maintain separation. Two-meter input/output and 903-MHz input/output connections were made using SMA chassis connectors soldered to the brass walls. The completed unit makes a nice RF-tight assembly that can be easily mounted to a chassis.

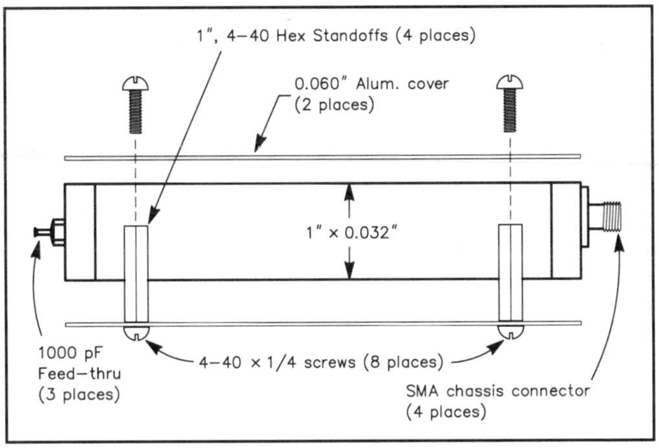

Fig 1

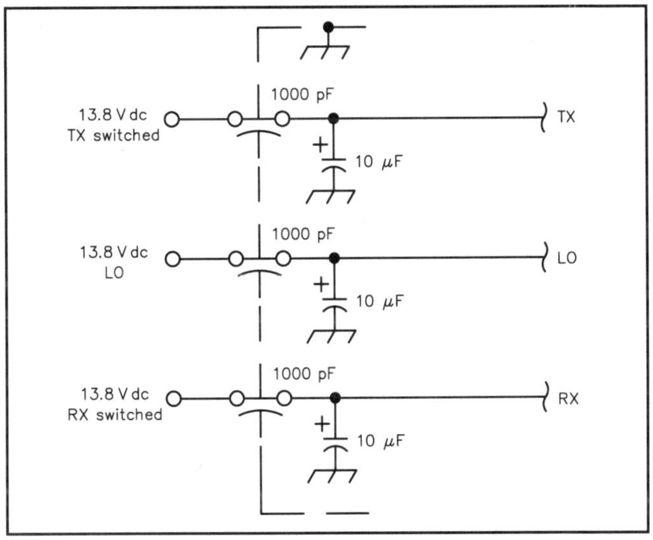

Fig 2

Supplying dc Power

Refer to Fig 2. All power is routed through 1000-pF, 50-V feedthrough capacitors. These are soldered directly to the brass walls of the box to insure a low inductance RF path. I used three feedthrough capacitors: one for the local oscillator, one for the transmitter, and one for the receiver. The 10-µF, 35-V Tantalum capacitors are used for additional bypassing of the dc supplies. I routed all of the wiring on the unetched side of the circuit board. Placing it on this side helps prevent any problems with RF floating on the dc bus.

Final Testing and Modifications

While testing the transverter on a spectrum analyzer, I found I was getting excessive spurious outputs from the transmitter. I tried squeezing and expanding the turns of L5 and L6 but this didn't get me any closer to a cleaner output. The only other thing I could do was move the bypass capacitors closer to the MMIC bias resistors. The easiest stage to work on was the final output MMIC. I moved the bypass capacitors as close to the bias resistor as possible and rechecked the output on the spectrum analyzer. I found that the gain had increased slightly and the spurs had decreased.

I heated up the soldering iron and unsoldered the remaining bypass capacitors in the transmitter chain. Not wanting to do half a job, I also moved the other bypass capacitors on the board closer to their intended source. Checking the output again showed an increase of approximately 3 mW over that previously obtained. This brought the transverter's output to approximately 18 to 19 mW, where it should be. The spectrum analyzer also revealed that the spurs had decreased considerably. Some of the spurs had dropped 10-20 dB. With a signal generator for an input signal, I checked the receiver for any improvements but saw no real evidence of any change. [KH6CP/1 comments: The chip caps may have caused the spurious signals. They are "better" than disk ceramics, resulting in sharper parallel resonances that prevent good bypassing. Jim Davey has recommended series resistors to help reduce these resonances. Moving them around varies the inductance of the parallel resonant circuit. See *Microwave Update 93*, p 58.]

Summary

I hope I have helped someone with these ideas. If you're just considering building one of the KK7B transverters, spend a little money and use chip caps throughout the transverter. Save those disk ceramics for something on HF or dc. If you're getting insufficient output, low gain, or a large spurious output from your transverter it might help to check the bypass capacitors and reposition them.

The "sandwiched" case idea has been used on a variety of projects around my shack. It has been ideal for packaging preamps, low-level power amps, and converter subassemblies. Whenever I need a housing for a project I get the size I need since the circuit board determines the dimensions. Besides this, it doesn't cost much in material when compared to a diecast aluminum box.

Even though the output from this transverter is only about 18 mW (10 mW for the 1296-MHz version) it's enough to start with. With the variety of RF brick modules on the market, it's possible to have up to a 20-W transverter very easily. With that amount of output, band openings should become interesting.

Notes

[1] *SHF Systems 902K Preliminary Instruction Manual*, Down East Microwave, 1989.

[2] R. Campbell, "A Clean, Low-Cost Microwave Local Oscillator," *QST,* July 1989, pp 18-23.

[3] R. Campbell, "A Single-Board No-Tuning 23-Cm Transverter," *Proceedings of the Central States VHF Society 1989.* (Reprinted in this book.)

[4] R. Campbell and D. Hilliard, "A Single-Board 900-MHz Transverter With Printed Bandpass Filters," *Proceedings of Microwave Update '89.*

[5] J. Hinshaw, "Modular Transmit And Receive Converters For 902-MHz," *Ham Radio*, March 1987, pp 17-20, 23-26.

[6] *SHF Systems 902K Assembly and Operational Guide*, Down East Microwave, 1990.

A Single Board No-Tuning 23-cm Transverter

By Rick Campbell, KK7B

(This article, which originally appeared in the *Proceedings of the Central States VHF Society 1989*, is reprinted from *The ARRL Handbook for Radio Amateurs.*)

This transverter uses bandpass filtering with printed hairpin filters. It will work with any MMICs capable of operating at 1.3 GHz at the appropriate input and output levels. New MMIC types may be easily substituted simply by changing the MMIC and its bias resistor.

System Block Diagram

Fig 1 is the transverter block diagram and Fig 2 is the schematic diagram. Fig 3 shows the layout of the 5- × 7-inch G-10 board with all functional blocks labeled. In the transmit path, FL1 removes the image, LO (local oscillator) and higher-order spurious outputs. FL3 selects the desired LO harmonic. The LO signal is filtered again in FL4 before the LO is split and applied to the transmit and receive mixers.

The received signal passes through FL5, which attenuates out-of-band signals. After amplification, image noise added to the received signal is attenuated by FL6.

One noteworthy feature of this transverter is that the choice of LO drive frequency is left to the builder. The filters are narrow enough to permit use of outputs at 1152, 576, 384, 288 and 230.4 MHz. Fig 3 shows component values for 576-MHz drive from the LO board described in note 1. Other

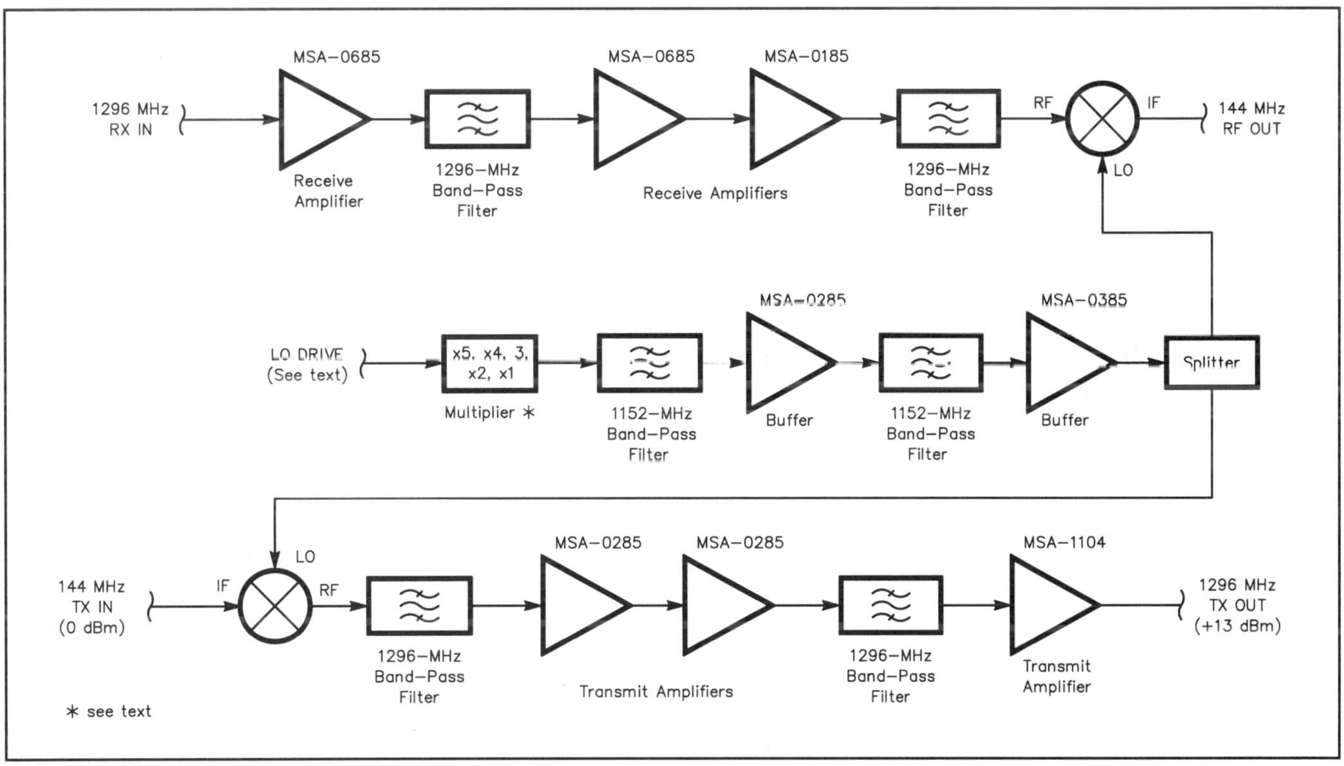

Fig 1—Block diagram of the single-board transverter. Extensive use of printed-circuit band-pass filters simplifies construction and alignment.

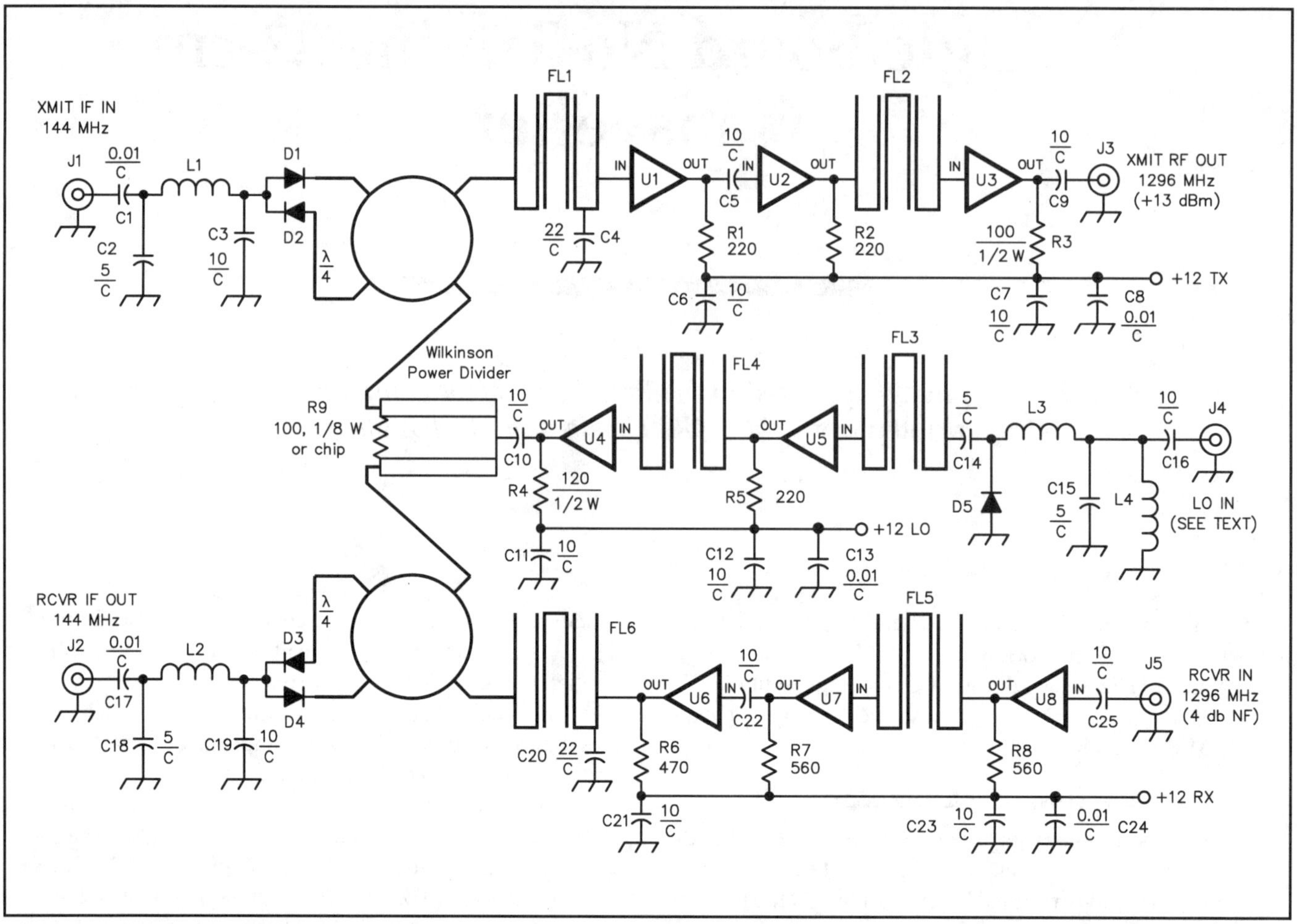

Fig 2—Schematic of the 1296-MHz transverter. All resistors and capacitors are chip types except as noted. Diodes D1 through D4 are HP 5082-2835. FL1 through FL6, the mixer tuned circuits and the Wilkinson power divider are etched on the circuit board.

LO frequencies may require minor component changes.

Circuit Performance

Careful measurements show a mixer conversion loss of 5.5 dB. The output for 1-dB compression is 0 dBm. LO suppression is about 25 dB. The mixers compare favorably with more-expensive packaged units.

Transmit output is about 13 dBm, which is suitable for some applications without further amplification. A major advantage of the MMIC output stage is that it offers an unconditionally stable, near-50-ohm source to the following stage or antenna.

Receive noise figure is less than 4 dB. The unconditionally stable 50-ohm load presented by the input MMIC is useful for direct connection to an antenna, and ideal as the stage following an external GaAsFET preamp.

The filters allow coverage of the entire 1240- to 1300-MHz band with a 144-MHz IF and appropriate LO drive. The bandpass filters used, although not mechanically critical, remove significant spurious responses. The passband response of one RF and one LO filter section is shown in Fig 4.

Construction

All components are surface mounted on the etched side of the board. The most tedious step is drilling the holes next to the MMIC mounting pads and soldering the "C" wires to provide a low-impedance ground. Plated-through holes would eliminate this step. After preparing the board with its "C" wires, mount the chip capacitors, inductors and connectors. Finally, install the MMICs and bias resistors.

The bottom of the hairpin U is a low-impedance point for signals in the filter passband. A bypass capacitor at that point has little effect on the filter passband but provides a ground path for VHF signals. Although 22-pF capacitors are specified, values from 12 to 30 pF gave good results in the prototypes.

The multiplier works best with drive levels of 0 to 10 dBm, without a bias resistor. For higher drive levels, the bias circuit described in Note 2 will provide more output. With 10-dBm drive at 230.4 MHz, output from FL3 is about –20 dBm, so about 33 dB of additional gain is needed. With 10-dBm drive at 576-MHz, output from FL3 is –3 dBm, so only 16 dB of gain is required.

Pads are included on the board for up to three MMICs. Choose the appropriate number of MMICs based on available LO drive level and frequency. FL3 and FL4 must be isolated. If only one MMIC is used it should be between them. If sufficient LO drive is already available at 1152 MHz, use a

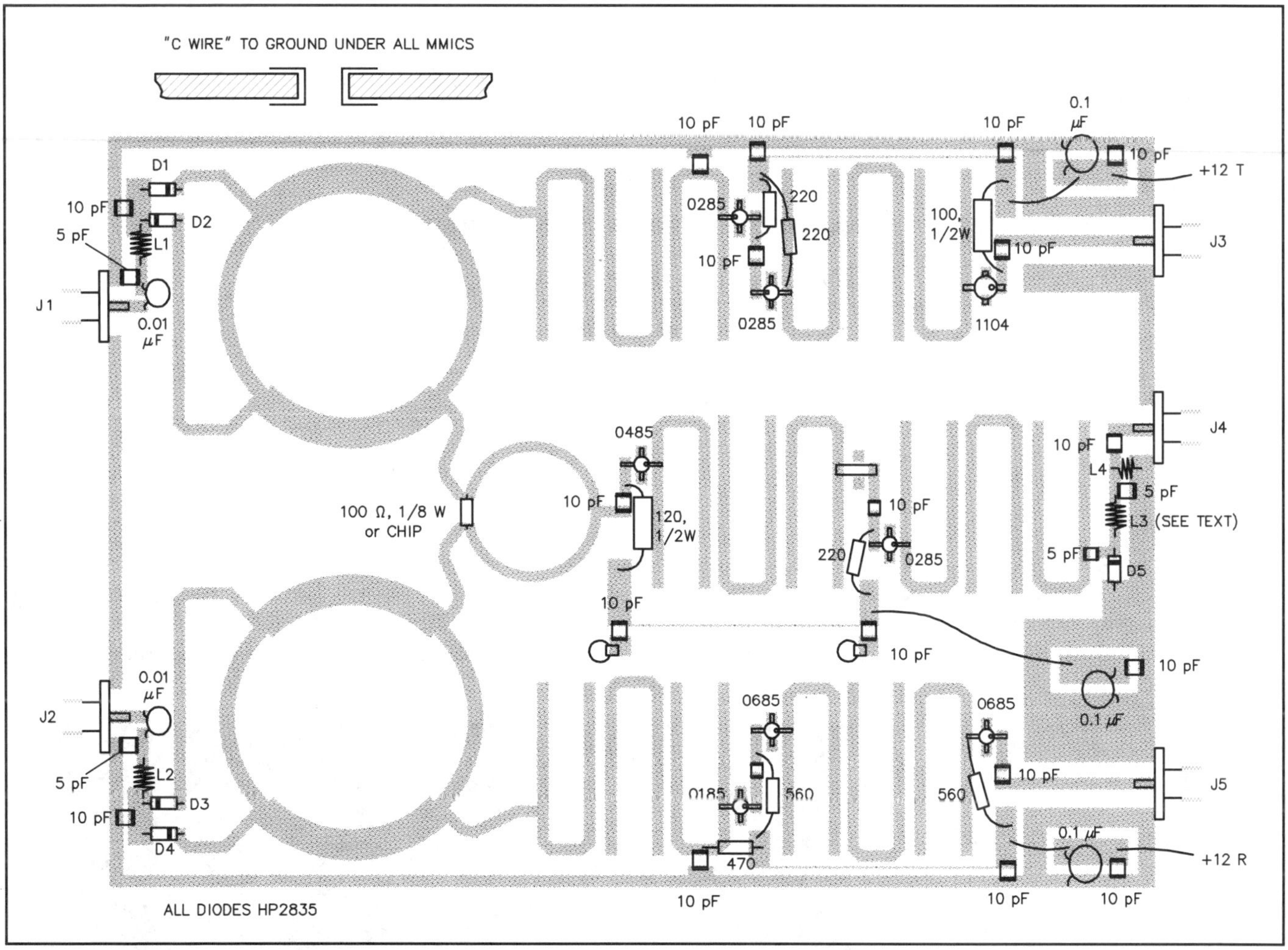

Fig 3—Part-placement diagram for the transverter. The "C" wires under each MMIC connect the component pad to the ground plane on the other side of the board.

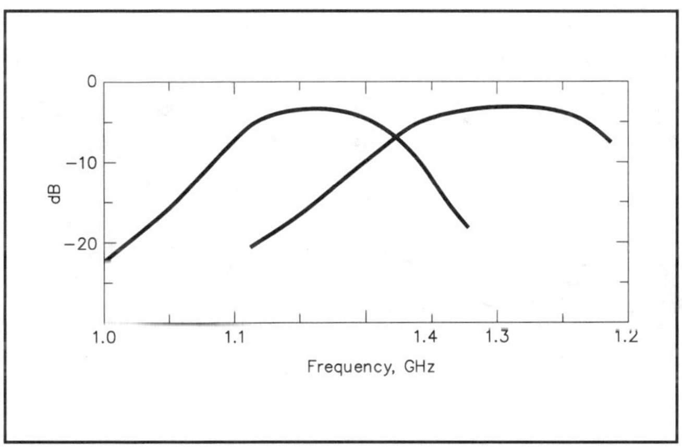

Fig 4—Passband response of one LO filter (A) and one RF filter (B) in the transverter. With appropriate choice of LO frequency, the transverter is suitable for use across the entire 23-cm band.

6-dB attenuator pad between FL3 and FL4.

Alignment

The tuning components in the LO multiplier are quite broad: network Q is about 1. The 576-MHz network will also work for lower frequency inputs, but the component values can be optimized for each drive frequency. L3 and L4 may be increased to 3 turns for 384 MHz, 4 turns for 288 MHz and 5 turns for 230.4 MHz if necessary. If LO drive is sufficient to saturate U4, however, mixer performance will be relatively unaffected by small changes in LO level, and tuning is unnecessary. A high-level MMIC is recommended for the LO-output stage.

Parts and Circuit Boards

Components are available from Microwave Components of Michigan, whose address appears elsewhere in this chapter. Etched and plated FR-4 circuit boards, complete kits and assembled transverters are available from Down East Microwave, whose address also appears elsewhere in this chapter.

The author acknowledges the assistance of Jim Davey, WA8NLC, John Miller of Michigan Technological University, Bob Dryden, W4OJK and John Molnar, WA3ETD.

Notes

[1]R. Campbell, "A Single-Board No-Tuning 23 cm Transverter," *Proceedings of the 23rd Conference of the Central States VHF Society* (Newington: ARRL, 1989), pp 44-52.

[2]R. Campbell, "A Clean Microwave Local Oscillator," *Proceedings of the 21st Conference of the Central States VHF Society* (Newington: ARRL, 1987), pp 51-57.

A No-Tune Transverter for 2304 MHz

By Jim Davey, WA8NLC
4664 Jefferson Township Place
Marietta, GA 30066
(From December 1992 *QST*)

Introduction

The circuit, layout and performance of the 2304-MHz transverter are similar to those of the 3456-MHz design described in the article that follows this one. It has the following features:

- 10-mW transmit output
- Receiver noise figure of 3.5-4.0 dB
- Same 2-meter IF requirements (1-mW drive level)
- Same size circuit board
- Same no-tune local oscillator
- Even easier to assemble than the 3456-MHz version

For construction details and background information, refer to the 3456-MHz article. Only the major features of the 2304-MHz transverter are covered in this article. A preliminary version of this transverter was described in *Proceedings of Microwave Update '89*.

Construction

The entire transverter can be assembled in a few hours, including the box. The brass-box type of enclosure is recommended for mechanical and electrical bonding of the connectors. A cover is optional. The 2304 transverter is indifferent to the presence of a cover. A few builders of the 3456-MHz transverter, however, have noticed increased spurious outputs

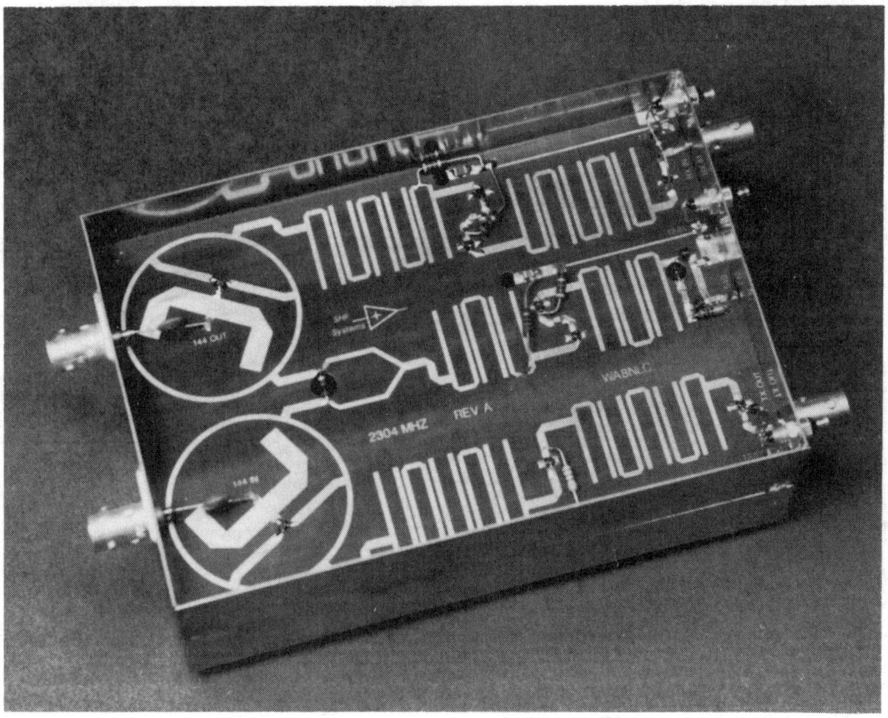

→

Fig 1—Schematic of the 2304-MHz transverter. See the text for parts sources. When building this transverter, follow good microwave construction techniques, as detailed in the 3456-MHz transverter article that follows this one. The Avantek MSA-0185 is equivalent to the Mini-Circuits MAR-1; the MSA-0685 is equivalent to the MAR-6, etc. (Lead labeling varies, however; follow the manufacturer's instructions.) Bias-resistor values are given for 13.6-V operation. All capacitors in the 2304-MHz signal path are 50- or 100-mil ceramic chips, except for the 0.01-μF coupling capacitors at the first IF input and output. Here, use ceramic discs, encapsulated chips or other low-loss capacitors. Use ¼-W carbon-composition or film resistors.

D1, D2—Hewlett-Packard HSMS-2822 surface-mount diode pair.
D3—Hewlett-Packard 5082-2835 Schottky diode.
FL1-FL6—Etched band-pass filter.
L1—4 turns #28 enameled wire, 0.075 in. ID, closewound.
U1, U2—Avantek MSA-0685 or Mini-Circuits MAR-6.
U3, U4—Avantek MSA-0185 or Mini-Circuits MAR-1.
U5, U7—Avantek MSA-0885 or Mini-Circuits MAR-8.
U6—Avantek MSA-0385 or Mini-Circuits MAR-3.

of the transmitter with a cover in place. When mounting components, keep the bias resistors flat against the board and away from the filter elements. Wrap a small piece of copper foil through the ground holes for the bypass capacitors, under each MMIC and at the ground end of the LO multiplier diode. I used SMA connectors for the 2304-MHz ports, but BNC connectors would be acceptable at this frequency.

As before, no matching is done at the IF ports. This has not been a problem, except if you want to measure the noise figure of the receiver. I have noticed that newer noise-figure

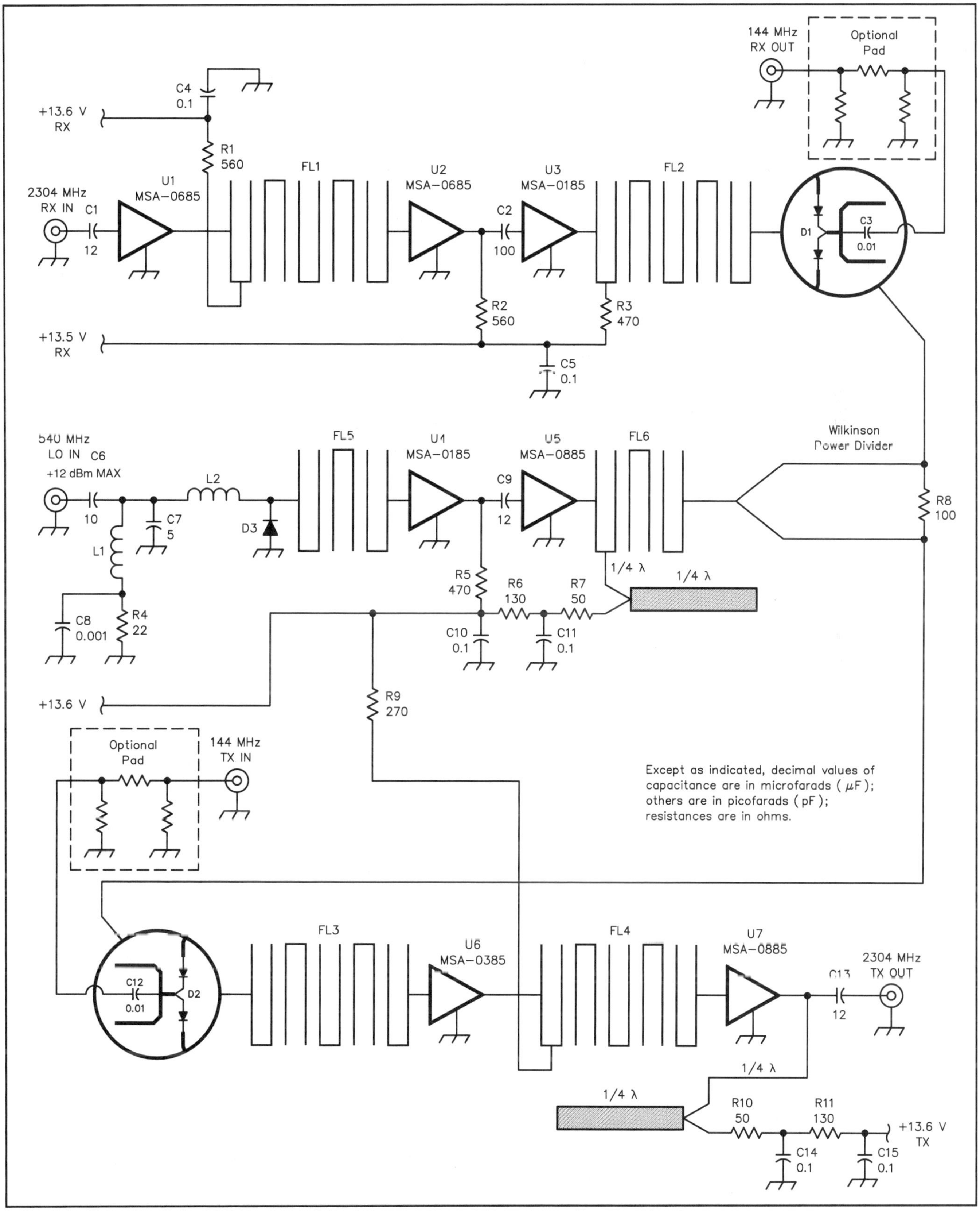

meters, like the HP 8970, have a problem with this arrangement due to the presence of other signals at the IF port, presumably the LO. If you want to check the noise figure, first run the receiver IF through a 2-meter converter to a lower IF, such as 28 MHz. When this is done, the problem with the test equipment should disappear.

Design

The transverter, shown schematically in Fig 1, is built on 0.032-in. woven Teflon board material with a dielectric constant of 2.5 and half-ounce copper foil. The key to the circuit's simplicity lies in the printed microstrip third- and fifth-order Chebyshev band-pass filters. These filters have excellent stopband rejection and low insertion loss. Midband insertion loss is just 2.1 dB, thanks to a low-loss Teflon PC-board substrate.

The center frequency and bandwidth of the receive filters were chosen to also allow receive-only operation on the OSCAR Mode S downlink at 2401 MHz using a 564.25-MHz LO. The transmitter-filter bandwidth is narrower, to help suppress unwanted mixing products. The transverter shouldn't be used for transmitting above 2325 MHz without an external cavity or interdigital filter that attenuates the LO signal by at least 10 dB.

Transmit Converter

Transmitter power output is 10 to 20 mW (10 to 13 dBm) and varies slightly with component variations in the Avantek MSA-0885 MMIC output stage. Although more power would help balance the transverter's capability with that of most home stations, power-amplifier stages are best implemented as external accessories. It's good design practice to limit the amount of gain in a single box to 30 dB or so, unless special efforts are taken to prevent feedback.

Receive Converter

The transverter's receive-converter noise figure (NF) is approximately 4.5 dB. This may seem high by today's standards, but it can be easily dropped to 1 dB or less by a modern GaAsFET preamplifier with a gain of 14 dB or more.

The transverter's first receive stage, an Avantek MSA-0685 MMIC, provides a 50-W input impedance to properly terminate an outboard preamp. The preamp can be mast mounted or connected directly to the transverter input.

Performance

When driven with 1 mW of 144-MHz RF, the transverter produces a clean 2304-MHz signal. The 1-dB compression point is 10 mW. All spurious products are down from the fundamental output by at least 50 dB. For a receive frequency of 2304.1 MHz, the measured image rejection is more than 70 dB and the measured NF is 4.3 dB.

Be careful not to exceed the transmit mixer's maximum drive rating of 0 dBm (1 mW). If you have extra transmit gain

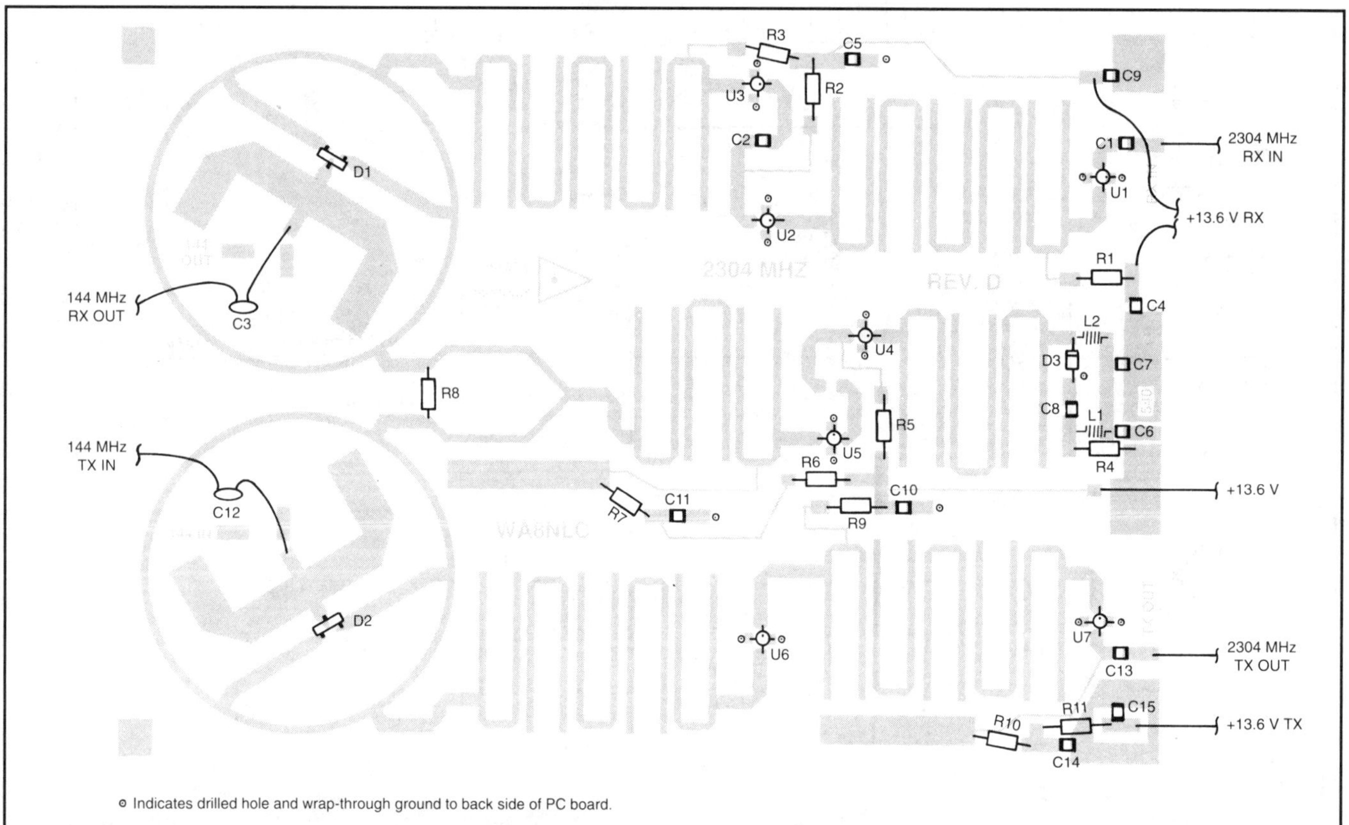

Fig 2—Part-placement guide for the 2304-MHz transverter (not shown actual size). All components mount on the etched side of the board. Feedthrough grounds, indicated by circles, *must* be installed and soldered top and bottom. Follow MMIC manufacturer's lead coding.

in your final configuration, reduce the transmitter drive.

Parts

Fig 2 is a part-placement drawing. Parts are available from Microwave Components of Michigan, PO Box 1697, Taylor, MI 48180; kits and assembled units are available from Down East Microwave, RR1 Box 2310, Troy, ME 04987, tel 207-948-3471, fax 207-948-5157. As in the 3456-MHz transverter, the success of the no-tune approach depends on how acccurately you can duplicate the printed filters. I strongly recommend you purchase the board from Down East Microwave, instead of trying to make your own.

A No-Tune Transverter for 3456 MHz

By Jim Davey, WA8NLC

(Reprinted from June 1989 *QST*.)

The development of easy-to-build equipment for the microwave bands has increased dramatically in the past few years. No longer do you have to have a small fortune invested in test equipment to get the satisfaction of building your own station for the bands above 1 GHz. Nor do you have to rely on the availability of surplus components gathered at hamfests—or be lucky enough to live in a high-tech part of the country where surplus is more plentiful—to home-brew your own equipment.

The recent increase in activity on the microwave bands has been well documented in *QST* and elsewhere. In the past two years alone, amateurs have conquered the difficult EME challenge on 3456, 5760 and 10368 MHz and set impressive distance records on nearly every microwave band through 47 GHz. UHF contest stations now regularly have 2304- and 3456-MHz equipment available. Our knowledge of the propagation characteristics of these bands has also benefited from the increased activity.

Commercially manufactured ham equipment from Europe has been available for a few years for the more popular microwave bands, and this equipment has helped to spawn activity. The 3456-MHz band, however, has not been supported by commercial manufacturers as of this writing. Let's not let the lack of commercial equipment stop us from having a little fun! Besides, you can get a lot of enjoyment and personal satisfaction from building your own equipment. I can personally attest to this: My own station uses no commercially manufactured equipment except for the IF rigs. This article describes a transmitting and receiving conversion module (transverter) for the 3456-MHz band. Any multimode 2-meter transceiver can be used as a tunable IF. The transverter has several features that make it ideal for the newcomer to 3456 MHz and the veteran looking for a simple loaner rig for grid-square expeditions:

- It doesn't require RF alignment or microwave test equipment for proper operation.
- The entire transverter, minus the 552-MHz local oscillator, is contained on one PC board, reducing the need for separate enclosures and expensive RF connectors and cables.
- The transmitter features 10-mW output, and the receiver features a 4-dB noise figure (NF)—performance sufficient for a lot of interesting work on this band.
- Inexpensive diodes and MMIC gain blocks are used to keep the cost low.
- An external receiving preamplifier, transmitter power amplifier and antenna relay can be added to make the unit a high-performance package for fixed or portable operation.
- The transverter can also be used as an IF for the higher microwave bands above 10 GHz.

Background

The straightforward design of this transverter is a result of two fairly recent developments in the amateur microwave field. First, the introduction of cascadable monolithic-microwave integrated circuits (MMICs) several years ago revolutionized the design of microwave amplifiers. Not only are MMICs inexpensive, but most are unconditionally stable and can be cascaded for increased gain or paralleled for greater power output. Al Ward, WB5LUA, authored excellent articles for *QST* concerning the use of these devices.[1] You are encouraged to review these articles for a more thorough treatment of the subject.

The second development contributing to the design of this transverter is my recent work on microstrip band-pass filters that provide the required selectivity for this system without the need for tuning adjustments. These filters were first introduced to amateurs in 1987 at the Microwave Update Conference in Estes Park, Colorado.[2] A later paper reported on further work to improve the SWR of the filters and compared the microstrip filters to other commonly used filters for the microwave bands.[3]

To avoid taxing your wallet or your patience, I used parts that are inexpensive and easy to get. You can get this transverter up and running for less than $200 including the LO. The idea here is to get you on the 3456-MHz band as easily as possible without relying on unique or hard-to-find components.

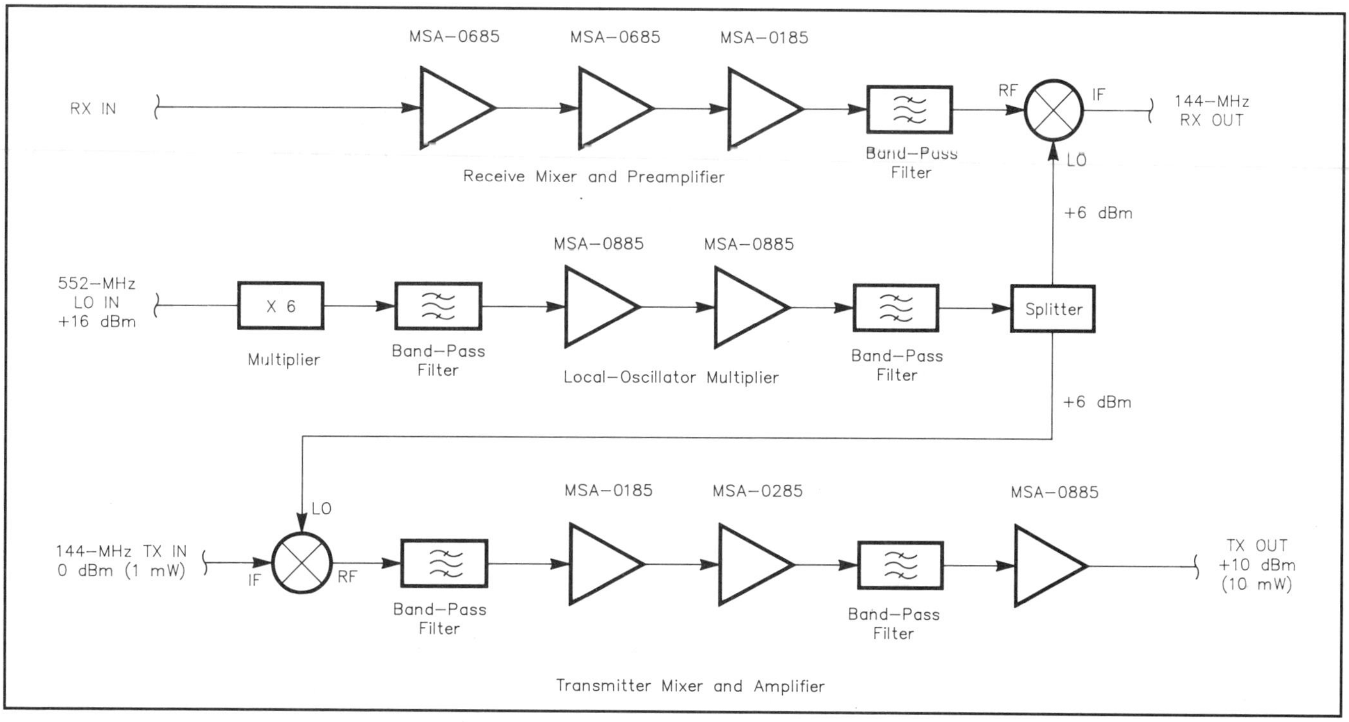

Fig 1—Block diagram of the 3456-MHz transverter.

Circuit Description

The transverter is divided into three basic sections: transmit mixer/amplifier, receive mixer/preamplifier and local-oscillator multiplier. See Fig 1. Each section is described in the following paragraphs.

Local-Oscillator Multiplier

The transverter requires an external 552-MHz local oscillator (LO) signal that is multiplied by six in the transverter to obtain 3312-MHz injection for the mixers. For the multiplier circuit, I used an idea developed by Rick Campbell, KK7B, who demonstrated that a simple diode multiplier and inexpensive MMIC gain stages can produce a clean microwave local oscillator. Although Rick's design placed each stage in a separate box, I was able to implement the entire multiplier on micro strip using my own printed filters. Rick used the printed-filter design idea to produce a clean 552-MHz LO module that I use as a companion to this transverter.[1]

The transverter requires +16 dBm (40 mW) of 552-MHz energy from an external LO. As shown in Fig 2, the first stage of the multiplier is a broadly resonant circuit that drives D1, a Schottky-diode comb generator. Following D1 is FL1, the first of two band-pass filters used to select the desired 3312-MHz output. Two stages of amplification (U1, U2) using Avantek MSA-0885 MMICs bring the level up to +11 dBm. A second filter, FL2, is used to further clean up the LO and reduce broadband noise that is generated in the amplifier stages. Following FL2, a 90° hybrid divider splits the LO signal into two equal outputs for injection into the transmit and receive mixers. The LO output level to each mixer is +6 dBm. Undesired products in the LO output are 35 dB below the carrier.

Transmit Mixer and Amplifier

The transmit mixer (Fig 3) is a 3/2 wavelength rat-race balanced design from an article by H. Paul Shuch, N6TX.[5] The original article described a 1296-MHz mixer etched on G-10 PC-board material. I developed the 3456-MHz mixer by resizing the line lengths and widths for the new frequency and for a Teflon-fiberglass substrate. To keep the mixer as efficient as possible and maintain high isolation between ports, I used a matched pair of diodes (D2, D3) in a surface-mount package. Isolation between the mixer's LO and RF ports is greater than 20 dB—about as good as any low-cost commercial mixer.

Conversion loss—about 9 dB—is a little higher (worse) than most commercial mixers. Impedance matching was not done at the IF port and did not appear to be necessary with the ICOM IC-202A transceiver I use as an IF rig.

I used three stages of amplification and two filters in the transmit-amplifier chain to reach the final 10 mW output level. The mixer RF port drives FL3, the first 5 pole band pass filter. Centered at 3500 MHz, FL3 strips image energy (3168 MHz) from the mixer output and also contributes about 20 dB of rejection at the LO frequency. Rejection at the LO frequency is important. With an LO injection level of +6 dBm and LO-to-RF-port isolation of 20 dB, the level of the LO signal at the RF port of the mixer is –14 dBm. For comparison, the desired signal (the sum frequency at 3456 MHz) is injected at 0 dBm (1 mW) and encounters a 9-dB loss through the mixer to emerge at –9 dBm, only 5 dB above the oscillator feedthrough.

Following the first filter are two amplifier stages (U3, U4) using an MSA-0185/0285 combination. With prototype versions of this transverter, I found that correct MMIC choice for these amplifier stages is critical because band-pass filters

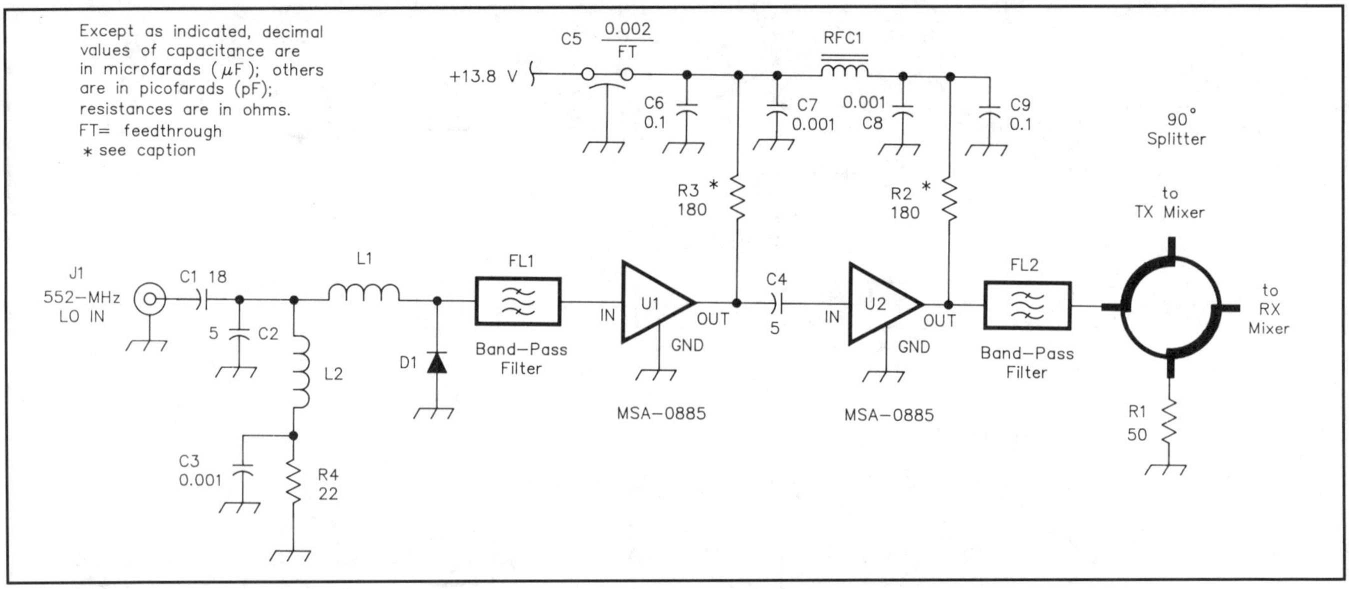

Fig 2—Schematic of the 3456-MHz transverter LO multiplier/amplifier.

C1—18-pF disc-ceramic or silver-mica capacitor.
C2, C4—5-pF porcelain chip capacitor.
C3, C7, C8—0.001-µF ceramic chip (preferred) or disc ceramic capacitor.
C5—0.002-µF feedthrough capacitor.
C6, C9—0.1-µF ceramic chip (preferred) or disc-ceramic capacitor.
D1—Hewlett-Packard 5082-2835 Schottky diode.
FL1, FL2—Band-pass filters printed on PC board.
J1—Female chassis-mount SMA connector.
L1—Inductor printed on PC board.
L2—3 turns no. 28 enam wire, 0.078-in. ID, close-wound.
R1—50-Ω chip resistor.
R2, R3—180-Ω, ¼- or ½-W carbon-film resistor. Note: This value is for 13.8-V operation. See text for information on operation at other voltages.
R4—22-Ω, 1/4-W carbon-film resistor.
RFC1—270 mH subminiature molded RF choke. A suitable alternative is 24 turns no. 28 enam wire on an FT-37-72 toroid core.
U1, U2—Avantek MSA-0885 or Mini-Circuits MAR-8 MMIC.

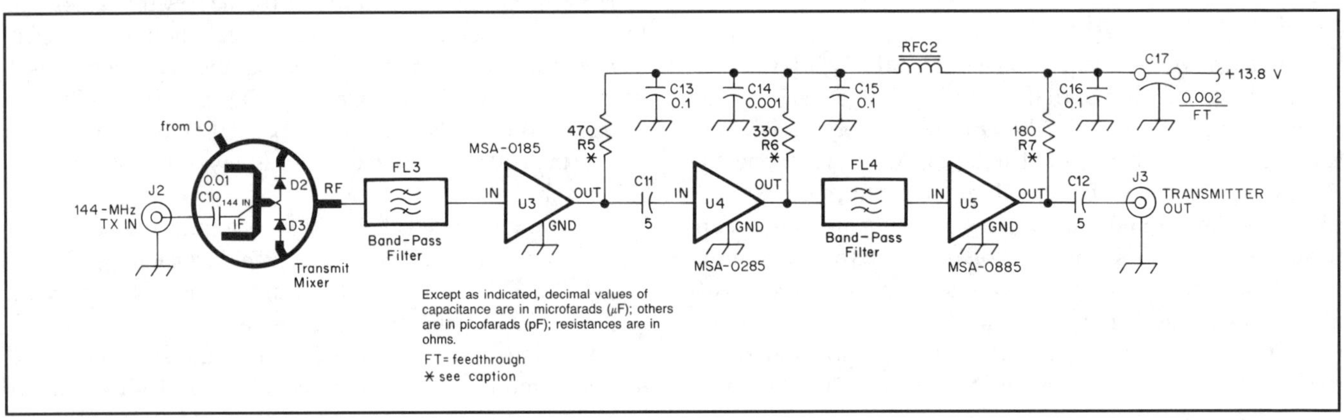

Fig 3—Schematic of the 3456-MHz transverter transmit mixer/amplifier.

C10—0.01-µF disc-ceramic capacitor.
C11, C12—5-pF porcelain chip capacitor.
C13, C15, C16—0.1-µF ceramic chip (preferred) or disc-ceramic capacitor.
C14—0.001-µF ceramic chip (preferred) or disc-ceramic capacitor.
C17—0.002-µF feedthrough capacitor.
D2, D3—Hewlett-Packard HSMS-2822 surface-mount diode pair. See text.
FL3, FL4—Band-pass filters printed on PC board.
J2—Female chassis-mount connector, builder's choice.
J3—Female chassis-mount SMA connector.
R5—470-Ω, ¼-W carbon-film resistor. Note: This value is for 13.8-V operation. See text for information on operation at other voltages.
R6—330-Ω, ¼-W carbon-film resistor. Note: This value is for 13.8-V operation. See text for information on operation at other voltages.
R7—180-Ω, ¼- or ½-W carbon-film resistor. Note: This value is for 13.8-V operation. See text for information on operation at other voltages.
RFC2—270 mH subminiature molded RF choke. A suitable alternative is 24 turns no. 28 enam wire on an FT772 toroid core.
U3—Avantek MSA-0185 or Mini-Circuits MAR-1 MMIC.
U4—Avantek MSA-0285 or Mini-Circuits MAR-2 MMIC.
U5—Avantek MSA-0885 or Mini-Circuits MAR-8 MMIC.

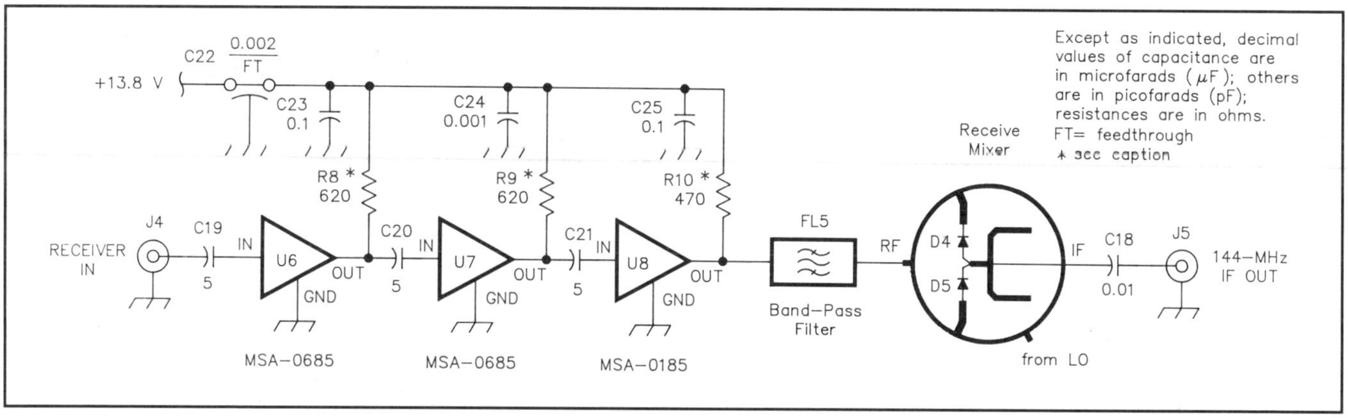

Fig 4—Schematic of the 3456-MHz transverter receive mixer/amplifier.

C18—0.01-μF disc-ceramic capacitor.
C19-C21—1-pF porcelain chip capacitor.
C22—0.002-μF feedthrough capacitor.
C23, C25—0.1-μF ceramic chip (preferred) or disc-ceramic capacitor.
C24—0.001-μF ceramic chip (preferred) or disc ceramic capacitor.
D4, D5—Hewlett-Packard HSMS-2822 surface-mount diode pair or matched pair of HP 5082-2835 Schottky diodes. See text.
FL5—Band-pass filter printed on PC board.
J4—Female chassis-mount connector builder's choice.
J5—Female chassis-mount SMA connector,
R8, R9,—620-Ω, ¼-W carbon-film resistor. Note: This value is for 13.8-V operation. See text for information on operation at other voltages.
R10—470-Ω, ¼-W carbon-film resistor. Note: This value is for 13.8-V operation. See text for information on operation at other voltages.
U8—Avantek MSA-0685 or Mini-Circuits MAR-1 MMIC.
U6, U7—Avantek MSA-0185 or Mini-Circuits MAR-6 MMIC.

(FL3, FL4) are present at the input and output of the amplifier strip. The filters are highly reactive out of band where their return loss is very low (in the range of 0 to 3 dB), so the MMICs used in conjunction with the filters must be unconditionally stable. Another consideration is that the microstrip filters were developed with good 50-ohm terminations on each end. The SWR at the input and output ports of the MMIC should be close to 50 ohms within the passband of the filters to maintain flat filter response and low insertion loss. An examination of the S-parameter data for the MSA-0185 and '0285 shows that they are excellent choices for this application.

FL4 was added in later prototypes to reduce mixing products above 3456 MHz that were present in the output. This filter, a modified version of the filters used in the LO multiplier chain, is centered at 3312 MHz. The combination of FL3 centered at 3500 MHz and FL4 centered at 3312 MHz creates a narrow window at 3456 MHz where the two response curves overlap, giving the effect of a much narrower filter. A single narrow-band, higher-Q filter could be used to accomplish this, but the stop-band rejection for one filter would be inferior to that of two filters. Technical references indicate that the stop-band rejection of a single microstrip filter is only about 40 dB because of surface-wave effects; two separate filters can achieve a total stop-band attenuation of 80 dB. FL3 and FL4 in tandem do a good job of cleaning up the output spectrum.

The final transmit amplifier is an Avantek MSA-0885, chosen for its gain and +10-dBm power-output capability at 3456 MHz. One disadvantage of the '0885 is that it is only conditionally stable: It can oscillate if supply line decoupling is inadequate or if it is terminated in highly reactive loads. U5 should be stable as long as the transmitter-output port is terminated in an impedance close to 50 ohms. If, however, the transverter is used "barefoot" in conjunction with an external antenna relay, I recommend that the dc power to the transmit amplifier section be removed during receive periods so the absence of an antenna load does not cause U5 to oscillate. My prototype units were stable without an output termination, but variations from device to device may cause U5 to oscillate in some transverters. Result: lots of noise in your receiver, making it impossible to hear weak signals!

Receive Mixer and Preamplifiers

The receive mixer (Fig 4) is identical to the transmit mixer. It is preceded by a 5-pole image-stripping filter, FL5, which is necessary to keep the noise energy at the image frequency from being converted to the IF and degrading receiver sensitivity. In a receive application, 20 to 25 dB of image rejection is all that is necessary to ensure that this does not happen. A 5 pole microstrip filter easily provides this much rejection at the 3168-MHz image frequency.

The receiver-preamplifier stages were chosen to have good SWR and low noise figure. Preceding FL5 is an MSA-0185 (U8) that terminates the filter well and has a noise figure of approximately 6.5 dB at 3456 MHz. The front end consists of two MSA-0685 stages (U6, U7). The overall receiver noise figure is about 4.4 dB. An outboard preamplifier can be added to make a state-of-the-art setup. Performance of the transverter is summarized in Table 1.

Construction Hints

The transverter is constructed on 0.031-inch-thick woven Teflon-glass substrate with a dielectric constant of 2.50. The board is double clad with ½-ounce copper. I get my board

Table 1

3456-MHz Transverter Performance

General

- Frequency range: 3456-3460 MHz
- IF range: 144-148 MHz
- Local oscillator: +16 dBm (40 mW) required at 552 MHz
- Power required: 13.8 V dc at 250 mA

Transmitter

- Output power: 10 mW (+10 dBm)
- Spurious rejection: –40 dBc
- IF drive level: 0 dBm max

Receiver

- Noise figure: 4.5 dB
- Image rejection: 25 dB
- Net gain: 15 dB

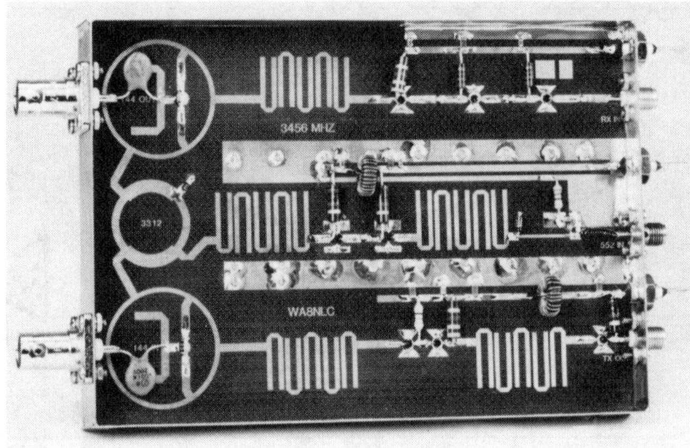

Fig 6—Photograph of a finished 3456-MHz transverter. This unit was built by Zack Lau, KH6CP, in the ARRL lab.

material from Taconics Plastics, Ltd, Petersburg, NY 12138, part number TLX-9-0310-R5/R5. A word of caution: The filters require that dimensional tolerances of +0.001 inch or better be maintained in the fabrication of the board. This is the price you must pay for microwave filters that require no adjustments. Because of the critical tolerances necessary and the many variables involved in the printing process, an etching pattern is not included in this article. You can purchase an etched board from Down East Microwave.[6]

Construction is as simple as populating the board with the components and mounting the connectors. A parts layout guide is shown in Fig 5. Lead dress can be seen in the photograph of the completed transverter (Fig 6). All of the parts needed to complete the project are available from amateur suppliers, and kits are available as well.

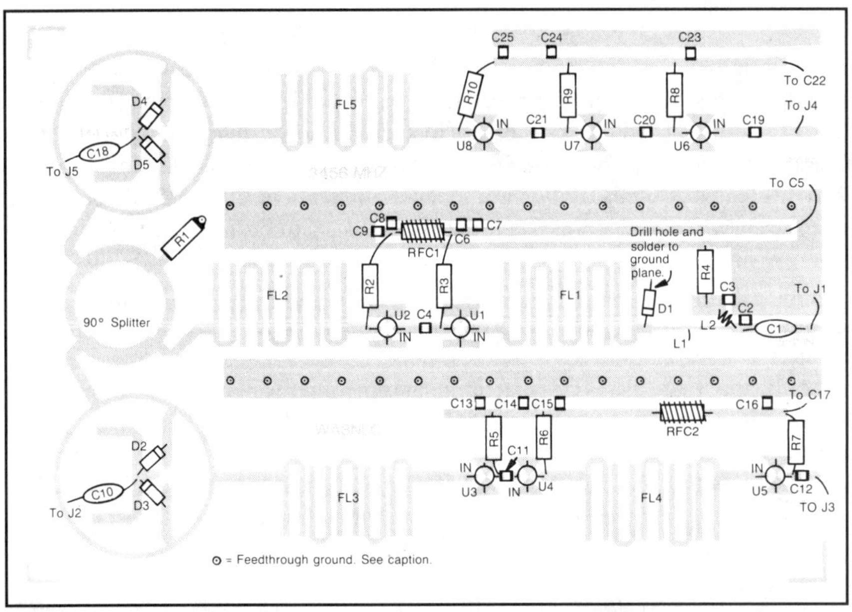

Fig 5—Parts-placement guide for the 3456-MHz transverter (not shown actual size). All components mount on the etched side of the board. Feedthrough grounds, indicated by circles, must be installed and soldered top and bottom; see text. See Fig 7 for mounting details for U1-U8.

Be sure to use high-quality porcelain-chip capacitors for coupling between stages. The remaining capacitors can be less expensive ceramic chips or disc ceramics. I used chip capacitors everywhere I could because they make for a neater layout and cost about the same as the equivalent disc-ceramic capacitors.

Mini-Circuits offers a line of MMICs equivalent to the specified Avantek parts. The corresponding part numbers are easy to determine. The Mini-Circuits MAR-8 is equivalent to the MSA-0885, the MAR-6 to the MSA-0685, and so on.

I strongly recommend that you enclose the board in a brass box to support the connectors and provide ground continuity to the top side of the board. I used strips of 0.032-inch-thick, ¾-inch-wide brass available at most hobby stores. The entire perimeter of the inside walls is soldered to the top and bottom of the board. This provides a ground connection to the component side in several places, as well as a ground for the input and output connectors.

The board material is soft and bends easily. It is quite tolerant of heat and rework if you misplace a part. I use a 27-watt pencil soldering iron with good results.

Through-the-board wires are required to tie components to ground, as shown on the parts layout. Tiny brass rivets can be used if you have them available, but I used no. 18 bus wire with good results. The ground leads of the MMICs are grounded to a small piece of copper foil under each lead. The foil is wrapped through a clearance hole under each MMIC lead and soldered to the ground plane below (see Fig 7). The cold end of the 50-ohm chip resistor (R1) on the LO power divider should be grounded in a similar fashion.

Keep the bias resistors and

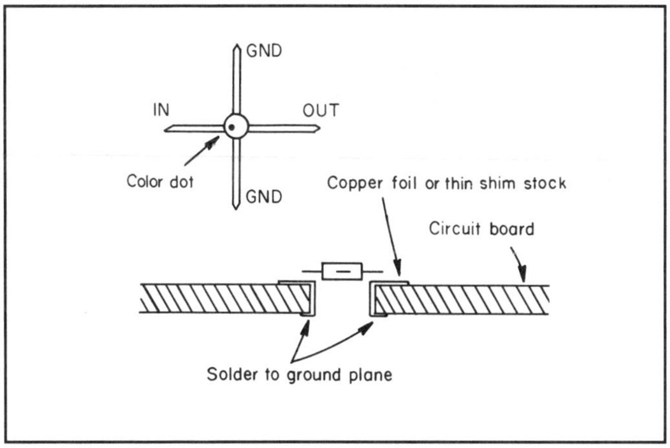

Fig 7—Mounting details for the MMICs.

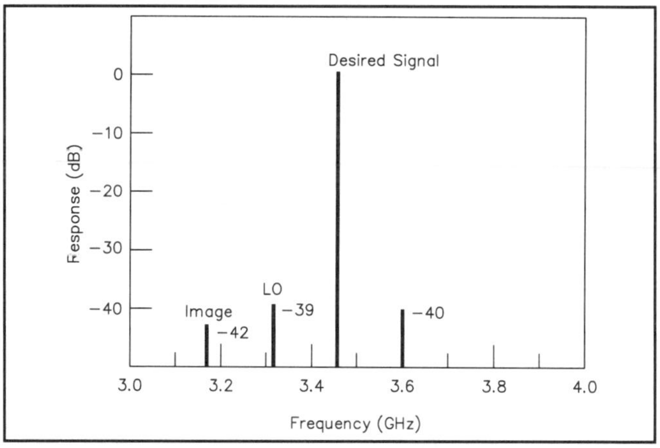

Fig 8—Output spectrum of the 3456-MHz transverter transmitter section.

chokes close to the board so they don't act as antennas. During testing of the prototypes, I noticed that there is some radiation of RF energy from the surface of the board. This is not a problem unless LO energy gets into the transmitter circuits and shows up in the output. Without any shields at all, the LO is more than 35 dB below the carrier.A spectrum analyzer display of the transmitter output is shown in Fig 8.

In the parts list, I have shown bias-resistor values for 13.8-V operation. If only 12-V operation is planned, the bias resistor values should be changed to maintain the correct operating current through each MMIC. Be sure to check resistor dissipation and use ½-W resistors as necessary. The bias current for each MMIC is as follows: MSA-0185, 17 mA; MSA-0285, 25 mA; MSA-0685, 15 mA; MSA-0885, 32 to 35 mA. On one portable expedition, I noticed that the receiver began oscillating after the transverter had been on for a while. I later found that the dry-cell battery had dropped more than 2 V, causing a corresponding drop in the bias current in each MMIC. When operated at lower than rated current, the MMICs are not unconditionally stable. If battery operation is planned, it is a good idea to regulate the power source to the transverter at 10 V or so and adjust the bias resistors accordingly.

For best performance, the mixer diodes should be a matched pair in a microstrip or surface-mount package. The recommended surface-mount pair is the Hewlett Packard HSMS-2822. This part is available from Down East Microwave.[4] If you have some microwave mixer diodes you want to try, go ahead and do it—just make sure the diodes are connected cathode to anode, as shown in Figs 3 and 4. The LO rejection can be measured with a power meter at the transmitter-output port by noting the difference between full carrier output and the residual oscillator level.

Accessories

You will need a few outboard accessories to complete your 3456-MHz station. First, a means of reducing the transmitter output power of your IF rig to 1 mW is required. A recent circuit published in *QEX* not only takes care of attenuating the transmitter, but also provides protected TR switching of the IF line.[7]

On the other end of the transverter, some sort of RF antenna relay is needed. Meeting this requirement may be a little more difficult, because good relays for this band are scarce. I have been fortunate enough to find small SMA-type relays at hamfests in the past, but larger relays with N connectors will probably work. Don't overlook relays equipped with TNC connectors. They are often good to well beyond 3456 MHz. You can take care of any connector mismatch in the jumper cables to the transverter. If you are unsure about the suitability of a relay for this frequency, ask a knowledgeable friend for advice. Remember, just because a relay has good microwave-quality connectors doesn't mean it will provide low loss and operate at an acceptable SWR at 3456 MHz.

If you can find one used, a circulator makes a neat TR switch. Isolation between the transmitter and receiver ports won't be as good with a circulator as with most relays, but at least you won't need the dc source necessary to operate a relay. If you can't find a circulator, you can use an isolator. I have converted many isolators to circulators by removing the isolator load resistor and installing a coaxial connector in its place.

Several types of antennas are popular at 3456 MHz. The most common 3456-MHz antenna is a small dish in the 2- to 4-foot-diameter range. A recent article in *QEX* shows how to build efficient feed systems for the 3.4, 5.7 and 10.3-GHz bands.[8] For respectable performance without the wind load of a dish, you can now buy loop Yagis for 3456 MHz.[9]

Feed line losses are severe at 3456 MHz. For home-station installations, and even for portable operations, a good Hardline, such as Andrew Heliax, is mandatory. Also, don't overlook the G-line.[10] I have used a G-line for a couple of years now, with very good results.

Station performance can be greatly improved with the addition of an outboard receive preamplifier and transmit power amplifier. A state-of-the-art receiving preamp that makes an excellent front end for this no-tune transverter was described in *QST*.[11] Preamps of this design have been duplicated by many amateurs; they deliver good performance without requiring tweaking on a noise-figure meter.

Several options are available for the transmitter. Some amateurs have found surplus traveling-wave tubes (TWTs) to be a great way of generating lots of power on this band. TWTs typically require only a milliwatt or so of drive for full output, so you will have to attenuate the output of the transverter by about 10 dB. I prefer solid-state, however. A receiving-type GaAsFET like the Avantek ATF-10135 can yield up to +17 dBm quite easily at low cost. Don Hilliard, WØPW, published several good ideas on how to bridge the gap between 10 mW and 1 W.[12] Don shows how, for about $30, you can break the 100-mW level with an Avantek ATF 21170 FET. I have built his circuit using a similar device (an Avantek ATF 25170) with good results.

Summary

What can you expect to work on the 3456-MHz band? A recent article in *QST* discussed the various modes of propagation at 2304 MHz.[13] Everything said about propagation there applies just as well to 3456 MHz from my experience. Stations using dishes have the advantage of increased antenna gains for the same physical size. For example, a 4-foot dish has 3-dB more gain at 3456 MHz than at 2304 MHz. The bottom line is that workable distances on 3456 MHz are on par with those on 2304 MHz. All it takes to prove this is a little more activity!

Notes

[1]A. Ward, "Monolithic Microwave Integrated Circuits," Part 1: *QST*, Feb 1987, pp 23-29, 32; Part 2: *QST*, Mar 1987, pp 22-28, 33.

[2]J. Davey, "Microstrip Bandpass Filters," *Proceedings of Microwave Update '87* (ARRL, 1987), pp 42-53.

[3]J. Davey, "Microwave Filter Update," *Proceedings of Microwave Update '88*, pp 1-8. This book is available from ARRL for $12 (plus $2.50 postage and handling, or $3.50 for insured parcel post or UPS) or from your local dealer.

[4]Complete parts kits for a suitable local oscillator, as well as assembled units, are available from Down East Microwave, Box 2310, RR 1, Troy, ME 04987, tel 207-948-3741.

[5]H. Shuch, "Rat-Race Balanced Mixer for 1296 MHz," *Ham Radio*, Jul 1977, pp 33-39.

[6]Etched boards, complete parts kits and assembled units are available from Down East Microwave (see Note 4).

[7]Z. Lau, "A VHF/UHF/Microwave Transverter IF Switch," *QEX*, Aug 1988, pp 3-4.

[8]D. Hilliard, "Antenna Ideas For 3.5, 5.8, and 10.4 GHz," *QEX*, Jan 1988, pp 3-5.

[9]Loop Yagis are available from Down East Microwave (see Note 4).

[10]R. Dryden, "G Lines for 1296, 2304 and Above," *Proceedings of Microwave Update '87*, pp 54-62. See Note 2.

[11]A. Ward, "Simple Low-Noise Microwave Preamplifiers," *QST*, May 1989, pp 31-36, 75.

[12]*Proceedings of Microwave Update '87*, pp 8-92. See Note 2.

[13]E. Pocock, "Getting From Here to There on 2304 MHz," *QST*, Nov 1988, pp 15-16.

A Single-Board, No-Tune Transverter for 5760 MHz

By Rick Campbell, KK7B

(From *QST*, October 1990.)

Single-board, no-tune transverter designs for the 902, 1296, 2304 and 3456-MHz bands have been published in recent years.[1-4] These boards follow a common theme: They use printed-circuit filters and inexpensive plastic monolithic microwave integrated circuit (MMIC) gain blocks to achieve good performance at low cost. The use of printed-circuit filters and broadband MMICs also eliminates the need for RF alignment or microwave test equipment for proper operation. Low cost and ease of assembly and operation have tempted many amateurs to experiment with the microwave bands.

This article describes a single-board transverter for 5760 MHz. Design and construction are similar to that of the 3.4-GHz transverter described by Jim Davey, WA8NLC, in June 1989 *QST*.[5] I won't repeat many of the design and construction techniques described in Jim's article, so a review of that information will be helpful if you're unfamiliar with the single-board transverter concept.[6] In addition to the transverter board described here, you'll need a local oscillator (LO), 1296-MHz IF radio and antenna. Information on completing your 5760-MHz station is presented later.

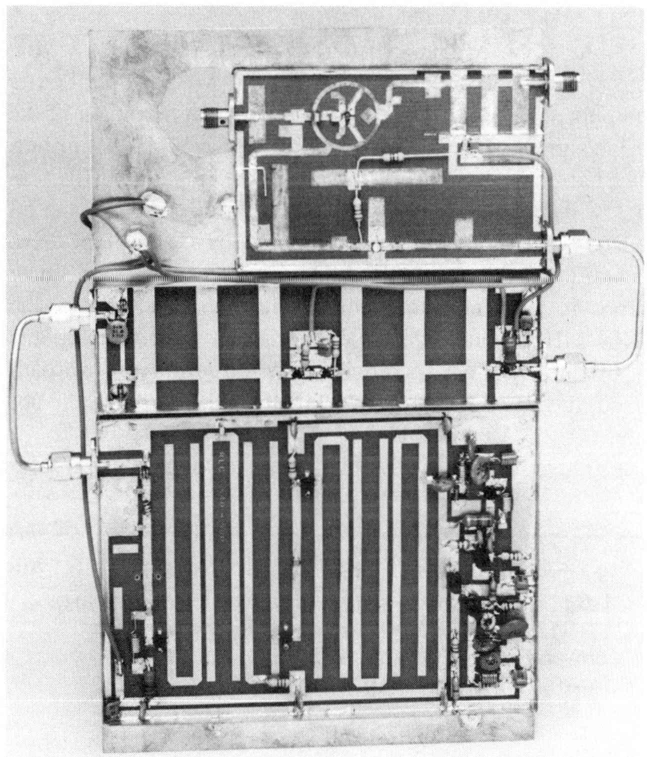

The KK7B 5760-MHz transverter (top board), along with its companion LO (bottom two boards) offers a no-tune approach to microwave operation.

Design Considerations

In attempting to push the single-board design concept to the 5760-MHz band, two difficulties were encountered:

• The performance of currently available silicon MMICs used in the transmitter and receiver gain stages rapidly deteriorates at this frequency.

• No-tune printed-circuit filters provide insufficient image rejection and LO rejection for use with a 144-MHz IF transceiver.

The problem of image and LO rejection was solved by designing the 5760-MHz board for an IF of 1296 MHz. A single three-section printed-circuit filter with 10% bandwidth provides image and LO rejection of more than 30 dB with a passband insertion loss of less than 1 dB. The problem of obtaining suitable devices for the gain stages was solved by integrating everything except the gain stages into a single circuit board, with each port matched to 50 ohms. Then you are free to use any external gain blocks that may become available.

The basic transverter board—without additional gain stages at 5760 MHz—provides a transmit signal of –6 dBm (250 mW) at the 1-dB compression point (–4 dBm saturated) and a receiver noise figure of 9 dB (assuming that the 1296-MHz IF rig has a 2-dB receiver noise figure). This is acceptable performance for line-of-sight contacts over distances of many miles when used with a small dish antenna. Leaving the gain stages off the transverter board solves another problem: You don't need to worry about finding a suitable 5760-MHz TR relay; the antenna connects to a common filter for transmit and receive.

There are several external gain blocks from which to choose. An excellent one is the 5760-MHz preamplifier described by Al Ward, WB5LUA, in May 1989 *QST*.[7] This preamplifier can also be biased for maximum power and used as a transmit amplifier. Another choice for a transmitter gain block is the Avantek MGA-64135 GaAs MMIC.[8]

Design Notes

Fig 1 shows the block diagram of the 5760-MHz transverter board, and the schematic is shown in Fig 2. There are three basic sections: LO doubler, mixer and bandpass filter.

A 4464-MHz LO signal is mixed with a 1296-MHz IF signal for operation at 5760 MHz. The transverter requires an external 2232-MHz LO; an on-board MMIC doubles this signal to 4464 MHz at +8.5 dBm for mixer injection.

Although you can use any 2232-MHz source, I highly recommend the no-tune LO system described in July 1989 *QST*.[9] The no-tune LO with a 93-MHz crystal delivers a 2232-MHz, +7-dBm signal.

An on-board doubler using an Avantek MSA-0835 MMIC provides an output of +8.5 dBm at 4464 MHz, with all other outputs more than 30-dB down. The second harmonic of the drive signal is obtained by overdriving the MMIC amplifier. The harmonic output is increased by reducing the MMIC bias below the value recommended for linear operation. In the reduced-bias condition, the MSA-0835 will oscillate if the drive signal is removed. A 2232-MHz drive signal of –8 dBm or more is sufficient to stabilize the MSA-0835, and the 4464-MHz output varies little for drive levels between +3 and +10 dBm. The MSA-0835 doubler should be driven by a broadband, flat 50-ohm source. The MMIC in the output of the no-tune 2232-MHz LO is ideal.

The output of the doubler passes through a 2232-MHz notch filter (L4 on the schematic) and a single-tuned circuit at 4464 MHz (L5 and L6). L4 is an open-circuit ¼-λ transmission line at 2232 MHz, so it appears as a short circuit at the drive frequency. L4 is an open-circuit ½-λ transmission line at 4464 MHz, so it appears as an open circuit at the desired second harmonic. The open circuited end of L4 is one-half wavelength away from the output of the MSA-0835 at 2232 MHz and one wavelength away at 4464 MHz, so the MSA-0835 does not see a short circuit at either frequency.

The MSA-0835 is a broadband amplifier, with a noise output from DC to more than 6 GHz. If the MSA-0835 noise output at 1296 MHz is passed to the mixer, some of it will appear at the IF port. This noise will directly add to the receiver noise figure. The broadband 4464-MHz single-tuned circuit (L5 and L6) was added to the artwork to stop the 1296-MHz noise component from getting to the mixer. Careful measurements with the single-tuned circuit indicate that the noise figure is approximately equal to the mixer conversion loss. Without L5 and L6, the noise figure is about 10-dB worse.

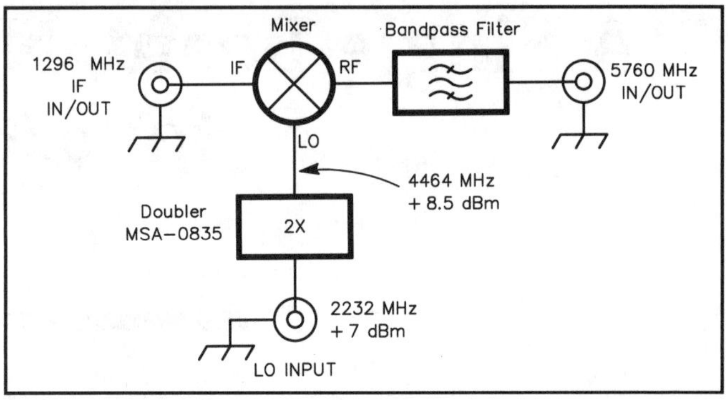

Fig 1—Block diagram of the 5760-MHz transverter.

Fig 2—Schematic of the 5760-MHz transverter. R1 and R2 are ¼-W carbon-film types.

C1-C10—Capacitive stub printed on circuit board.
C11-C13—5-pF, 50-mil-square microwave chip capacitor.
C14—0.01-mF chip capacitor.
C15—18-pF microwave chip capacitor.
D1—HP HSMS-2822 diode assembly.
L1-L3—¼-λ open-circuit stub etched on PC board.
L4—¼-λ stub at 2232 MHz (and ½-λ stub at 4464 MHz)

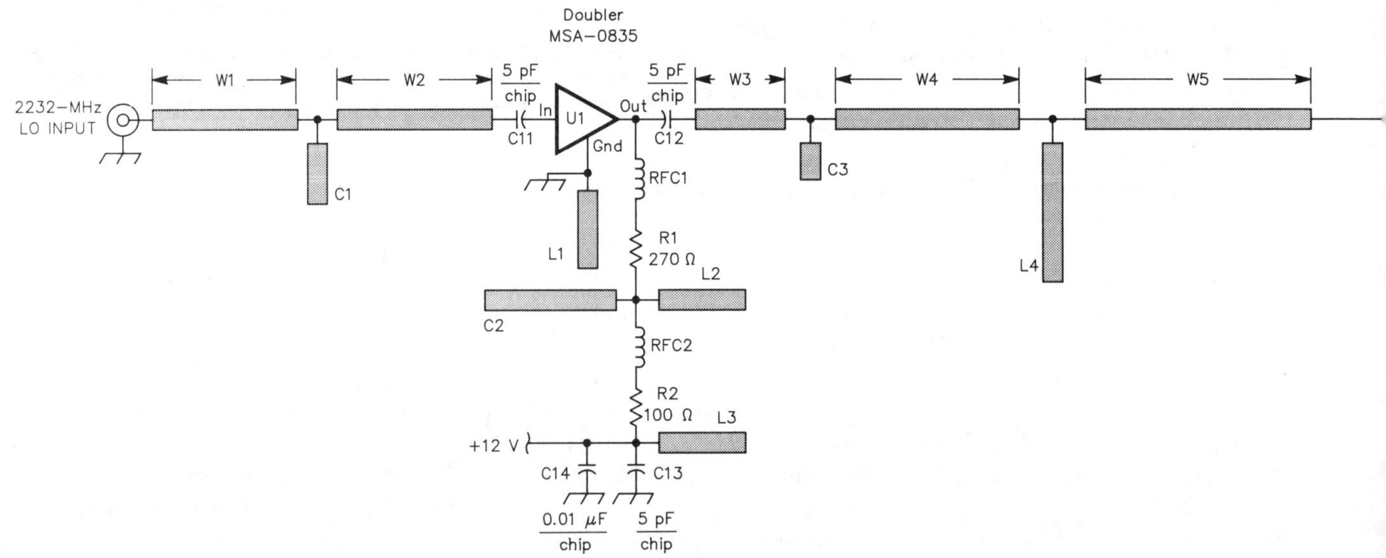

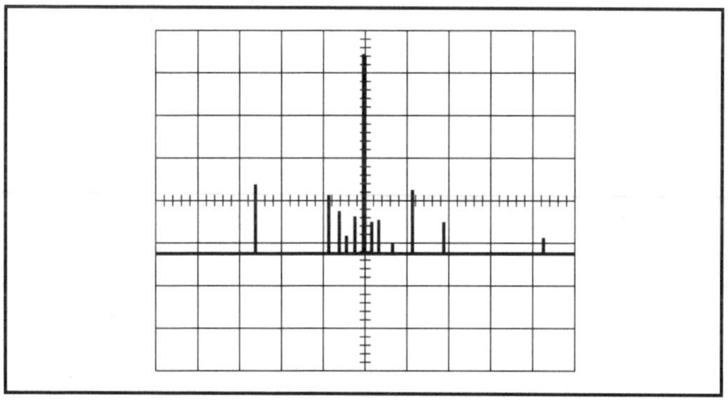

Fig 3—Output spectrum of the 5760-MHz transverter. Each vertical division is 10 dB; each horizontal division is 500 MHz. The pip at the center of the display is the 5760-MHz signal. All other outputs are more than 30-dB down.

When properly driven, this frequency doubler is well behaved, clean and stable. The six versions constructed to date have shown less than 0.5-dB variation in output level when driven from the same 2232-MHz no-tune LO.

The mixer is a standard design, except that the HP-2822 diode pair is used well above its specified frequency range. On the assumption that the imperfections of the diodes were reactive and relatively uniform from part to part, I built a standard 6/4-λ mixer, then added tuning "confetti" empirically to improve the conversion loss. The confetti was then added to the mixer artwork, and subsequent mixers show good uniformity.

The IF port is also empirically matched to 50 ohms at 1296 MHz. The bare mixer displays a conversion loss of only 6.5 dB at 5760 MHz, which is much better than many commercial mixers designed for broadband performance in this range.

The filter is an off-center-tapped 0.16-dB-ripple Chebyshev type designed using the procedure described by

Table 1
Transverter Specifications

General

- Frequency range: 5650-5925 MHz
- IF range: 1240-1300 MHz
- LO required: 2.18-2.32 GHz at +7 dBm nominal; +3 to +10 dBm acceptable
- Power requirements: 12 V dc at 10 mA; 8-15 V acceptable.

Transmitter

- Output power at 1-dB compression: −6 dBm.
- Saturated RF output: −4 dBm.
- LO signal at RF port: −36 dBm, max.
- IF drive level: approx 0 dBm.

Receiver

- Noise figure: 9 dB.
- Conversion loss: 7 dB.
- Image rejection: 30 dB.

Beebe.[10] It is exceptionally flat, has steep skirts and a loss of well under 1 dB in the passband.

After I integrated the doubler, mixer and filter onto a single board, the interconnecting line lengths and tuning confetti were varied to obtain best performance with 50-ohm pads on all ports. The output spectrum at the 1-dB compression point is shown in Fig 3. The image, which is 32-dB lower than the 5760-MHz output, is off the lower end of the spectrum-analyzer range. Table 1 lists the transverter specifications.

One cautionary note: Any metal box big enough to hold the transverter board will be a resonant cavity at a series of frequencies within the gain bandwidth of the MMIC. Imag-

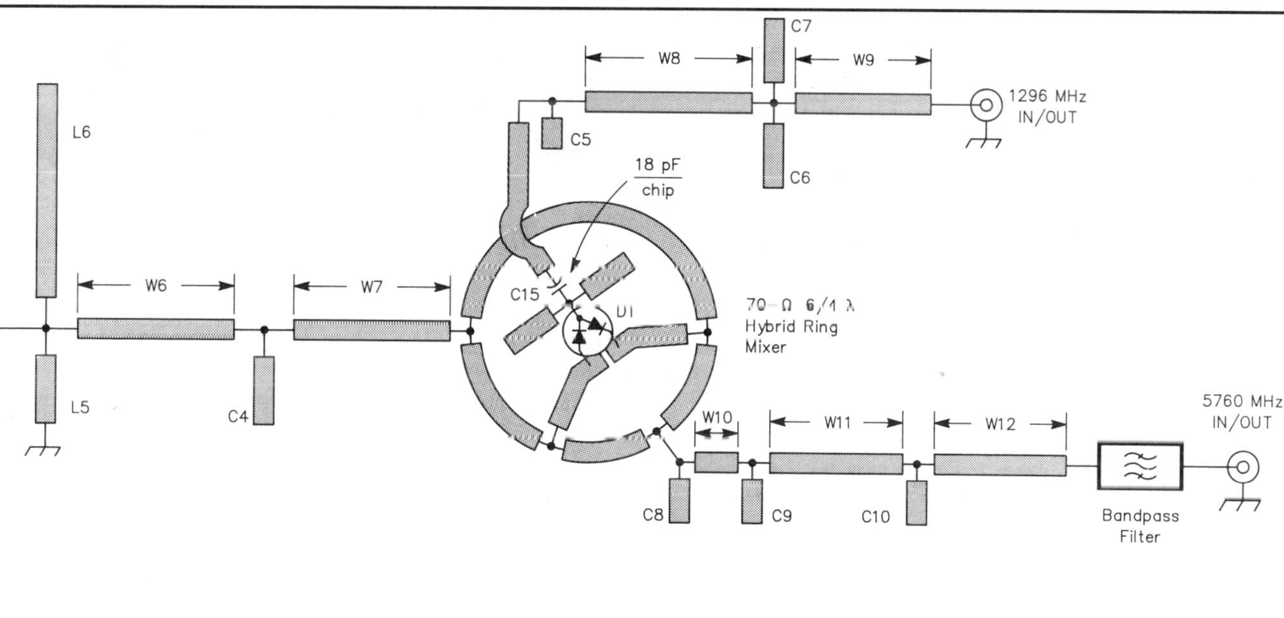

etched on PC board (see text).
L5, L6—Broad tuned circuit at 4464 MHz (and short circuit at 1296 MHz) etched on PC board (see text).
RFC1, RFC2—Inductance of R1 and R2 leads (not critical).
U1—Avantek MSA-0835 MMIC.
W1-W12—Tuned lengths of 50-Ω microstrip transmission line etched on PC board.

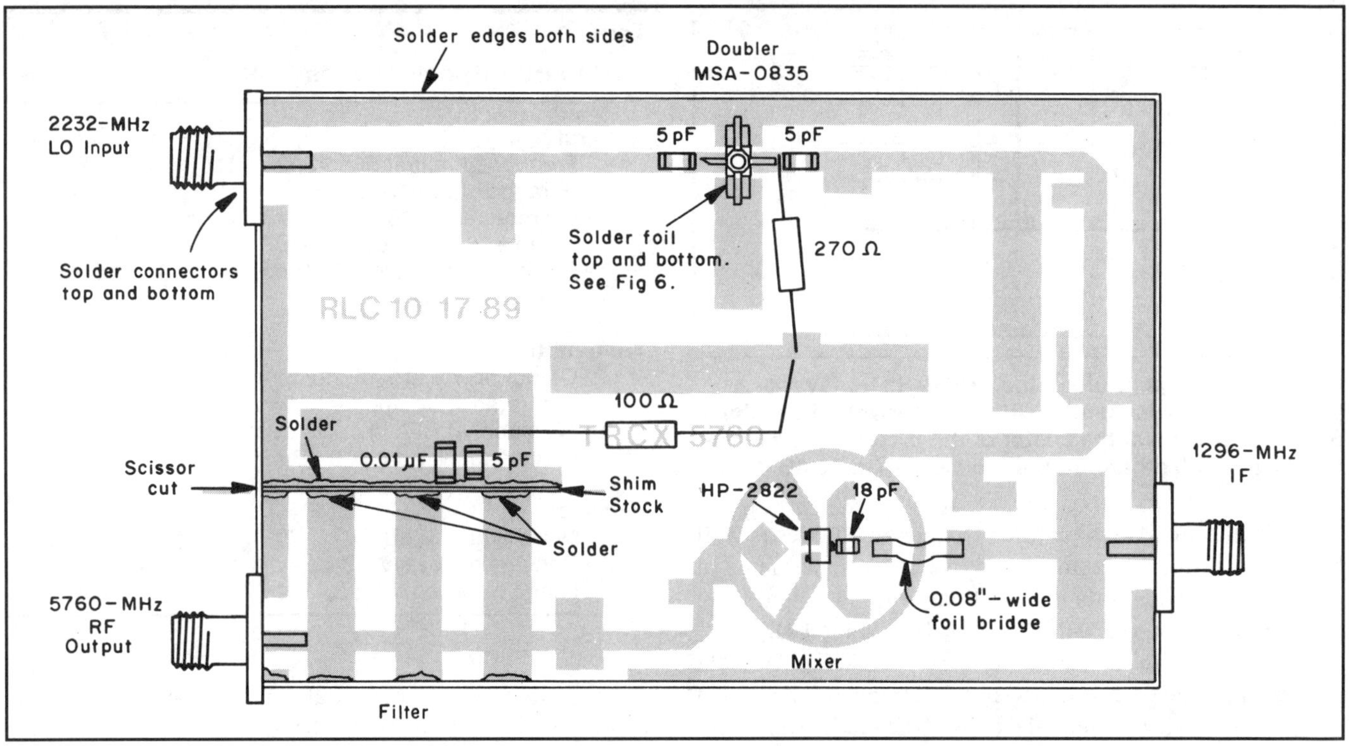

Fig 4—Part-placement diagram for the 5760-MHz transverter board (not shown actual size). All components mount on the etched side of the board. See Fig 6 for mounting details for the MSA-0835 MMIC.

ine building a 40-meter linear amplifier in a box the size of a football field! No-tune transverters and LOs that work perfectly on the bench often exhibit spurious oscillations when enclosed in a metal box. The transverter board will work fine if it is simply wrapped with plastic-bubble packaging material or nonconductive foam and mounted near the antenna feed. For permanent installations, a plastic food container works well.

If a shielded metal box is necessary, I recommend shielding individual stages with thin brass or copper shim stock and then enclosing the shielded individual stages in a larger box. Spurious oscillations in the no-tune LO or lower-frequency transverters may often be cured by replacing "hot" MMICs like the MSA-0835, MSA-0685 and MAR-6 with well-behaved parts like the MSA-0235, MSA-0285 and MAR-2.

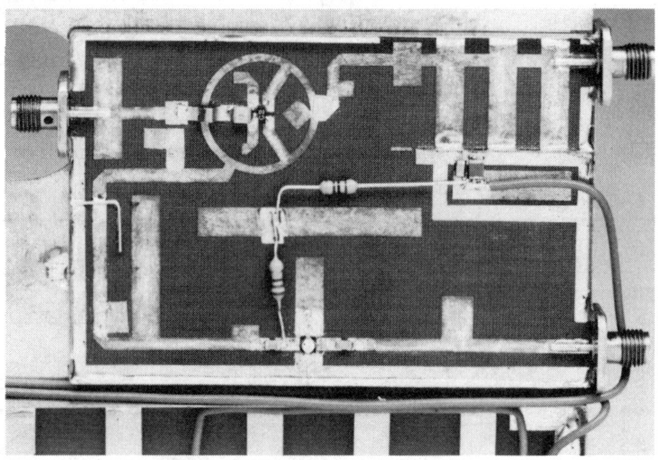

Fig 5—A finished 5760-MHz transverter.

Construction

The transverter is constructed on 0.031-inch-thick Teflon-glass substrate with a dielectric constant of 2.5. The board is clad with ½-ounce copper on both sides. One side is etched; the other is unetched and acts as a ground plane. The material I used is made by Taconics Plastics, Ltd, Petersburg, NY 12138, and the part number is TLX-9-0310-R5/R5. The filter requires that dimensional tolerances of ±0.001 inch or better be maintained in the fabrication of the board. (Close construction tolerances are essential for microwave filters that require no adjustments.) Because of the critical tolerances necessary and the many variables involved in the *QST* printing process, an etching pattern is not included in this article. If you are interested in making your own board, send an SASE to the ARRL Technical Department for a dimensioned copy of the artwork.[11] Or, if you wish, you can purchase an etched board from Down East Microwave.[12]

Part placement is shown in Fig 4. All components mount on the etched side of the board, and construction is conventional for microwave circuits. Additional details may be seen in Fig 5.

Brass sides enclose the board to support the SMA connectors and provide continuity between the grounds on the top and bottom of the board. The sides are made from strips of 0.032-inch-thick, ¼-inch-wide copper or brass shim stock available at hobby stores. The perimeter of the inside walls is soldered to the top and bottom of the board. This provides a ground connection to the component side in several places, as well as a ground for the connectors.

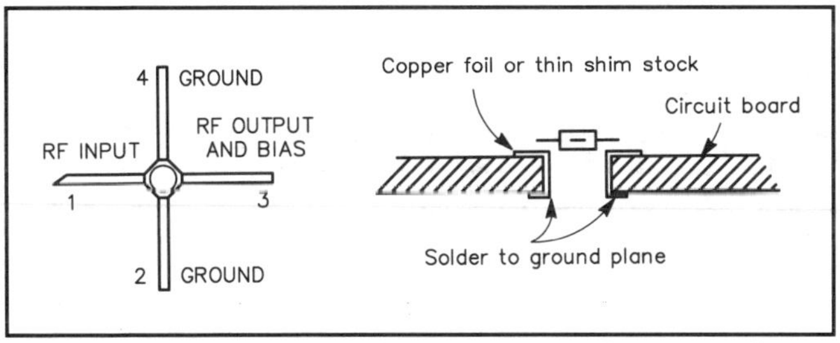

Fig 6—Mounting details for the MSA-0835.

Additional grounding is needed for the MSA-0835. Cut a hole in the board as shown in Fig 6, and wrap copper foil through the hole to connect grounds on the top and bottom of the board. Then solder the foil on both sides.

Both ends of the filter elements must be grounded. The ends of the elements along the edge of the board are grounded simply by soldering them to the vertical board edge strips, which must also be soldered on the bottom. The ends of the elements toward the center of the board are grounded by making a straight scissor cut as indicated by the arrows, and inserting a strip of thin copper shim stock the full length of the slit. Then solder the shim stock to the PC-board copper on both sides of the slit, on the top and bottom of the board.

Additional copper foil is used to make a bridge between the 1296-MHz IF connector and the mixer. Cut the foil to 0.08-inch wide and solder one end to the printed trace from the IF connector. Then bend the foil into an arch over the printed mixer ring and solder it to the pad leading to the 18-pF coupling capacitor.

Completing Your 5760-MHz Station

In addition to the transverter and LO boards, you'll need an IF radio and antenna. There are two convenient approaches to the IF rig:

- Use a 1296-MHz multimode transceiver (ICOM, Kenwood and Yaesu make suitable radios).
- Build a 1296-MHz no-tune transverter board[13] to use as a first IF, and use a 2-meter multimode radio as the IF transceiver.

In either case, it's worth a look at Zack (KH6CP) Lau's transverter control circuit described in August 1988 *QEX*.[14]

Selecting an antenna for 5760 MHz is easy: Nothing matches the performance of a small dish. I've had good success with a 19-inch dish that mounts easily on a small camera tripod. It shows a 10-degree beamwidth and about 25 dB gain with a simple feed. Don Hilliard, WØPW,[15] and Tom Hill, WA3RMX,[16] have presented some excellent antenna ideas for 5760 MHz.

Conclusion

This transverter board, used with the no-tune 2232-MHz LO, provides a straightforward approach to a transceive system on the 5760-MHz band. Although it's not as attractive as the previously described transverters for 902 through 3456 MHz in terms of output level, noise figure and the use of a 144-MHz IF, it is by far the simplest board in the family. Construction of a high-performance transceive system should be relatively easy for an experimenter ready to move up to the 5760-MHz band.

Acknowledgments

I thank Tony Bickel, K5PJR, Merle Cox, W7YOZ, and Jim Davey, WA8NLC, for inspiration and encouragement, and Mark Schreiner, NK8Q, Dave Erickson, N9JBI, and Jim Berner, K8OSF, for help in evaluating the prototype 5760-MHz transverters.

Notes

[1] R. Campbell and D. Hilliard, "A Single Board 900 MHz Transverter with Printed Bandpass Filters," *Proceedings of Microwave Update '89*, pp 1-8. (Reprinted in this book.)

[2] R. Campbell, "A Single-Board No-Tuning 23-cm Transverter," *Proceedings of the 23rd Conference of the Central States VHF Society*, pp 44-52. (Reprinted in this book.)

[3] J. Davey, "A No-Tune Transverter for 3456 MHz," *QST*, Jun 1989, pp 21-26. (Reprinted in this book.)

[4] J. Davey, "No-Tune Transverter for 2304 MHz," *Proceedings of Microwave Update '89* (ARRL, 1989), pp 30-34. (Reprinted in this book.)

[5] See Note 3.

[6] Copies of this article are available from the ARRL Technical Department Secretary for $3 and an SASE.

[7] A. Ward, "Simple Low-Noise Microwave Preamplifiers," *QST*, May 1989, pp 31-36, 75.

[8] Contact Avantek, 3175 Bowers Ave, Santa Clara, CA 95054, tel 408-727-0700, for the name of your local dealer.

[9] R. Campbell, "A Clean, Low-Cost Microwave Local Oscillator," *QST*, Jul 1989, pp 15-21.

[10] G. Beebe, "Analysis of a Class of Microstrip Bandpass Filters," MSEE thesis, Michigan Technological University, Feb 1988.

[11] Send a no. 10 SASE to the ARRL Technical Department Secretary; request the Oct 1990 *QST* 5760-MHz transverter template.

[12] Etched circuit boards, parts kits and assembled and tested circuit boards for this project are available from SHF Systems through Down East Microwave, 954 Rte 519, Frenchtown, NJ 08825, tel 908-996-3584, fax 908-946-3072. Circuit boards for the other transverters and the no-tune 2232-MHz LO referred to in the text are available from the same source.

[13] See Note 12. Also see Z. Lau, "Product Review: SHF Systems 1240K 1296-MHz Transverter Kit," Feb 1990 *QST*, pp 33-34.

[14] *QEX*, Aug 1988, pp 3-4.

[15] D. Hilliard, "Antenna Ideas for 3.5, 5.8, and 10.4 GHz," *QEX*, Jan 1988, pp 3-5.

[16] T. Hill, "A Triband Microwave Dish Feed," *QST*, Aug 1990, pp 23-27.

10-GHz Gunnplexer Communications

By ARRL Staff

The M/A-COM Gunnplexer transceiver shown in Fig 1 will fit conveniently in your hand and is operated from a single 12-V power supply, either ac line operated or batteries! This makes the Gunnplexer ideal for fixed or portable operation.

The Gunnplexer is most often used in wide-band FM systems and is ideal for audio, full-color video and data transfer. With suitable peripheral equipment, systems for full-duplex audio, color video and up to several megabit data transfer are possible. Such high data rates should allow direct computer-to-computer memory transfer that would not be practical (or legal) on other lower-frequency amateur bands.

The Gunnplexer

The heart of the Gunnplexer is a Gunn diode oscillator, named after its inventor, John Gunn of IBM. Refer to the cutaway drawing of the Gunnplexer, Fig 2, for this discussion.

The Gunn diode is mounted with a varactor diode in a resonant cavity. When a regulated voltage is applied to the Gunn diode it oscillates; the frequency of oscillation is determined by the capacitance of the varactor diode and two mechanical tuning screws. The mechanical tuning screws can be likened to coarse tuning controls and are factory set for the appropriate tuning range. The voltage applied to the varactor diode (1 to 20 V dc) tunes the frequency electronically a minimum of 60 MHz. Power is coupled out of the cavity through a small iris that has been designed as somewhat of a compromise between maximum power output and isolation from changes in diode impedance and load.

The Gunn oscillator is also used to provide the local-oscillator signal for the detector diode. A ferrite circulator couples an appropriate amount of energy into the low-noise Schottky mixer diode and isolates the transmitter and receiver. Because the Gunn oscillator functions as both the transmitter and the receiver local oscillator, the IF at each end must be at the same frequency. Furthermore, the frequencies of the Gunnplexers must be separated by the IF. This is illustrated in Fig 3. Intermediate frequencies of 30 MHz are more or less standard for audio work in the US. Both 45 and 70 MHz are used for video and high-speed data work.

Fig 1

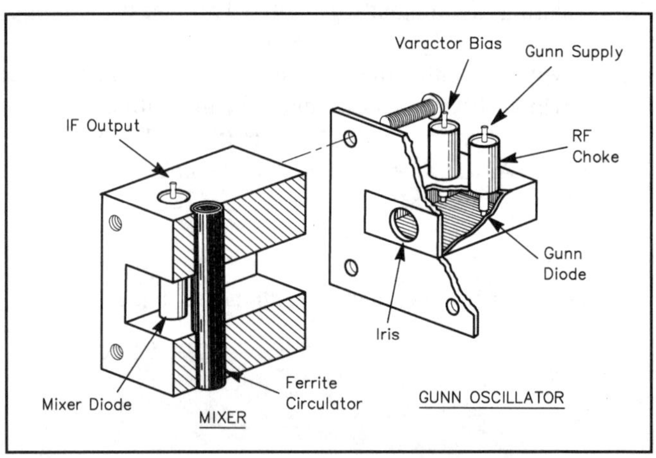

Fig 2

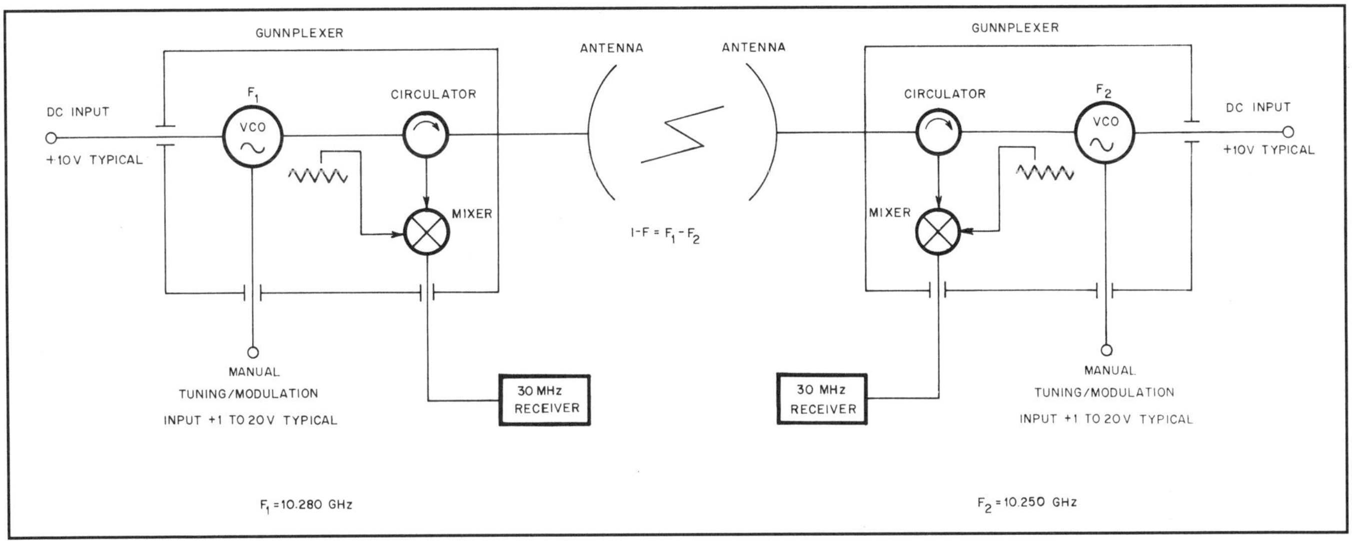

Fig 3

As can be seen from Fig 3, the Gunnplexer communications system is full duplex. In other words, both parties can talk and listen at the same time, without throwing switches. This is something that may take a while to get used to as most amateurs are programmed for VOX or PTT operation. In short, it is the ultimate break-in system!

One detail of the Gunnplexer that does require some specific attention is frequency control. The Gunnplexer has a frequency stability specification of –350 kHz frequency change per degree Celsius increase. This does not pose much of a problem with wide-bandwidth applications such as video or data transfer. For relatively narrowband audio work (200 kHz and less), however, some form of AFC, a phase-lock or other frequency-control scheme is required. In most cases simple AFC circuitry is sufficient and quite simple to implement. The electrical characteristics of the Gunnplexer are given in Table 1.

Bare Minimum Audio Communications System

The simplest of communications systems using the Gunnplexer transceivers can be formed with two 88- to 108-MHz FM receivers, two Gunnplexers, two microphones and associated amplifiers, and two sources of 12 V dc. A diagram of such a system is shown in Fig 4. Some of the low-cost FM converters and receivers for automotive use make good IF strips for this communications system. The AFC signal developed in the converter or receiver can be routed to the Gunnplexer varactor diode to lock the two units together. The microphone amplifier can be a single 741 op amp as shown in the diagram.

There are two shortcomings with this system. The first involves the use of the FM broadcast band as the IF. If mountaintop DXing is planned, it is likely that strong FM broadcast stations will be received no matter how short the lead between the detector and the FM converter or receiver is made. It will be necessary to select that part of the band where there are no strong signals present.

Table 1
Gunnplexer Specifications @ T_A = 25° C

Electrical Characteristics	
RF Center Frequency	10.250 GHz[1]
Tuning	
Mechanical	± 50 MHz
Electronic	60 MHz min.
Linearity	1 to 40%
Frequency Stability	–350 kHz/°C max.
RF Power vs Temperature and Tuning Voltage 6 dB max.	
Frequency Pushing	15 MHz/V max.
Input Requirements	
Dc Gunn Voltage Range	+ 8.0 to 10.0-V dc[2]
Maximum Operating Current	500 mA
Tuning Voltage	+1 to +20 volts
Noise Figure[3]	<12 dB
RF Output Power[1]	
Model	*P out (mW)*
MA87141-1	10 min. 15 typ.
MA87141-2	20 min. 25 typ.
MA87141-3	35 min. 40 typ.

Notes
[1]Tuning voltage set at 4.0 volts.
[2]Operating voltage specified on each unit.
[3]1.5 dB NF at 30 MHz.

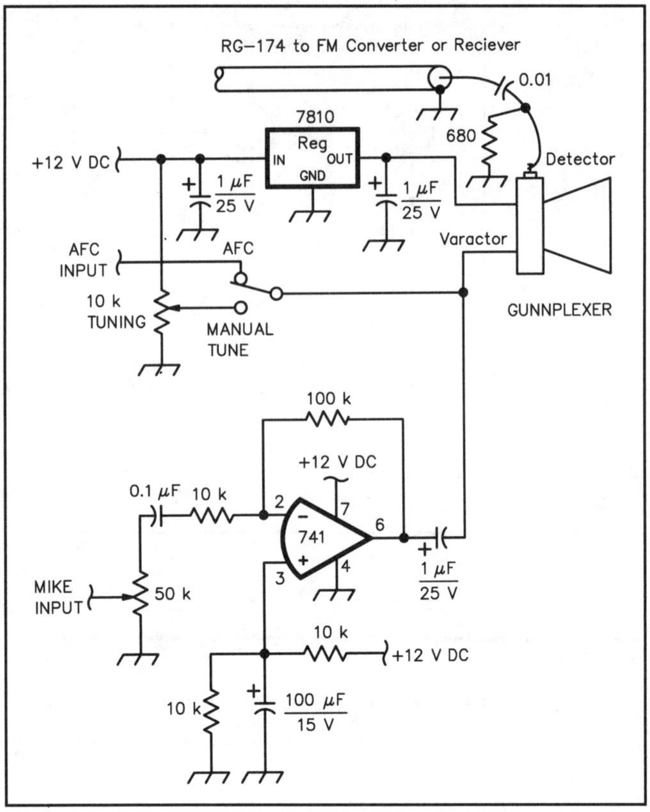

Fig 4

The second item involves AFC. Since it is possible to tune the Gunnplexers over a 60-MHz range with the varactor diode tuning, it is possible to tune them on either side of each other. This means that the single-polarity AFC system in the receiver or convertor will work for only one combination. If the units are operated so that the AFC polarity is incorrect, the AFC will push the received signal out of the receiver passband. If this happens, simply tune the two Gunnplexers to produce the other IF signal. AFC lock should then be obtained.

Modifications for the SSB Electronics 10 GHz Transverter

By Kent Britain, WA5VJB

(From *Microwave Update '91*)

The SSB Electronics 10-GHz Transverter is a truly state-of-the-art piece of equipment. They were the first commercial 10-GHz transverter. There are, however, several areas where these units can be improved.

Power Leads

All three boxes use insulated pins for the 12-V power connections. Replacing the pin with a feedthrough capacitor makes the box RF tight. This really helps in the high RF environment found on many hill tops. I also suggest adding a diode from the power lead to ground in each box. Use a 1N4001 or a 15-V, 1-W Zener. This adds reverse polarity protection for those little accidents that occur in the field (Fig 1).

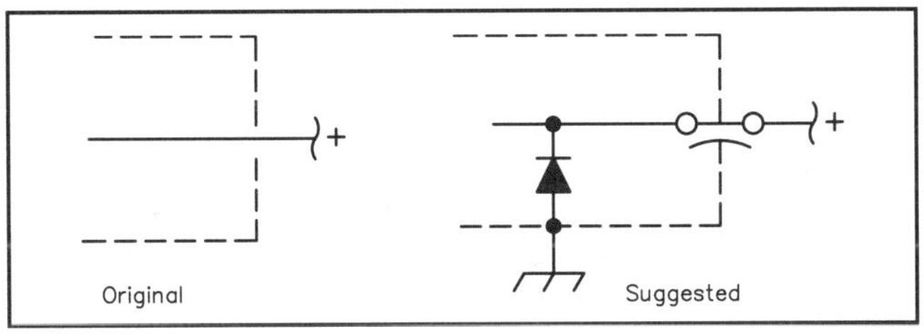

Fig 1—Reverse-polarity protection.

Image Filtering

The basic unit has very little rejection of the 10.080-GHz image. While there are several filters in the transmitter, there is only one filter in the receiver. With very careful tuning, I was only able to get 8 dB of image rejection. Sweeping with a signal generator, I was also able to get responses on six frequencies between 10 and 11 GHz.

These extra images have resulted in the publication of some very optimistic noise figure measurements. Typically, the noise figure runs about 3-½ dB for the basic unit. An external preamp and 10-GHz filter will really clean up the receiver. After adding a pair of ½-in. plumbing end cap "cheap" filters, all spurious responses were more than 40 dB down. WA5TNY and W7CNK are using waveguide filters ahead of their transverters (Fig 2).

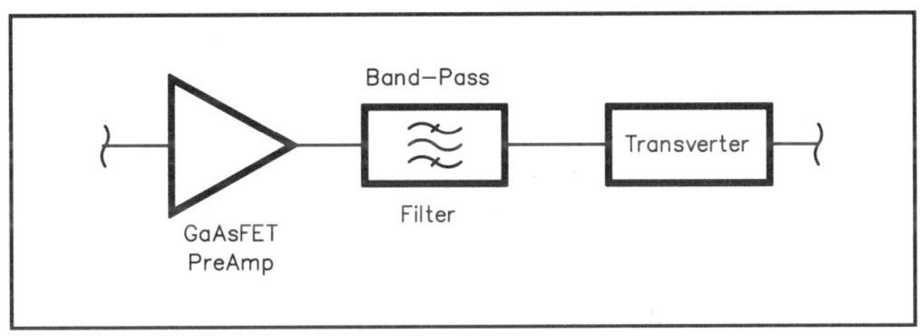

Fig 2—An external preamp and filter will remove receiver spurs.

The Local Oscillator

The LO module puts out 6 to 7 mW at 2.556 GHz. If the LO output power ever drops below 5 mW, the 2.556 to 10.224 GHz multiplier stops multiplying. The LO tuning can be quite critical in the 2.556-GHz filter stage. The filter depends on self resonance in the three trimmer caps to form the filter.

If you simply tune the LO for max output, you'll see a pretty dirty output on a spectrum analyzer. Tune for cleanest response, and you may not get the 6 mW the next stage needs. Several stations have solved this problem by completely replacing the LO module. Using a Frequency West Brick crystaled for 2.556 GHz, or even 10.224 GHz (bypassing the quadrupler stage) has been popular.

Another solution is to add a filter-amp stage between the LO and receiver modules. Replacing the jumper cable with an MSA-0404, MAR-4, or MSA-1105 MMIC, followed by a plumbing cap or interdigital filter can really clean up the output and produce plenty of drive for the next stage.

Buffer Amp

The output of the 10-GHz GaAsFET receiver mixer goes directly to the IF output. The slightest accidental transmission back into the receiver is *fatal* to this GaAsFET. Many of the new solid-state switching 2-meter rigs can actually start transmitting faster than a relay can close. Installing a buffer amp and resistive pad on the IF output can protect the receiver while providing enough gain to drive secondary receivers, moon/sun-noise meters, or long IF cable runs. On my transverter an 18-dB pad is permanently mounted on the IF

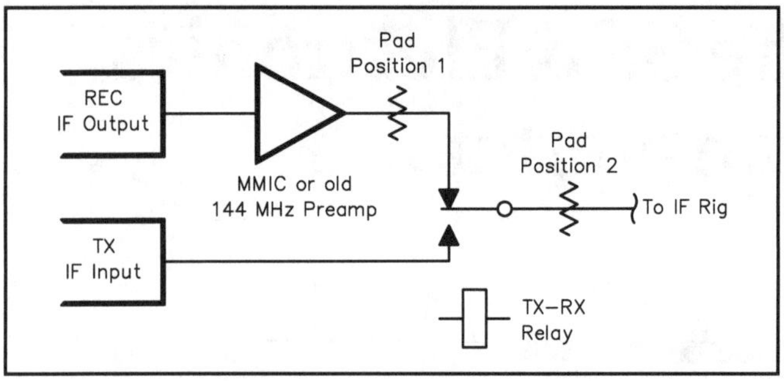

Fig 3—A buffer amp and attenuator protect the receiver from the IF rig transmitter.

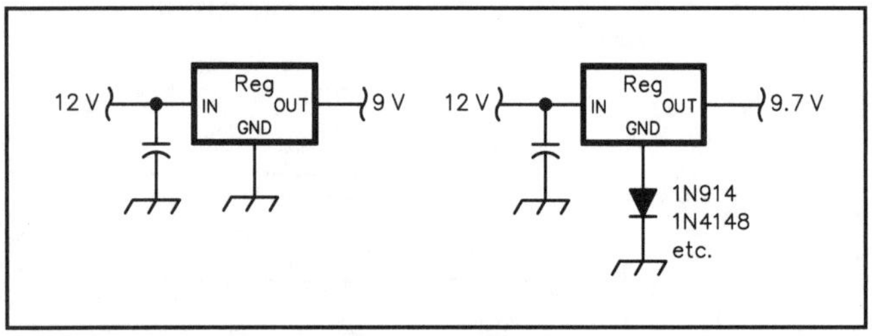

Fig 4—Adding a diode in the GND leg of the regulator raises LO supply voltage, improving stability and increasing drive.

connection and is used during both transmitting and receiving (Fig 3).

Transmitter Module

The output of the transmitter usually exceeds SSB's specs, and is quite clean. The only problem was driving a TWT to saturation. When a TWT is driven to saturation to get that last watt out for EME, the TWT is often driven to 15 dB of gain compression. This makes any spurs or images 15 dB stronger at the TWT output. An extra 10.368-GHz filter between the transverter and the TWT keeps all those precious watts on 10.368 MHz.

Increasing LO Output and Stability

Allen Katz, K2UYH, was also having problems with LO drive and poor stability. His solution was to add a diode in the ground leg of the 9-V regulator in the LO module. This increased the oscillator and multipliers operating voltage from 9 to 9.7 V. Allen says this really helped (Fig 4).

Conclusion

The 10-GHz SSB Transverter is an excellent piece of gear. Most of these modifications attach between units. The mods replace the normal jumper cables, so modifications to the SSB boxes themselves are not necessary.

SSB/CW Equipment Concepts for 24 and 47 GHz

By Tom Hill, WA3RMX

(From *Microwave Update '89*)

INTRODUCTION

As interest and activity increases on the microwave bands, the relentless push for ever-higher frequencies leads us up into the so-called Millimeter Wave bands. In the past, most work at these frequencies has been done using wideband FM, as the stability of the available components did not allow for narrowband work. Today the technology for better frequency stability and phase noise is becoming more available within the amateur community, and we are beginning to take some advantage of it. This article shows the techniques and equipment I have used at 24 and 47 GHz. It is not a construction article, as it does not contain complete construction details, but it does present the concepts I used in my implementation of Millimeter Wave SSB and CW. These ideas may be useful as you venture higher in frequency.

MODULATION MODE

If we exclude, for now, the more exotic synchronous detection and DSP signal recovery methods as being beyond the scope of this project, then we are left with CW as the

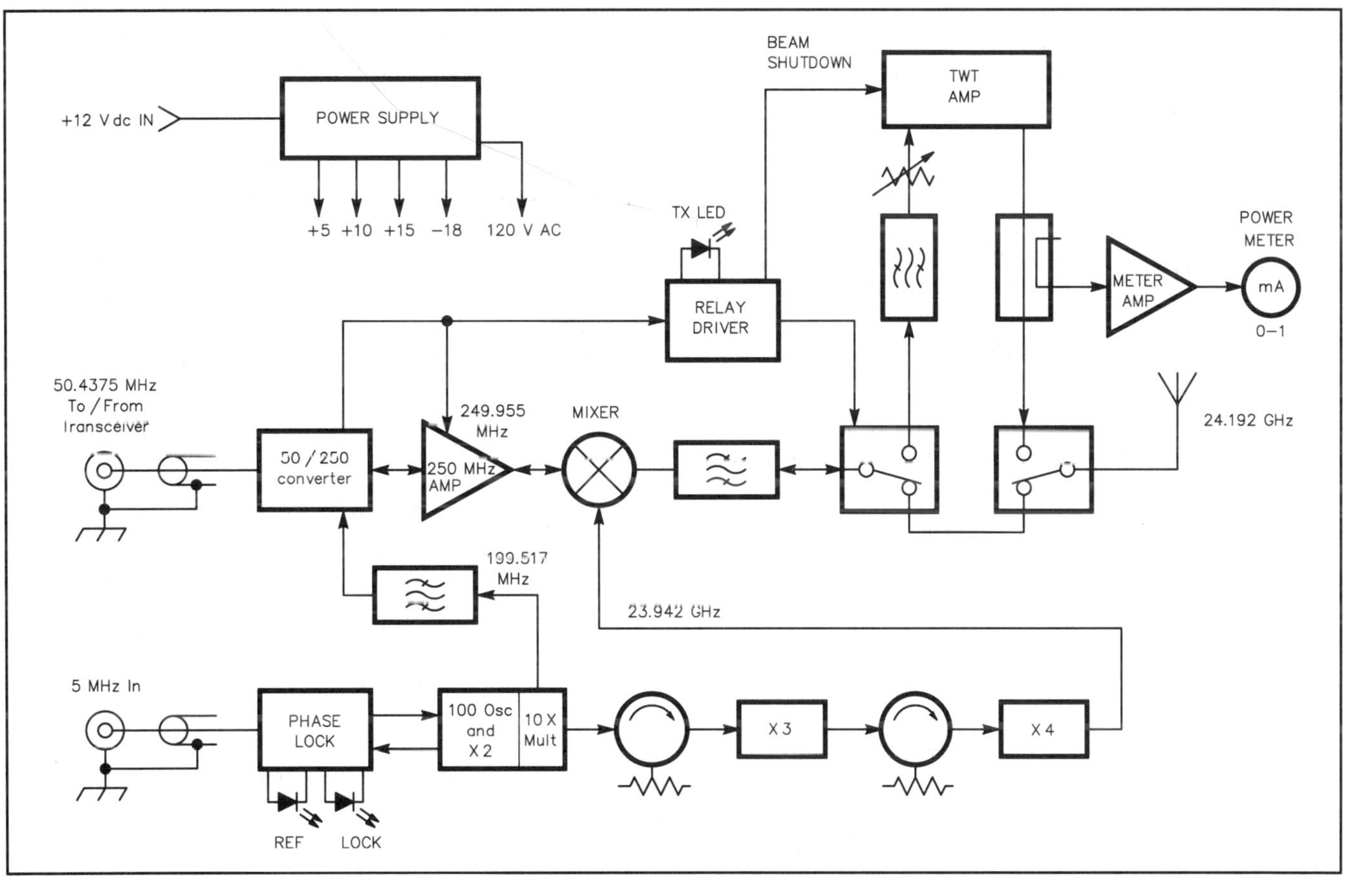

Fig 1—Block diagram of 24-GHz transverter with 6-meter IF.

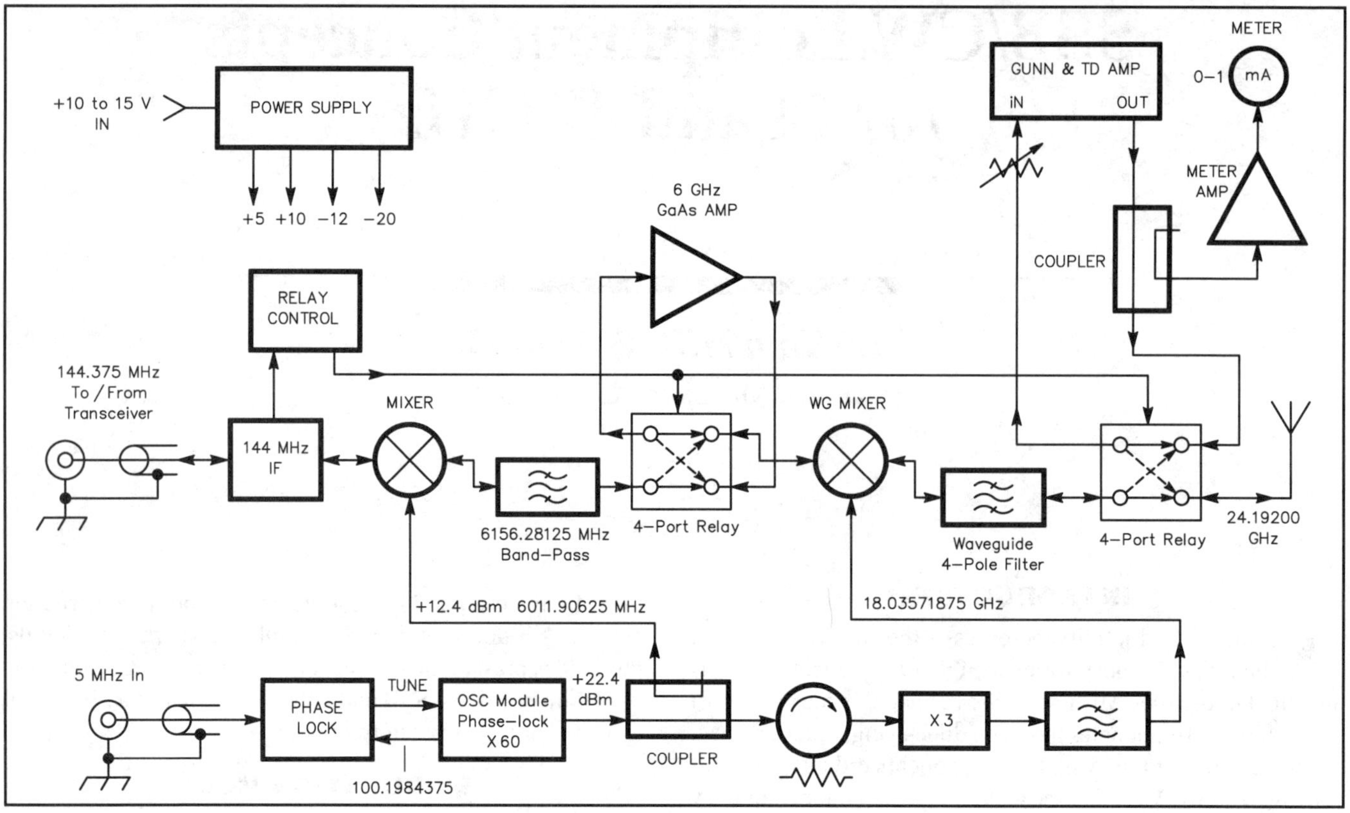

Fig 2—Block diagram of 24-GHz transverter with 2-meter IF.

modulation/detection scheme most likely to recover the weakest signal. It is important to consider that weak-signal aspect, as the atmospheric losses at these frequencies are fierce and the attainable transmitter powers are miniscule at best. The receiver for CW will do SSB for free, so this is a good choice when the signals are strong.

A linear transverter allows the most modulation flexibility, and has only a small penalty in output below using the LO chain itself as a CW transmitter. Even this limitation can be overcome today, since surplus TWTs (Traveling Wave Tubes) are available for 24 and 47 GHz.

THE TRANSVERTERS

24 GHz

My first millimeter transverter was for 24 GHz. The block diagram is shown in Fig 1. It was built as an extension of a previously done 10-GHz rig. I used what I had available at the time, which was an NEC digital telephone-link radio. The original local oscillator chain was the useful part, and I modified the last multiplier to work at 4X instead of the original 2X. With a 6-meter last IF, the existing LO chain provided for a mid IF at 250 MHz, which was barely high enough for the waveguide filter to lower the LO feedthrough at the RF output to an acceptable level.

The TWT amplifier is contained within the same cabinet with the rest of the system. The various blocks are essentially the same as the later 47-GHz rig, and so for a more complete description, refer to the circuit descriptions which are for the 47-GHz rig.

With the growing availability of the 6-GHz phase-locked oscillator modules (CMI, CTI, and others), it seems desirable to use these as they provide a simpler, smaller, lighter, and more cost-effective solution. I have designed such a rig, and the block diagram for this is shown in Fig 2.

47 GHz

There are two separate rigs for 47 GHz. This is so that I will have someone to talk to. The two differences between these two are the different IF—6 meters for one, and 2 meters for the other—and the fact that the one has the TWT built in, while the second has the TWT amplifier built as an external accessory. Since they are essentially the same, I will present only the one with the separate TWT. The block diagram for the transverter is Fig 3, and the amplifier is shown in Fig 4. In this version, the last IF is 2 meters, and the high IF is near 2 GHz.

DUAL CONVERSION

The choice of IF is often influenced by what SSB radio you may have available. On the lower bands this will likely be the only IF used. But with 24 GHz and up it is usually desirable to have a high IF as well, with a dual-conversion rig the result. This is necessary because the filter used at the operating frequency must remove both the image frequency and the LO. Since it is often difficult to obtain fully balanced mixers for these frequencies, the LO power coming out of the mixer along with the desired signal may be 10-dB stronger than the desired signal. This places quite a strain on the design of the output filter if the IF is low. The original 24-GHz rig used a 250-MHz high IF, which is quite marginal. Although the image conver-

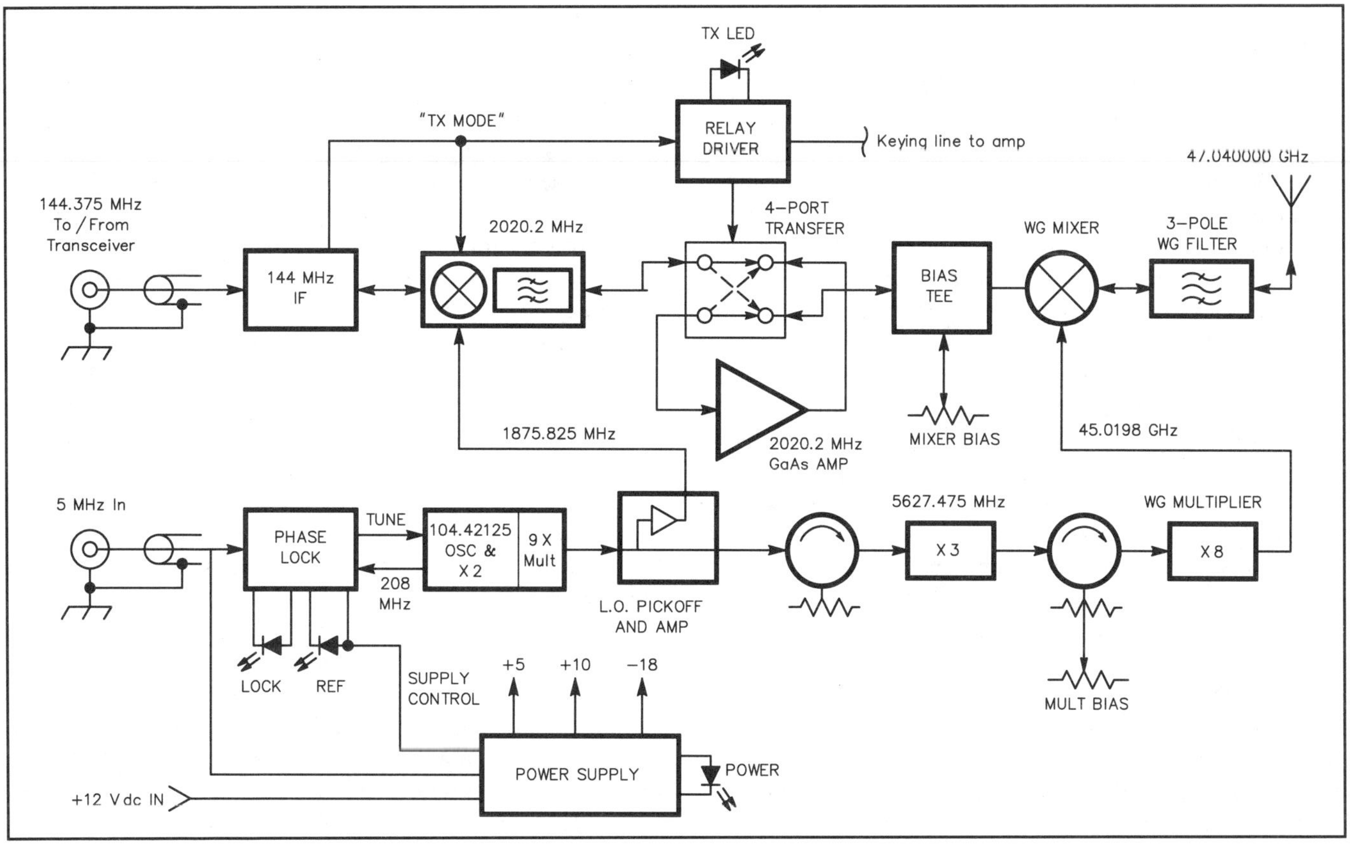

Fig 3—Block diagram of 47-GHz transverter with 2-meter IF.

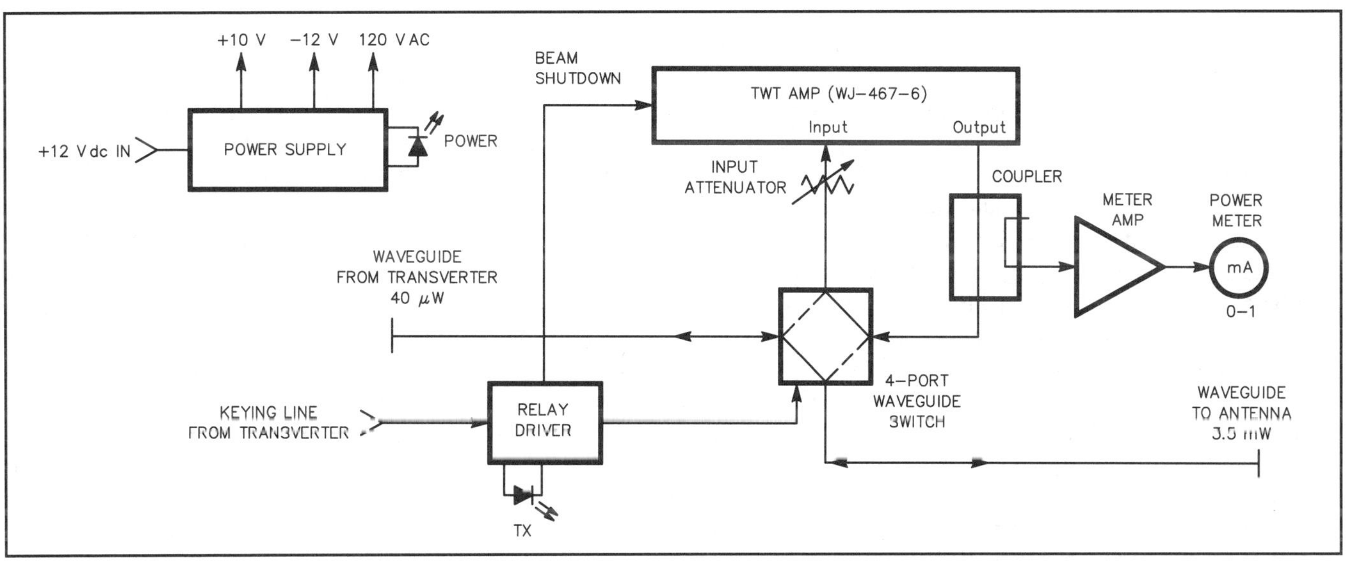

Fig 4—Block diagram of 47-GHz TWT Amplifier.

sion is about 70 dB below the carrier, the LO is only 20 dB down! I therefore put a second filter in the input to the TWT so that when I go to high power (4 mW) the LO is about 35 or 40 dB down. I didn't want the extra 2 dB loss of the second filter in the receive path if I didn't need it. The new 24-GHz rig uses a 6-GHz high IF which nicely gets around these problems. The 47-GHz rigs use 2-GHz high IFs which, while not as good as 6 GHz, they made use of the oscillator/multiplier assemblies that I had available at the time.

The LO scheme used in all these transverters has another feature. I always make the frequencies come out so that all conversions are done with multiples of the same crystal. This means that if I can control the frequency and residual FM of this one oscillator, there will be no uncertainty added by any other oscillators. The importance of this will be further explained later in the section on Frequency Stability. This also means that I have fewer modules to build if I can just tap off the various multipliers as I go.

CLEAN LOCAL OSCILLATOR

Residual FM

If the LO is varying quickly (or FM-ing) it will add this FM to received signal before it passes it to your IF rig. Free-running oscillators, such as Gunn oscillators, tend to have poor residual FM. Note that even 100-kHz-wide FM communications are difficult with Gunn-diode LOs, and narrowband operation is impossible. The stability of a crystal is necessary.

You can indeed phase-lock such a Gunn to a crystal, but the loop calculations need to be done carefully to ensure that the loop response is wide enough to get all that noise and FM clear out to several hundred kHz. Either a locked oscillator or a multiplied crystal can be made to work, but I have found the multiplied approach to be most convenient with the parts I have available.

When you are multiplying the signal from an oscillator by a factor of 470 or more, then any FM-ing or drift of that oscillator will be multiplied by the same factor. This means that if a copiable CW or SSB signal is to result, the crystal oscillator must be very good to start with.

Phase Noise

Any oscillator has some slight amount of fast "jitter," which is called phase noise. This usually manifests itself as a rise in the broadband noise as you measure closer and closer to the carrier itself. This causes two basic problems.

One is that if it is strong enough and wide enough to actually place noise into the mixer at the intended receive frequency, the apparent noise figure of the receiver will be degraded. This is not usually a problem with reasonably high IF frequencies, but it used to sometimes show up in klystron rigs with 10-MHz or lower IFs.

The other problem is that even the relatively low phase noise of a crystal oscillator will begin to show up when it is multiplied as many times as you must to get to 24 GHz or higher. Even if the multiplier adds no noise of its own, the noise will increase in amplitude by a factor of 20 log(N), where N is the multiplication factor, and the answer given is in decibels. This computes to an increase of 53.4 dB in the phase-noise level of 100-MHz oscillator that is multiplied up to 47 GHz. Since the phase noise is much stronger as you get closer to the carrier, the noise that is in the area of 300 Hz to 3000 Hz (the SSB voice area) must be particularly low to start with, or the incoming signal will be drowned out by the receiver's own audio noise!

One of the easiest ways to get better phase noise is to run the crystal oscillator with a high level of power in the crystal itself, in order to get the crystal to help filter out the noise of the active device. This is exactly the worst thing that you can do to a crystal oscillator if you want good frequency stability (particularly long term).

Frequency Stability

The frequency of the transverter must stay put or you will have to chase the signals all over the band. For CW to work in a narrow bandwidth, the signals must be inside this bandwidth. The intelligibility of SSB signals is severely degraded if the frequency is wandering around. If you want to reduce the variables that are all working to prevent you from making your mmWave contact, a good place to start is with a rock-solid signal.

A point of comparison is that a typical "high accuracy" crystal oscillator in a two-meter FM rig has about 5 parts per million accuracy, for about 700 Hz possible error. This transcribes into about 235 kHz of accuracy at 47 GHz. Two rigs using this accuracy of oscillators trying to make contact could be almost *half a megahertz apart!* On the other hand, a high-accuracy, ovenized oscillator can be sometimes found surplus with one part in a billion (1×10^{-9}) frequency drift per day. This gives my transverter an accuracy of 47 Hz at 47 GHz. The 2- meter SSB radio I use as the IF is probably not that good.

This accuracy is only achievable if all of the oscillators in the transverter are locked to the same reference. This is why the idea of using a LO chain that has taps for the different conversions rather than separate oscillators is so important. This saves building separate phase-locks for each one. The one does it all.

REFERENCE LOCK

The answer that I use is not new here (it is in several commercially available pieces of test equipment).[1] Run the main crystal oscillator for best phase noise, and lock it to an accurate external reference. I use a frequency near 100 MHz as it is fairly easy to get acceptable phase noise here. The exact frequency used for each rig will be determined by the aforementioned process of getting all the IFs to come out right. The crystals are reasonably affordable, but are not accurate enough for my purposes.

The accuracy comes from phase-locking the main crystal oscillator to the ovenized reference oscillator also previously mentioned. See Fig 5 for the schematic of the phase lock for the 47-GHz rig. The lock is accomplished with a purposely narrow loop bandwidth. Usually a phase-lock loop has a wide enough bandwidth to reduce the phase noise of the locked oscillator to that of the reference oscillator. However in this case the locked oscillator is considerably cleaner than the reference oscillator. The reference has fantastic accuracy, but not very good phase noise. The loop is cut for about a 1-Hz bandwidth, which allows the main oscillator to determine the phase noise further out than 1 Hz from the carrier.

This scheme also allows me to have only one expensive and hard-to-get reference and put it with whichever rig I am using at the time. This reference oscillator is in its own box, with batteries, and is left on 24 hours a day all year long. This ensures maximum frequency accuracy, as the drift of these references decreases the longer you leave them on.

CIRCUIT DESCRIPTIONS

Refer to Fig 3, the 47-GHz block diagram, for the overall signal flow as each of the individual modules is described.

144-MHz IF Module

The 144-MHz IF module is shown in Fig 6. The SSB rig I used is a Mizuho hand-held, which puts out some voltage through the antenna connector when in the transmit mode. This is sensed here to key the parts of the transverter which are

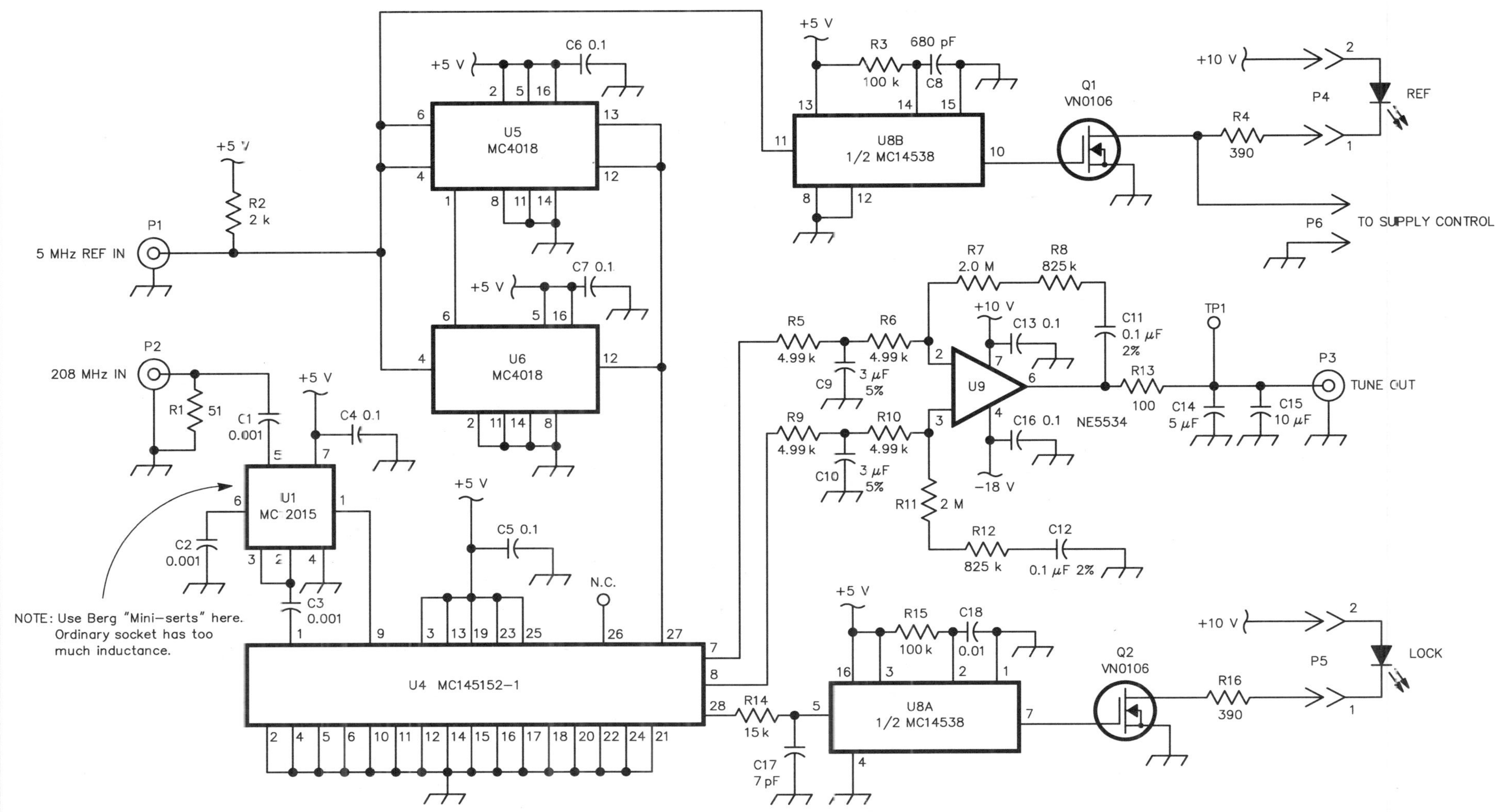

Fig 5—Schematic of the 47-GHz Phase Lock.

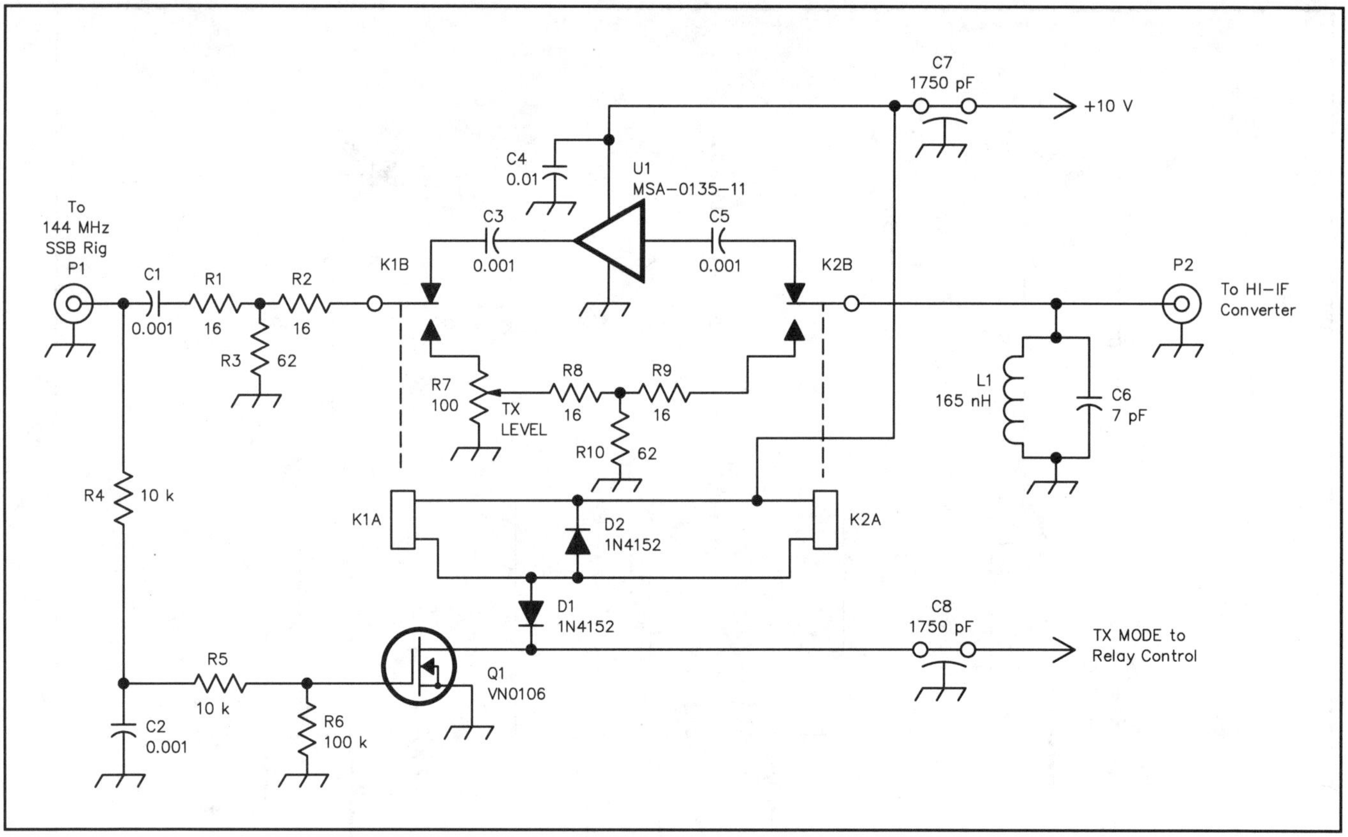

Fig 6—Schematic of the 144-MHz IF Module.

not inherently bi-directional. This module attenuates the 200 mW 2-meter transmit signal to the level which the rest of the rig can tolerate. The gains are set up so that the millimeter wave mixer saturates first, so the "TX LEVEL" pot can adjust the power out for the maximum possible without too much distortion. On receive, the signal from the high IF is amplified by U1 to overcome some of the 2-GHz converter loss and the pad in the 2-meter line.

2-GHz Converter

Fig 7 shows the 2-GHz converter. This is a surplus assembly originally intended for 2072 MHz, which I retuned to 2020 MHz. It is balanced only for the LO, but that is not a problem with the frequency spacings involved here. The 4-pole TEM (coaxial) mode filter is more than adequate to get rid of the image of the 144-MHz IF.

2-GHz GaAs Amplifier

The 2-GHz amp, Fig 8, is a single GaAsFET, which is tuned for lowest noise figure by adjusting the positions of the taps on the input inductor. The noise figure is indeed important here, since this is the first active stage on receive, and therefore the noise figure here will directly add to the system noise figure. This unit achieved 1.7-dB noise figure with 11.5-dB gain. Another dB could be achieved here with a bit more work on the input section, but this was simple and effective. The output is a simple broadband transformer that gives excellent stability.

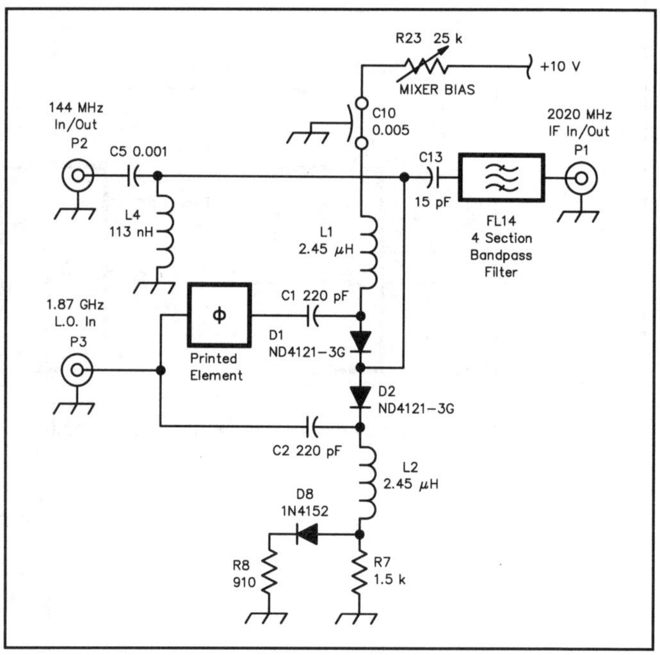

Fig 7—Schematic of the 2-GHz Converter Module.

Waveguide Mixer

The waveguide mixer is a converted surplus unit. There is a basic sketch of it in Fig 9. It originally had the back end filled with a terminating resistor for flatness, and the LO came in the coax. This was used for harmonic mixing with the LO

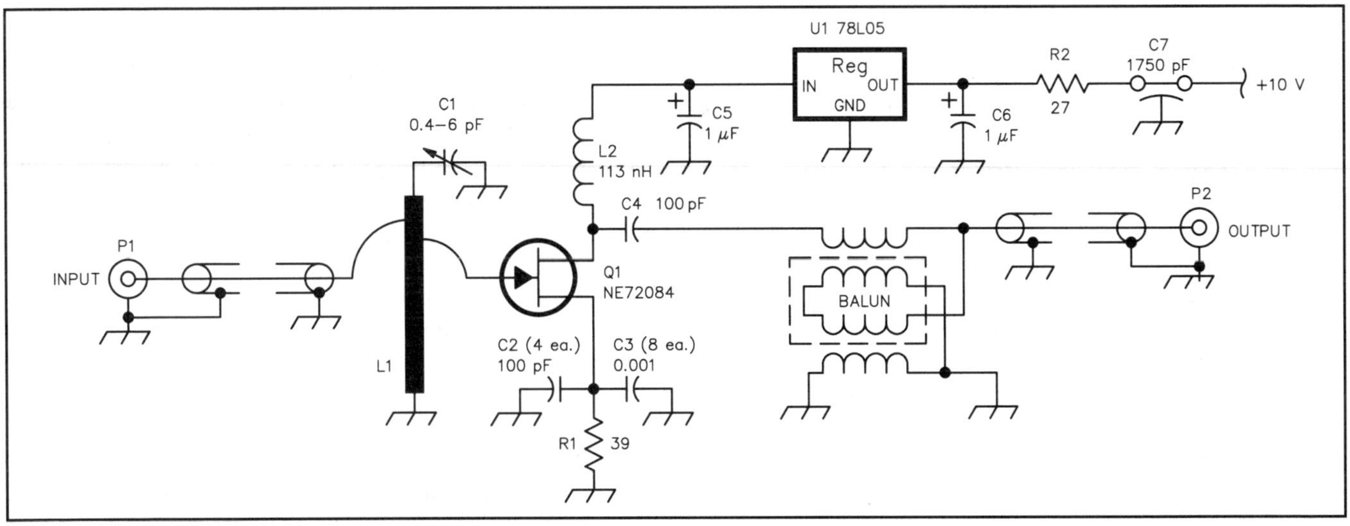

Fig 8—Schematic of the 2-GHz IF GaAsFET Amplifier.

at 2 to 6 GHz. I am using fundamental mixing for much better conversion loss, so I opened up the back and inject the LO here. I installed three tuning screws on each side of the mixer diode, and with a *whole lot* of tuning I got not only a reasonable match, but also there is some reflection of the LO back towards the diode from the RF side and vice-versa, which improves conversion loss over the plain diode. Fig 10 shows the bias tee that gets the dc bias into the mixer which is required by the single diode to optimize conversion loss.

LO Chain

The LO chain is mostly a surplus digital telephone transmitter. The original oscillator and multiplier is simply re-tuned to the new operating frequency. The crystal oscillator had to be very carefully tuned for best residual FM. There is a double and a power amp that feeds a snap diode to get the 1875 MHz, at about +20 dBm. The oscillator also needed a varactor added to it for the tuning voltage that allows the phase lock to work. This has a 4-pole R-C filter added to its tune line to filter out any op-amp noise, 60-Hz noise, etc, so as to improve the locked residual FM. After the snap diode output, I added an LO pickoff to tap a small amount of power off and amplify it to drive the 2-GHz converter. This is a simple resistor and Avantek MSA amplifier. The varactor tripler and the two isolators around it are stock from the surplus telephone gear.

The waveguide multiplier is a bit more difficult. It is another of the surplus mixers (same as the waveguide previously discussed), with a snap diode mounted in it, except that after removing the resistive terminator I closed off the end to reflect the power back and help increase the output power. This also has the same tuning screws and is *even more difficult* to tune. Some positions of the tuning screws return some of the other unused harmonics to the diode to re-convert for more output. They perform a similar function to the idler circuits in a varactor multiplier. In fact, this is almost acting as much like a varactor as it is a snap, since the operating frequency is above the operating range of this diode. The diode is a Metellics mesa snap diode, part number MMD-840-C11, with a 35 pS snap time. They have faster ones now, so I might experiment more

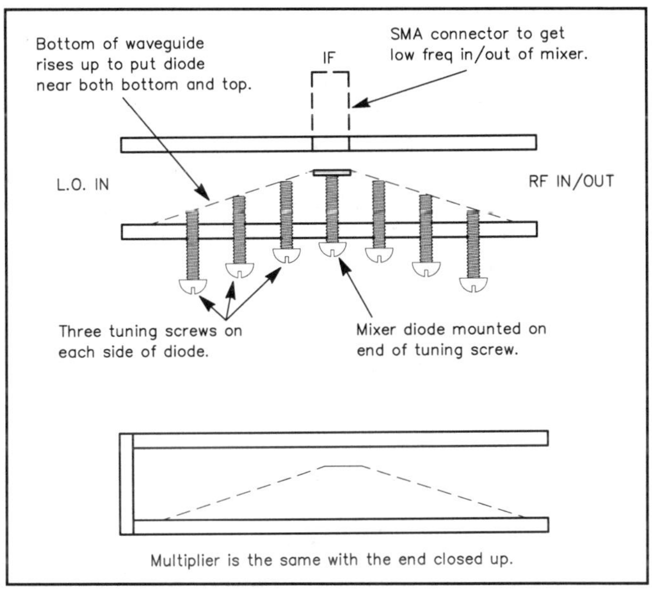

Fig 9—47-GHz waveguide mixer.

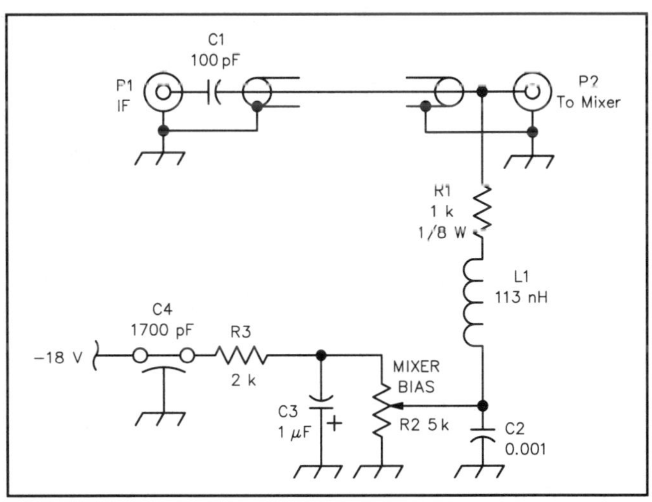

Fig 10—Mixer bias tee.

in the future. The multiplier requires bias too.

Phase Lock

Fig 5 is the phase lock that locks the 100-MHz oscillator to the external reference. This is a Motorola PLL chip, U4, that can control a dual-modulus prescaler chip, U1. The 200 MHz from the oscillator and multiplier is fed to the pre-scaler, and the 5-MHz reference is divided by the two MC4018 dividers, U5 and U6, before going to the PLL chip. These extra multiply and divide operations were required to get the correct lock ratio for the 104.421296296 MHz required frequency, while maintaining a high enough frequency at the phase detector to keep sidebands from being inserted on the carrier. U9 is the loop amplifier, and I had to use the "pole splitting" technique of adding C9 and C10 to reduce the phase-lock sidebands. U8A detects the lock condition so as to let the operator know if the frequency is right, and U8B senses the presence or absence of the reference.

Power Supplies

The whole unit is required to operate from batteries for mountaintopping, so it must operate from 10.5 volts to about 15 volts dc. This is accomplished with two supplies. Fig 11 shows the linear low-overhead regulator that supplies +10 volts to most of the RF circuits. The surplus oscillator/multiplier on the other hand, required –18 volts at about 300 mA (of all things)! This is provided by a small switcher that is shown in Fig 12. This was designed to not need any transformer or inductor, so it is easily reproducible.

It has its own free-running oscillator, but the *very* fast transitions of the switcher FETs do cause some VHF birdies. There is a fair amount of filtering to reduce this, but there is still a very tiny amount left. If the free-running oscillator were to drift to just the right frequency, then one of its harmonics might be barely audible in the 2-meter IF. It would drift around a bit, causing the operator to maybe mistake it for a weak signal. In order to prevent this, U2 in the supply divides the reference frequency down to the switcher frequency, and if the phase-lock module tells the supply that a reference is present, the divider output is used for the switcher drive. The switcher is a simple "rail-to-rail" topology, with a linear post-regulator for overall simplicity, if not the ultimate in efficiency.

The relay control module runs all the transmit/receive switching functions and gets its input from the detector in the 144 IF module.

POWER AMPLIFIER

The power amplifier is a traveling wave tube. For one of the 47-GHz rigs it is contained in its own separate housing, so that I can leave it behind if I don't need it. In this case the definition of "power" is a bit different than you may be used to. The most powerful amplifier I could find was a surplus receiving type of TWT. This amplifier was made by Watkins Johnson, model number WJ-467-6. It is a self-contained amp, with tube and power supply in one compact unit. Fig 4 is the block diagram of the amplifier I built.

TWT

The TWT that I managed to find is a unit that is intended to operate from 26.5 GHz to 40 GHz, with 35 dB of gain and 4-mW minimum output (most of them do close to 10 mW out). 47 GHz is well outside of its operating range, by 54% of its original bandwidth. The tube as received was not useful at this frequency. I opened it up and readjusted the helix voltage

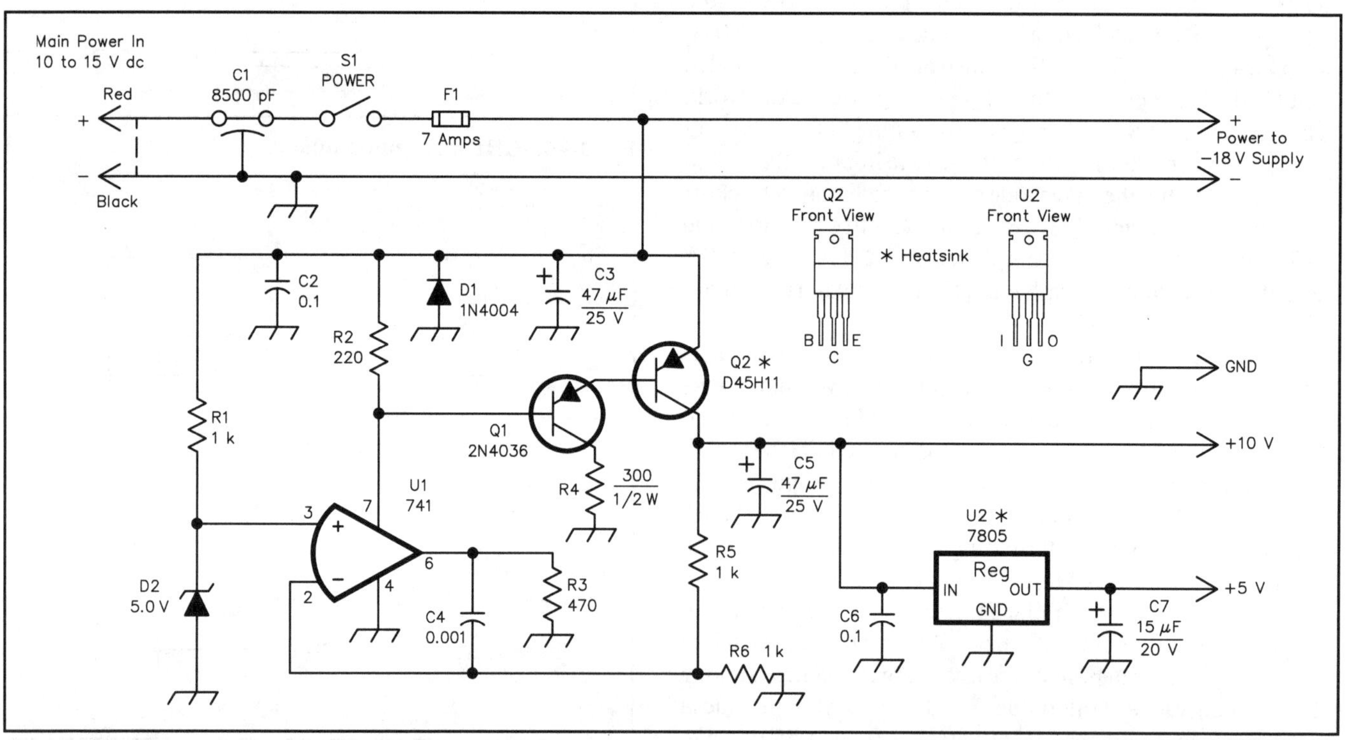

Fig 11—Ten- and 5-volt power supplies.

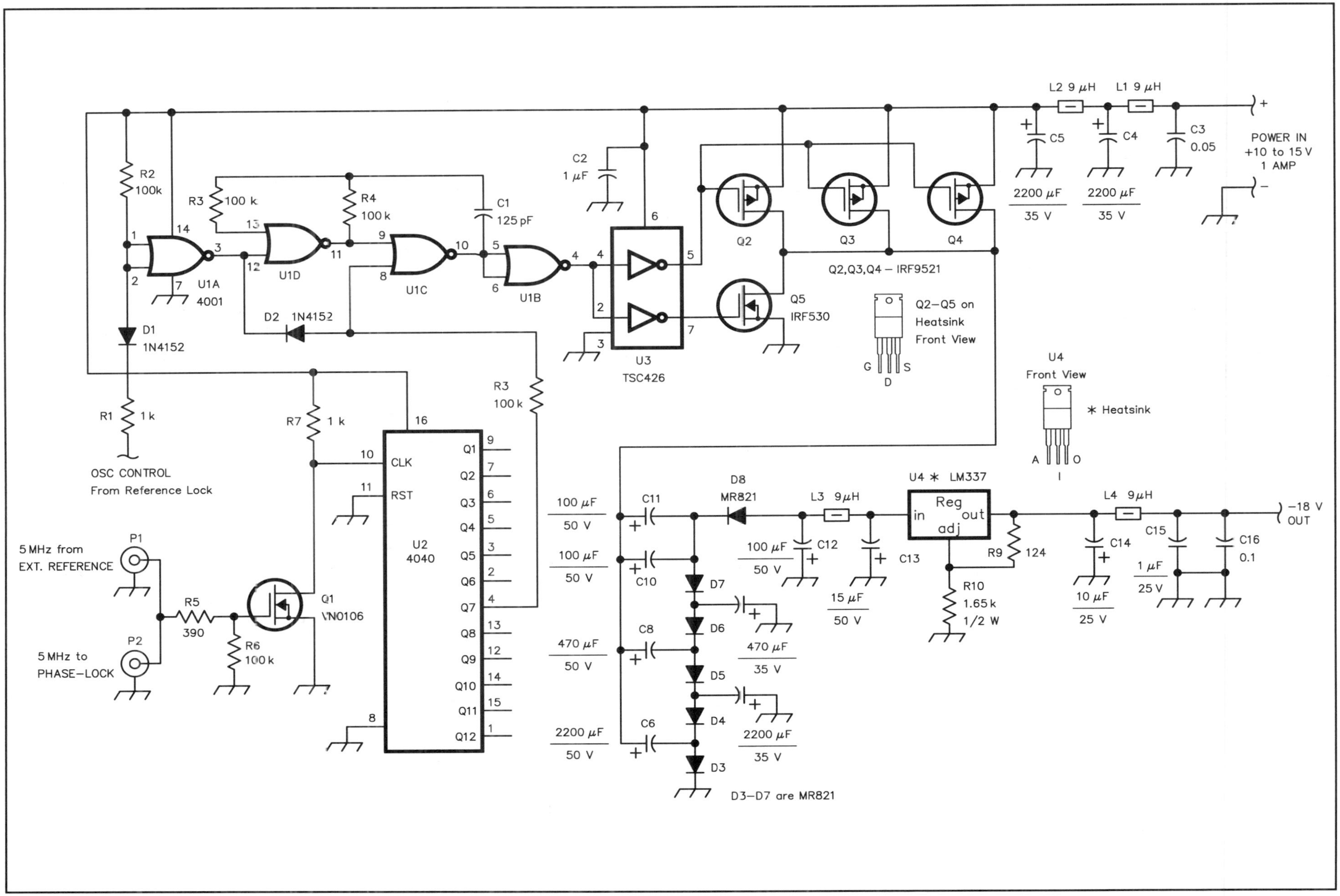

Fig 12—Negative 18-volt switching supply.

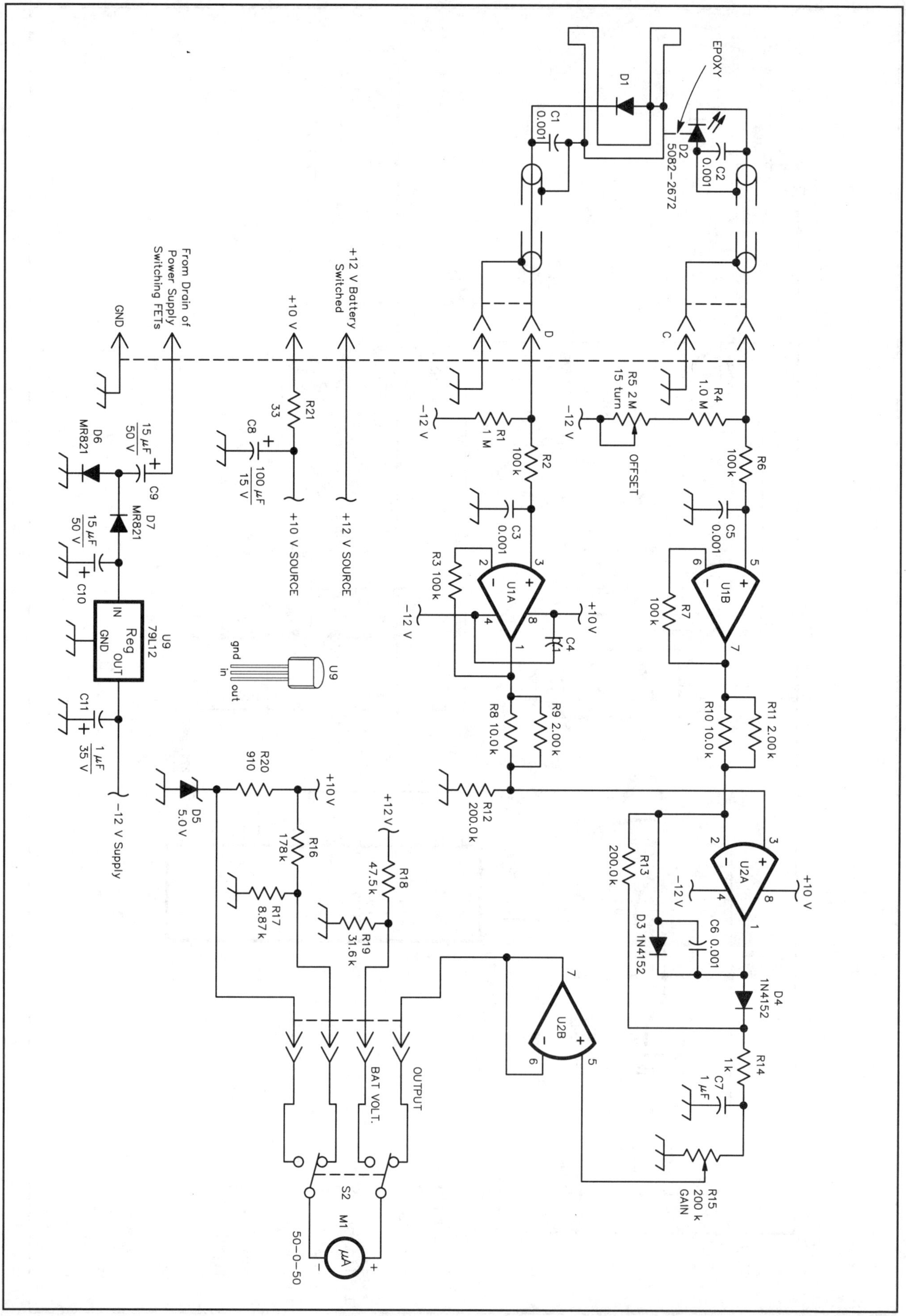

Fig 13—Power detector and meter driver.

lower until the tube attained useful gain at 47 GHz. This, however, lowered its power output to well below one milliwatt. To fix this, I lowered the grid bias so as to raise the collector current until I got the output power back up. The tube now shows 24-dB gain with 4.5-mW power out.

I also added an external connection to the TWT module that controls the grid voltage in order to shut down the beam. This is necessary so that the tube is inactive during receive. The tube's noise figure is slightly worse than the barefoot transverter, so it would only hurt if I were to try to use it during receive. The TWT is therefore bypassed in receive.

The TWT requires 120 VAC for its primary power, so the switching power supply that runs the amplifier was designed to run at 380 Hz. It drives a transformer to get the necessary voltage for the TWT.

Attenuator

There is also a waveguide attenuator that I built by boring a small hole in the wide side of a piece of waveguide, and pushing in a piece of microwave absorber. It is adjusted by pushing the absorber to different depths in the guide. This attenuator is needed to prevent overdriving the TWT. If it is overdriven it will not only distort, but it will even reduce its power out as it gets driven harder.

Power Meter

In order to be able to set the attenuator correctly (a one-time operation), a power meter is helpful. It is also quite useful to keep an eye on the general health of the transmitter. I built a coupler out of two pieces of waveguide soldered next to each other with a hole bored between them. This taps off a small bit of signal to a diode which then feeds a peak-reading meter. The peak detector circuit allows reading of SSB signals. The schematic of the waveguide diode mount and peak detector is Fig 13.

OBTAINING PARTS

The parts for this project are proportionately harder to find than "ordinary" microwave parts, as the frequency is that much higher. I have found many of them (including one of the TWTs) at the Dayton HamVention. There are an assortment of places in California with *real* expensive parts (the waveguide switch was $100, an 18" broken piece of flex guide was $50). Occasionally I find millimeter parts in ordinary local hamfests. It took me over two years to collect enough parts to begin serious construction on the 47-GHz rig.

Another note worth mentioning here is that I never located much waveguide for any band that includes 47 GHz. I wound up using all parts designed for the 26.5-40 GHz band. This means that I must keep all discontinuities in the guide to a minimum, to prevent mode-hopping. This would result in excessive losses.

OPERATION

The 47-GHz rig described here was used with an 18-inch dish at one end of a world record DX contact that stood for 4 years. The other rig had a 29-inch dish with it. Both dishes are equipped with rifle 'scopes to aid with aiming. This contact was between WA3RMX at Crater Lake Oregon, and K7AUO on top of Mt. Ashland, Oregon, for a distance of 65.37 miles.[2]

CONCLUSION

I hope these ideas will spur on more hams to venture into the Millimeter Wave arena. Modify whatever you can get a hold of, and get on the air! With the current push to reassign our bands away, how long before someone finds these useful?

Above 10 GHz the calling frequencies are (in GHz):

- 24.192
- 47.040
- 80.640[3]
- 120.000
- 144.000
- 245.760

I expect to hear more people on these bands soon!

Notes

[1]This scheme is used in Tektronix spectrum analyzer models 494AP and 2756P, as well as in other places.

[2]For a complete description of this contact see *QST*, Dec 88, p 87 "The New Frontier" column by Bob Atkins, KA1GT.

[3]Although this frequency does not strictly follow the rule of the lowest reasonable multiple of 96 MHz, I am hoping we will use this multiple instead, since in this particular band the atmospheric losses decrease with increasing frequency.

A Solid-State Laser Transceiver

By Roger Wagner, K6LMN

(From *Proceedings of the 37th Annual West Coast VHF/UHF Conference*)

INTRODUCTION

Welcome to the world of Amateur Radio lightwave communications! Here, we are communicating voice/CW/data/video, line-of-sight over unguided coherent light beams at frequencies of 350 to 680 THz (terahertz or million megahertz) or about 3 million times the frequency of the 2-meter band. This translates in wavelength to 700 nM (deep visible red) to 400 nM (visible violet). Of course, infrared and ultraviolet can be used as well. The FCC has given hams and non-hams any of the frequencies above 300 GHz, where extremely deep infrared begins.

In this article, I have developed the concept of an all solid-state CW/phone transceiver, such as is used on the UHF bands. The units are slightly larger than a hand-held transceiver, complete with batteries. The power output is only 5 mW, and the receiving "antenna" is a 3-inch telescope. As on the lower bands, range is a function of power output and antenna size. Communicating over laser beams sure is a lot of fun and the band is certainly not crowded! A simple laser diode/LED AM-voice modulator is just a couple of amplifier chips and a transistor or two. A simple receiver is a small solar cell followed by an amplifier chip. Most parts are available from Radio Shack.

Unfortunately, there has been a shortage of good and practical articles in the ham magazines on the use of semiconductor lasers and associated solid-state photodetectors. Most of the authors in this field got their start with gas lasers. I hope this article will shed some light on laser diodes, and encourage your investigation in this new world of communication.

BACKGROUND

After some 35 years of HF through UHF operation I wanted to try something new—lightwave communications. I started out using automobile quartz-iodine headlights. Believe me, you can throw a light beam a long way with one of these. My goal was voice/CW communications and I quickly found out the heated filament didn't modulate well at frequencies above a few hundred hertz. Next, I began experimenting with Radio Shack's high-powered red LED. It offers an output of some 2000 millicandelas for a cost of less than $2. I built up a couple of voice-modulated AM transceivers, using solar panels as photodetectors. Range was several blocks and fidelity was excellent. Then, I experimented with various lenses to collimate the LED's wide-angle beam. Range improved considerably since the beamwidth was now down to 2°. I do a lot of VHF/UHF contesting and I was now ready to generate some extra points for ARRL VHF/UHF contests. To my surprise though, only coherent light sources, ie, lasers, are permitted.

I purchased several laser diodes, complete with collimators, and added the laser assemblies to my transceivers. Now I'm legal for VHF/UHF contesting. My two transceivers were used for the 1990 ARRL June VHF contest from Mt. Pinos by the N6CA group. I also used them during the 1992 ARRL January VHF Sweepstakes. My range so far is only 5 miles. I think longer distances are possible, but I have great difficulty pointing the extremely sharp beam (0.02° beamwidth).

SAFETY

Lasers are not toys! Never point a laser at eye level at close range. The beam emitted by a laser diode is harmful if aimed into the unprotected human eye, especially at night, when the iris is wide open. Never point the laser into known aircraft flight paths.

LASER BASICS

Laser is an acronym for Light Amplification by Stimulated Emission of Radiation. A laser generally consists of an active medium that can amplify light, and reflectors that return a portion of the light back to the medium. A small portion of the amplified light that bounces back and forth escapes into a beam with these characteristics:

- Highly directional
- Single or narrow spectrum of wavelengths
- Phase coherent
- Continuous for relatively long periods

The medium can be any of the following:

- Solid-state (ruby, YAG)
- Liquid (dye laser)
- Gas (Helium-Neon, Argon)
- Semiconductor (laser diode)

The wavelength of the emitted light is a function of the medium and mirror spacing. All of these lasers have their applications, but only two are presently used for free-space communications by hams.

HELIUM-NEON LASERS

What about using commonly available surplus HeNe and HeCd laser tubes? Since the HeNe (gas) lasers have been around for some time, they are cheap and plentiful on the surplus market. They cost between $40 and $100, including high-voltage power supply. Gas lasers do the job well if you are interested in just pointing a narrow beam of light at something. To me, they aren't very useful for communication, as they are difficult to fully modulate. In addition, they require high-voltage power supplies (over 1000 Vdc). You can achieve 10-15% AM-voice modulation by modulating the power supply. For CW, a chopper wheel (such as a compact fan), in the beam path will produce square-wave modulation at about 1000 Hz. Now, just key the power supply on and off to transmit 100% modulated MCW.

ARRL contest rules state that lightwave-frequency transmitters must generate coherent light, and be capable of communicating at least 1 km. The receiver must have at least one stage of amplification. As LEDs and headlights don't generate coherent light, they can't be used for contest QSOs. Serious ham DX is still being worked with gas lasers and mechanical chopper wheels.

SEMICONDUCTOR LASERS

Semiconductor lasers (laser diodes) are now coming to the surplus market at reasonable prices. They are used in lecture pointers, bar-code readers, CD players and laser printers. Why use a laser diode for communication? The primary advantage of laser diodes over HeNe lasers are:

- Greater efficiency
- Physically rugged
- Compact size
- Light weight
- Long life (over 50,000 hours)
- Easy to modulate up to 1 GHz
- Simple power supply (3-12 Vdc)

The typical laser diode looks like a TO 5 case transistor with a glass window (Fig 1). Unfortunately, low-cost laser diodes are at near-visible infrared wavelengths of around 750-820 nM. Visible red (670 nM) laser diodes are now available, but at much higher cost. A small collimator lens is required to narrow the beam from about 10×30 degrees to 1-2 milliradians (mrad). With an adjustable collimator the beam can be focused to any width required. A HeNe laser has a typical divergence of 1 mrad, without external optics.

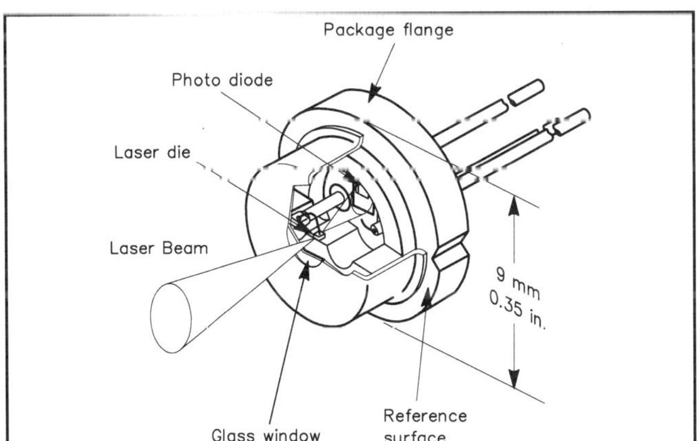

Fig 1—A typical laser diode.

For the same optical power output, the HeNe laser appears almost 10 times brighter to the eye than the typical laser diode. The HeNe wavelength of 633 nM is closer to the eye's most sensitive wavelength (green, 540 nM). For communications, the red and near infrared (670-850 nM) semiconductor lasers are a better match to commonly available silicon photodetectors, which have a sensitivity peak at near-infrared wavelengths. Laser-diode efficiency is typically 20%, compared to 2% for HeNe lasers. Above the lasing threshold, laser diode efficiency approaches 80%. If you are contemplating portable, battery operation, this feature is significant. Table 1 is a comparison of several available laser diodes.

OPTICAL RANGE

The maximum line-of-sight (LOS) range of a free-space, all solid-state laser communications system can be calculated by:

$$R = \sqrt{\frac{(Po \times Ar \times Tor \times Tot \times Ta)}{Pt \times D}} \qquad \text{(Eq 1)}$$

where

R = LOS range in meters between transmitter and receiver

Po = Peak power output from laser (watts)

Ar = Area of the receiver lens or mirror in square meters

Tor = Transmissivity of receiver optics, including filters

Tot = Transmissivity of transmitter collimator and beam-shaping optics

Ta = Atmospheric transmittance

Pt = Threshold power sensitivity (0-dB signal + noise : noise ratio) (watts)

D = Collimated laser-beam divergence (radians)

This equation assumes that the apertures of the transmitter optics have been matched to the beamwidth of the laser diode for minimum loss. Otherwise, another term must be included in the numerator.[1]

The receiver's minimum detectable power term (Pt) needs some explanation. This discussion pertains only to solid-state photodetectors. The minimum incident power on a photodiode required to generate a photocurrent equal to the total photodiode noise current is defined as the noise equivalent power, or NEP. The NEP is dependent on the square root of the receiver bandwidth, which is usually limited by the audio passband. The NEP is given in watts/ $\sqrt{Hz}$. The noise generated by the diode is a combination of shot noise and Johnson noise. The shot-noise current is a function of the dark leakage current; Johnson noise is related to the diode's internal resistances. Both are device and temperature dependent. Photodiode NEP can be expressed by:

$$NEP = \frac{In}{Rs} \qquad \text{(Eq 2)}$$

where

NEP = noise equivalent power (watts/ $\sqrt{Hz}$)
In = photodiode noise current (A/$\sqrt{Hz}$)
Rs = photodiode sensitivity (A/W)

Typical silicon PIN photodiode figures are In = 2.6^{-14} A/ $\sqrt{Hz}$, and Rs = 0.5 A/W at 670 nM. NEP then = 5.2^{-14} W/ $\sqrt{Hz}$ A low-noise op amp (OPA27) following the photodiode has a much greater noise current of 4^{-13}A/ $\sqrt{Hz}$. The geometric sum of the diode and op amp noise currents is then 4.01^{-13A}A/ $\sqrt{Hz}$, yielding NEP = 8.02^{-13} W/ $\sqrt{Hz}$.

Let's predict the range for my laser transceivers:

Po = 0.003 watt max.
Ap = 0.0046 meter^2 (3-inch lens diam)
Tor = 0.95 (estimate for plastic Fresnel lens, no filters)
Tot = 0.9 (est)
Ta = 0.9 (extremely clear night at mountaintop altitudes)
Pt (2.7-kHz receiver bandwidth) = 4.18^{-11} W
D = 0.001-radian circular beam

Range = 500 km, or 311 miles for a signal barely detectable in the noise. This is the theoretical maximum; no consideration is given for modulation type or atmospheric effects, such as air turbulence or transmitter-to-receiver beam misalignment. The theoretical VHF radio LOS range between a transmitter and receiver, both at 9000 ft (unobstructed path) is about 268 miles. The optical range should be a bit less. The beam diameter at long range is approximately:

Beam diam = Full-width Beam Divergence (radians) × Range

Thus, beam diameter at 1 mile = 0.001 radian × 5280 ft = 5 ft.

TRANSMITTER BEAM SHAPING AND OPTICS

Unlike gas lasers, laser diodes alone emit a divergent asymmetric beam, due to diffraction effects in the diode's asymmetrical laser cavity. In addition, unlike the long cavity of the gas laser, the semiconductor laser cavity is extremely short (sub-millimeter) and the short cavity does not allow for a collimated beam. A typical near-visible laser diode without external optics emits an asymmetric (elliptic), divergent cone of light of approximately 11 × 37 degrees, 3-dB beamwidth. Referring to Fig 1, the ratio of the orthogonal beamwidths is called the aspect ratio. For these diodes it is approximately 3.5:1. The aspect ratio for the typical visible laser diode is nearly 5:1. Laser diodes also have a small amount of astigmatism. External optics are thus required to transform the output of a laser diode, to give a corrected, circularized, collimated light beam. The ideal collimated beam has no divergence. In other words, it is a parallel beam of light out to infinity distance.

When a simple lens is used to collimate this asymmetric output, a narrow beam with an elliptic cross-section results. A slightly elliptical beam is acceptable for amateur communications. It is, however, desirable to correct this distortion. An anamorphic prism pair or cylindrical-lens telescope can be used. Prisms are easier to align and can be adjusted to correct different amounts of ellipticity. An elliptically shaped fixed iris can also be used to circularize the beam, but at a considerable loss in beam power level.

There are two common ways of achieving beam collimation for laser diodes in free-space operation. A simple plano-convex singlet lens is a very economical solution to laser-diode focusing and collimation. A well-designed doublet lens is usually superior to a corresponding singlet, because of the much higher degree of aberration correction which is possible. There are also available aspheric lenses, which have shorter focal lengths than equivalent plano or bi-convex lenses. The corrected aspheric lens has a much higher numerical aperture (to capture maximum emitted light) than the corresponding spherical singlet, while having less aberration.

Beam astigmatism is caused by the light sources for the orthogonal axes being at slightly different distances. In the simpler, gain-guided devices like visible laser diodes, this distance can be 40-50 microns. In the case of index-guided lasers (typical infrared), the astigmatism is typically less than 5 microns. Astigmatism often needs correction if a small spot size or ultra-precise collimation are required. This is particularly true in the case of visible lasers. Cylinder-ground lenses are available to correct astigmatism. By slight rotation around the optical axis, the lens can correct astigmatism without introducing additional aberration. The astigmatic correction lens, if required, should be placed after the collimator lens. Sony laser diodes use an integral slanted window over the diode chip to achieve very low astigmatism, without external corrective lenses.

For these corrective lenses, we are not talking about large-diameter optics like those used in telescopes. Typical collimating lenses have diameters of about ¼ inch. With good optics, a circular beam of less than 0.2 mrad divergence can be achieved. A simple plano-convex lens can easily converge a near-visible infrared laser diode to a 1 × 3 mrad beam. Antireflection coatings should be used to prevent light scattering, and to reduce path losses. A simple coated plano-convex collimator lens has a transmissivity over 95%.

LASER TRANSMITTERS

Now we have the transmitted beam shaped, what about modulating it? There are several types of modulation suitable for communication with laser diodes. As on the lower bands, the type of modulation used depends on whether you wish to transmit voice, data, Morse code or video. Fig 2A shows a diagram of a simple laser transmitter and its analogy to an HF AM-voice transmitter. At lightwave frequencies, the FCC does not dictate the mode of modulation or maximum bandwidth. Here, simplicity and low cost may dictate the modulation method. A continuous (CW) or unmodulated light beam carries no information and the receiver detector can only detect the presence or absence of the light beam.

Three basic methods of modulating the laser diode are:

- Modulated CW (MCW), using an external light chopper
- Amplitude modulation for voice/CW/data/video
- Pulse modulation

MCW

MCW is used by the HeNe laser gang. It is simple and cheap, but is suitable only for Morse code. To generate a

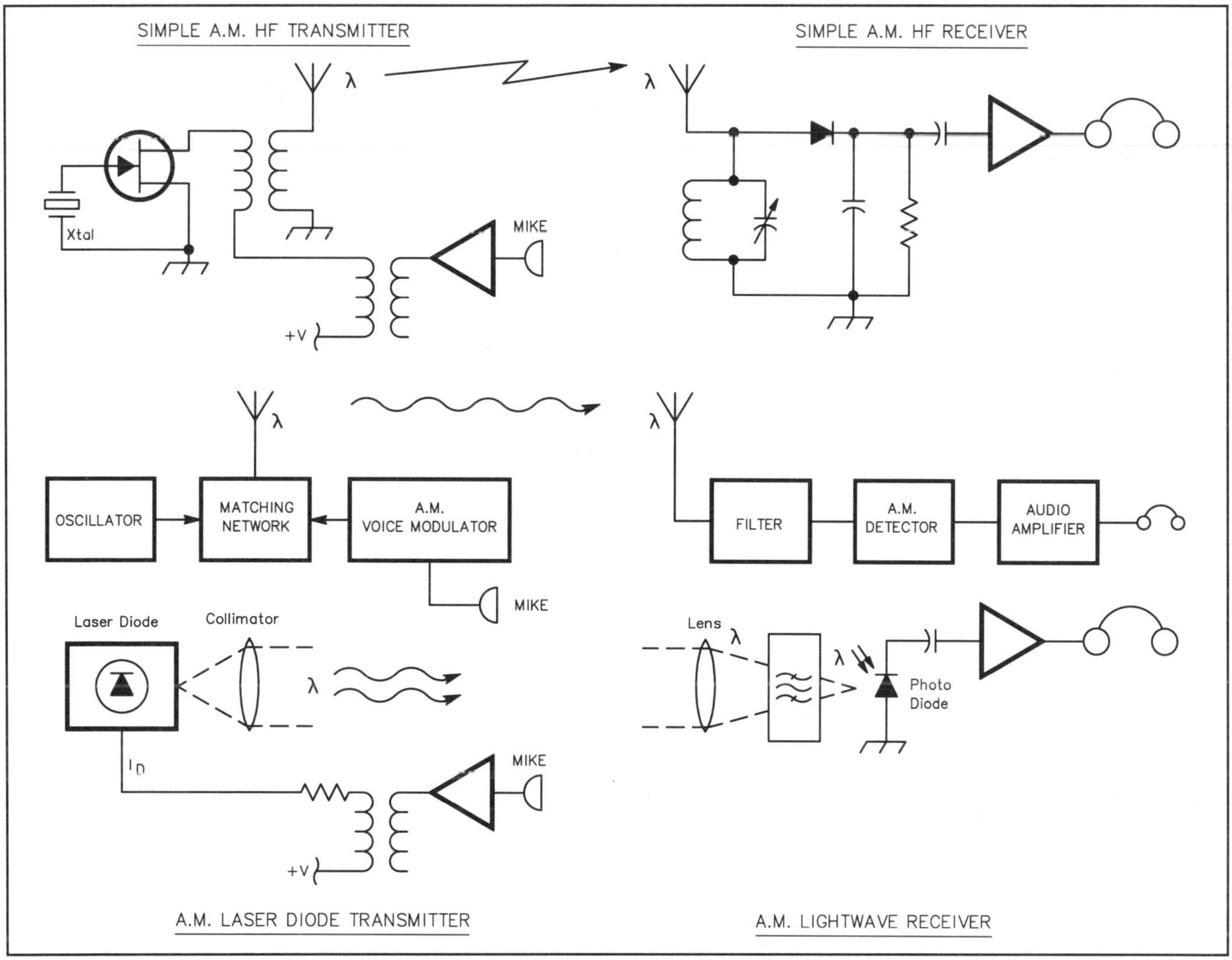

Fig 2—Comparison of HF and lightwave transmitters and receivers.

modulated light beam, all that is needed is a motor-driven wheel (fan) spinning in front of the light beam, to interrupt it at an audio rate. A solenoid-operated shutter, liquid-crystal shutter, or even your hand, can be used to key the light beam. Liquid-crystal shutters are sluggish, especially at the low temperatures often found at high elevations. They have a transmissivity of only 70% when clear. Of course, you can also key the laser power supply, or modulate a laser diode supply.

Amplitude Modulation

Amplitude or intensity modulation is rather simple to implement with laser diodes, and is suitable for voice, data, Morse code and video. Another plus is that AM is compatible with the receivers used for gas lasers. This feature is handy during contests, as it allows contacts with stations using either laser type. For AM, the laser diode output is established at the 50% output level with no modulation present. The modulating waveform then modulates the laser from the bias point up to 100% and down to 0% output. For Morse code, the sinewave tone oscillator can be keyed. The average light power output in all cases is 50% of the peak laser output. A 30-100 kHz subcarrier can be amplitude modulated, and itself used to modulate the laser. This method complicates the receiver slightly, but offers a major advantage: this receiver will be nearly immune to stray light modulated by ac power like street lights. For television transmission, couple the camera baseband output through a wideband stepdown transformer to the laser diode. For best picture quality, drive level and bias may have to be more carefully adjusted than for low-fidelity voice transmission.

Another modulation scheme involves coupling a few milliwatts of HF SSB into the laser diode. For a simple receiver, just hook up a wideband photodetector, such as a PIN diode (biased for maximum bandwidth) to the HF receiver antenna input. A low noise, high frequency preamp following the photodetector will improve receiver performance. Speaking of exotic modulation techniques, for coherent CW, the laser diode transmitter tone oscillator can be phase-locked to a master frequency such as WWV. At the receiving end, the detector can be referenced to the same master frequency. With an extremely narrow receiver baseband bandwidth of 2 to 10 Hz, a tremendous system gain over voice modulation can be achieved. The sloppy tone generated by mechanical chopper wheels typically used with gas lasers will not permit the narrowband DX work possible with laser diodes in such a system.

Pulse Modulation

Pulse modulation is quite suitable for Morse code operation, but can be used for voice and data modulation as well. All that is necessary is to pulse the laser on to full power output and then pulse it off, with a squarewave. The light output is a pulse or squarewave. Using pulse modulation, you don't have to worry about laser diode thresholds and linearity of light output. For Morse code, drive the laser diode with a squarewave current of 0-75 mA peak at audio frequencies and key the oscillator to form characters. Use a simple transistor switch driven by a square-wave oscillator like the 555 IC. The simple receiver can be any HeNe-system photodetector and receiver.

For voice and data modulation, either pulse-width modulation (PWM) or pulse-position modulation (PPM) can be used at a subcarrier frequency. For PWM, the subcarrier frequency is normally 30-100 kHz. The audio modulation input varies the pulse width or duty cycle. At the receiver, the recovered pulse train passes through a low-pass filter before the audio is recovered. Because the carrier frequency is so far above the audio range, this system isn't compatible with most HeNe laser receivers.

For pulse-position modulation, the pulse train average frequency is again set at 30-100 kHz. This time though, the pulse phase or frequency is varied by the audio input signal. The receiver detector is a little more complicated: a phase or frequency detector is required. The benefit? A static-free, hi-fi communications system.

A word of caution: All laser-diode manufacturers strongly recommend slow turn-on to avoid transients that may destroy the device. If you plan to use high-speed pulses, verify the drive waveform with a high-frequency oscilloscope. You should also take precautions to avoid spurious oscillations in the driver. A 100-pF capacitor may be connected in parallel with the diode. Keep the leads short. Test to be sure the additional capacitor doesn't cause your driver stage to oscillate. Another good idea is to add a small silicon diode in parallel with but in reverse polarity to the laser diode. A 1N4001 is suitable for frequencies up to 100 kHz. Do not exceed the manufacturer's specified maximum power ratings.

Laser Diode Operation

Before getting into the actual driver circuits, we must understand the laser diode's operation. The shape of the laser diode voltage-versus-current curve is quite similar to that of a silicon diode.

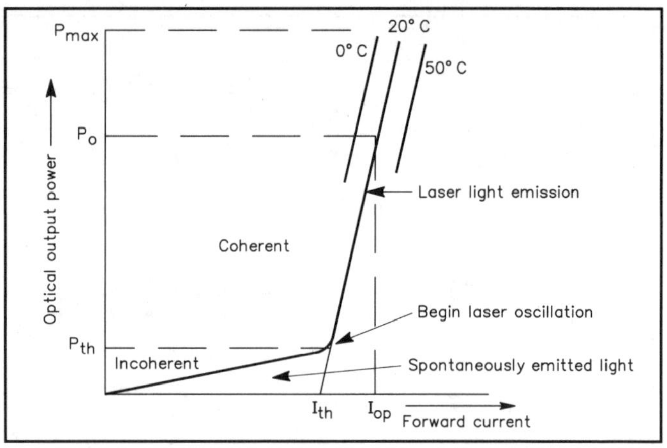

Fig 3—Laser diode forward-current optical output characteristics.

The major difference is the threshold voltage: 0.6 V for silicon diodes, but 1.8-2.2 V for laser diodes. The laser diode should always be mounted in a heatsink. Laser diodes have a Zener diode-like curve. The voltage drop remains fairly constant with increases in current. The laser-diode curve also resembles that of an LED, but its light output is nearly proportional to drive current above the lasing threshold. The lasing threshold of a laser diode is constant for a given temperature; as temperature increases, so does threshold current.

Drive current should be adjustable. The transmitter design should regulate drive current according to temperature, unless the diode is always going to be operated at room temperature. Laser-diode drive is a variable current, not voltage. The bias circuit must accommodate a 1.5-2.5 V swing, but it is diode current that modulates the light output. A calibrated laser light meter is a very helpful tool for adjusting the transmitter.

Laser Drive Circuits

A simple wideband laser-diode bias circuit can be a simple resistor and negative temperature coefficient (NTC) thermistor

Table 1
Comparison of Laser Diodes

Model	*Mfr*	*CW Pwr MW Max*	*Wavelength Peak #*	*Beamwidth*	*Ith*	*Iop mA*	*Lo mA*	*Notes Cost $*
LT022MC	Sharp	5	780	11	33	50	65	Surplus
LT024MD	Sharp	30	780	10	29	55	85	
SL151U	Sony	5	670	11	30	75	85	Low astig
SL151V	Sony	5	670	11	30	75	85	
SLD201U3	Sony	50	780	14	28	80	120	
SLD304V	Sony	1000	810	13	28	450	1400	Cost >$2500!
TOLD9200	Toshiba	3	670	7	34	76	85	
TOLD9201	Toshiba	5	670	10	35	80	90	
TOLD9211	Toshiba	5	670	8	33	50	60	Low astig
TOLD9410	Toshiba	3	650	7	35	70	80	Low wavelength
NDL3210	NEC	6	675	9	34	50	60	
LN9705	Panasonic	5	788	10	35	40	50	
SDL5311	Spectra Diode Labs	100	830	10	30	35	170	Cost $800

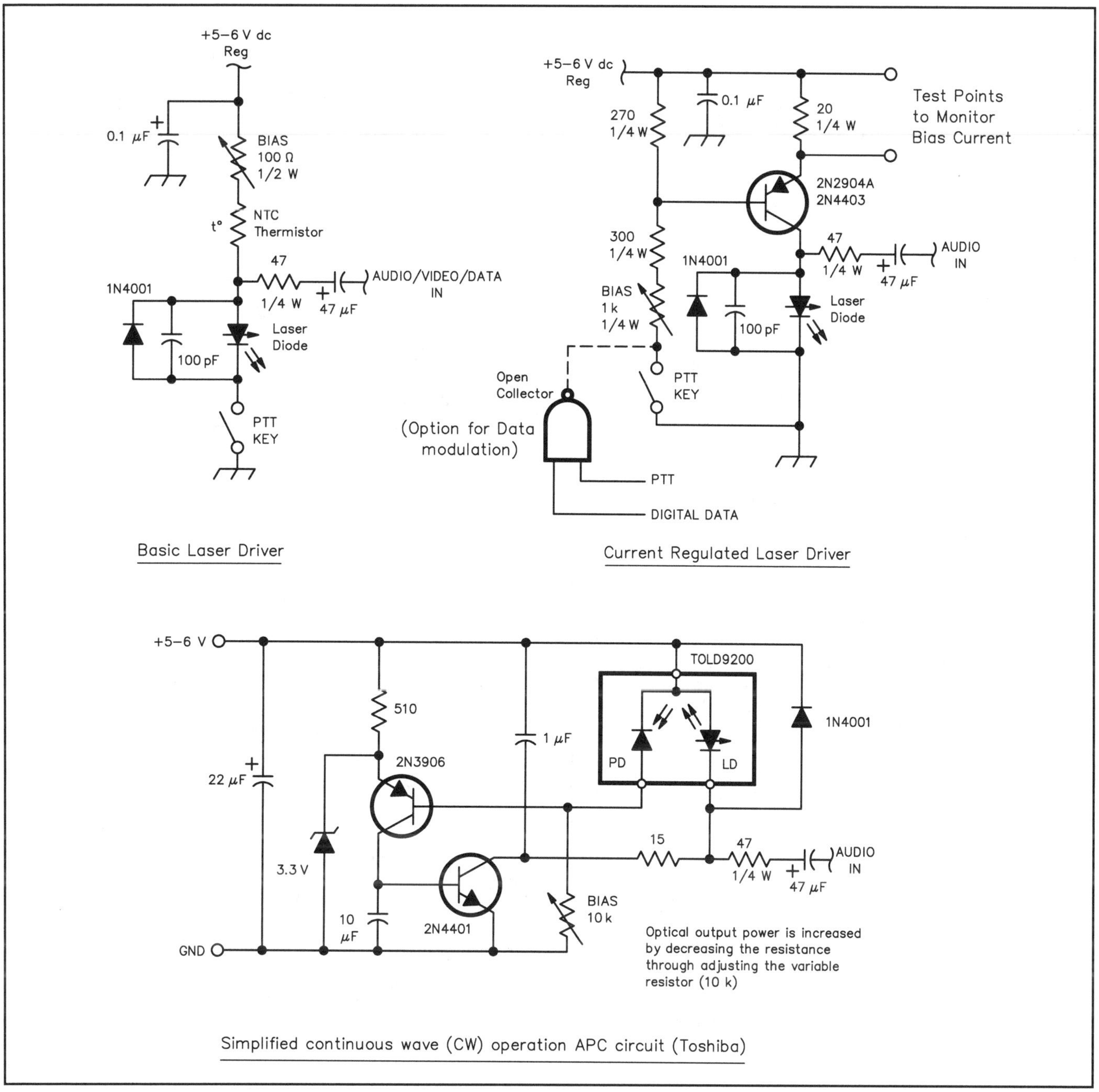

Fig 4—Laser diode driver circuits.

from the laser diode anode or cathode to a regulated voltage source of 5-12 Vdc. The modulating signal can then be capacitor or transformer coupled to the laser diode. A better method is to create a constant-current generator with a transistor or op amp (Fig 4). Temperature compensation can be easily included in this circuit. I do not use the monitor PIN photodiode built into the laser diode. It is commonly used in a feedback circuit (APC) to automatically stabilize the laser diode output over the temperature range. If you don't use the monitor diode, conduct initial transmitter testing with an inexpensive high-current red LED. Electrostatic discharge can destroy a laser diode. Ground yourself and your soldering iron when handling them.

LASER RECEIVERS AND OPTICS

The schematic of a direct-detection (non-superhet) free-space, optical receiver and its analog in a basic HF receiver appears in Fig 2. The receiver "antenna" is a telescope used to gather as much of the lightwave signal as possible. It can be a reflecting (mirror) or refracting (lens) telescope. Its main purpose is to focus the incoming light on the photo detector. From there, the light signal is converted or amplified, and converted to an electrical signal. The optical filter is analogous to the front-end tuned circuit, and is used to remove unwanted signals. The best overall choice for a telescope is a refractor using a Fresnel lens. This lens, composed of concentric engraved

Table 2
Comparison of Photodetectors

Device	*NEP @ 670 nM $W\sqrt{Hz}$*	*Gain cm²*	*Detect Area V dc*	*Power Supply*	*Relative Size*	*Comments*
PMT	2-60^{-16}	>10000	>4	>1 kV	large	fragile, fast
PIN	0.2-9^{-14}	none	0.01-2	0-100	small	ultra fast
APD	1-9^{-14}	200	0.01-2	0.2-2 kV	small	ultra fast
Photo diode	0.1-1.0^{14}	none	0.01-big	none	small-big	medium speed
Photo transistor	?	>1000	0.01	<50	very small	medium speed
Photo resistor	?	none	1-5	1-100	medium	slow

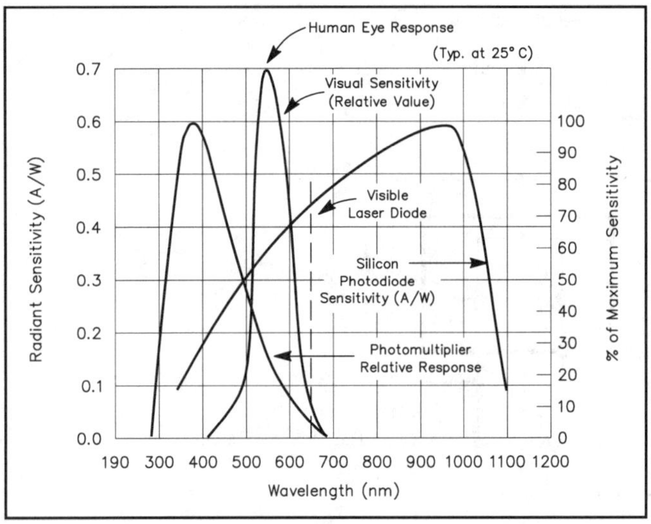

Fig 5—Photodetector responses at various wavelengths.

rings and usually made of plastic, is light, thin and has high transmissivity. It is also surprisingly cheap for such large diameters. For example, a surplus 12-in. diameter Fresnel lens cost me $4 at a flea market. The focused image is not as good as a conventional glass lens, but the low cost and light weight of the plastic Fresnel lens are definite advantages.

I highly recommend filters selected for the wavelengths to be received. A narrow interference-type filter reduces stray light and allows laser communications in daylight. When working laser-diode stations, remember that, unlike gas lasers, some diode lasers may drift 10 nM up and down in wavelength with temperature and drive level. Variations from device to device may result in additional difference in wavelength of plus or minus 20-nM. The interference-type filter is sensitive to beam angle, and should be placed where the incoming beam is most collimated, usually in front of the lens. Loss is about 3 dB. Colored plastic and glass filters are cheap, but usually broadband, with losses of only 1 dB. Don't forget a lens cap for the telescope, to protect the lens and avoid frying the detector if the telescope is accidentally pointed at the sun!

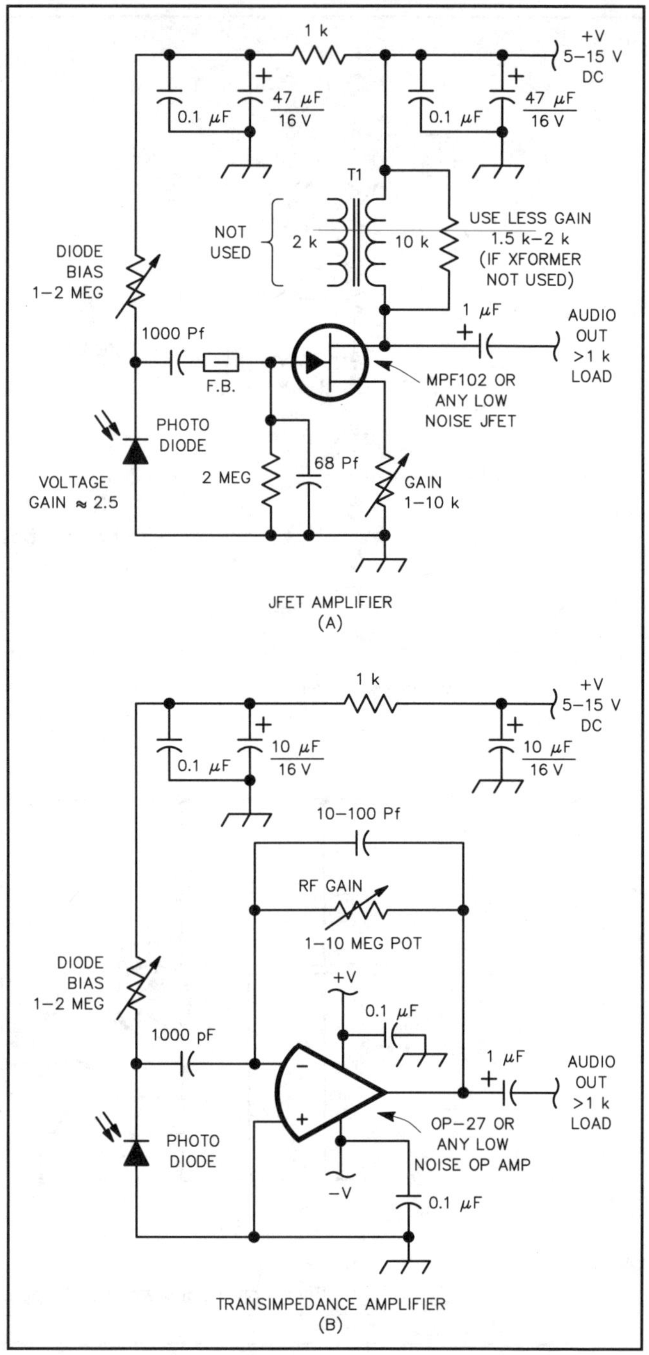

Fig 6—Low-noise photodiode preamplifiers.

The receiver detector can be any photodetector:

- Photomultiplier tube (PMT)
- PIN photodiode
- Avalanche photodiode (APD)
- Silicon photodiode and solar cell
- Phototransistor
- Photoresistor

Table 2 is a brief comparison of typical detectors. Because of its superior sensitivity and built-in gain, the PMT is the king of detectors for serious DX work. The typical PMT is blue sensitive. Its sensitivity drops off at the red and near-infrared (Fig 5). As you can see in the figure, the silicon PIN photodiode, followed by a low-noise, high-gain preamp is a

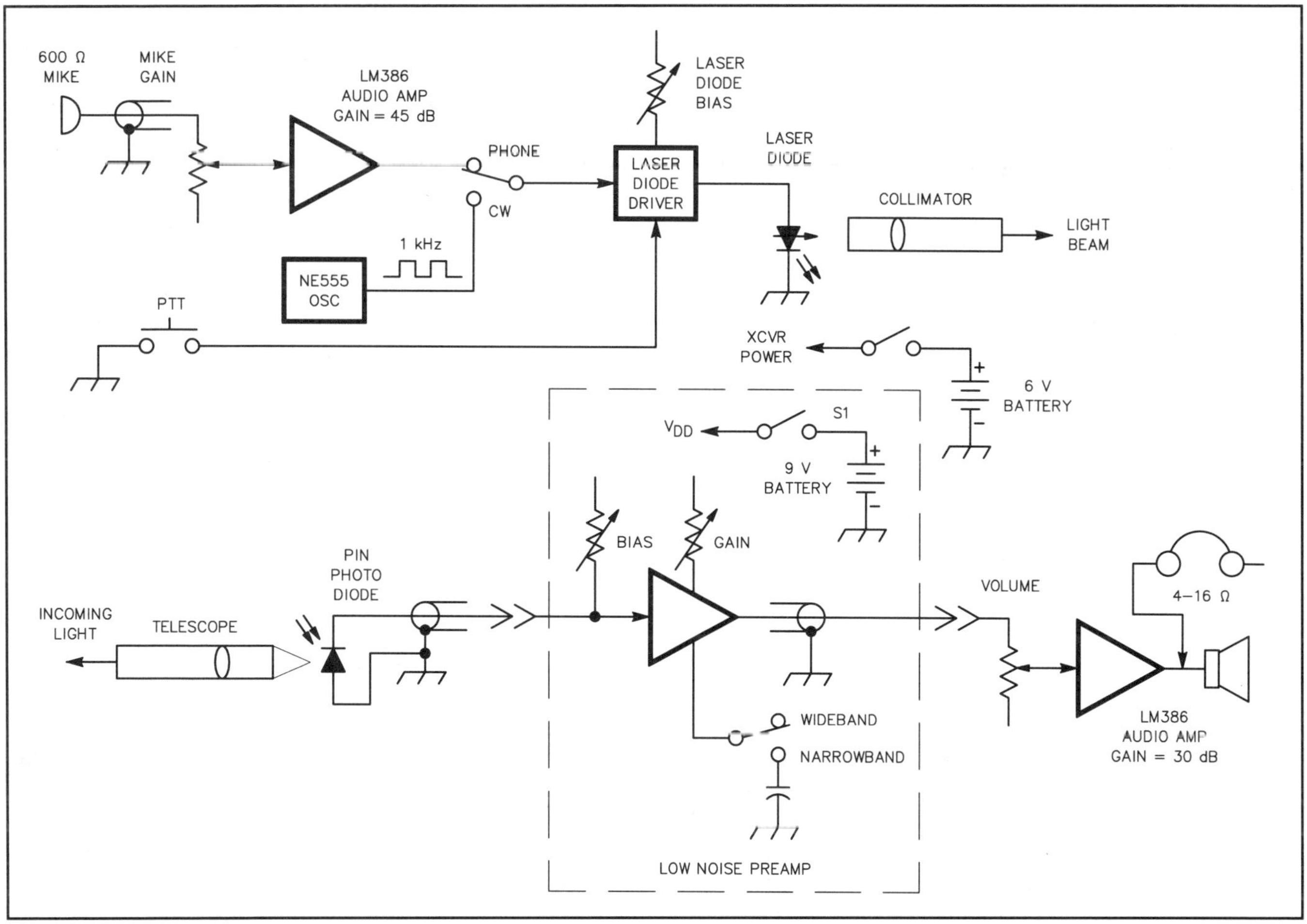

Fig 7—Block diagram of K6LMN laser transceiver.

practical second choice. It is the perfect mate for the red and near-visible infrared lasers. Its sensitivity peaks in the near-visible infrared region, and drops only about 20% at 670 nM. For lowest noise, the diode should not be back biased; for wide bandwidth (out to 1 GHz) though, the PIN diode requires 1-100 Vdc back bias. When avalanche photodiodes drop in price they will be more attractive, as they offer built-in amplification, like the PMT.

Typical silicon solar cells and photodiodes show peak sensitivity at 800 nM. A single, small solar cell chip at the focus of the telescope offers an inexpensive photodetector for the beginner. For a few dollars more, the highly sensitive PIN diode is a better choice for serious work.

Fig. 6 shows schematics of two common photodiode amplifiers. The transimpedance amplifier (B) is most useful for wide-bandwidth applications, and is commonly used with fiberoptic detectors. I like the JFET approach for a simple photodiode audio preamp, because of its low noise, resistance to self oscillation, and simplicity. A variable-frequency, variable-bandwidth filter should follow the preamp. A good choice might be one of the aftermarket audio filters. For optimum S+N/N ratio, limit the bandwidth as much as possible.

LIGHTWAVE TRANSCEIVER DESCRIPTION

Had enough theory? The K6LMN laser-diode transceiver provides push-to-talk, full-duplex voice/data/CW communication capability at visible light wavelengths in free space. The transceiver is a lightweight, compact (6- × 4- × 2-in.) package, with a self-contained power supply (4 AA cells). This unit is the heart of a complete lightwave station, including an "antenna" mounted on a 12- × 12-inch platform equipped for tripod or table mounting. A low-power red LED (665 nM) in parallel with the laser diode provides visual monitoring of the modulation. Fig 7 shows a block diagram of the station.

In addition to AM phone, the transceiver features MCW mode. With the key down, the laser is switched rapidly between 0 and 5-mW (full-power) output, by a 1-kHz square wave. The square wave is generated by a 555 IC. Average power output is thus 2.5 mW. An unmodulated CW mode may be added, to aid in visual sighting of the beam.

The laser transmitter output or lightwave emitter is a 5-mW peak, 670-nM (448 THz) laser-diode, integrated with a lens collimator. The collimator is adjustable, for divergence tweaking. A coherent and extremely narrow divergence (about 0.3-mrad high × 1.0-mrad wide, after the collimator) red light beam is the result. A die-cast aluminum case from a surplus laser printer serves as the diode heat sink. The laser-diode assembly is mounted on the transceiver cabinet. A low-astigmatism Toshiba 9211 laser diode is the emitter. It features a stable, single-transverse mode (clean spectrum) and low lasing-current threshold. A PIN photodiode integrated with the laser diode can be used to monitor and stabilize the power output, although I don't use this feature.

A 386 audio amplifier IC amplifies the 600-Ω mike input. Its high-level audio output amplitude modulates the current to the laser diode. The laser diode is purposely biased at the threshold of lasing, to reduce power consumption and the possibility of destroying the diode on voice peaks. The resulting modulation is slightly non-linear, but adequate for amateur voice/data communication. Only a small amount of temperature compensation is provided, and the bias point is not tightly regulated. An internal variable resistor sets the bias.

With the transmitter unkeyed, the current drawn from the internal batteries is about 10 mA. In transmit mode (no modulation), current consumption rises to 50 mA. On voice peaks, the current may reach 110 mA. The alkaline AA batteries used have a rated capacity of 1500 mAH. I use PTT, instead of operating full duplex, to conserve the batteries. Depending on path length and atmospheric clarity, it may take 15-30 minutes of continuous transmission and searching for the other station to complete a QSO.

For receiving, an "antenna" jack is provided, to accept audio input from any low-level-output photo detector. No dc bias is provided at this jack to power an active photodetector, such as a phototransistor or PMT. This transistor is designed to operate with solar cell detectors or sensitive photodiodes, followed by a preamp. The detector input circuit is protected against levels exceeding 1.5-V peak-to-peak, but less than 100 mA maximum. The input impedance is 10 kΩ. The detector input is amplified by another 386 IC. The output can drive 4-16 Ω external speakers or headphones, and a speaker jack is provided on the control panel.

The receiving antenna, which is mounted on the platform adjacent to the laser, is a super-cheap design. It is a 3-inch refracting telescope made of a free mailing tube, a $4 plastic Fresnel lens and a $5 silicon photodetector purchased at a swapmeet. The telescope mount is a bit flimsy, and needs refinement. The lens focal length is 12 in. The photodetector is a United Detector Technology model PIN 10RP438-1 silicon PIN diode, with a 1-cm^2 detection area. A broadband red bandpass optical filter in front of the detector filters non-red stray light, but can accommodate HeNe and near-visible infrared lasers. The photodetector is followed by a high-gain, low-noise JFET preamp mounted in a separate, shielded box. The photodiode is back biased with a variable voltage of 0-9 Vdc. A pot on the preamp front panel adjusts the bias for day or night operation with weak or strong signals. Another pot controls preamp gain. A front-panel switch sets the preamp bandwidth for voice or Morse code. Preamp output is connected to the antenna or detector input jack on the transceiver. My other transceiver uses a 15-V solar panel for an antenna, and has no preamp.

This transceiver is very simple to operate, with a minimum of controls. Although the transmit and receive channels are independent, crosstalk within the unit provides sidetone. The transceiver can be self tested by bouncing the beam off a nearby reflecting object. For sighting, I look along the transceiver case edge and point at the other station. Instead of using a spotting scope boresighted with the laser, I use a marked, vertical white pole placed several yards in front of the transceiver along the expected path. A heavy-duty, rock-steady tripod is an absolute necessity for serious DX work. Moving the beam a mere 0.06° moves the beam off axis 5 feet in one mile!

For phone operation, plug in a mike, flip the mode switch to Phone and push the PTT switch. Because the transceivers are full duplex, there is no transmit/receive changeover delay. The bandwidth of the voice channel is about 2.7 kHz (300-3000 Hz), which should be usable for data transmission.

THE FUTURE

I can see into the near future, where a lightwave repeater is a possibility. Bandwidth, as indicated by laboratory work being done with fiberoptic cables, is over 40 GHz. While the extremely sharp beam is suitable for point-to-point links, it doesn't lend itself to wide-area repeater use. The answer may be an array of semi-collimated laser diodes for the transmitter, and an array of photodiodes for the receiver.

Bibliography

[1]R. Atkins, "Laser Communication Systems," *Ham Radio*, Mar 1990, pp 18-21. See also F. Mimms, *Lightwave Communications* (Indianapolis: Howard W. Sams, 1982).

[2]R. Atkins, "Optical Receiver," *Ham Radio*, Apr 1990, pp 14, 17-19.

[3]M. Forbes, "Designing Circuits With LEDs," *QEX*, Jul 1990, pp 15.

[4]L. Foltzer, "Small Aperture IR Optical Links Using LED Light Sources," *QEX*, Aug 1990, pp 9-14.

[5]B. Bergeron, "A Laser Communications Primer," —*Part 1*, *QST*, Sept 1990, pp 19-24; —*Part 2*, *QST*, Oct 1990, pp 22-26.

4

Power Supplies and Control Circuits

Contents

An RF and DC Switching Notebook

By Dave Mascaro, WA3JUF

Here is a collection of circuits I've used to switch and key transverters, preamps, power amps and IF rigs.

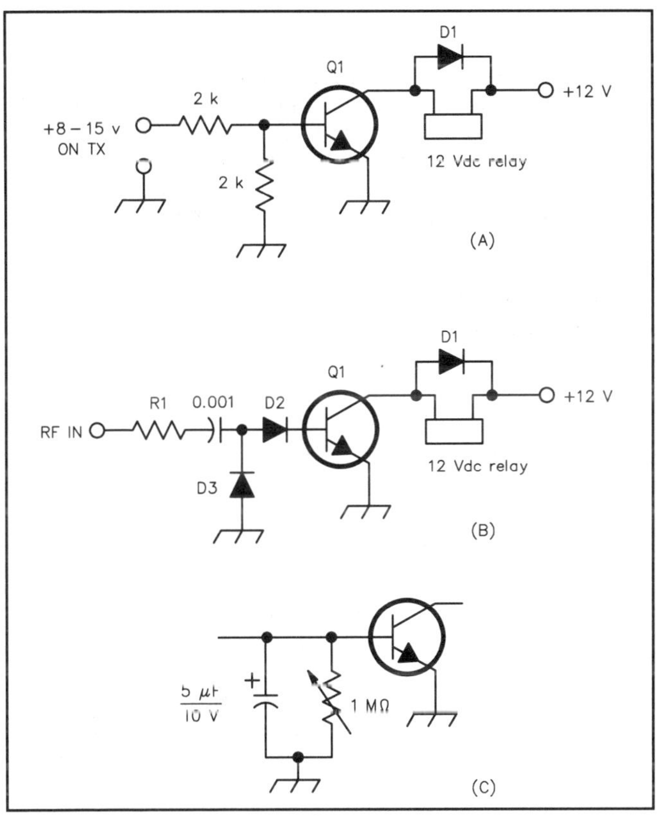

Fig 1—Relay drivers for relays with 12-V dc coils. In A, the relay is keyed from a low-current 8-15 V dc supply. In B, the relay is keyed by rectified RF. The value of R1 may range from 1000 to 10 kΩ, and must be determined. For SSB operation, the circuit in C keeps the relay from dropping out between syllables. Resistors are ¼-watt film or composition.

D1—1N4001 or equiv. (50 PIV, 1 amp).
D2, D3—Silicon switching diodes, 1N914, 1N4148 or equiv.
Q1—MPS-A13, 2N2222A or equiv.

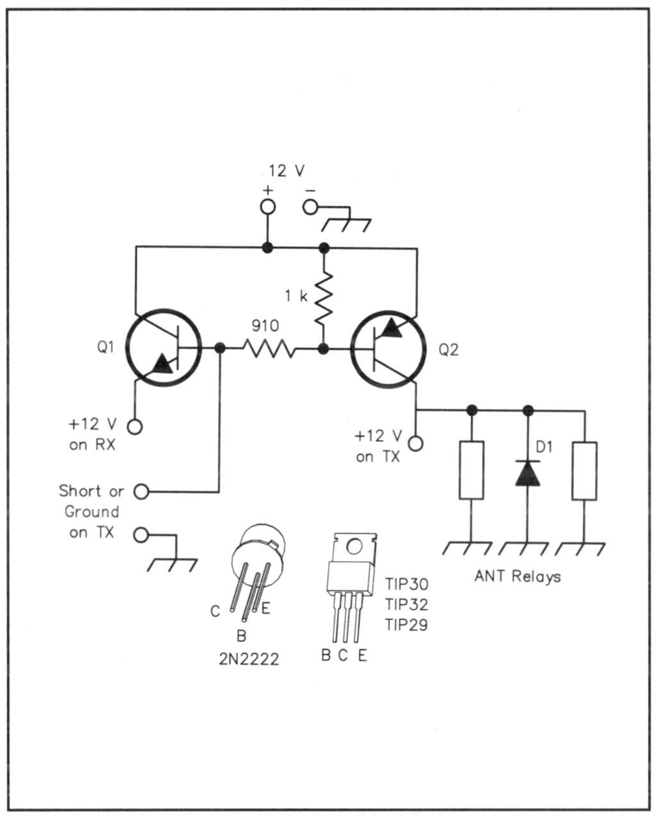

Fig 2—Transverter keying from the normally open (N.O.) contacts of the IF rig. Resistors are ¼-watt film or composition.

D1—1N4001 or equiv.
Q1—2N2222A, TIP29 or equiv.
Q2—TIP30, TIP32 or equiv.

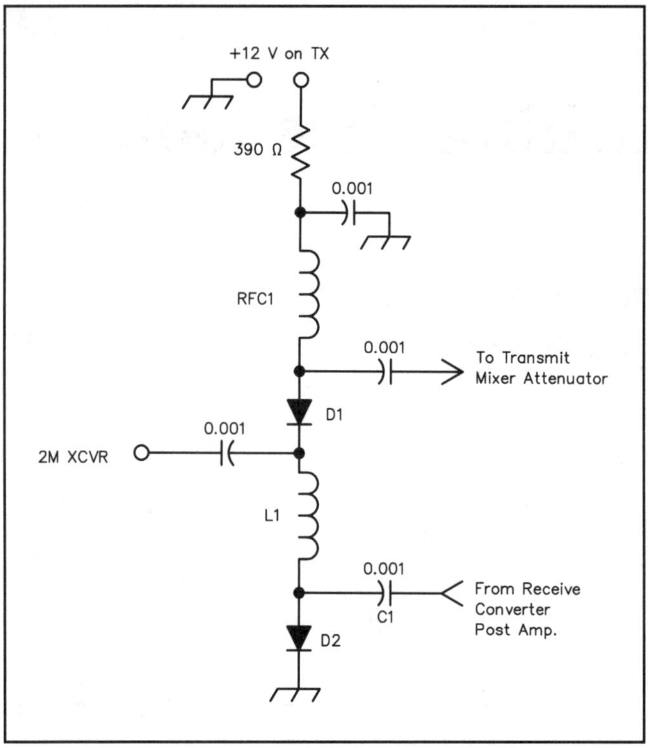

Fig 3—Transverter-RF switching of the IF. If the receive post amplifier has a dc-blocking capacitor, C1 is unnecessary.

D1, D2—PIN diodes, Phillips BA182, Motorola MPN3401 or equiv.
L1—For 2 meters: 12 turns #26 AWG, 0.1-in. dia., or λ/4 length of miniature coax at frequency being switched.
RFC1—For 2 meters: 12 turns #24 AWG, 1/16-in. dia.

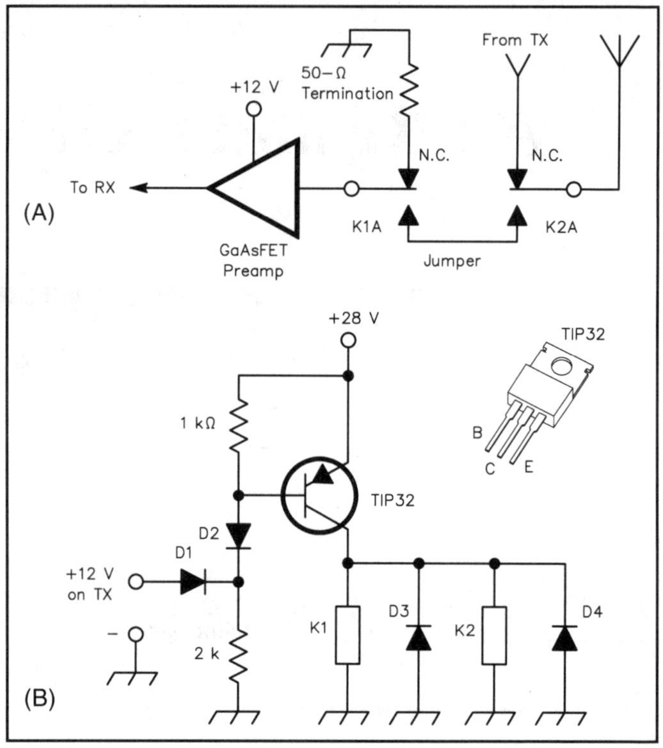

Fig 5—Switching a mast-mounted preamp. RF connections are shown at A. Use a double-male type-N adaptor for the common connection between the relays. B shows the control wiring. When control or relay power is off, the relays are in the TRANSMIT position, and the preamp input sees the 50-Ω termination. This scheme protects the preamp when the station is off the air. The preamp should be powered by a dedicated supply or NiCd batteries.

D1-D4—1N4001 or equiv.
K1, K2: Transco-type dual-coil relays.

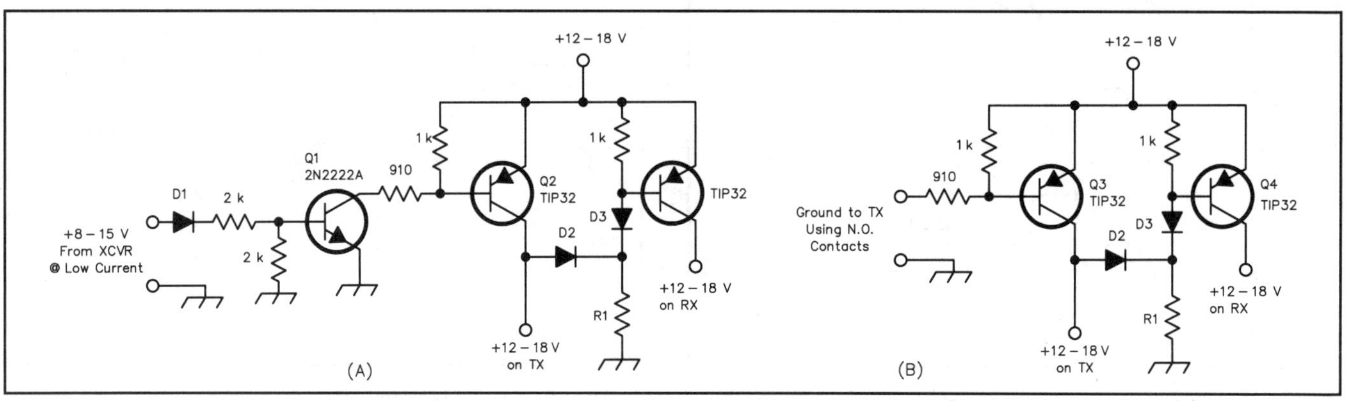

Fig 4—Transverter keying/switching. As drawn, these circuits will switch the supply between the receive and transmit sides of the transverter. At A, the circuit is switched by a +12 V signal from the IF transceiver. At B, the circuit is switched by a N.O. relay contact in the transceiver. Either circuit may be used to energize 2-coil Transco-type relays by changing R1 to 2 kΩ. Also, increase the supply from 12-18 V to 28-30 V. Resistors are 1/4-watt film or composition.

D1—Silicon diode, 1N914 or 1N4001.
D2, D3—1N4001 or equiv.
Q1—2N2222A or equiv.
Q2-Q4—TIP32 or equiv.
R1—For 12-V switching, 560 Ω; for 28-V switching, 2 kΩ.

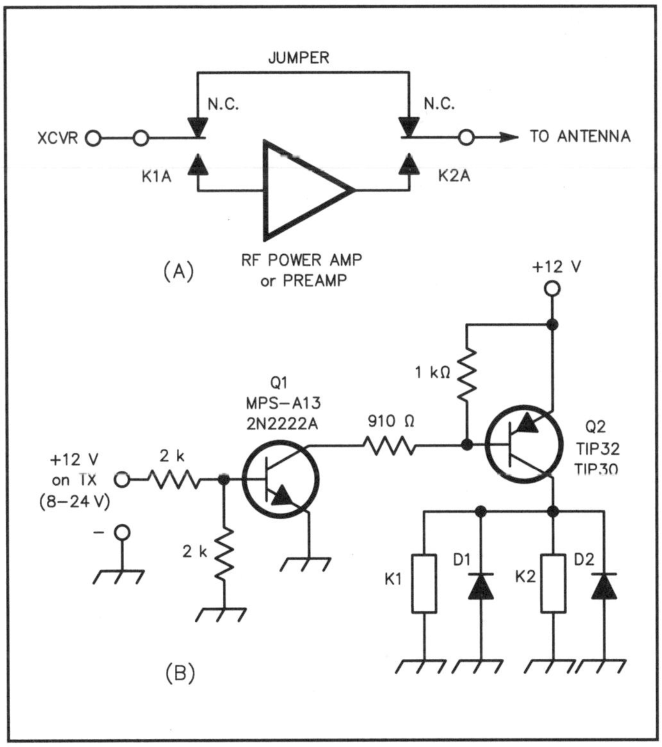

Fig 6—Switching external power amplifiers. These circuits are useful for switching "brick" amplifiers connected to transceivers. While "hard" keying with dc (A) is preferred, the RF-sensing circuit at B may be used. Resistors are ¼-watt film or composition.

D1, D2—1N4001 or equiv.
K1, K2—12 V dc miniature coaxial relays.
Q1—MPS-A13, 2N2222A or equiv.
Q2—TIP30, TIP32 or equiv.

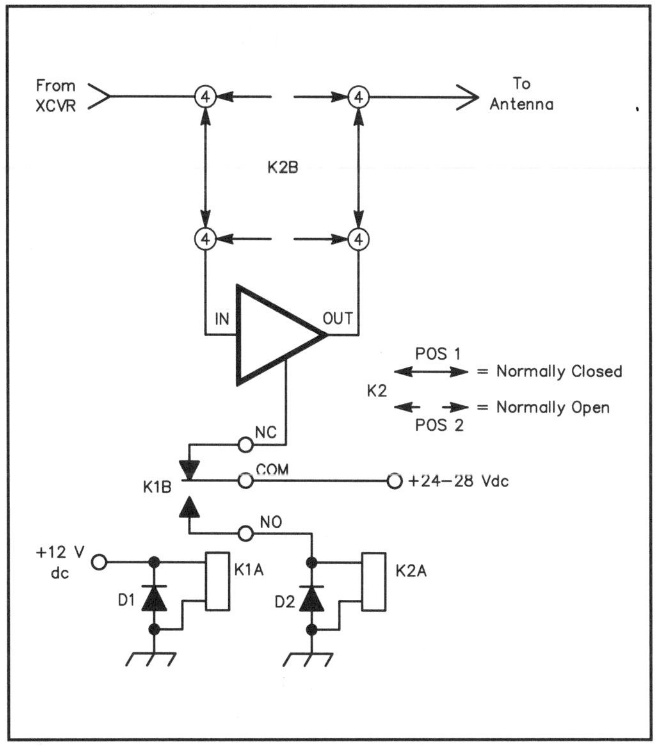

Fig 7—Using a single Transco transfer relay to select/bypass an amplifier. In this example, the amplifier is also powered from 28 V dc. K1 also removes amplifier supply voltage when the amplifier is bypassed, because the amplifier input is connected to its output by K2. An amplifier that operates from another supply voltage can be switched by substituting a DPDT relay for K1. One set of contacts is used to switch the amplifier, the other to switch K2.

D1, D2—1N4001 or equiv.
K1—Control relay, SPDT, 12-V dc coil.
K2—Transco 310C00200 transfer relay.

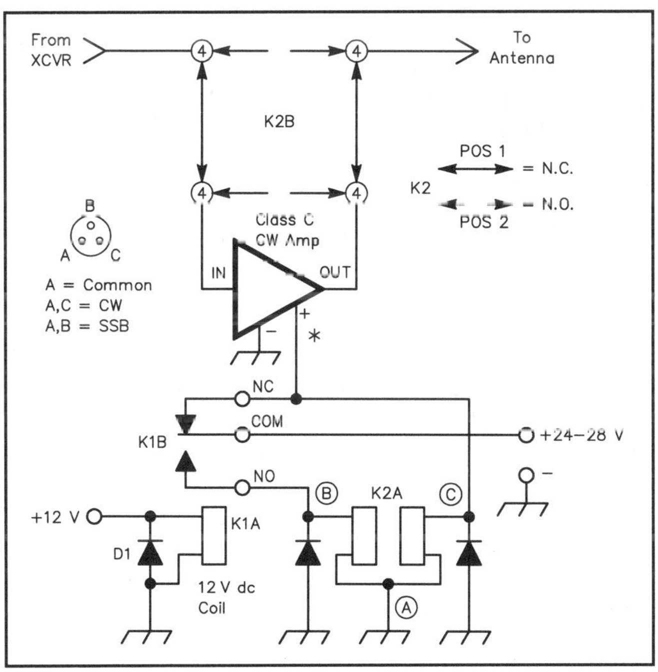

Fig 8—Using a single General Communications transfer relay to select/bypass an amplifier. K2 has two coils, unlike the Transco relay depicted in Fig 7, which has only one. See the caption of Fig 7 for information on using amplifiers operating from a supply voltage other than 28 V dc.

D1-D3—1N4001 or equiv.
K1—Control relay, SPDT, 12-V dc coil.
K2—General Communications 2NRP1 transfer relay.

How To Use PIN Diodes In A 2-Meter IF Switch

By Greg Raven, KF5N

It is rare to see PIN diode RF switches in homebrew amateur equipment. Most experimenters have probably tried building a PIN diode switch, with limited success. Excellent switches can be built that are comparable to mechanical switches in both insertion loss and isolation. A PIN diode switch is physically small and lightweight, so they are great for portable equipment. Most of the RF relays that hams have obtained via surplus channels require 28 volts, which is incompatible with the 12-volt systems commonly used. Any voltage can be used with a PIN diode switch with proper design of the bias network. This article describes the design details of a PIN diode switch that is useful for IF switching in microwave transceivers.

To RF, the PIN diode looks like a small resistor when forward biased, and like a small capacitor with zero or reverse bias. A simple narrow-band switch that has good performance is shown in Fig 1. In transmit mode, a positive bias voltage is applied to the resistor. Both diodes are forward biased.

Note that L1 and C1 form a parallel resonant circuit at the frequency of operation. The reactance of L1 is chosen to be 50 ohms. This means that this resonance is very low Q, due to the external loading, making the circuit relatively broadband. The shunt PIN diode is across the path to the receive terminal, providing very good isolation. In practice, the parasitic inductance of the diode package limits the isolation. Isolation can be improved by adding a second inductor (L2) in parallel with D2, the series diode. L2 is chosen to resonate with the *off* capacitance of the diode. This value is given on the manufacturer's data sheet. C6 blocks dc from the inductor.

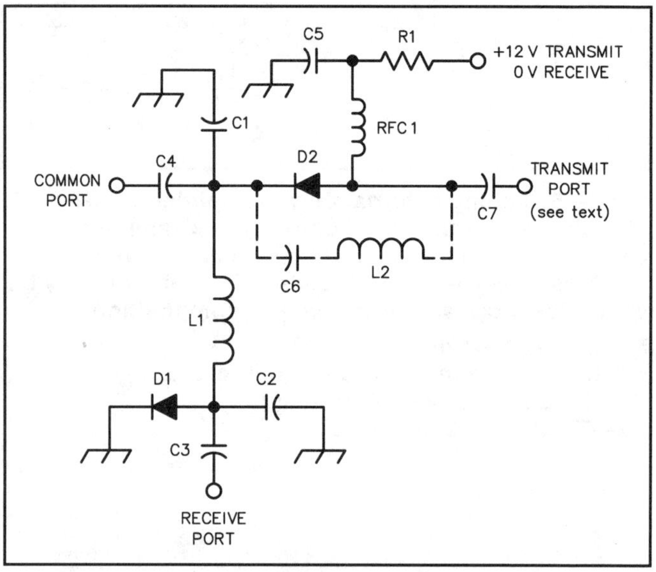

Fig 1—Schematic of a PIN-diode T/R switch. See the text for information on calculating C1 and L1. C6 and L2 may be necessary in some cases; see text for details.

C1-C7—Ceramic chip capacitors. C3-C7 are series-resonant type.
D1, D2—PIN diodes. See text.
L1, L2—See text.
R1—Calculated; see text.
RFC1—Reactance should exceed 500 Ω at operating frequency.

In the receive mode, 0 volts bias is applied to the diodes. C1, L1 and C2, plus the capacitance of the shunt PIN form a quarter-wavelength section of lumped transmission line. The pi-section equivalent of a ¼-λ transmission line simply requires the reactance,of the shunt capacitors and the series inductor equal 50 Ω. The capacitance of the shunt PIN diode is "parasitically absorbed" into C2.

In some set-ups that use a class-A final amplifier that remains active during receive, excess thermal noise will be present at the transmit port. This noise will be conducted into the receiver via the capacitance of the series diode D2. If this is a problem, the transmit to receive port isolation can be improved by using L2/C6 to resonate with the off-capacitance of D2.

Resonating the series diode becomes more important with higher-frequency designs or higher-power diodes, which tend to have more *off* capacitance. The impedance of the transmitter in the *off* state can actually increase the insertion loss of the switch in the receive mode, if the isolation is not high enough. (The transmitter tends to load the switch.)

If the transmitter-to-receiver isolation in transmit mode

is not good enough, another shunt PIN diode can be added. It should be incorporated into another ¼-λ transmission-line section. This extra diode typically improves isolation from about 30 dB to more than 50 dB. The disadvantage is the additional loss that the diode and other components add to the receive path. This loss should not be a problem in most applications.

Design Example

Choose a PIN diode app pF ˙ate to the power level and frequency of your application. I just happened to have a few surplus Unitrode UM9401 diodes. This diode is perfect for this kind of antenna switch. The capacitance is a relatively low 1.5 pF, while the forward c istance is about an ohm.

Choose C1 and L1, using a design frequency of 144 MHz.

$$C1 = \frac{1}{50(2\pi)144 \times 10^6} = 22.1 \text{ pF}$$ (use a 22-pF chip cap)

$$L1 = \frac{50}{(2\pi)144 \times 10^6} = 55.3 \text{ nH}$$ (use an airwound coil, 20 gauge close spaced, 0.25 inch inside diameter, 4 turns.)

C2 = 22 pF (the shunt capacitance of D2 is small enough to ignore. No parasitic absorption of its capacitance is needed.)

$$R1 = \frac{12 - 1.5}{0.05} = 210\ \Omega$$ (Switch designed for 12 V. Assume a voltage drop of 0.75 V for each diode. Use a diode bias current of 50 mA.)

That is all there is to it. The coupling capacitors C3, C4 and C5 are series resonant. I typically use 470 pF chip caps at 2 meters.

The circuit was breadboarded "deadbug" style, using a piece of PC board. SMA connectors were used for the antenna, receive, and transmitter ports. The entire circuit is only 1.5 inches square. The circuit took about 15 minutes to build. It was measured on a network analyzer with the following results:

Transmit port to antenna port insertion loss	:	0.15 dB
Receive port to antenna port insertion loss	:	0.17 dB
Transmit to receive isolation (switch on)	:	29.2 dB
Current drain in transmit mode	:	50.0 mA
Current drain in receive mode	:	0.0 mA

This switch should be able to handle up to 50 watts in the transmit mode. The intermodulation performance of this switch in the receive mode is excellent. The circuit is useful to at least 900 MHz. For 1296 MHz and above, transmission lines are substituted for the lumped elements. I have built PIN switches for the 900-MHz band, and I am sure they will work at 1296 MHz and higher. With the right devices, low-loss switches can be made to switch kilowatt power levels.

Three Useful Circuits

By Rick Fogle, WA5TNY

LATCHING RELAY DRIVER CIRCUIT

The circuit of Fig 1 drives a latching relay that has only one coil. When the logic level at point A changes state, a pulse is applied to the coil of K1. The polarity of the pulse depends on whether the logic level is going from positive to negative, or negative to positive. The pulse duration depends on the coil resistance and the value of C1. Pulse duration should be long enough to drive the relay to its opposite state.

VOLTAGE DOUBLER

The circuit in Fig 2 operates a small Transco-type relay (SMA connectors) with a 28-V coil from a 12-V dc supply. U1 generates square waves. Together with D1, D2, C1 and C3, it forms a half-wave, voltage-doubling power supply.

VOLTAGE INVERTER

Fig 3 shows a circuit that uses a square-wave generator to provide a negative supply for FET or step-recovery diode (SRD) bias from a positive 12-V source. The output voltage is approximately equal to the input voltage, but opposite in polarity.

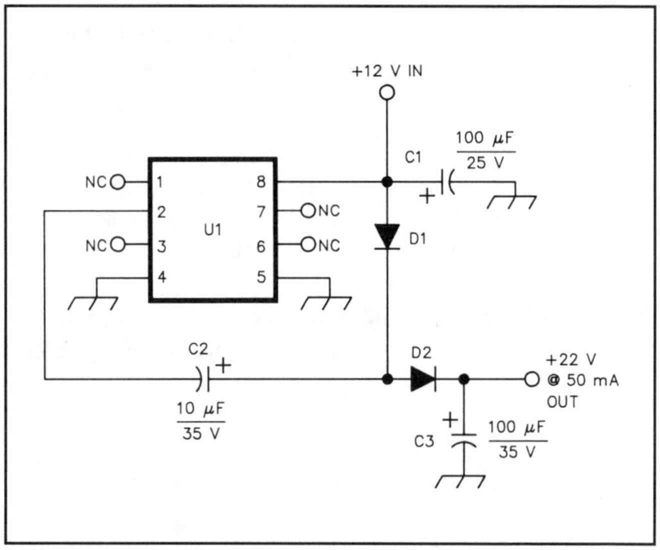

Fig 2—Voltage doubler to operate 28-V relay coils from 12-V supply.

D1, D2—1N4001 or equiv.

U1—Siliconix Si7661CJ square-wave generator.

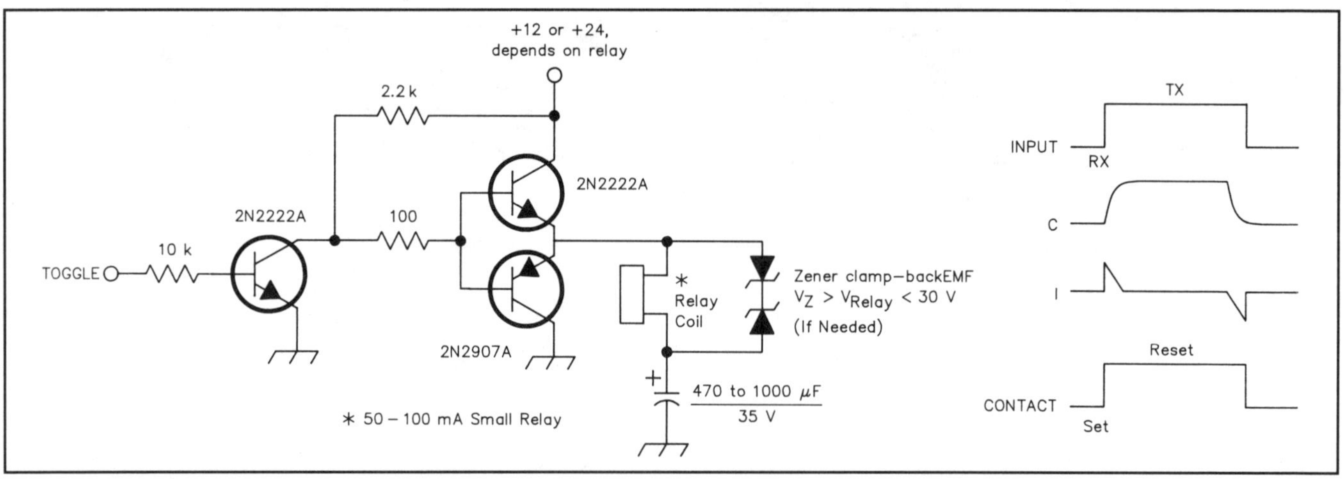

Fig 1—Latching relay driver circuit. Vcc is either 12 or 24 V dc, depending on the coil voltage of K1. The circuit shown will handle coil currents of 50-100 mA. Resistors are ¼-watt film or composition.

C1—Electrolytic capacitor, 470-1000 μF, 35 V.

D1, D2—Zener diodes; breakdown voltage should be greater than the coil voltage of K1 but less than 30 V.

K1—12- or 24-V latching relay.

Q1, Q2—NPN switching transistor, 2N2222A or equiv.

Q3—PNP switching transistor, 2N2907A or equiv.

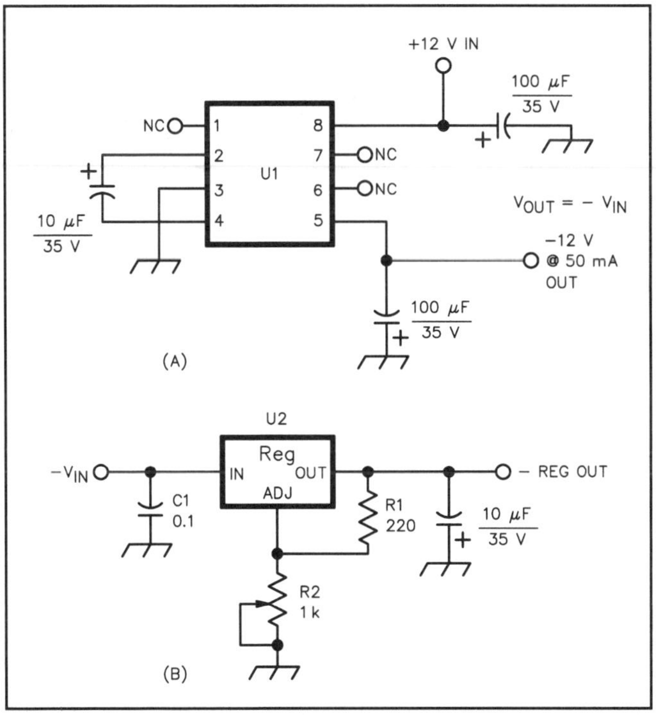

Fig 3—At A, the voltage inverter. Its output is not regulated. If regulation is needed, use the circuit shown at B. Resistors are ¼-watt film or composition.

C1—Ceramic disk, mounted close to U2.
U1—Siliconix Si7661CJ square-wave generator.
U2—National LM337L negative-voltage regulator, or equiv.

Power Supply for GaAsFET Amplifier

By Zack Lau, KH6CP

(From *QEX*, March 1991)

I have used this circuit to supply numerous GaAsFET low-level amplifiers with good results. It's an active supply to compensate for device and temperature variations. It fits in well with the no-tune concept of microwave circuitry. After all the work of coming up with RF circuitry that doesn't need tuning, why should you need a set of trimmers to adjust dc bias parameters?

Circuit Description

This circuit isn't original—a similar circuit appeared in an article by Al Ward, WB5LUA.[1] Unfortunately, the author didn't supply a printed circuit-board pattern, which is the need this article fulfills. In addition to Al's article, you may want to refer to Avantek application note AN-S003.

The negative bias generator is a simple 555 oscillator circuit feeding a diode rectifier. The use of high-speed rectifier diodes permits a higher oscillator frequency, allowing smaller filter capacitors. A high-frequency supply comes up to operating voltage faster, reducing the stress on the GaAsFETs. Common switching diodes (1N914/1N4148) will work if you don't exceed their limited current capabilities. Three-terminal regulators provide stiff voltages for the active bias circuitry that don't exceed the FET limits.

The active circuit uses R3 to set the desired GaAsFET drain current. R1 and R2 act as a voltage divider that sets the drain voltage. Don't forget that in addition to the emitter base voltage drop, there is often a drain resistor that adds yet another voltage drop.

The Hand-Waving Explanation of Circuit Operation

The circuit is based on a PNP transistor which tries to maintain 0.7 V across the base-emitter junction. If this voltage

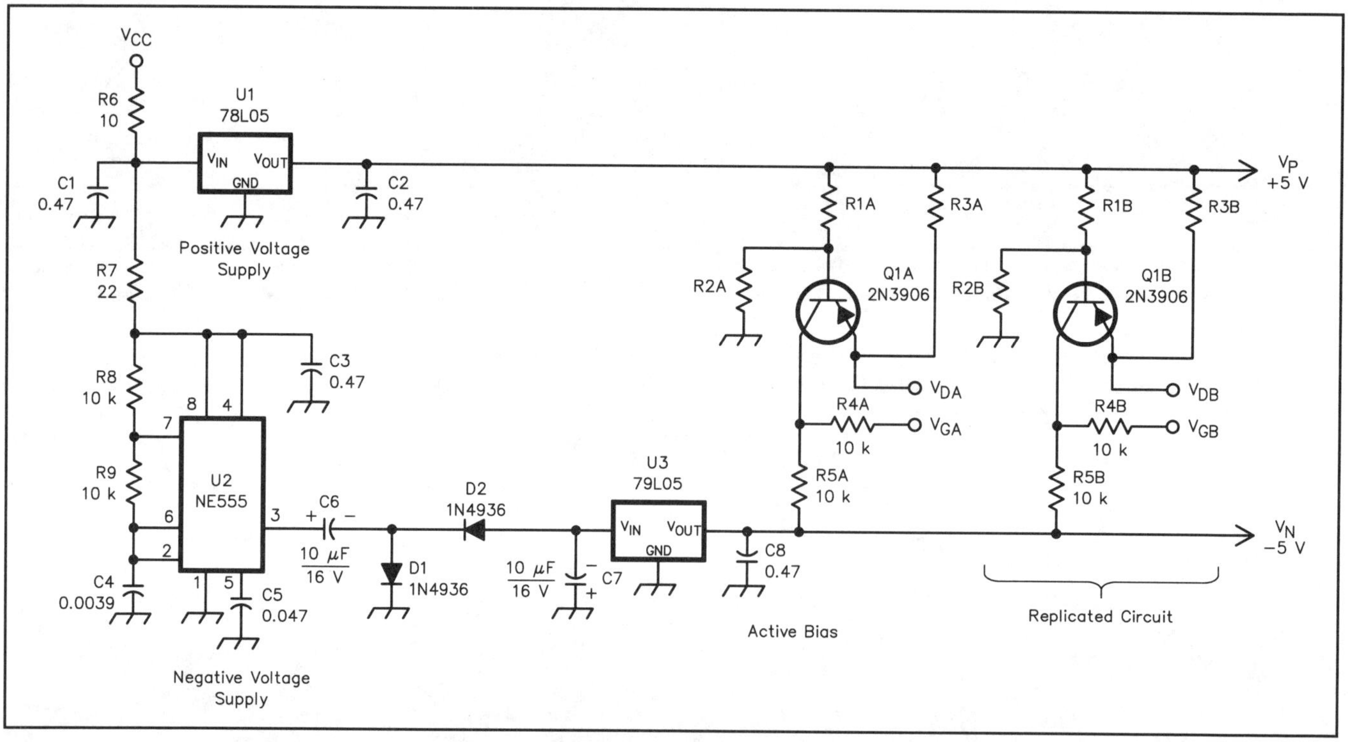

Fig 1—Schematic of the GaAsFET amplifier power supply.

is too low, the current across the base is reduced, making the FET gate voltage more negative. The more negative gate-source voltage reduces the drain current, which increases the emitter voltage. Since the base voltage is approximately fixed, the base-emitter voltage is therefore increased until it's approximately the voltage drop for a saturated transistor. Similarly, too low a drain current increases V_{be}, increasing the current across the base, making the FET gate voltage more positive, which increases the drain current.

Choosing Values for R1, R2, R3

$$R3 = (V_p - V_d) / I_d \qquad \text{(Eq 1)}$$

$$R1 = R2\,(V_{be}+V_p - V_{dd})/(V_{dd} - V_{be}) \text{ approx.} \qquad \text{(Eq 2)}$$

V_p = The positive supply bus

V_{dd}= The drain voltage supplied to the GaAsFET amplifier (don't forget any voltage drops from series resistors)

I_d = The drain current

V_{be} = The base-emitter drop (0.7 V, but temperature sensitive)

I have had pretty good luck using Eq 2 with R1 and R2 values in the low to mid-kilohm range. These equations seem to be close enough for picking standard resistor values.

Experimental Results

The resistance values are assumed to be those marked on 5% ¼-W resistors. The voltages were measured with a Fluke 77 multimeter.

R1	*R2*	*R3*	V_{dd}	V_p
2 k	3.9 k	15	3.92	5.03
2.2 k	4.3 k	51	3.99	5.04

Construction

The printed circuit board is designed to allow a variable number of devices to be powered. If you only need to power a few devices, you might wish to cut the board short. With a bit more work, more than 6 devices can be accommodated by replicating the active bias pattern and extending the board. It may be necessary to heat sink or replace the 78L05 with a bigger device if you supply too many FETs. The pads are purposely made big enough for swaged terminals, which are ideal for this application if you have access to the expensive swaging tool ($300+ new).

Reliability

Since the negative supply comes on shortly after the positive supply, questions about damaging the FETs invariably arise. So far, out of over 20 devices operating from 2304 to 10368 MHz, only one failure has been noted. I suspect this reused 72084 failed as a result of excessive heat or static dis-

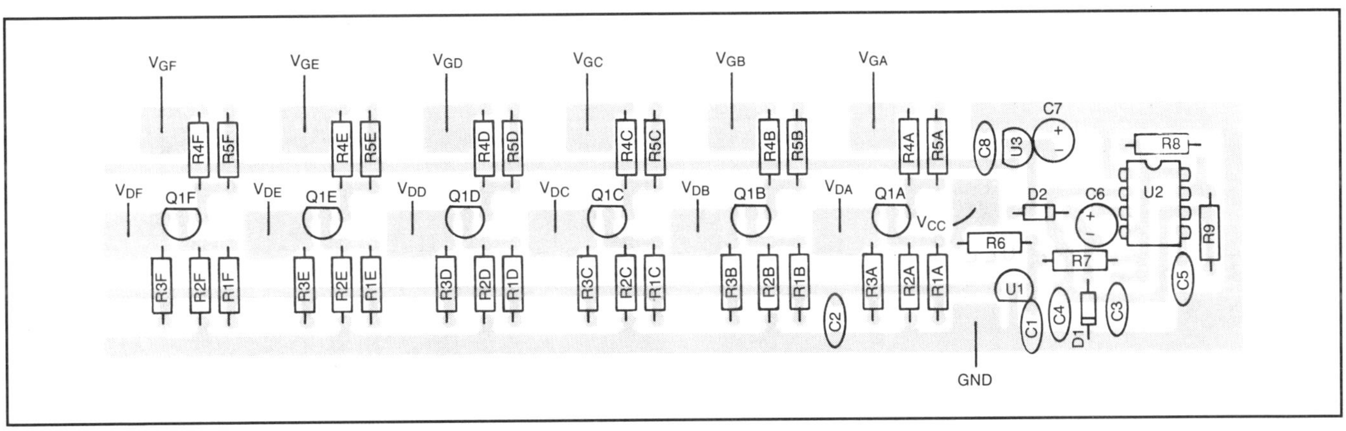

Fig 2—Component layout of the power supply.

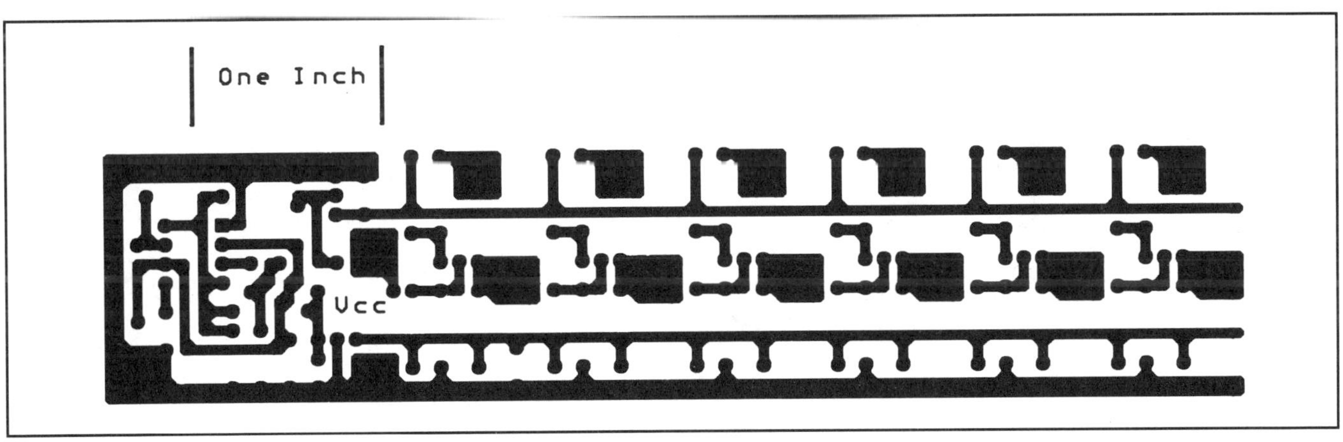

Fig 3—Etching pattern for the power supply.

charge, as it's not always easy to remove devices from circuits that didn't work as well as hoped. Actually, my biggest problem with microwave devices are the leads breaking off while attempting to reuse them!

There is a trade-off between current limiting and voltage limiting. By raising the open-circuit voltage, you can increase the drain series resistance, improving the current limiting abilities of your supply. However, if the voltage is too high, you can exceed the V_{DS} rating of the device. There are situations where a more sophisticated supply is needed to avoid exceeding device ratings. An example might be a 5-volt maximum V_{DS} FET that needs a drain bias of 40 mA at 4 V, as well as 51 Ω of series resistance for RF stability!

[1] A. Ward, "Simple Low-Noise Microwave Preamplifiers," *QST*, May 1989, pp 31-36, 75.

[2] Avantek AN-A, as well as other Avantek applications notes, are available from your local Avantek distributor. If you do not know the location of your local distributor, contact Avantek, 3175 Bowers Ave, Santa Clara, CA 95054-3292, tel 408-727-0700.

5

Oscillators and Multipliers

Contents

A Clean, Low-Cost Microwave Local Oscillator

By Rick Campbell, KK7B

(From *QST*, July 1989)

Obtaining a suitable local oscillator (LO) has traditionally been a major obstacle in amateur microwave work. This article describes a straightforward, inexpensive, easily constructed microwave LO. The oscillator may be combined with a simple low-noise preamp[1], an image filter[2] and an off-the-shelf doubly balanced mixer to build a complete high-performance receiving converter for OSCAR Mode S. This LO can also be used as a building block in a transverter for the 2304- or 3456-MHz bands.

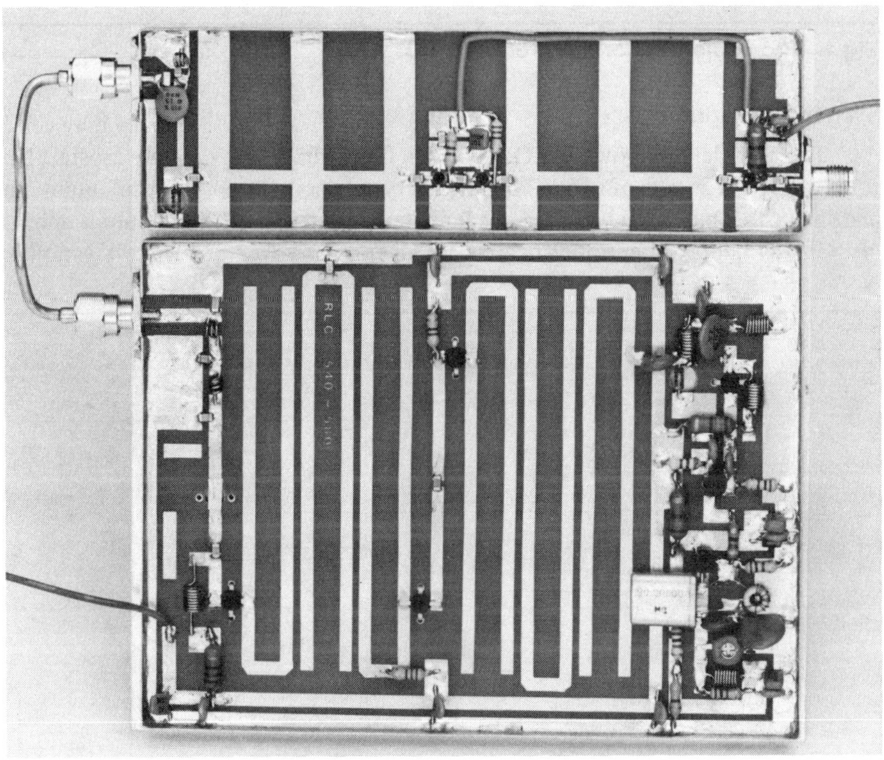

The complete microwave LO is built on two PC boards. The larger (bottom) board provides a signal anywhere from 540 to 580 MHz, depending on the crystal frequency. The smaller board is a ×4 multiplier that provides an output from 2160 to 2320 MHz, depending on input frequency. Used separately or together, these boards have a wide variety of UHF and microwave applications.

All of the critical microwave circuitry in this LO is readily taken care of by a pair of fiberglass-epoxy (G-10) PC boards. The remaining parts include noncritical chip capacitors for interstage coupling and bypassing, standard ¼- and ½-W bias resistors, inexpensive, plastic-cased monolithic-microwave integrated-circuit (MMIC) amplifiers, a pair of 99-cent diodes, a few hand-wound inductors, disc-ceramic capacitors, and a 90-MHz, 5th-overtone crystal oscillator. PC board manufacturing tolerances, component variations, and construction tolerances have all been allowed for in the design. There are no RF tuning adjustments except for the 90-MHz oscillator tank circuit.

Design Goals

This project began with a list of design goals:

1) No tuning adjustments should be required.
2) All frequency-sensitive elements are printed on G-10 board.
3) Use inexpensive, readily available components.
4) Offer sufficient output to drive a standard-level mixer.
5) Use a single 12-V power supply.
6) All spurious outputs are more than 40 dB down.
7) Have electrical, mechanical and thermal stability consistent with portable CW operation on mountaintops in bad weather.

These goals have been met, with one minor exception: The 90-MHz crystal oscillator tank circuit must be tuned to make the oscillator start reliably. This adjustment can be made by listening for the crystal-oscillator output on an FM-broadcast radio. The electrical, mechanical and thermal stability are impressive. One of these LOs was still operating after an airline baggage-handling event left the aluminum transverter case so badly bent that the top had to be removed with a hammer!

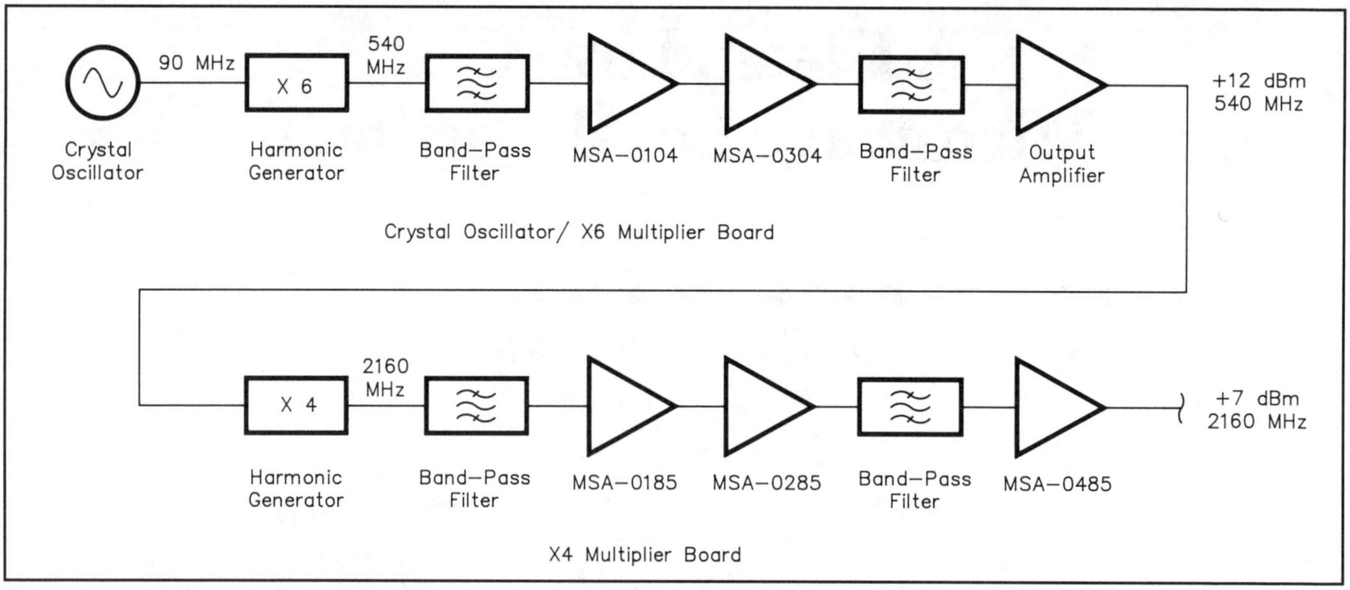

Fig 1—Block diagram of the 2160-MHz LO.

System Description

The complete microwave LO, shown in block-diagram form in Fig 1, consists of two PC boards: a crystal oscillator and times 6 (×6) multiplier board; and a ×4 multiplier board. The crystal oscillator/×6 multiplier board can generate any frequency between 540 and 580 MHz; simply choose the appropriate crystal. The output level depends on the device chosen for the output amplifier. An Avantek MSA-0404 is used for the output amplifier in the version described here. (See this article's Amplifiers section for more details.)

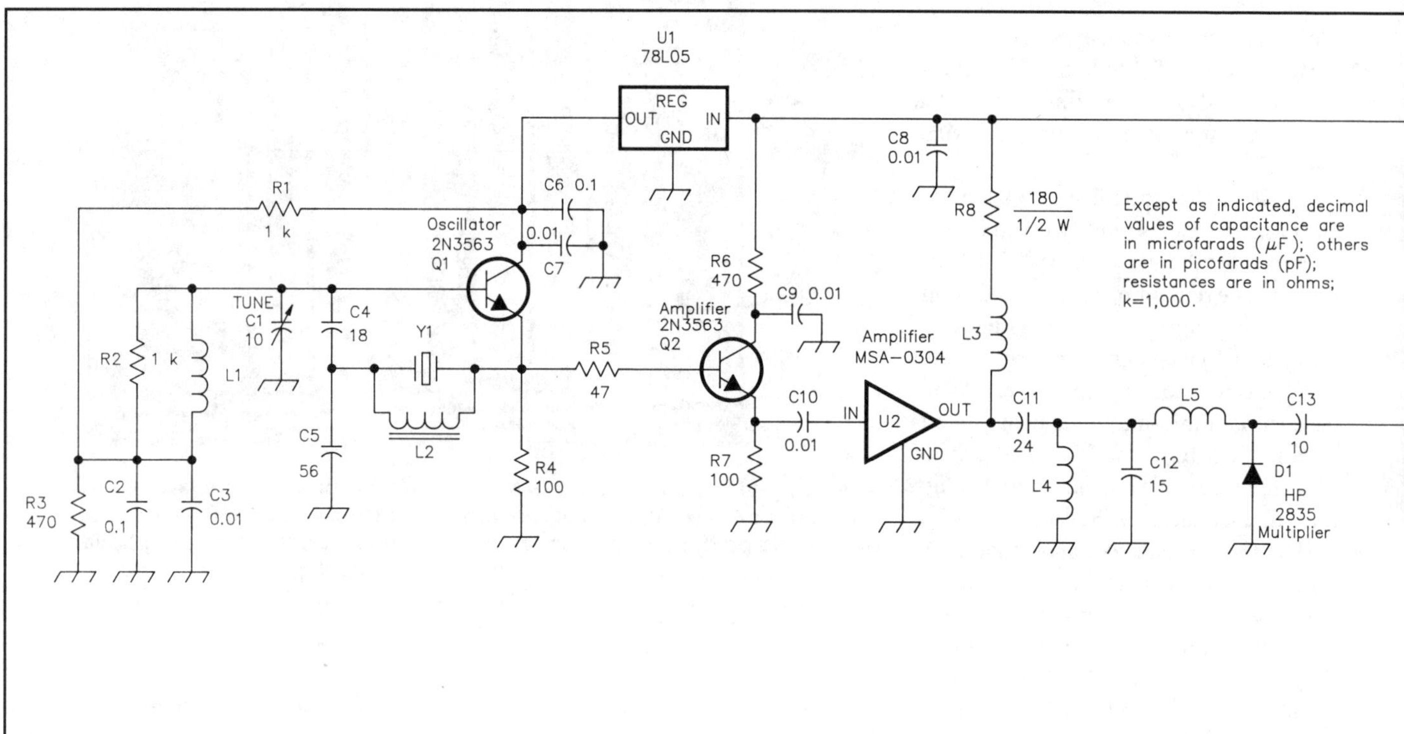

Fig 2—Schematic diagram of the crystal oscillator/×6 multiplier board. Resistors are ¼-W carbon-film types unless otherwise indicated. Capacitors are 50- or 100-V disc-ceramic types unless otherwise noted.

C1—8- to 10-pF trimmer capacitor. Ceramic-piston trimmer preferred; standard ceramic trimmer acceptable.

D1—Schottky diode; Hewlett-Packard 2835, 2800, 2811 or equiv. See text.

J1—SMA female chassis-mount connector preferred. See text.

FL1, FL2—Band-pass filters printed on PC board.

L1, L3, L4, L5, L6—8 turns no. 28 enam wire, 0.1 inch ID, closely wound.

L2—10 turns no. 32 enam wire on T-25-6 toroid core, or 0.33 μH miniature RF choke.

L7, L8—3 turns no. 28 enam wire, 0.0625-inch ID, spaced 1 wire diam.

The ×4 multiplier board can be used for any output frequency between 2140 and 2360 MHz. The harmonic-generator components are sufficiently broadly tuned that the board works equally well as a ×3 or ×5 multiplier. Any input level between +7 and +13 dBm is fine, and inputs as low as 0 dBm may be used, at reduced output levels.

These two boards can be used independently—in fact, they were developed for two separate projects. The 540- to 580-MHz board was developed at the suggestion of Jim Davey, WA8NLC, who wanted a simple 552-MHz driver for his single-board 3456-MHz transverter.[3] The 2140- to 2360-MHz multiplier board was developed as part of a no-tune 2304-MHz transverter that was described in the *Proceedings of Microwave Update '88*.[4]

The local-oscillator system shown in Fig 1 has four functional blocks: the 5th-overtone crystal oscillator; the Schottky-diode harmonic generators; the printed band-pass filters; and the MMIC amplifiers. Each of these blocks is described in the following sections. A schematic of the crystal oscillator/×6 multiplier board is shown in Fig 2, and Fig 3 shows a schematic of the ×4 multiplier board.

Crystal Oscillator

The crystal oscillator generates the signal that is subsequently multiplied into the microwave region. In this design, the 90-MHz crystal-oscillator signal is multiplied by 24 to produce the final output signal (2160 MHz). Any long-term drift or "warblies" on the 90-MHz oscillator will be 24 times worse at the output frequency. Common crystal-oscillator circuits that work well in a 144-MHz, or even 432-MHz, receiving converter may be unacceptable when the output frequency is multiplied into the microwave region.

The Butler emitter-follower circuit shown here was originally suggested to me by Al Ward, WB5LUA, and modified to the present circuit by Jim Davey, WA8NLC. (I don't waste much time arguing with those two—when they express an opinion, they generally turn out to be right.) This oscillator will free-run on the tank-circuit frequency if the crystal and its shunt inductor (L2) are replaced with a 47-Ω resistor. This characteristic is especially useful if you want the tank circuit to operate at another frequency. After initially testing the prototype oscillator with inexpensive 2N5770 transistors, I tried replacing them with some 20-year-old pullout 2N3563s, some MPS3563s, a pair of 2N5179s that I found on the floor under the bench, and some new AT-42085 microwave transistors from Avantek. All of these devices worked in this circuit. I also discovered that the value of R1, which sets the operating points of Q1 and Q2, can be varied to change the power output. A 1-kΩ resistor was fine for all the transistors except the AT-42085s. The output power from the AT-42085s was about

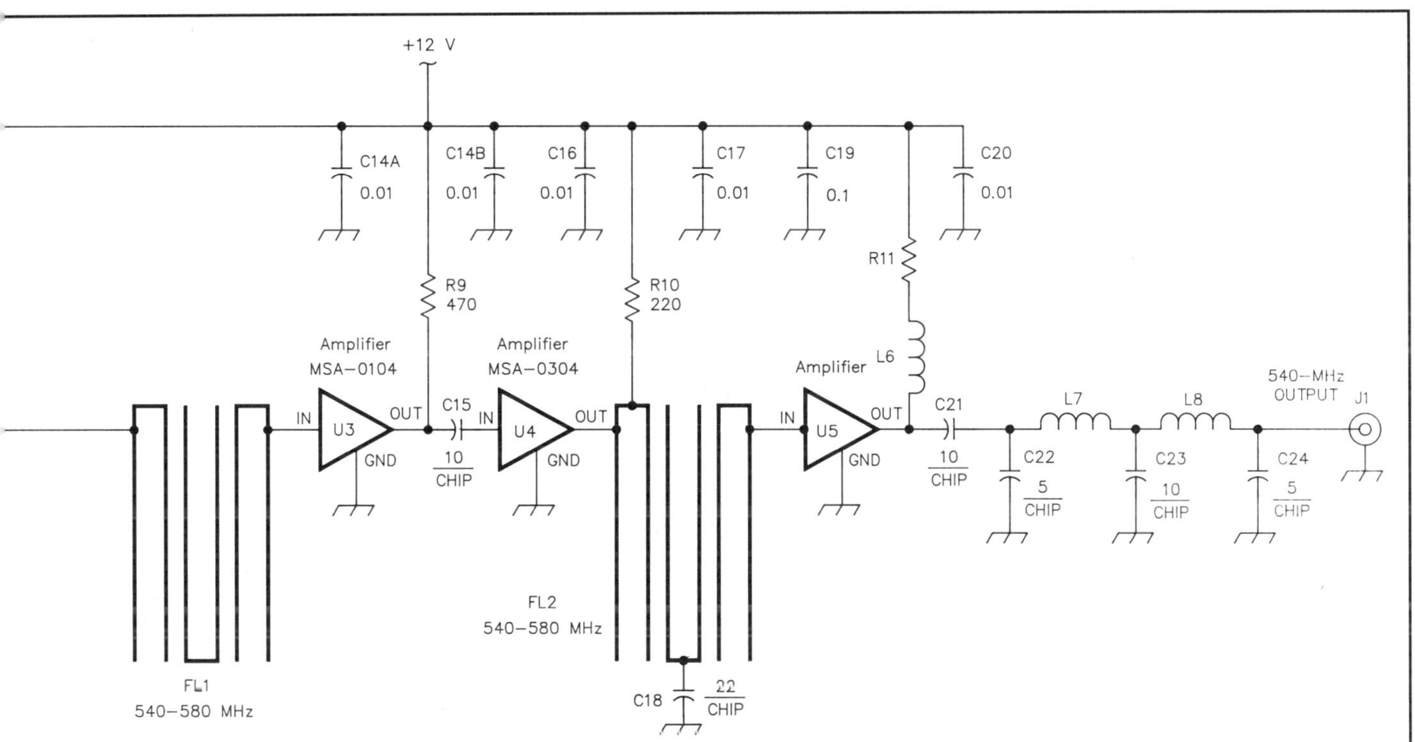

Q1, Q2—2N3563, MPS3563, 2N5179 or equiv. See text.
R11—If U5 is an MSA-0404, use 120-Ω, ½-W resistor. If U5 is an MSA-1104, use a 100-Ω, ½-W resistor. See text.
U1—5-V, 100-mA, 3-terminal regulator.
U2—MSA-0304 MMIC preferred. MSA-0404, MSA-0385, MSA-0485, MAR-3 or MAR-4 also usable. See text.
U3—MSA-0104 MMIC preferred. MSA-0185, MSA-0685, MAR-1 or MAR-6 also usable. See text.
U4—MSA-0304 preferred. MSA-0285, MSA-0385 or MAR-2 also usable. See text.
U5—For +12 dBm out, use MSA-0404. For +16 dBm out, use MSA-1104. See text.
Y1—90-MHz, 5th-overtone, series-resonant crystal.

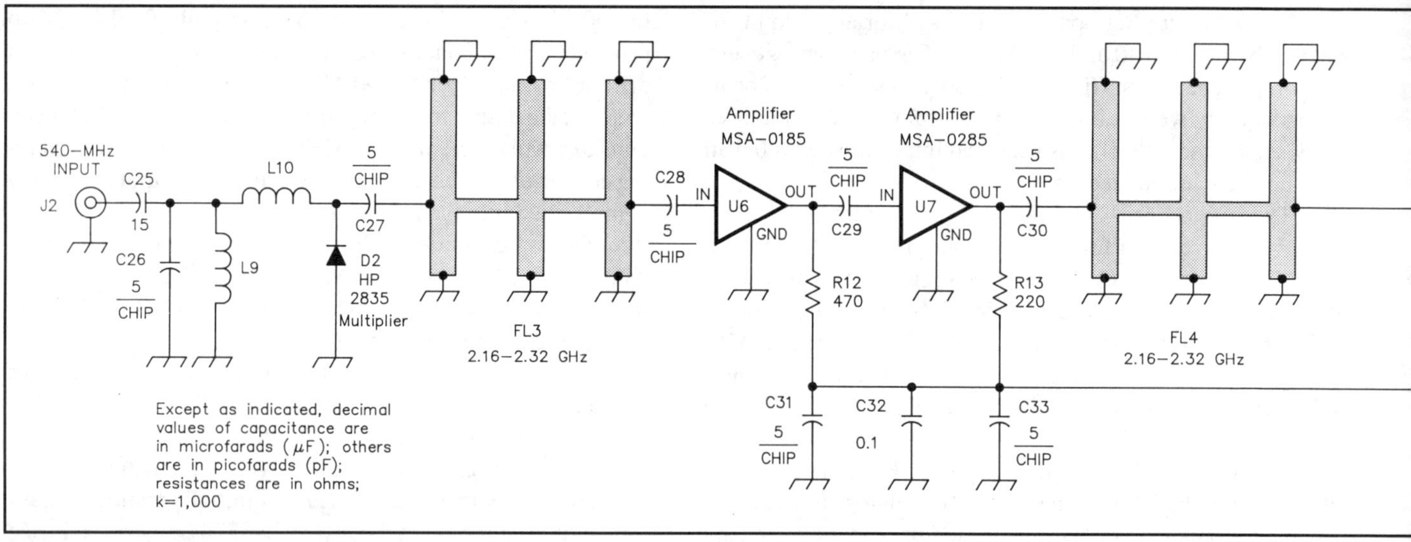

Fig 3—Schematic diagram of the ×4 multiplier board. Resistors are ¼-W carbon-film types unless otherwise indicated. Capacitors are 50- or 100-V disc-ceramic types unless otherwise noted.

D2—Schottky diode; Hewlett-Packard 2835, 2800, 2811 or equiv. See text.
J2, J3—SMA female chassis-mount connector preferred. See text.
FL3, FL4—Band-pass filters printed on PC board.
L9—3 turns no. 28 enam wire, 0.0625-inch ID, spaced 1 wire diam.
L10—Inductor printed on PC board.
U6—MSA-0185 or MAR-1 MMIC preferred. See text.
U7—MSA-0285 or MAR-2 MMIC preferred. See text.
U8—MSA-0485 or MAR-4 MMIC preferred. See text.

+6 dBm—a little too much drive for the MSA-0304 buffer (U2).

The LO shown in the photo on the first page of this article varies slightly from the schematic in Fig 2. The photo shows a Zener-diode regulator in Q1's collector circuit. When this board is used in a setup with a battery supply, the difference in voltage when switching from receive to transmit may be enough to cause an observable frequency shift. This problem is eliminated by using a 3-terminal, 5-V regulator (U1), as shown in Fig 2.

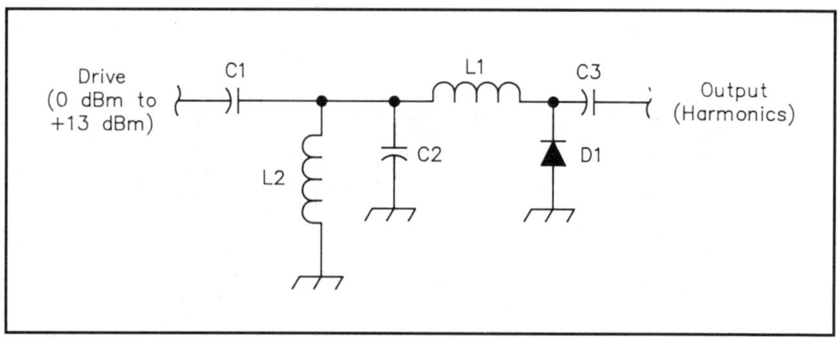

Fig 4—Schematic diagram of the basic harmonic generator used on both boards. See text for discussion.

Harmonic Generator

Harmonic generation is easy—or, at least, not generating harmonics is very difficult. Solid-state power amplifiers must be low-pass filtered to get rid of the harmonics generated by the nonlinear characteristics of the transistors. In fact, if you do anything to a sine wave—clip it, drive a class-C amplifier with it, half-wave rectify it—anything that distorts its perfectly sinusoidal shape, the resultant signal will contain harmonics.

Harmonic generation is difficult only if you want to do it with high efficiency. High efficiency in microwave-LO harmonic-generator stages was important in the 1960s and 70s because amplifying a low-level microwave signal to the +7 dBm level required for many diode-ring mixers was expensive. Back then, tuning up efficient multipliers that used expensive step-recovery or varactor diodes required hours in front of a spectrum analyzer tweaking a handful of $5 piston trimmers to within a half turn of oblivion. And you had to do it all again if the drive level, load or temperature changed.

Now that unconditionally stable, broadband MMIC amplifiers are available for less than a dollar, multiplier efficiency is a minor consideration. By taking advantage of readily available modern components, we can build a broadband multiplier—with no RF-tuning adjustments—that is unconditionally stable with variations in temperature, load and drive level.

The harmonic generator shown in Fig 4 is just a half-wave rectifier with a simple low-pass filter (L1) feeding in the fundamental, and a simple high-pass filter (C3) picking off the harmonics. A half-wave rectifier based on an ideal diode generates only odd harmonics. A Schottky diode (D1 of Fig 4) has an offset voltage of a few hundred millivolts, so the conduction angle is less than 180 degrees. Odd and even harmonic levels are approximately equal for drive levels up to about +10 dBm. This basic harmonic generator is used on both boards. For higher drive levels, a bias circuit with

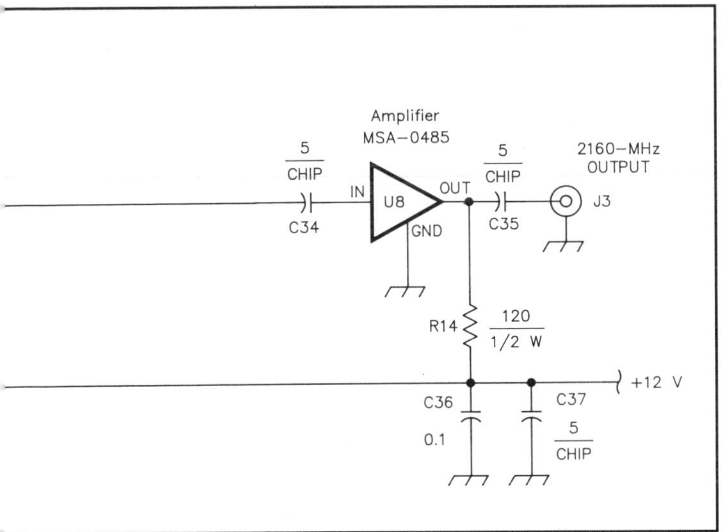

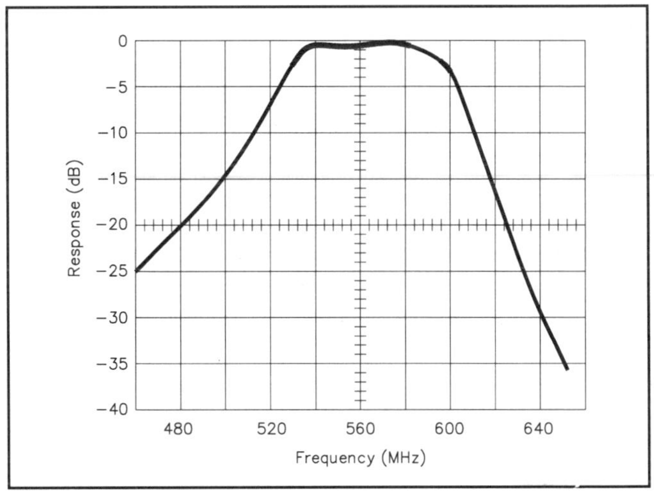

Fig 5—Frequency response of the 540- to 580-MHz hairpin filters printed on the crystal oscillator/×6 multiplier board.

a trimmer potentiometer is recommended.[5]

Other diodes will work in this circuit. Schottky diodes like the HP-2800 and HP-2811 are suitable. A silicon switching diode like the 1N4148 works fine, and even provides more output than the specified Schottky diode, but it has one major disadvantage: It may oscillate! If you don't believe this, compare the circuit of Fig 4 with that of a parametric amplifier circuit in an early VHF manual. Better yet, connect the input of a 1N4148 multiplier to a signal generator and the output to a spectrum analyzer and try it. At some combination of input frequency and drive level, the output noise floor will rise considerably, and the output spectrum will contain many discrete output signals—typically, subharmonics of the drive signal modulating the desired output. This circuit needs to operate reliably from a motorcycle battery on a mountaintop in a rainstorm, so the use of switching diodes is discouraged.

Filters

The filter selects the desired harmonic output from the harmonic generator. In the past, amateur-built frequency multipliers usually were limited to multiplication factors of 2 or 3, because of the difficulty of tuning to the correct harmonic. With fixed-tuned filters having steep skirts and flat tops, it is easy to build multipliers of much higher order. The theory behind hairpin filters (FL1 and FL2 in the 540- to 580-MHz board) and off-center-tapped half-wave filters (FL3 and FL4 in the ×4 multiplier board) is covered in the amateur and professional literature.[6-9] Only the practical aspects are mentioned here.

The primary design goals for these filters were low cost and reproducibility without tuning adjustments. To achieve the first goal, I've specified G-10 board. As a result, the filters are more lossy than equivalent designs on a more expensive substrate, such as Teflon fiberglass. The loss for each of the three-element sections used here is about 3 dB. Because a 10-dB-gain MMIC capable of compensating for this loss costs only about $1, and because better substrates may cost $100 a square foot, G-10 is an attractive trade-off.

To achieve reproducibility without tuning adjustments, the filters are made broadband (with lots of low-Q resonators), rather than narrowband (with only a few high-Q resonators). The resulting passband characteristic, shown in Fig 5, looks more like that of an SSB filter in an HF rig than something out of an LO chain! This filter characteristic is fundamentally different from a single-tuned circuit that must be tuned exactly on frequency—a signal anywhere in the 40-MHz-wide passband (540 to 580 MHz) passes through, but the undesired 5th and 7th harmonics 90 MHz above and below the passband are greatly attenuated by the steep filter skirts.

There are two major advantages to using flat-passband filters in a crystal-controlled LO. One is that the frequency may be moved anywhere in the passband by simply changing the crystal. The other is that allowance for manufacturing tolerances and variations in circuit-board material can be designed in before the circuit is built, rather than having to be tuned out afterward. For example, if the PC-board manufacturer is a little sloppy and the production boards are 1% smaller than the prototype, the passband will be 5 MHz higher. The desired signal will still get through, and the undesired signals will still be attenuated.

Even a change in circuit-board material from G-10 to its fire-retardant variant, FR-4 (which has a slightly different dielectric constant), results only in a well-behaved upward shift in the passband. The FR-4 boards provide a few decibels more output at 576 MHz, and the G-10 boards are a few decibels better at 540 MHz. The passband shape is the same for both G-10 and FR-4 materials.

In an attempt to discover how tolerant the 540- to 580-MHz hairpin filters are, I reduced the length of several hairpin resonators by one millimeter. The output dropped about 2 dB, worst case, and the spurious outputs remained acceptably low. I also sprayed a complete 2160-MHz LO with a thick coat of clear Krylon, and could not detect any change in the module's output. These filters work better than double-tuned circuits; require no tuning; are tolerant of manufacturing, device and construction variations; allow a range of frequencies to be generated from a single board layout; and cost no more than the PC-board material required to make them.

Amplifiers

The MMIC amplifiers serve two important functions: They raise the level of the desired harmonic, and they provide broadband, 50-Ω interstage terminations. Many suitable amplifiers are available, and substitutions require only a little care concerning output level. The MMICs most popular with hams are available from Avantek and Mini-Circuits Labs. The Avantek line has part numbers starting with MSA-; the MSA-0404 is an example. Mini-Circuits parts numbers start with MAR-; the MAR-4 is an example. The devices used here were chosen because they were closest to the top of the pile of parts on my workbench. Drive to D1 should be about +10 dBm—an MSA-0304 or '0404 works well for U2. The desired harmonic coming out of FL1 is at about –17 dBm, so a device with a 50-Ω input and more than 10 dB of gain is useful for U3. The MSA-0104, MSA-0185, MSA-0685 and MAR-6 are good choices.

The second interstage amplifier, U4, must deliver about +10 dBm and provide a 50-Q load for FL2. MSA-0304, MSA-0285 and MAR-3 devices have all been used in this stage. The final stage, U5, may be omitted if +6 dBm is enough output. If U5 is not used, also omit R11, L6, C20 and C21. A good plan, however, is to use a high-power device that will not saturate, and then add a resistive attenuator to reduce the output to the desired level. An MSA-0404 will provide +12 dBm output, and an MSA-1104 works well for +16 dBm out.

It's not really necessary to operate any of these amplifiers linearly, but the output stage should be kept at or below its 1-dB compression point, or the spurious outputs will rise to an unacceptable level. MMIC selection is left up to the builder. Just about anything will work—and who knows what will be available next year for 49 cents? If you'd rather not think about which parts to use, just use those specified in the schematics. If you use devices different from those specified, be sure to use an appropriate bias resistor, as specified in the manufacturers' data sheets.

Construction

Both boards are etched on 0.062-inch-thick, double-clad, G-10 PC board material. All components are surface mounted on the etched side; the other side is unetched and acts as a ground plane. The components necessary to complete this project are available from several suppliers, and etched PC boards and complete parts kits are available as well.[10]

The first construction step for the crystal oscillator/×6 multiplier board shown in Fig 6 is to drill the component holes and file the edges flat. The filter elements need not be grounded, so the exact size of the board is not critical. The edges of the upper and lower ground foils must be connected all the way around the board, though. A plated-through hole or shorting wire every half inch or so works fine. Copper tape wrapped the whole length of each edge and soldered top and bottom also works well. The method I recommend is to solder brass or tin walls all the way around the circuit board, making sure to solder to both the top and bottom of the circuit board. This results in a nice, rigid box with solderable shield walls suitable for mounting feedthrough capacitors and the output connector.

The next step is to add the bridges across the unused breaks on the crystal oscillator/×6 multiplier circuit board (between C21 and C22), and a piece of copper tape to short the cold end of multiplier diode D1 to the ground plane. The MMIC ground leads are bent at right angles to the device body, passed through holes drilled in the board, and soldered directly to the

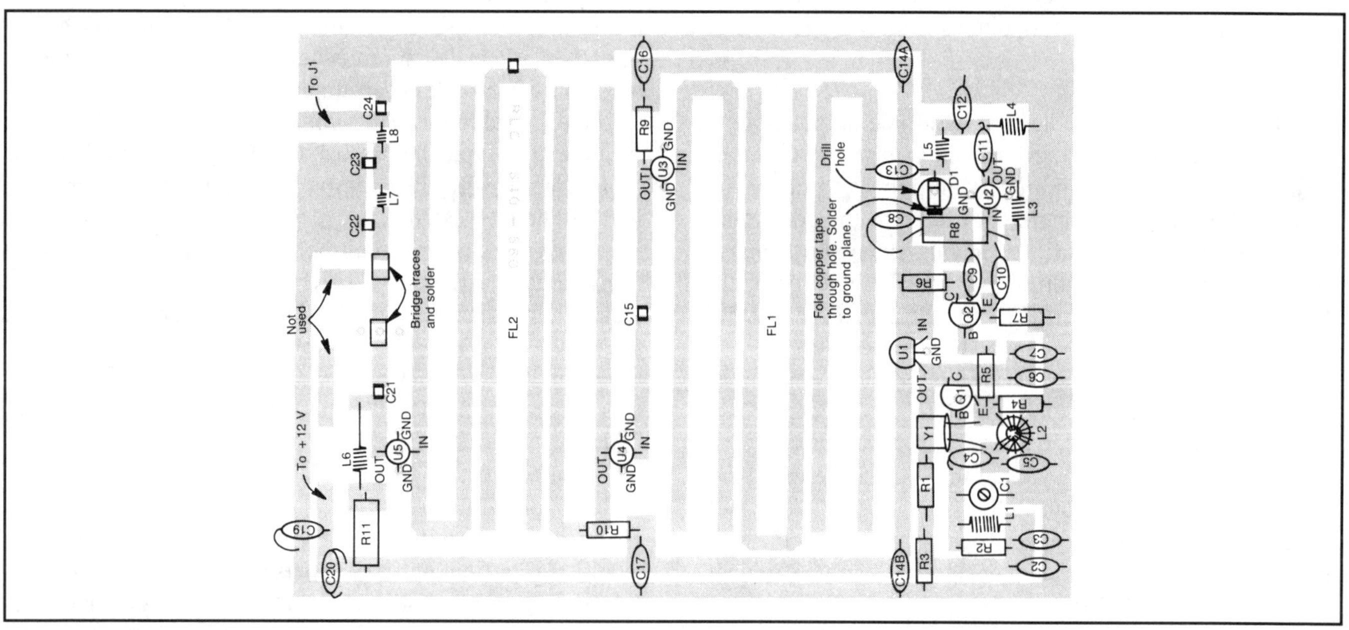

Fig 6—Part-placement diagram for the crystal oscillator/×6 multiplier board (not shown actual size). All components mount on the etched side of the board.

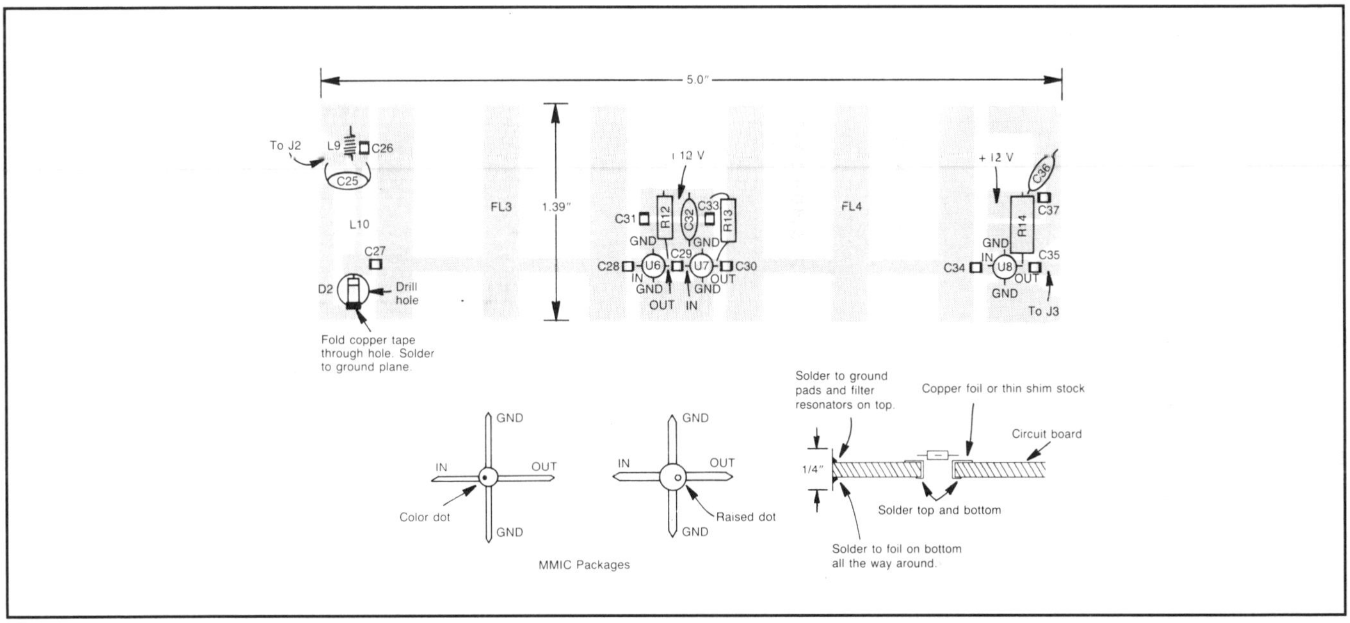

Fig 7—Part-placement diagram for the ×4 multiplier board (not shown actual size). All components mount on the etched side of the board.

ground plane. The 5-V regulator's ground lead also goes through the board and is soldered to the ground plane. I usually put the chip capacitors on first, followed by the inductors, resistors, disc-ceramic capacitors, diode, transistors, 5-V regulator IC and MMICs, in that order. The crystal can be installed as shown on the parts-placement diagram by sticking it to the board with a small piece of double-sided foam tape.

I prefer to use SMA end-launch connectors, even at the output of the crystal oscillator/×6 multiplier board,

because they are small and easy to use. (I also use an SMA connector at the output of the ×4 multiplier board because it's an excellent microwave connector. By standardizing on connectors, I don't have to keep switching adapters on my power meter.)

After you've carefully checked all your mounted components against the parts-placement diagram, you can apply 12 V to the board. Tune C1 until you hear the 90-MHz signal in a nearby FM broadcast radio. Then turn the power supply on and off a few times to make sure the oscillator starts reliably. The crystal oscillator/×6 multiplier board is now complete.

The filter topology on the ×4 multiplier board (Fig 7) differs from that of the ungrounded hairpins on the crystal oscillator/×6 multiplier board. The width of the ×4 multiplier board determines the resonant frequency of the shorted half-wave filter elements. The correct length for all the half-wave resonators is obtained by cutting the circuit board precisely to the width shown in the drawing, and then soldering the board's brass wall to the ground plane on the bottom, and to the end of each resonator on the top of the circuit board. Plated-through holes, ground wires and copper tape wrapped around the board edges will not work with this layout.

I obtain the correct board dimensions by scribing a line on the top of the circuit board at exactly the correct place. Then I cut the board slightly oversize. Next, I lay a large, flat file on my workbench and work the circuit board back and forth until the board edge is filed to the scribed line. This results in a nice square edge as shown in Fig 7. Only the width of the ×4 multiplier board is critical. Because FL3 and FL4 form a bandpass filter with a flat passband response, construction errors of up to about 0.032 inch do not significantly affect the output level at 2160 MHz.

After soldering the side and end walls to the ×4 multiplier board, add copper tape to ground the MMICs and multiplier diode as shown in Fig 7. Then add the chip capacitors, disc-ceramic capacitors, inductor, diode, MMICs, bias resistors and SMA connectors. No adjustments need be made to the ×4 multiplier board

Performance

Fig 8 shows the output spectrum of the crystal oscillator/×6 multiplier on G-10 board with a 90-MHz crystal and an MSA- 0404 output device. The plot is from dc to 1 GHz. The largest spur, at 450 MHz, is 70 dB below the +12-dBm, 540-MHz output. The 360-MHz and 630-MHz spurs are just barely visible at about 75 dB below the 540-MHz output. The harmonics at 1.08 GHz and 1.62 GHz (not shown) are more than 55 dB below the 540-MHz output, and are not measurable because of the limited dynamic range of this spectrum analyzer. This is a clean LO!

Different crystal frequencies result in different spurious-output levels. Worst-case spurious outputs are about –45 dB for any output between 540 and 580 MHz.

Once the board is built and tested, frequency stability can be enhanced considerably by thermally insulating the crystal oscillator. I usually tape a small piece of sponge packing foam over the oscillator and then package the entire system in a box to keep rain and cold mountain breezes out.

Fig 9 shows the output spectrum of the ×4 multiplier board

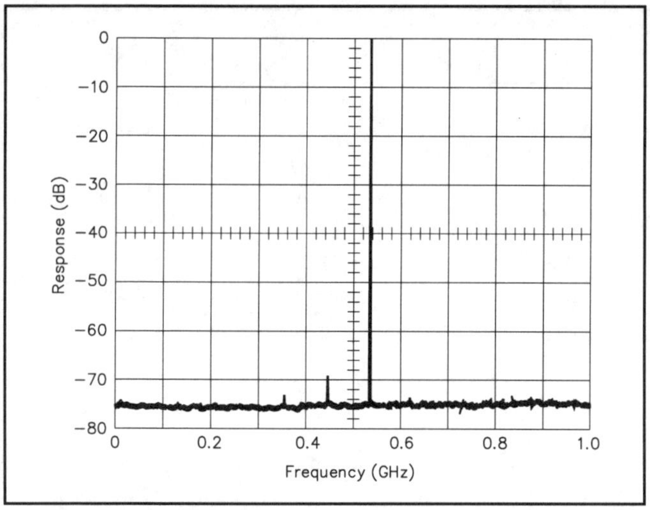

Fig 8—Output spectrum of the crystal oscillator/×6 multiplier board. The desired signal is +12 dBm at 540 MHz.

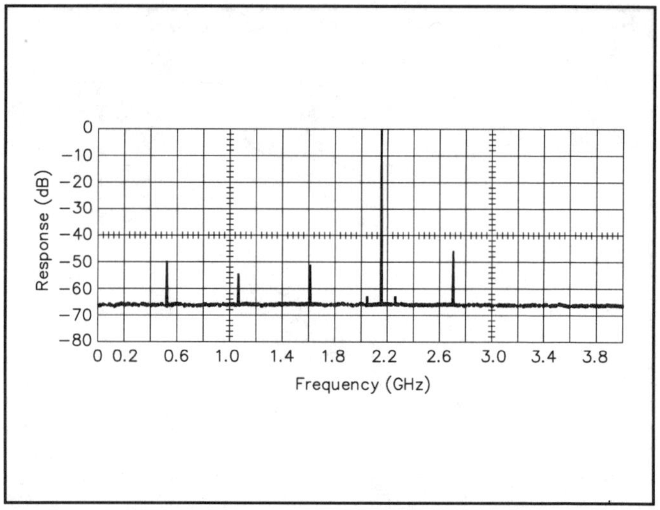

Fig 9—Output spectrum of the ×4 multiplier board. The desired signal is +8 dBm at 2160 MHz.

driven with the signal shown in Fig 8. The plot covers dc to 3.7 GHz. The largest spur, at 2.70 GHz, is 45 dB below the +8-dBm, 2.16-GHz output. The lower frequency spurs, at 0.54, 1.08 and 1.62 GHz, are more than 50 dB down. The second harmonic, at 4.320 GHz (not shown), is only about 25 dB down. However, harmonic spurs are not too important on LO outputs, since the mixer generates harmonics of the LO signal anyway. If the output of the ×4 multiplier board is used to drive an antenna or another multiplier stage, then a filter (as described in reference 5) may be added.

A word about LO drive level is in order here. Many engineers, both amateur and mercenary, agonize because they have a +7-dBm mixer and only +6 dBm of LO drive. It's true that 1-mW (0 dBm) is probably not enough drive for a 5-mW (+7-dBm) mixer, and that 100 mW (+20 dBm) is too much—but there is some latitude. A few decibels either way won't make any measurable difference in most systems. If you really don't want to get on the air, having only +5-dBm drive for your +7-dBm mixer is as good an excuse as any. But the guy down the street with 10 countries worked via moonbounce is probably running +3 dBm into the RF port of an unknown surplus mixer.

Applications

These two boards have been used for a surprising number of applications. I use the 2160-MHz LO described here in a pair of 2304-MHz transverters with 144-MHz IFs, and in a 3456-MHz transverter with a 1296-MHz IF. Simply change the crystal to 96 MHz, and you have a low-power CW transmitter for 2304 MHz. A 94-MHz crystal will provide a 2256-MHz LO for OSCAR Mode S. The ×4 multiplier board may be used with a suitably modified 70-cm FM exciter to generate 2304-MHz FM (460.8 × 5 = 2304 MHz).

The crystal oscillator/×6 multiplier board is even more versatile. With a 96-MHz crystal providing 576-MHz output, it serves as the LO for my single-board 1296-MHz transverter. It may be used with a ×4 multiplier for 2304-MHz CW output, a ×6 multiplier for 3456-MHz output, a ×10 multiplier for 5760-MHz output or a ×18 multiplier for 10.368-GHz output. If an untuned Schottky diode multiplier is used, it will provide useful signal levels for receiver and filter alignment on the calling frequencies of all the amateur bands from 2.3 through 10 GHz. With 92-MHz crystal and an MSA-1104 providing +16 dBm at 552 MHz, it can serve as the LO board for the WA8NLC single-board 3456-MHz transverter described in June 1989 *QST*.

Acknowledgments

This work would not have been possible without the forum for amateur microwave technology exchange provided by Don Hilliard, WØPW, in the form of the Microwave Update Conferences. The speakers and attendees at Microwave Updates '85 through '88 provided much of the basic information and all of my motivation for pursuing this work. In particular, I thank Al Ward, WB5LUA, for the wealth of information on MMICs he has made available to the amateur community, and Jim Davey, WA8NLC, for bringing us hairpin filters and encouraging me to keep pushing the state of the art.

Notes

[1]A. Ward, "Simple Low-Noise Microwave Preamplifiers," *QST*, May 1989, pp 31-36.

[2]R. Campbell, "2.3 GHz Transverters," *Proceedings of Microwave Update '88* (Newington: ARRL, 1988), pp 9-32.

[3]J. Davey, "A No-Tune Transverter for 3456 MHz," *QST*, Jun 1989, pp 21-26. *Proceedings of the 21st Conference of the Central States VHF Society* (Newington: ARRL, 1987), pp 51-57. (Reprinted in this book.)

[4]See note 2.

[5]R. Campbell, "A Clean Microwave Local Oscillator," published in *Proceedings of the 1296 and 2304 Conference*, Estes Park, CO, Sep 1985. Reprinted in *Proceedings of the 21st Conference of the Central States VHF Society*. Newington: ARRL, 1987), pp 51-57.

[6]J. Wong, "Microstrip Tapped-Line Filter Design," *IEEE Transactions on Microwave Theory and Techniques*, Vol MTT-27, No. 1, Jan 1979, pp 45-51.

[7]G. Beebe, "Analysis of a Class of Microstrip Bandpass Filters," MSEE thesis, Michigan Technological University, Feb 1988.

[8]J. Davey, "Microstrip Bandpass Filters," *Proceedings of Microwave Update '87* (Newington: ARRL, 1987), pp 42-53.

[9]J. Davey, "Microwave Filter Update," *Proceedings of Microwave Update '88*, pp 1-8. See note 2.

[10]Most of the parts for this project are available from Microwave Components of Michigan. Etched PC boards and parts kits are available from Down East Microwave.

Phase Lock Control Circuit For Use With Brick Type Oscillators

By C. L. Houghton, WB6IGP and Kerry Bane, N6IZW

This project was started to overcome the inherent stability problems in the California Microwave and Frequency West "brick" type oscillators. We have found that the basic brick oscillator's crystal (operates in the 100-MHz range) can vary 1 kHz or more (short term stability). This system if left unchecked can vary in output frequency up to 40 kHz or more (at 10 GHz) when left on for long-term operation. This drift is due to the brick's internal 100-MHz oscillator temperature and aging characteristics.

Several modifications to the brick crystal oscillator have improved the basic stability of this high-frequency control oscillator. These include: (1) Better temperature control circuitry for the crystal housing, and (2) Changing some of the capacitors in the oscillator to temperature coefficient types, which improved overall stability. This improved system stability, but the improvements still missed the mark. Something several magnitudes better was sorely needed.

We designed the phase-lock control circuit to improve upon existing stability problems looking for a high state of accuracy. What we developed was a circuit that controls the brick's crystal oscillator, by phase locking this crystal to a high accuracy external reference source (Fig 1). This improved the basic brick oscillator to what we feel is the ultimate, with very simple circuitry. Part of the design goal was to minimize modifications to the brick oscillator. The phase-lock circuitry would be external to the brick, and would control the 100-MHz oscillator by comparing it to a very high-accuracy standard. In practice, we were able to stabilize the crystal oscillator to about 100 Hz at X band.

Kerry, N6IZW, did the initial work and came up with the design while testing two different prototypes. The first test circuit used two separate high stability low frequency oscillators to divide the high frequency brick oscillator control crystal (a fractional value) down to a common PLL frequency. This system worked well, but with two different oscillators the circuit was cumbersome and might be difficult for others to duplicate. After quite a bit of number crunching on a computer Kerry discovered that a single 10-MHz standard could be used in the system eliminating the second oscillator.

The original design was modified adding an additional 74HC163 programmable divider and providing pinouts for any other possible divide by ratio to be programmed in. Our system used a single 10-MHz high stability oscillator as the system reference, although other frequency combinations are possible. We have included a program that was used in our system allowing other oscillator combinations to be compared to your brick oscillator frequency. First a little about our system.

The 10-GHz frequency that we have standardized on for SSB work is 10.368 GHz. The local oscillator system uses a 10.223 GHz (Brick Frequency) for a lower side mix incorporating our Frequency West Brick oscillators. The fractional crystal frequency for a Frequency West brick operating at 10.223 GHz is 100.2254902 MHz. This fractional frequency is quite a challenge to phase lock but I believe you will like the scheme used to do the job.

The single oscillator approach mixed the tenth harmonic of the 10-MHz oscillator with the 100.225-MHz signal from the brick oscillator leaving a difference of 0.2254902 MHz. The computer program (attached) will give the relationship between this number and 10 MHz. In this case it was found that dividing 10 MHz by 1020 and 0.2254902 MHz by 23 produces a very close match to 0.01 Hz. Note other combinations are possible but this was the simplest choice. This new frequency (9.803913043 kHz) primary reference is fed to a 4046 phase detector along with the crystals divided down input. The output of the 4046 PLL controls a varactor located in the brick's crystal adjust circuitry trimming the crystal to exactly the proper frequency.

System error using this system of phase-lock control results in errors of less than 1/100th of a cycle at 100 MHz, (actually it's 0.008526 Hz.). Using the system, the 100.2254902-MHz brick oscillator will lock up in about 2 minutes after applying power up the circuit and brick. The circuit that was finalized is quite simple in keeping with the original design goals. The harmonic mixer is nothing more than a 2N918 with its collector circuit tuned to the low frequency product (0.2254902 MHz) in this example. The output of the mixer is coupled into a VN10KM VMOS FET, amplified and squared up in two sections of a 74HC04.

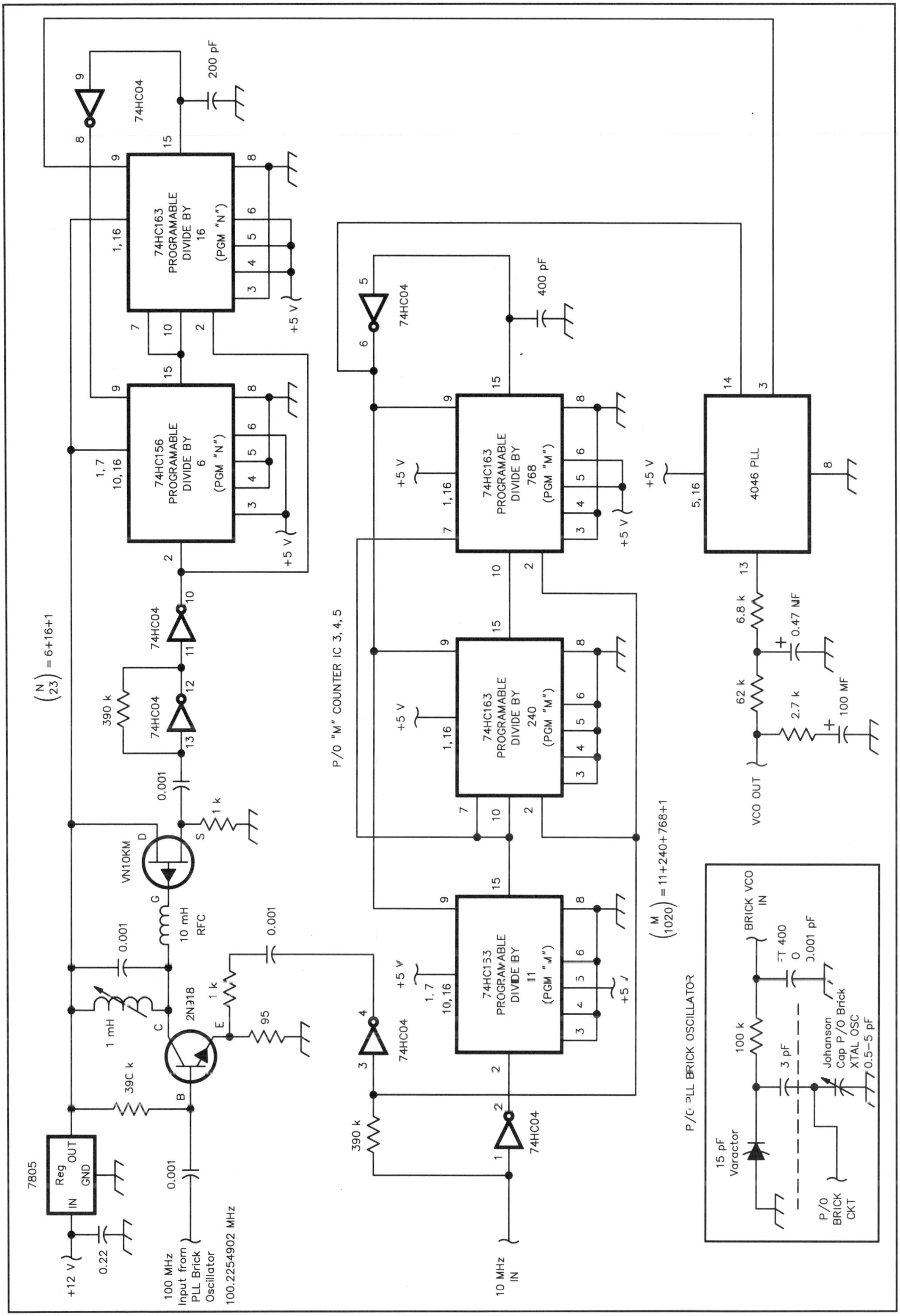

Fig 1—10-GHz phase-locked "brick" oscillator stability circuit designed by N6IZW and WB6IGP.

This signal is then coupled into a programmable divide by a string of 74HC163s. Note that all programming pins are brought out, allowing custom programming for different division ratios. The 10-MHz oscillator input is squared up in a 74HC04 inverter and feeds two signal paths. One is the harmonic mixer and the other is the programmable M divider for the master reference. The 4046 PLL compares the output of the M and N dividers and adjusts the control voltage for the varactor in the brick. This varactor is in series with a small 3-pF capacitor which is tied in parallel to the existing Johanson capacitor in the Frequency West brick.

The test results of this system, using identical phase-locked circuits, at either end of the test path, were excellent. From a cold start, as soon as the high-stability main-reference oscillators were stabilized, we attempted communications on 10-GHz SSB. We selected 14.01 MHz for both SSB IFs. The RIT control was the only adjustment needed to clarify SSB operations at both ends of the test path, as both signals were in each other's passband on 2 meters. The error was about 100 Hz.

An off-shoot modification to the basic high-stability circuit is just the opposite but can be very useful. Our 2-meter radios are SSB hand-helds. Frequency setting is done by small BCD switches. Searching for other stations proved cumbersome. The solution to the problem was to disable the high-stability circuitry. In its place, we installed a pot to vary the voltage on the brick's varactor. This gave us about 100 kHz of frequency tuning (at 10 GHz). Fine control was obtained by using a 10-turn pot. This gave us a method to search and scan quite a bit of frequency in short order, instead of flicking BCD switches. Once you verify what frequency the other station is using, you can return to high stability and let the other station adjust to your frequency.

Perspective

The use of high-stability oscillators brings greater opportunities for weak-signal work. If you know where a signal is without question, you can start some interesting applications. One such possibility is to copy a repetitive signal being transmitted over a very long path that normally is not received. With special techniques, it is possible to reach under the noise floor and pull these signals out. Very high-stability oscillators and computers are necessary to further these techniques.

Brick Oscillator Number Generation Program

The program listed in Table 1 generates the proper division ratio for other combinations of brick oscillator frequencies and reference frequency input. Additionally, the program will define the M and N counter division numbers and give the data bits 0 or 1 to be preset into each counter. Program step 190 is set at 1 to start division comparisons. This number can be raised if you find that higher order numbers are needed. In our case we set up to 500 or so for most work with a 10-MHz input reference in step 190.

We hope you will adapt this circuit to your phase-locked oscillators, as the improvement in stability is quite noticeable. After all, 100-Hz stability at 10 GHz with simple circuitry is something to brag about!

Table 1

BASIC Program Listing

```
10  DEFDBL A-Z
20  DEFINT I
30  DIM P(13)
40  REM THIS PROGRAM CALCULATES THE DIVIDE
     BY "M" AND "N" COUNTER
50  INPUT "ENTER BRICK FREQUENCY",BF
60  INPUT "ENTER REFERENCE OSCILLATOR
     FREQUENCY",RF
70  REM RH IS HARMONIC NUMBER BEING COM-
     PARED
80  REM HD IS DIFFERENCE BETWEEN RH AND BF
90  REM M IS THE NUMBER BY WHICH REFERENCE IS
     DIVIDED
100 REM N IS THE NUMBER BY WHICH DIFFERENCE
     IS DIVIDED
110 REM S & U SET PRECISION OF DIVISION MATCH
120 RH=CINT(BF/RF)
130 PRINT "BRICK OSC FREQ= ";BF:"MHZ"
140 PRINT "REFERENCE OSC FREQ= ";RF;"MHZ"
150 HD=ABS(BF-RF*RH)
160 PRINT "DIFF FREQ IS ";HD;"MHZ"
170 IF RF*RH>BF THEN PRINT "PHASE SEN REV"
180 PRINT " M ";" N"; " CALC DIFF FREQ IN MHZ"
190 M=1
200 N=M*ABS(HD)/RF
210 S=SBS(RF*INT(N)/M-HD)
220 IF S<.00001 THEN PRINT M; INT(N);RF*INT(N)/M
230 IF S<.00001 THEN 300
240 U=ABS(RF*INT(N+1)/M-HD
250 IF U<.00001 THEN PRINT M;INT(N+1);RF*INT(N+1)/
     M
260 IF U<.00001 THEN GOTO 295
270 M=M+1
280 GOTO 200
290 REM THIS CALC M CNTR PRESET FOR M=1
295 N=INT(N)+1
300 D=2048
301 PRINT
305 M=M-1
320 FOR I=12 TO 1 STEP -1
330 IF M<D GOTO 360
340 M=M-D
350 P(I)=0 GOTO 370
360 P(I)=1
370 D=D/2
380 NEXT I
385 PRINT "M CNTR PRESET LSB ";
390 FOR I= 1 TO 12
400 PRINT P(I);
410 NEXT I
480 PRINT
485 REM THIS CALC N CNTR PRESETS FOR N-1 CNT
490 D=128
495 N=INT(N)-1
500 FOR I=8 TO 1 STEP -1
510 IF N<D THEN GOTO 540
520 N=N-D
530 P(I)=0 : GOTO 570
540 P(I)=1
570 D=D/2
580 NEXT I
```

Frequency Multipliers Using Silicon MMICs

By Jim Davey, WA8NLC

Introduction

Two recent articles describe the use of MMICs as frequency doublers.[1,2] The advantages cited are high efficiency and unconditional stability. MMICs can also be used for higher order multiplication. In low-power applications, the MMIC multiplier offers several advantages over Schottky-diode multipliers including lower drive power and less current for the total oscillator system.

Background

My work with no-tune transverters for 2304 and 3456 MHz[3] required the use of ×4 and ×6 multiplication on the board. This was done by driving a Schottky diode with a high-level source to generate harmonics. Filters and MMIC amplifiers were added to select the desired harmonic and amplify the output to a usable level. The success of MMIC doublers cited in the references intrigued me. How would a MMIC multiplier perform in a ×4 or ×6 stage? Would the MMIC offer any advantage over the Schottky diode?

This paper describes a series of experiments I performed with silicon MMICs to see what could be done in a ×4 and higher mode. The experiments were performed on local oscillators for 2160 and 3312 MHz. The results show that MMICs work well at multiplication orders up through at least ×6, with less drive power than the Schottky diode. Efficiency was enhanced by selecting an optimum MMIC-to-filter spacing.

Test Methodology

Experimental data was obtained by following this general procedure:

1. Measured the relative strengths of the harmonics in the output of the MMIC.

2. Varied the drive level to see over what range of input power the MMIC delivered the highest harmonic levels.

3. Added an output filter to the MMIC to select one of the higher harmonics. Checked for instabilities resulting from the VSWR of the filter.

4. Using one harmonic (6th), varied the device bias current to see what correlation it had to the level of the 6th harmonic.

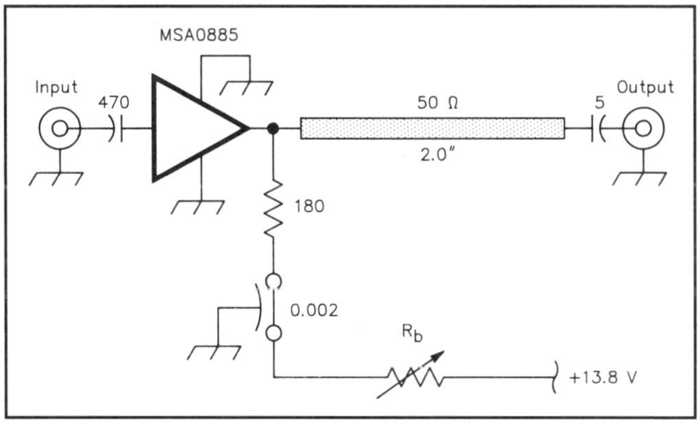

Fig 1—Schematic of the 540-MHz test fixture.

5. Varied the electrical distance between the MMIC and the filter to see if the "filter gain" phenomenon also worked for active devices. Recorded output level data for several values of bias at each spacing.

Test Details

I chose to limit the test to multiples of 540 MHz, since I had an extra 540 MHz no-tune oscillator available that could deliver +14 dBm. For the active device, I used an Avantek MSA 0885 (Fig 1). The gain of the 0885 is 26 dB at 540 MHz with 1-dB compression at an output level of +14 dBm. Saturation should begin at an input of –12 dBm.

The MMIC was mounted on a small 50-Ω test fixture made of 0.031-inch TFE-glass board, to allow bias and dc isolation capacitors to be added. SMA connectors were edge-soldered to the microstrip for input and output connections. The output was measured on an HP 8551B spectrum analyzer and HP 431B power meter.

In the first tests I looked at the effects of drive and bias on the 4th and higher harmonics. Initially, a wide range of input levels was tried to see what input level the MMIC preferred for best harmonic output and spectral purity. At the same time, the device bias was varied to test the effect of bias current at each level of input power. In general, for each combination of input power and harmonic number, I found a unique bias condition that gave the best output. The bias used in this test were lower than the specification value to prevent excessive current in the MSA 0885 when driven at +10 dBm.

The practical upper limit for input drive appears to be about +10 dBm. Above that point no additional output power was observed with any level of bias. Overall efficiency dropped as drive levels approached +10 dBm, too. Nearly as much output was available at lower drive levels if the bias was adjusted to optimize the output level. For this reason I decided to conduct the remaining experiments at a lower input level of 0 dBm. This level is about 12 dB higher than the level required to cause +l dB compression of the fundamental in the MMIC.

Next, I filtered the output and tried to optimize the 4th and 6th harmonic. Microstrip bandpass filters were attached to the test fixture and the drive and bias were varied while watching for greatest output. The optimum drive and bias conditions found in these tests did not correlate well with the unfiltered test data. The reasons for this were apparent after the next test was performed.

"Filter Gain" Test

A Hewlett-Packard application note[4] describes some tests HP made on a step-recovery diode multiplier/filter setup. It showed that each harmonic can be peaked by locating the diode at specific electrical distances from the reflective-type filter. At these locations the reflected energies in the other harmonics are returned to the diode in proper phase to convert more of the unwanted harmonic power into power at the desired harmonic. Of particular interest was the fact that the optimized filtered output level at any chosen harmonic was higher than the unfiltered level in the broadband output, the added power coming from the other harmonics. If this principle also applied to the active MMIC multiplier, higher efficiency may be achievable than the unfiltered output would suggest.

To test this principle, I chose the 6th harmonic at 3312 MHz and constructed a test fixture of a 3312 bandpass filter (10% BW, 1.5 dB loss) preceded by a long 50-Ω input line. The input line was made over 360 electrical degrees long, to be able to see the whole range of phase conditions. The line was marked off in steps of 0.1 inch, which corresponded to about 15 degrees on this dielectric. The 50 Ω line was driven by an Avantek MSA 0885.

Data was then taken at four levels of bias from 12 mA to 29 mA at each distance to the filter, down to 15 degrees. The data was then plotted, revealing that for each bias condition, there are spacings that definitely enhance the efficiency of the multiplier. When plotted on top of each other, it was evident that for most bias conditions, 210 to 230 degrees gave the highest output. I could not correlate these results to the HP application note because HP's tests were performed on 8th and higher harmonics. I believe, however, that the effect is adequately demonstrated for active multipliers.

The one variable I held constant in these tests was the input power level. It is possible that combinations of input power and bias could be found that would give even greater outputs. In addition, the type of filter used may have an effect on the overall results. The hairpin filter used in these tests looks like an open circuit to the highest level signal in the output, the fundamental at 540 MHz. If another type of filter is used that appears as a short circuit, such as an interdigital, the phase of the reflection will be altered and different results would be expected.

The "filter gain" effect is seen in a comparison of the data. The level of the 6th harmonic in the unfiltered test was –24 dBm for an input of 0 dBm. Under the same bias conditions, the level of the 6th harmonic was –13 dBm when passed through a bandpass filter located 220 degrees away. The apparent "gain" is 11 dB. This is in the range of the HP test results.

Comparison to Schottky-Diode Multiplier

The Schottky-diode multiplier layout in Fig 2 is the one used in my 3456-MHz transverter. In that design, the diode spacing to the filter is between 15 and 30 degrees. Diode bias and input matching are optimized for maximum output. Total conversion loss for the diode is 29 dB.

Applying the results of the above test, the multiplier chain shown in Fig 3 could be realized. Signal levels are kept below 0 dBm until the last amplifier. Shielding and isolation require-

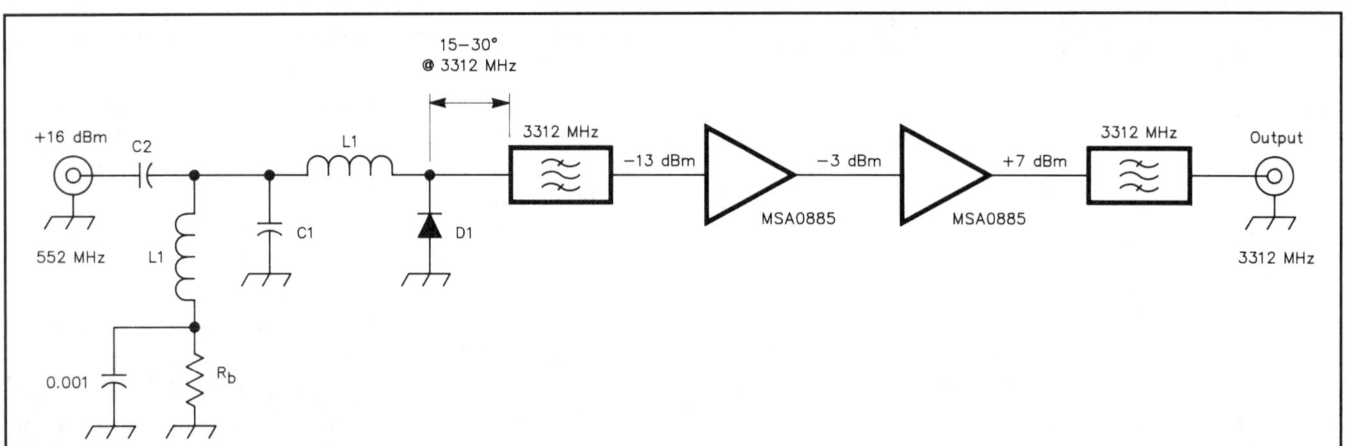

Fig 2—Block diagram of a Schottky-diode multiplier for use in a 3456-MHz transverter. L1-C1 and L2-C2 resonate at 552 MHz. D1 is the Schottky diode.

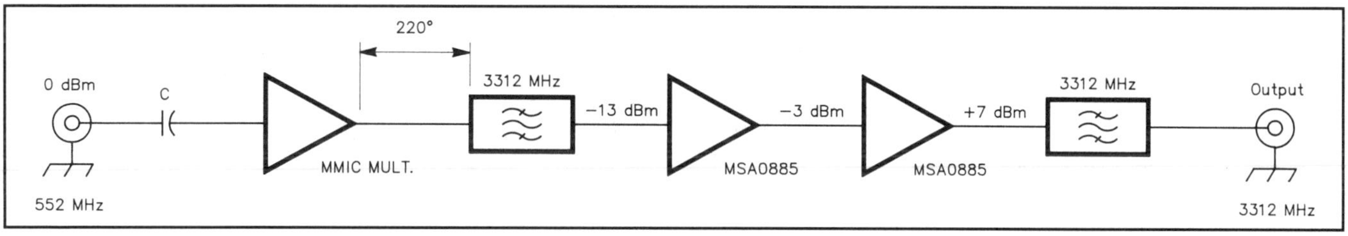

Fig 3—Block diagram of a multiplier for the same application, but using a MMIC instead of a Schottky diode.

ments would be greatly reduced with this design.

Although not usually an important design consideration in amateur equipment, the total current required in the MMIC version is lower. The parts count is about the same when the MMICs bias and dc blocking components are considered.

Summary

Simple frequency multipliers can be made with inexpensive silicon MMICs well into the microwave range that offer an alternate to Schottky diode multipliers. Drive requirements are lower with the MMIC design. The principle of "filter gain" can be effectively used to achieve maximum efficiency in MMIC multipliers. For each set of operating conditions, there is an optimum MMIC-to-filter spacing for best results. Experiment!

Notes

[1]J. Hinshaw, "MMIC Multiplier Chains for the 902-MHz Band," *Ham Radio*, Feb 1987, pp 72-79.

[2]J. Hinshaw, "MMIC Active Multipliers," *RF Design*, June, 1988, pp 64-68.

[3]J. Davey, "A No-Tune Transverter for 3456 MHz", *QST*, June 1989, pp 21-26.

[4]"Comb Generator Simplifies Multiplier Design," *Hewlett-Packard Application Note 983*, July 1981.

Weak-Signal Sources for the Microwave Bands

By Paul Wade, N1BWT

Introduction

There are a few areas of the country where microwave experimenters have the luxury of beacons or local hams to provide a signal source for checking out microwave equipment. The rest of us need a local weak-signal source for tuneup, and to be able to verify that things are working when no signals are heard. Relying on harmonics of VHF equipment can be very misleading—I learned this the hard way on top of Burke Mountain in northern Vermont during the 1990 UHF contest.

These weak-signal sources provide a local signal source that is strong enough to be heard at a reasonable distance, perhaps a few hundred feet, but not strong enough to ride through directly into the receiver, so the entire system, including preamps, coax switches, feed line, and antenna, may be verified. The crystal frequency can be chosen so that none of the harmonics fall in the IF band; most of those listed in the table on the schematic avoid all ham bands.

The design is based on the no-tune transverter[1] and oscillator[2] work of KK7B, and is very similar to the LO section of the 903 transverter. The only significant differences are in scaling of the filters for 903, 1152, and 1296 MHz; also, I flipped the artwork to avoid bending the base leads of common transistors under the can. Fig 1 is a schematic of the source.

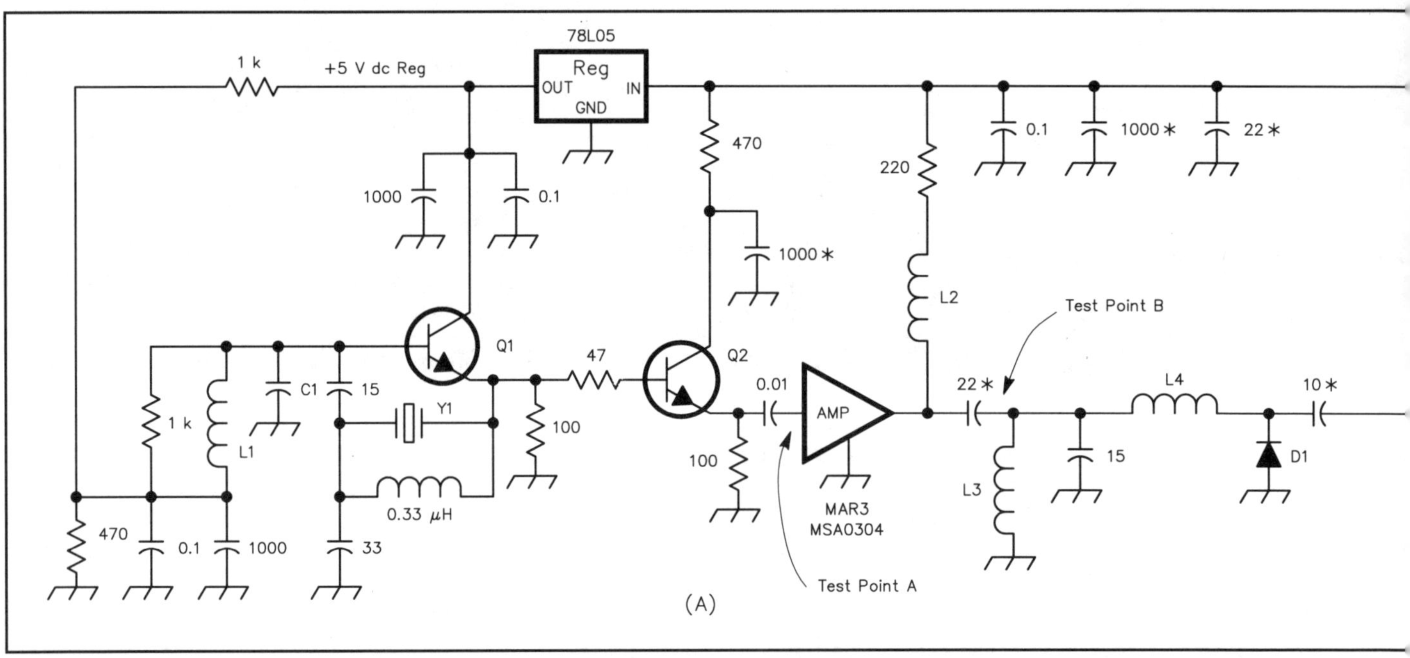

Fig 1—Schematic diagram of the N1BWT weak-signal source. The inset at B is an optional amplifier. See text. Capacitors marked * are chip capacitors. The source can be built in a Radio Shack 270-238 enclosure.

C1—10 pF trimmer.
D1—Schottky diode, Hewlett-Packard HP2835 or equiv.
L1—Below 90 MHz: 11 turns no. 24 enam wire, 0.1-inch dia. Above 90 MHz: Same, except 10 turns.
L2-L5—8 turns no. 28 enam wire, 0.0625-inch ID, spaced 1 wire diam.
Q1—2N5179, 2N4124 or equiv.
Y1—Crystal; see text and Table 1 for details.

Tuneup

Tuneup is straightforward, and easiest if the source is built in stages, following the schematic from left to right. Most of the component values aren't critical—the ones mentioned below require some attention. The first step is to install all parts up to Test Point A, and measure the output at the test point. Adjust trimmer C1 for max output, checking to see that the crystal is running on the right overtone. Most of the crystal frequencies can be checked by listening on an FM broadcast radio (if it isn't near the right frequency, replace the crystal with a 47-Ω resister and adjust L1 until C1 tunes through the desired frequency). I don't like to leave trimmers in the circuit, so after peaking C1, I estimate its capacitance and replace it with a fixed capacitor, then squeeze and stretch L1 to repeak the output. The second stage is to add components up to Test Point B and check the output there. I typically see one milowatt (0 dBm) at Test Point A and 10+ milliwatts (10 to 12 dBm) at Test Point B.

Now it is time to add the multiplier and next two amplifiers. There isn't enough output after the multiplier to measure unless you have a spectrum analyzer (that's cheating!), so the amplifiers give enough signal to measure, typically –10 to –20 dBm, depending on the number of times the crystal frequency is multiplied. This is plenty of output for a weak-signal source, but if you feel a need to peak it, squeezing and stretching L4 sometimes makes a difference.

The optional amplifier is not recommended for 903 and 1296, but may be useful at 1152 to get to the higher bands. To install, drill a hole in the output stripline for the MMIC and make a cut with a knife for the output blocking cap. With this amplifier there is enough gain in the box to make things touchy; an external amp might be better.

Table 1
Suitable Crystal Frequencies for Weak-Signal Sources

Output Frequencies (MHz)		
903.1	*1152*	*1296.1*
72.26	76.802	81.01
82.1	82.287	86.41
90.31	88.617	92.58
100.344	96.002	99.70
112.89	104.73	108.01

Board

Boards and kits are available from Down East Microwave.

Higher Bands

All the standard calling frequencies above 2 GHz are multiples of 1152 MHz (+100 kHz), so this is the starting point. I made this one with the optional amplifier, which provides measurable (about –30 dBm) harmonic output at 2304 and 3456 MHz to use as a weak signal, preferably after filtering. Output at 5760 and 10368 was not detectable, so I used a very simple harmonic generator and high-pass filter: a waveguide diode detector and a waveguide-to-coax adapter, both surplus, bolted together (Fig 2). The 1152-MHz output

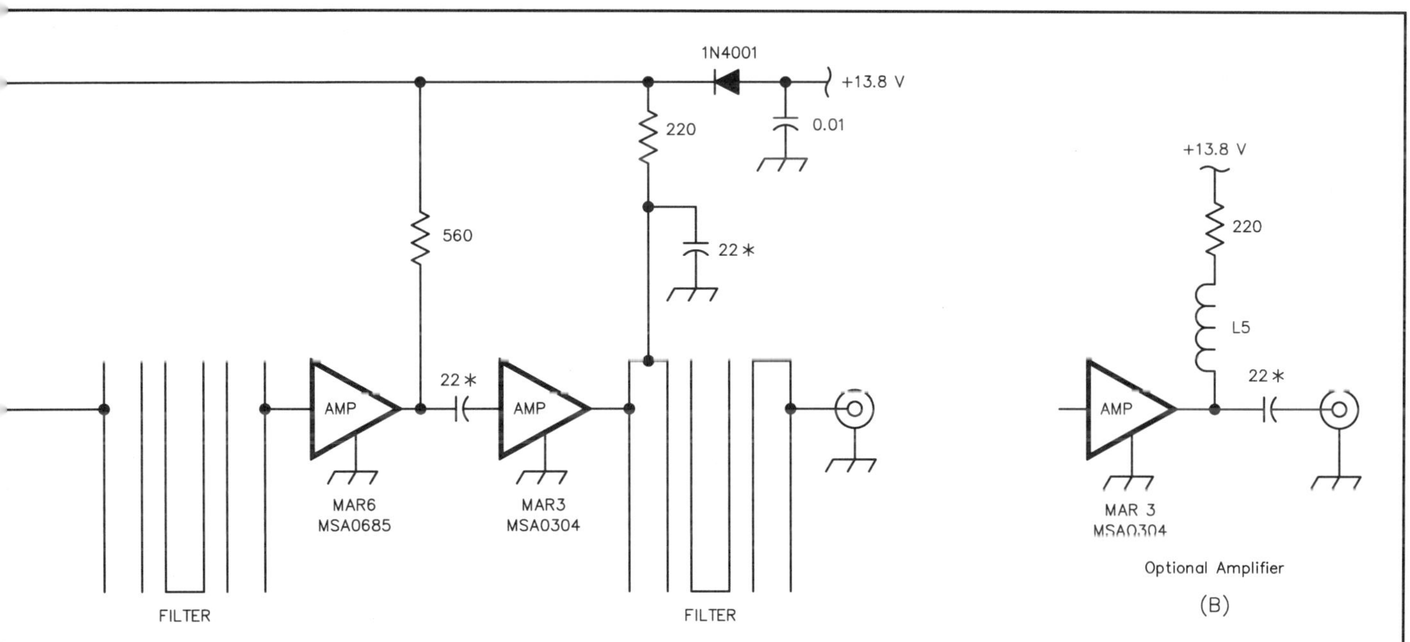

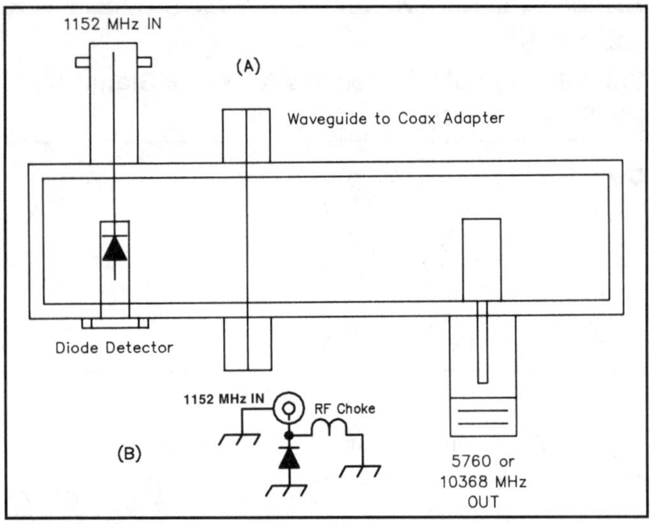

Fig 2—Waveguide harmonic generator for use with weak-signal source. At A, mechanical construction. B shows a schematic diagram of the harmonic generator, including an external RF choke required for use at low drive levels.

drives into the detector BNC output, and 10 GHz comes out the coax end of the waveguide to coax adapter. Drive of 8 to 10 dBm (from an external MMIC amplifier) produces an output of around –50 dBm at 10.368 GHz, but 0-dBm drive produced no output; apparently this is not enough to produce harmonics from a 1N23 diode. The waveguide section does rudimentary filtering, with all harmonics in the waveguide passband appearing; a real filter could be inserted here.

I haven't tried this with C-band waveguide components, but there's no reason it shouldn't work just as well for 5760 MHz.

Notes

[1]R. Campbell, KK7B, and D. Hillard, WØPW, "A Single Board 900-MHz Transverter with Printed Bandpass Filters," *Proceedings of Microwave Update 1989*, ARRL, 1989, p 1.

[2]R. Campbell, KK7B, "A Clean Low-Cost Microwave Local Oscillator," *QST*, July 1989, p 15.

Phase-Locked Microwave Sources

By Greg McIntire, AA5C

Introduction

The many phase-locked microwave sources circulating in surplus have been a real boon to microwave enthusiasts, particularly those building equipment in the 3456, 5760, and 10368-MHz bands. Their stable operation greatly simplifies the frequency variable when making microwave DX shots. Building around a good source simplifies much when working with microwaves.

This article briefly describes the principles of operation of these sources, and provides information and procedures that should allow you to get your LO up and running.

Principles of Operation

Most phase-locked microwave sources are based on the block diagrams shown in Figs 1 and 2, although there are many variations of the same theme. A microwave cavity oscillator serves as the fundamental RF oscillator for the source. The output of this oscillator is either the direct RF output or it drives a multiplier block, from which the output RF is obtained. A comb of harmonics extending into the frequency range of the fundamental oscillator is generated from a 5th- or 7th-overtone reference crystal oscillator. The fundamental oscillator is then phase locked to a harmonic of the reference oscillator.

The phase difference between the reference harmonics and the fundamental signal is obtained in a phase detector, basically a mixer. When the fundamental and one of the reference harmonics are equal in frequency, a pure dc level is obtained from the phase detector. An ac signal rides atop the dc level when the signals differ in frequency. The output of the

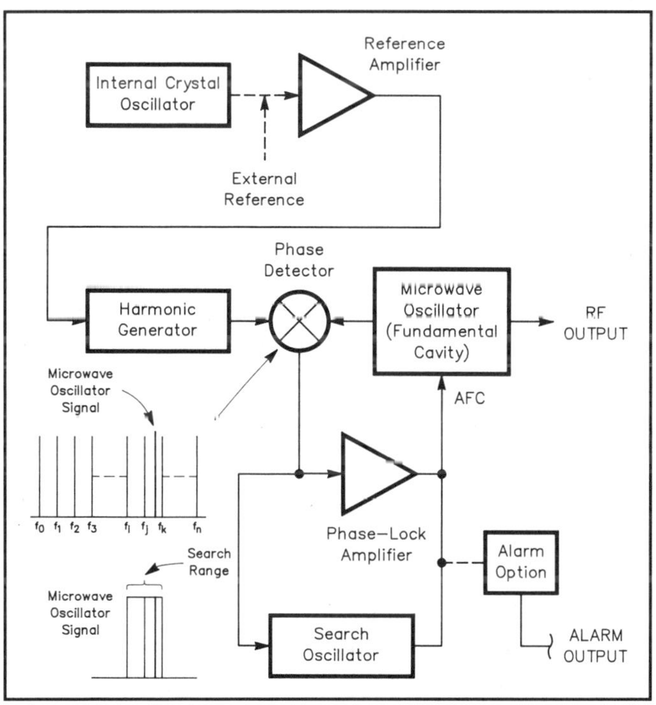

Fig 1—Block diagram of a phase-locked, mechanically tunable microwave signal source.

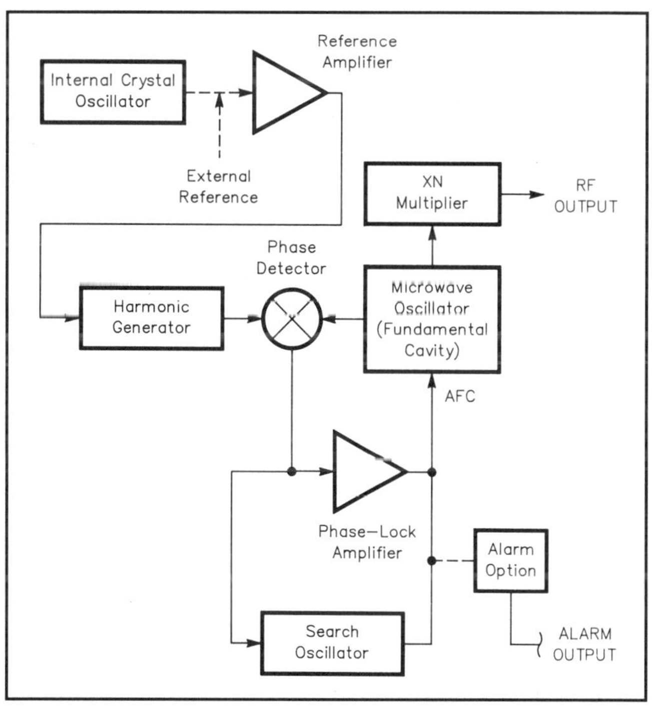

Fig 2—Block diagram of a phase-locked, mechanically tunable cavity-plus-multiplier microwave signal source.

phase detector is automatic-frequency control (AFC) of the fundamental cavity, typically via a varactor diode inside the cavity.

A search oscillator is used to initially acquire phase lock. The search oscillator slews the fundamental oscillator frequency, via AFC, so that it sweeps across the reference harmonics. Once lock is obtained, the search circuit is disabled. The search oscillator automatically kicks in if lock is broken on most units. Some require manual initiation of a new search.

The multiplier block, when used, consists of a step recovery diode (SRD) and an interdigital filter. A relatively high level signal from the fundamental cavity, typically +27 dBm, hits the SRD and generates the comb of harmonics. The interdigital filter selects the desired harmonic.

The main advantages of this type of source include:

1. High spectral purity output
2. High rejection of reference sidebands, typically 120 dB
3. Long term stability is that of the reference oscillator

Models

I'm aware of at least three companies that manufacture the more common "brick" sources for the 3-18 GHz range: Communications Techniques Inc. (CTI), Frequency West, and California Microwave. Sources from these companies are based on the same scheme and most are mechanically and electrically interchangeable. Non-"brick" sources I have worked with include 6- and 11-GHz units from Collins (Rockwell International). Almost all of the units operate from negative power supplies, common in the telecommunications industry.

The CTI, Frequency West, and California Microwave sources use a fundamental cavity operating in the 1-2 GHz range. A fifth overtone crystal in the 90-120 MHz range is the reference. All of the units I've seen have multiplier blocks. Depending on the fundamental cavity range, the sources can be used directly without multipliers for 1152 and 2160 MHz high level LOs (most have threads matching a female chassis mount BNC connector). A representative drawing of interconnections and adjustment points for a Frequency-West source, is shown in Fig 3. Options and variations I've come across include:

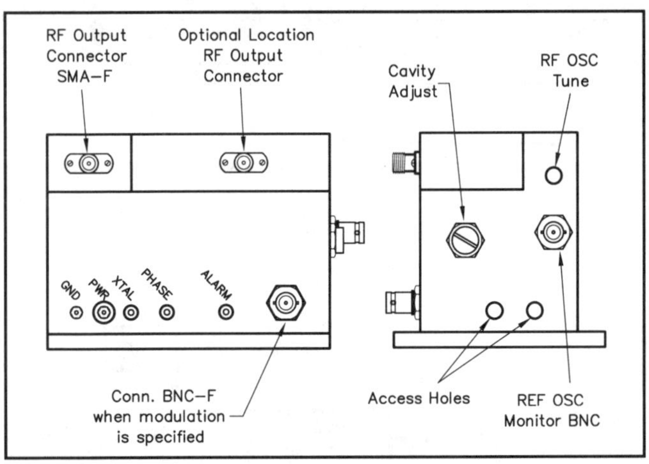

Fig 3—Locations of connectors and adjustment points on a typical Frequency West "brick."

- Output power level: 5, 10, or 20 mW
- Output connector: SMA, TNC or WR-75 output (WR-75 on some 10-15 GHz sources)
- Alarm (Y/N) and output type: relay contacts on older units, an open emitter transistor on newer ones
- Modulator input (Y/N): typically via a BNC-F connector
- "IF" output for external phase-lock control circuit
- External reference: reference oscillator not in basic unit
- Supply polarity: positive or negative (generally –19 or –20 V dc)
- Precision reference: the reference oscillator is phase locked to a precision reference, usually 20 MHz
- Multiplier: from 3 to 17 depending on manufacturer and frequency

Collins designed and built its own 6- and 10-GHz phase-locked sources. The Collins sources both use a fundamental cavity operating on the intended RF output frequency, ie, no multipliers are used.

The fundamental cavity in the Collins 6-GHz units tunes 5.9-6.4 GHz without modification. Units with P/N 622-5829-001 have two output SMA connectors, a tuning control and a bolt on the cavity. Units with P/N 622-5828-001 have a single SMA output connector, a tuning control, and no bolt on the cavity. Both 6-GHz units operate from +12 and –12 V dc, and have the following characteristics:

- Supplies: +12 V @ 110 mA, red lead
–12 V @ 450 mA cold, 180 mA warm
Common is black lead
- Output: +7 dBm via SMA connector
- Alarm: open emitter 2N2907, gray lead, on indicates unlocked
- Phase lock: blue lead, –6 Vdc indicates unlock—settles near 0.1 Vdc when locked
- Crystal: fifth overtone, 100-116 MHz
- Multiplier scheme: Fo=(F × Xtal × M)–25 MHz(M=25, 26, 27, 28 . . . so output falls within cavity range)
- Reference Output: SMA, 2× crystal frequency

A 3579.545-kHz color-burst crystal is multiplied by seven to provide the 25-MHz offset oscillator. I presume this was used for some sort of subcarrier, but have not investigated further.

The Collins 10-GHz sources were designed for the high end of the 10-GHz or the low end of 11-GHz range. I've put them on 10,224 and 10,368 MHz successfully, but the cavity output power is dropping off at the lower frequencies, which makes phase-locking more difficult. A tank-type reference oscillator is phase-locked to 4 times the reference crystal frequency. The reference oscillator is phase locked to the fundamental cavity. The Collins part number for these sources is 622-2983-002 and characteristics include:

- Power Supply: –20 Vdc @ 1.2 A cold and 0.8 A warm
- Output: +7 to +10 dBm via SMA connector
- Alarms: Cavity—yellow lead; reference—green lead; both open emitter 2N2907

• Phase lock: white lead, 0 to –18 Vdc swept during power up

• Sweep button: manual initiation of sweep when unlocked

• Reference indicator LED: on indicates unlocked reference

• Oscillator indicator LED: on indicates unlocked fundamental

• Crystal: fifth overtone 105-120 MHz

• Multiplier scheme: (crystal × 4) × 22, 23, 24 . . . in range of cavity oscillator

• Reference output: SMA connector, 4× crystal

• RF monitor port: coupled output, SMA connector

• Test points: a) crystal oscillator level (peak for –4 to –5 V, b) reference AFC (tune for about –3 to –5 V with reference LED off), 3) oscillator RF power (should read –0.4 to –0.6 V) paralleled with brown lead, 4) oscillator AFC (same as phase-lock indicator)

Figuring Out What You Have

Visual Check

The first thing you should tweak when you get a source is NOTHING! A little bit of investigation before you do anything can save a lot of time later. Get any model number and frequency information from the source. For the brick sources, the model number should be on a label atop the multiplier block. It's pretty obvious if this label has been removed. If intact, you can generally look up the model number to get the multiplication factor, crystal range, and power level. The output frequency and crystal frequency are often also on the unit.

Generally, you'll find that the fundamental cavity of a Frequency West brick tunes about 300 MHz. You can multiply this up to determine potential output tuning range. The range of the multiplier block generally limits the tuning range.

Operational Check

Next, remove the side cover and see if there is a reference crystal in the unit. If so, get the crystal frequency. Now apply regulated power, usually –19 or –20 V dc. First verify that you have output RF. Check the output level and frequency (if you can measure the RF frequency directly). If you can't measure the RF frequency directly, you'll absolutely need to know the multiplication factor.

The pickup probe on the multiplier block fits in a hole in the fundamental oscillator cavity. You can remove the multiplier block and a female chassis-mount BNC connector will directly thread into the threaded hole. Measure the frequency here (1-2 GHz range) if you can't measure the RF frequency at the output directly. Remember, the signal level here is about ½ watt! Different loading than the multiplier block provides can force the source to lose lock. Padding sometimes helps attain phase lock. Check the Φ terminal for a dip when lock is achieved. (Reference oscillator operation can be verified by a voltage peak on the XTAL test point when tuning the reference adjust control.) This verifies operation of the phase lock circuitry, reference oscillator and fundamental cavity.

After you've checked for phase lock, check the range of the fundamental cavity. The range is usually great enough so that three multiples of the reference oscillator fall within the range of the fundamental cavity, ie, the 10th, 11th and 12th. You'll need this information to help pick the reference crystal frequency. Now you should know whether the unit is worth modifying, and the multiplication scheme.

Ordering the Reference Crystal

Run through several multiplication schemes for the complete source. Pick a crystal frequency that falls in the middle of the cavity range when multiplied. Reference crystal frequencies from 95-113 MHz are common for the brick style oscillators.

All of the units I've seen use a TO-5 style crystal case. These have gotten rather expensive from many vendors, about $50 each. A recent order to International Crystal resulted in a $13 price for a TO-5 packaged crystal, however. An HC-18 style crystal holder can be used with a little extra effort. The cost of this style crystal holder typically ranges from $12-20. To use holder, you have to unsolder the machined oven from the oscillator printed circuit board. You also have to remove the lip at the bottom of the hole the crystal fits in, by drilling it out (first remove the ceramic disk with the socket pins for the TO-5 type case). Then install the crystal directly on the board and reinstall the oven over the crystal.

The Collins sources use a plug-in oven. The crystal leads can be plugged directly into the socket pins on the board. Placing the oven over the HC-18 style case is awkward, but will provide some temperature control.

Crystal specifications (ideal) are:

Mode: Fifth overtone
Oven: 75° C
Shunt capacitance: 4.5 ± 0.75 pF
ESR: 60 Ω max
Frequency tolerance: ±0.0005% @ 75° C
Temperature stability: ± 0.002% from 70-89° C
Spurious response ratio: 2:1

Sources for crystals include:

Bliley (814) 838-3571 (TO-5)
2545 W. Grandview
P.O. Box 3428
Eric, PA 18508

EG&G Cinox (TO-5)
(513) 542-5555
4914 Gray Road
Cincinnati, OH 45232

International Crystal (both)
10 North Lee
Oklahoma City, OK 73102

JAN Crystals (HC-18)
P.O. Box 06017
Fort Myers, FL 33906-6017

Crystek (HC-18)
2351/2371 Crystal Drive
P.O. Box 06135
Fort Myers, FL 33906-6135

Modification Procedure

CTI, Frequency West and California Microwave Sources

The technique I've generally followed to put a brick

source on a new frequency is to install the new reference crystal and then "walk" the multiplier block and fundamental cavity to the new frequency. Going directly for the final output frequency may result in power inadequate for any indication, when tuning the multiplier block.

Crystal Installation

Install the crystal and apply power. A XTAL test point is available on all of the sources except those designed for external reference. Monitor the voltage at this point and adjust the crystal tune control for peak voltage, generally –1.5 to –2.0 V. A check of the reference frequency can be made on the SMA connector. Final adjustment for frequency should be done once the source is installed in your system.

"Walking" the Fundamental Cavity and Multiplier Block

There are several styles of interdigital filter adjustments on the multiplier blocks. Some permit adjustment from above, using small lock nuts. The best way to adjust these is with a tiny, hollow socket, which will pass a hex wrench to the socket on the setscrew. (Although more difficult, the job can be done with long-nose pliers or hemostats!) With a socket tool, you can hold the locking nut close to final position while adjusting the setscrew. This arrangement is convenient, in that you don't have to remove the block for filter adjustment. You still have to remove the block to adjust the output connector coupling.

Another construction style has locking setscrews accessible from the top, with adjusting setscrews on the side facing the brick. You must remove the multiplier block to adjust the screws. Fortunately, by design, the multiplier block will sit "backwards" on the brick for adjustment. Multiplier blocks with waveguide flanges won't seat completely for adjustment, however. Tweaking these is a trial and error process the best I can tell . . . remove the block to tweak and replace to see how well you did.

Walking the source to the new frequency involves adjusting the fundamental cavity tuning screw to slew the cavity a hundred megahertz or so, then tweaking the multiplier block to maintain signal. Tweak the output connector probe adjustment (usually a slotted "screw") and all filter setscrews for maximum signal. The adjusting setscrews nearest the cavity probe are the most critical. Then slew the fundamental frequency further toward the desired frequency and retweak. Monitor the 0 test point when you think you're close to final frequency You should be able to achieve lock before making the final multiplier block adjustments. Sometimes, however, the multiplier block needs to be close to final frequency to provide a good match for the fundamental cavity in order to achieve phase lock. When you achieve lock, make a final pass on the filter and probe tuning screws.

Generally, I've not lost too many dB in output power from the original power level, depending on how far the desired frequency is from the design frequency. One LO I put on 3456.350 MHz was originally set for about 4.0 GHz. When finished, the tuning screws were fully engaged to get maximum power, which was about 0 dBm. Don't despair if this is your case. An outboard MMIC amplifier easily brings up the signal level for driving a mixer. Remember, we are generally using the sources well outside their design range and losing a few dB can be expected.

If you're feeling adventuresome, one tweak which usually makes a significant difference is the depth the multiplier block probe inserts into the fundamental cavity. The probe is threaded. Some units use nuts to lock the threads. Unfortunately, the lead from a ⅛-W coupling resistor is wrapped around the probe underneath the lock nut, in these units. I can assure you this resistor is fragile! Other units tack solder the probe in place. Although the solder wicks into the threaded sleeve, I've been able to remove it, make the adjustment for maximum power, and resolder. This adjustment also affects the match of the probe to the cavity for both styles. You should simultaneously monitor for phase lock and peak power.

Collins 6-GHz Sources

Remove any existing reference crystal and slew the cavity oscillator to the new frequency. There is no multiplier block/filter to tune on these sources since the fundamental cavity is tuned to the RF frequency. Tuning the cavity below the 5.9-GHz range requires capacitive loading. The units with a bolt allow the most tuning range. A Teflon rod is used for the tuning adjustment on these sources. I've soldered Teflon rods from other units to the cavity side of the brass bolt and thus inserted a second Teflon rod into the cavity. Some experimenting is usually required to trim for the 5616-MHz range. The additional loading biases the cavity and reduces the tuning range to less than 200 MHz.

Install the reference oscillator crystal and tweak until you see a signal on the reference oscillator test point. These units will slew the fundamental oscillator over a broad range, so you should see the source lock up without further action on your part if you're at all close on the fundamental frequency. If not, tune the fundamental in increasing increments each way until you obtain lock.

Final trim of the reference crystal should be done after you've installed the source in your system with its own regulator. Remember, the signal at the reference SMA test connector is twice the crystal frequency.

Collins 10-GHz Sources

Like the 6-GHz Collins sources, no multiplier is used in these units. Remove the reference crystal and slew the fundamental cavity to the desired frequency. You may want to monitor the cavity power test point and get a feel for how much it drops off as you tune to the ends of the cavity range.

Next install the reference crystal. Monitor voltage at the test point for the crystal oscillator and adjust for maximum voltage. Then adjust the reference tuning capacitor (REF). A red LED on the unit is quite useful in indicating when the reference tank oscillator locks to the reference crystal oscillator. It will go out when lock is obtained. The reference oscillator operates in the 425-450 MHz range and locks to the fourth multiple of the crystal. Now adjust the reference tuning while monitoring the REF AFC test point. Tune for about –4 volts. You can tune for lower voltage but spectral purity of the reference signal, and thus lock stability, becomes a problem. The "cleanliness" of the reference signal is very sensitive to the

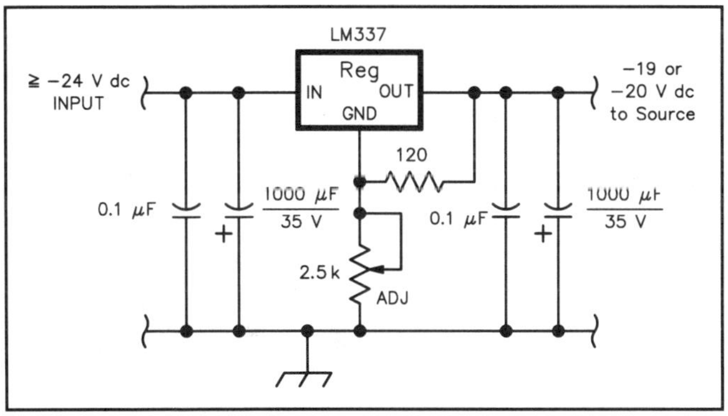

Fig 4—Schematic diagram of a regulator for "brick" signal sources that operate from negative supplies.

REF TUNE adjustment. I've sometimes had to make this adjustment using a spectrum analyzer to get a clean reference signal. Consider trying this if you have trouble obtaining lock.

The source may have already locked at this point. If not, press the red "SWP" button. When the fundamental cavity is close in frequency, and a sufficient sampled signal from the cavity is available, lock should occur. Monitor the "OSC AFC" test point. You should see the voltage swing close to the power supply rails (0 to –18 volts) and then settle in mid-range when lock occurs. The "OSC" red LED will go out when the cavity is phase-locked to the reference.

System Installation

A most critical aspect of installing a phase-locked source in your system is power supply regulation. Small variations in supply voltage result in "FMing." I strongly recommend using a dedicated regulator close to the source. My favorite is an adjustable negative regulator based on the LM337 IC (Fig 4). I substitute a 47-Ω, 2-W resistor for the source, to get the supply close to (less than) final voltage before connecting to the source. Tweak to the final level when you've connected the source.

Source output isolation is important for sources, particularly for those without multiplier blocks or if you're keying the output; for example, a PIN diode switch for beacon applications. Isolators directly on the source output are the best answer. In keying applications, 20 dB of isolation has been sufficient to prevent FMing or loss of lock.

After you've used your source for six months or so, recheck your output frequency. Most of the variation due to crystal aging occurs within this period.

6

Filters

Contents

Interdigital Bandpass Filters for Amateur VHF/UHF Applications

By Reed E. Fisher, W2CQH

[Adapted from *QST*, March 1968]

Multiple-section stripline interdigital filters are easily constructed in a few hours, and will work the first time with little or no adjustment. Their low midband insertion loss is comparable with the best circular coaxial filter and their out-of-band attenuation properties are just what modern filter theory predicts.

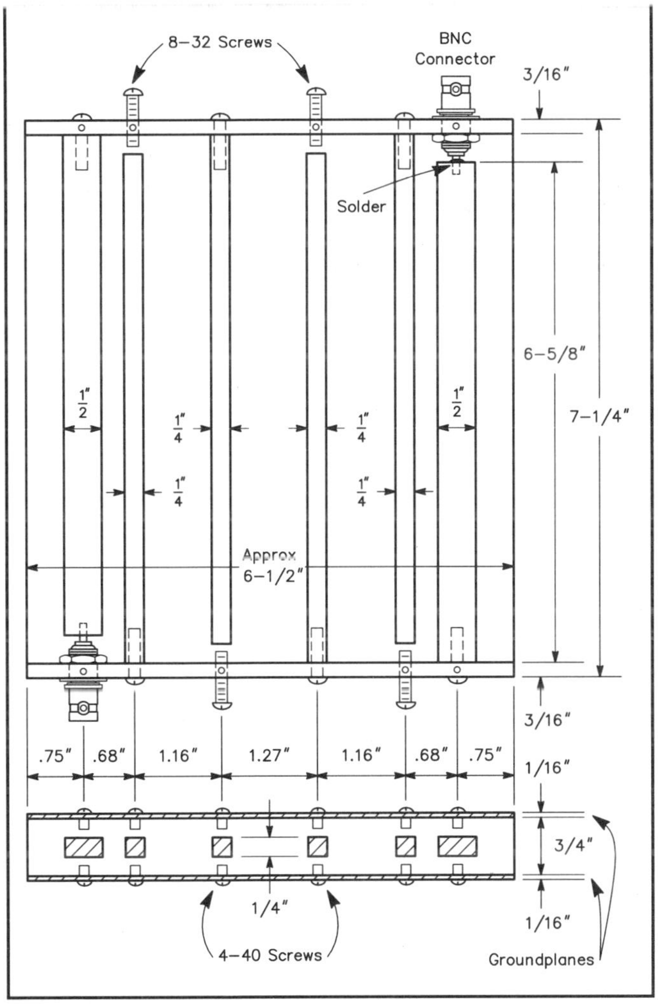

Fig 1—Mechanical details of the 432-MHz filter.

This article discusses the construction of two basic models which, when scaled to other frequencies, will satisfy most VHF/UHF filtering requirements.[1] Both filters are constructed of stock size brass and aluminum, which can be purchased at most supply houses. No milling is required, although a small lathe is useful for squaring the resonator ends.

The first filter, Fig 1, is a four-resonator model centered near 432 MHz, with a 3-dB bandwidth of 2%. The measured insertion loss versus frequency in a 50 Ω system is shown in Fig 2. The filter has a Butterworth response, which means there are no loss ripples in the passband. It was used to reject the oscillator and image frequencies of my 28-432 MHz transmitting converter.

The filter consists of six interdigitated rectangular rods centrally located between two ground planes. The four ¼-in. square open-ended rods approximately ¼-wavelength long constitute the high-Q resonators. The two larger rods, whose open ends are soldered to BNC coaxial connectors, are low-Q

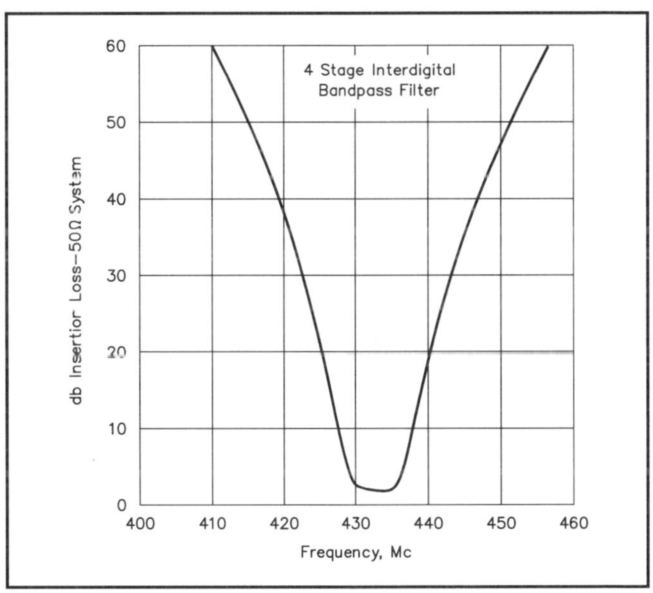

Fig 2—Bandwidth and insertion loss of the 432-MHz filter.

coupling sections. One end of each rod is drilled and tapped for an 8-32 machine screw, so that it may be securely attached to an end wall. The top and bottom ground planes are 1/16-in. brass or aluminum, fastened to the drilled and tapped end walls by several 4-40 or 6-32 machine screws. Ground-plane screws should be located near the center line of each rod, as large RF currents flow in this region. Note that no side walls are required. The ground planes overlap the coupling rod by an amount sufficient to prevent any radiation loss.

In the first model I built, the rods were plain stock brass and the ground planes were 1/16-in. thick aluminum. This filter has a 1.4-dB midband (432 MHz) insertion loss. A second model, constructed by W2CCY, with silver-plated brass rods and ground planes, exhibited a 0.5-dB insertion loss. Tuning screws were included in the first model. I later found that, if all four resonators are made precisely the same length, subsequent tuning is unnecessary.

The open ends of the resonators are loaded by "fringing capacitance." This loading means their lengths are always less than 1/4 λ. You'd find it difficult to determine this capacitance, and hence accurately predict how much to shorten the rods. Since the resonators are easily removed, it is easy enough to remove and prune them to length after you initially test the filter.

When the 3-dB bandwidth of a filter is made larger, the midband insertion loss decreases. A three-resonator filter centered near 1296 MHz, with an 8.5% 3-dB bandwidth, is shown in Fig 3. This filter has a 0.4-dB insertion loss, using plain brass rod construction. This filter was a "four-hour special" that worked the first time with no tuning.

These filters can be scaled to any other frequency by changing the rod length, but keeping the center-to-center and ground-plane spacings the same.[2] For example, the rods would be approximately 20-in. long in a filter designed for 144 MHz. The 3-dB bandwidth would remain at 2% or 2.9 MHz.

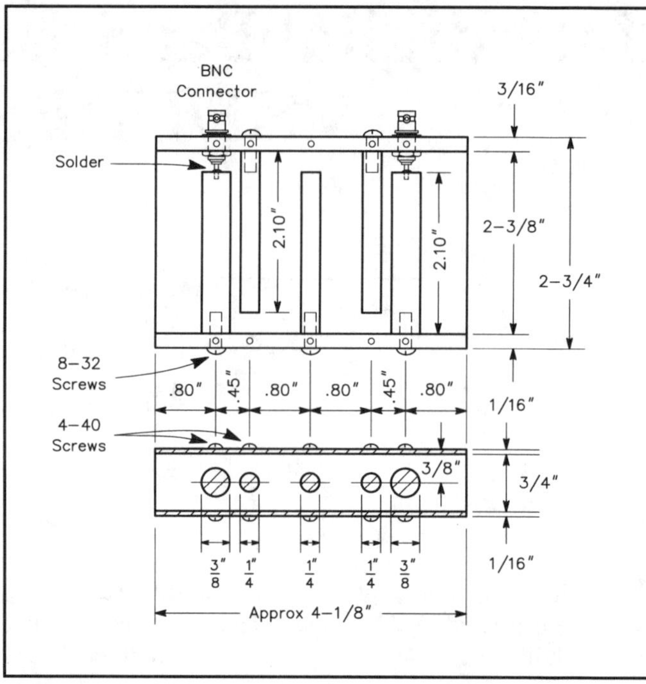

Fig 3—Structural details of the 1296-MHz filter are similar to those of the 432-MHz model, except that three circuits and cylindrical conductors are used.

Notes

[1]The theory of interdigital filters is described in G. L. Matthaei, "Interdigital Band-Pass Filters," *IRE Trans. on Microwave Theory and Techniques*, vol. MTT-10, Nov 1962, pp 479-491 and W. S. Metcalf, "Graphs Speed Design of Interdigital Filters," *Microwaves*, Feb 1967, pp 91-95.

[2]For more information on scaling, see "Interdigital Filter For 2304 MHz," which appears later in this chapter.

A Half-Wave Transmission Line Filter for 902 MHz

By Don Hilliard, WØPW

This filter is based on a 1296-MHz filter described by Joe Cadwallader, K6ZMW.[1] The filter is mechanically simple; Fig 1 gives almost all the information needed to duplicate it. My filter tunes to 902 MHz with the capacitor plates about 0.1 in. apart. Its –3-dB bandwidth is about 17.5 MHz; –10-dB bandwidth is about 49 MHz. Insertion loss is only about 0.6 dB.

The trough line is made from 5 pieces of 1/16-in. double-side printed-circuit board, soldered together. You can also use sheet brass or copper. The capacitor plates are #10 brass washers. One plate is soldered to the inner conductor (1/4-in. copper or brass tubing); the other is soldered to the 10-32 brass tuning screw (Fig 2). The coupling loops are #14 AWG solid copper wire (bare).

I used BNC connectors for the 902-MHz filter. For 1296 MHz and above, I prefer type-N connectors.

[1]J. Cadwallader, "1296 Transverter," *Ham Radio*, July 77, pp 10-17.

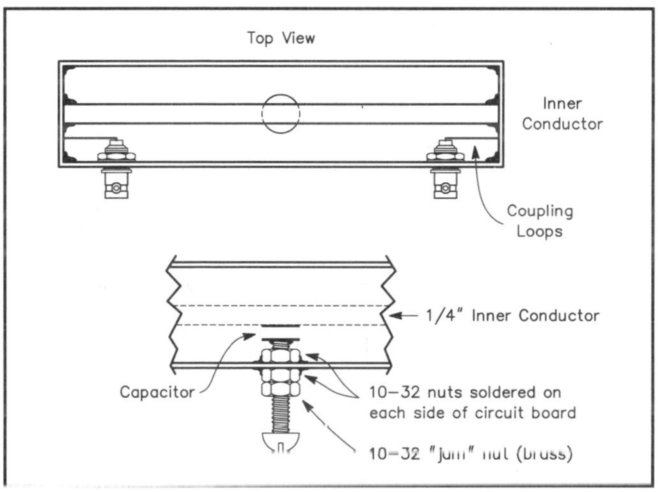

Fig 1—Mechanical details of the 902-MHz filter.

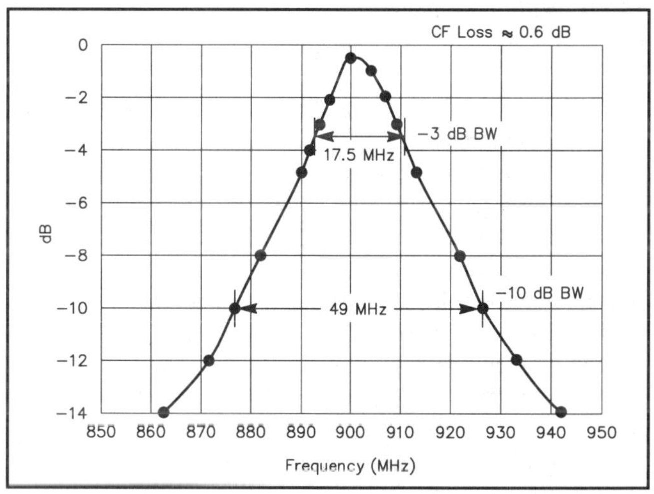

Fig 2—Details of the tuning capacitor.

A Simple Cavity Filter for 2304 MHz

By Bob Atkins, KA1GT

(From *QST*, September 1989)

The filter shown in Fig 1 is designed for 2- to 2.6-GHz use, but will tune down to around 1.5 GHz. I have used this filter to:

- Suppress the 2160-MHz LO component in the output of a 2304-MHz transmit mixer
- Reduce the image frequency response (2016 MHz) of a 2304-MHz receiver
- Clean up a 2556-MHz LO chain output before a x4 multiplier to 10.224 GHz (the LO for a 10-GHz transceiver).

With tight coupling (connectors close to the center conductor), the loss at the tuned frequency is less than 0.5 dB, and the rejection 150 MHz from the tuned frequency is about 15 dB. With loose coupling, the insertion loss is higher (as much as 3 dB), but the rejection 150 MHz off frequency increases to around 30 dB. Whether low loss or high rejection is important depends on the application; a nice feature of this filter is that coupling is adjustable by simply threading the connectors in and out. This filter should be capable of handling high power, so it should be usable for cleaning up the output of a power amplifier.

The best construction material for this filter is copper. Brass is acceptable, though performance suffers. Thick-walled tubing can be used for the cavity or, alternatively, a ⅞-inch hole can be drilled in suitable bar stock. Soldering should be done on the outside of the cavity wherever possible, because solder inside the cavity lowers its Q.

From electrical and mechanical standpoints, SMA connectors are preferred, but BNC (or better still, TNC) connectors could also be used. The tuning screw (Fig 1—top) should have a fine pitch to make tuning easier, and its end should be filed flat and smooth. Better still would be a DRO (dielectric-resonance oscillator) tuning piston or cavity-tuning assembly, such as those available from a number of commercial microwave-tuning-component companies. The inside of the filter should be polished to a smooth, mirror-like finish for maximum Q. Silver plating is a good idea if brass is used, but probably would make little difference in the performance of a copper cavity.

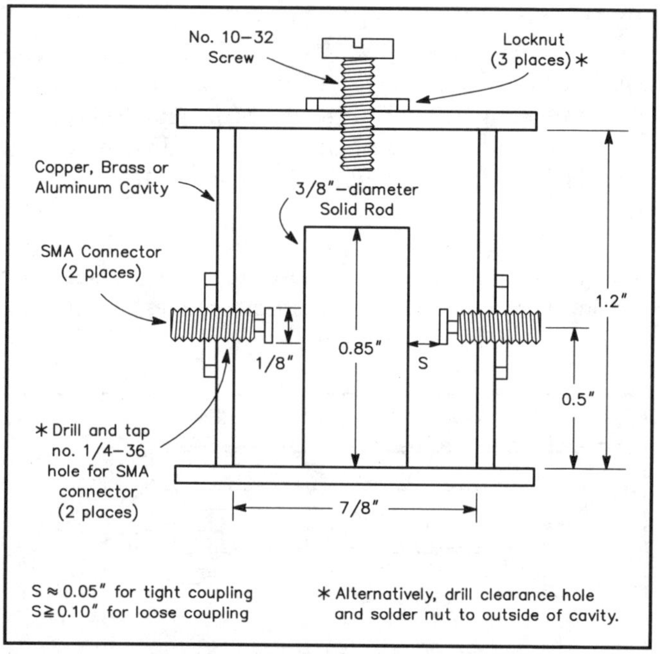

Fig 1—2- to 2.6-GHz band-pass cavity filter has several practical applications. The construction material of choice is copper, but brass or aluminum are also usable.

An Interdigital Filter for 2304 MHz

By Don Hilliard, WØPW

When I scaled the interdigital filters described by Reed Fisher, W2CQH,[1] I found that proportionally adjusting resonator lengths resulted in filters that didn't resonate at the expected frequency. At higher frequencies, of course, small changes in resonator length drastically affected the center frequency. The calculated resonator length for a 2304-MHz filter was 1.181 in. Not wanting to make the resonators too short, I cut them to 1.187 in. A filter built with these resonators resonated at 2090 MHz.

Scaling from *this* frequency yielded a calculated resonator length of 1.076 in. for a 2304-MHz filter. Again, not wishing to make the resonators too short, I cut them to 1.1 in. This filter resonated at 2324 MHz! The "fringing capacitance" Reed mentions in his article was apparently more of a factor than expected. Nonetheless, measured insertion loss at 2304 MHz was only 0.3 dB. I could have used tuning screws to lower the resonant frequency, but I considered this loss acceptable. On the other hand, loss at 2324 MHz appeared to be only 0.1 dB. Although I questioned this value, I measured it three times with the same result.

Fig 1 shows the layout and dimensions of my filter. The resonator rods are ¼-in. dia round brass; coupling rods are ⅜-in. dia round brass. I recommend you use type-N connectors for frequencies above 1296 MHz. A short length of brass tubing sleeved over the connector center pin connects the pin to the coupling resonator. See the article by W2CQH for complete construction details.

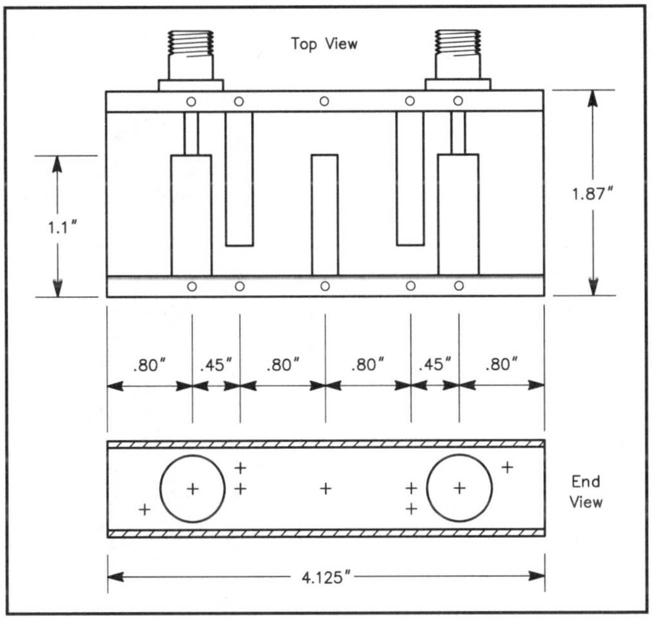

Fig 1—Mechanical details of the 2304-MHz interdigital filter.

[1] R. Fisher, "Interdigital Bandpass Filters For Amateur V.H.F./U.H.F Applications," *QST*, Mar 1968, pp 32-33 (reprinted in this chapter).

Cheap Microwave Filters From Copper Plumbing Caps

By Kent Britain, WA5VJB

[From Proceedings of Microwave Update '88]

These filters for our 3456, 5760 and 10,000 MHz bands are based on filters used in the DJ6EP 5760-MHz transverter. When I first saw these filters, I was impressed with their simplicity. But Roman's design uses Teflon PC board and small pins through 50-Ω stripline for coupling. I wanted to find a simple way to duplicate them with commonly available materials. I ended up using ½-, ¾-, and 1-in. copper end caps, sweat soldered onto PC board (Fig 1).

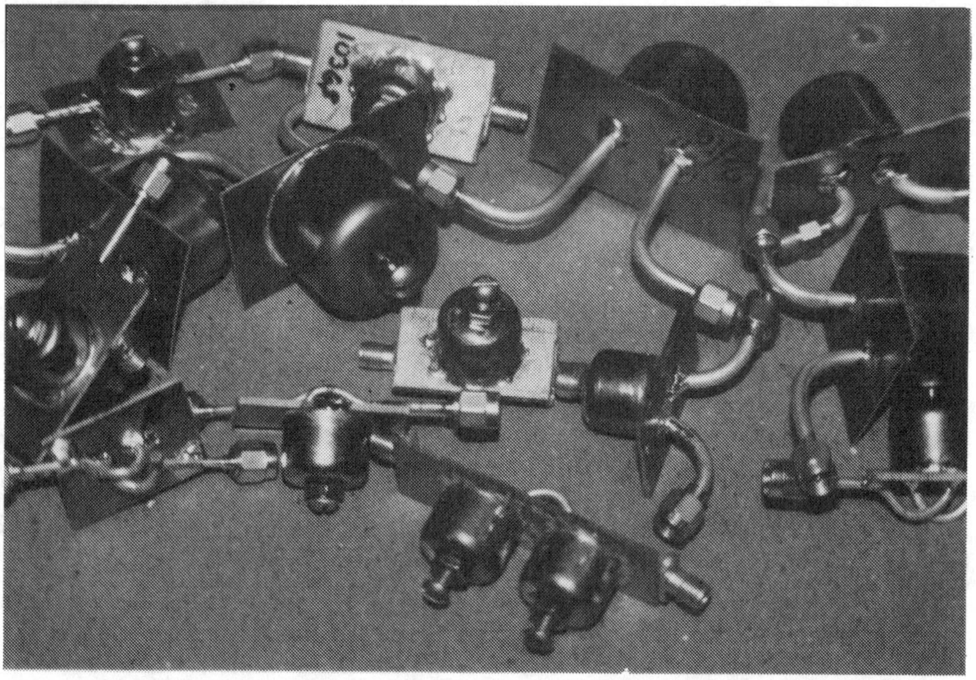

Fig 1—Almost too good to be true. Copper caps used to seal plumbing tubing can be used as cavities for microwave filters. The lower unit has been taken apart to expose the coupling probes, which are the center conductor and dielectric of semi-rigid coaxial cable.

The first batch of filters I built all worked well. Next, I built several with intentional errors, lots of sloppy solder, misaligned probes, unequal probes, off-center tuning screws, etc. Loss went up a bit on a few of them, but they all worked. These filters are very forgiving!

The length of the probe determines the coupling and, therefore, the Q of the filter. Keep the probes as short as you can, consistent with how much loss you can stand in your system. I did build some multi-stage versions to get a tighter bandwidth, but it really wasn't worth the effort. To improve Q, just use shorter probes.

You can drill and tap the hole for the tuning screw, but I found it much easier to drill a slightly undersized hole and force a steel screw (same thread size) through. The locking nut is tightened down after you have adjusted the tuning.

3456-MHz Filter

The 3456-MHz version is based on a 1-in. copper plumbing end cap (Fig 2). The hollow filter resonates between 6 and 7 GHz. The tuning screw pulls the filter down to 3.4 GHz at a depth of about ⅝-in. In a 3456-MHz station using a 144-MHz IF, a filter using 3/16-in. probes would give 25-dB rejection of the LO and better than 30-dB rejection of the image, with 2 dB of loss. Fig 3 shows the effects of probe depth on insertion loss and bandwidth.

5760-MHz Filters

The 1-in. filters will also tune to 5760 MHz when the tuning is about ⅜ in. into the cavity. With a ⅛-in. long probe, a 5760-MHz transverter would see about 20-dB rejection of the LO and almost 30-dB rejection of the image, with a 144-MHz IF.

A 5760-MHz filter made from a ¾-in. cap has similar performance to the 1-in. filter, but with less insertion loss than the 1-in. filter (Fig 4).

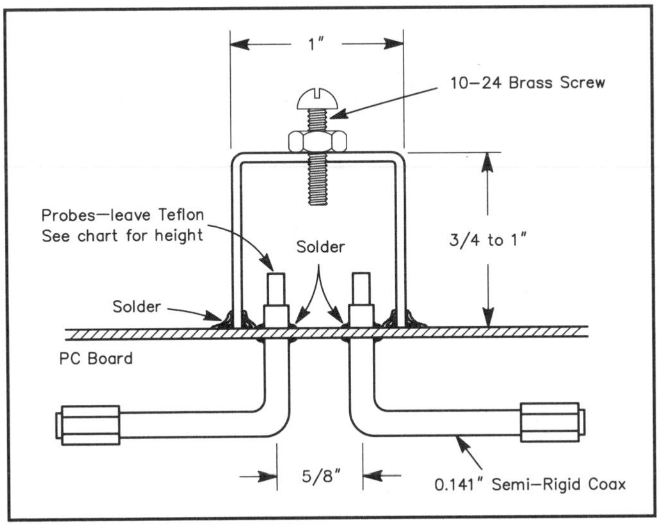

Fig 2—Mechanical details of the 3456-MHz filter.

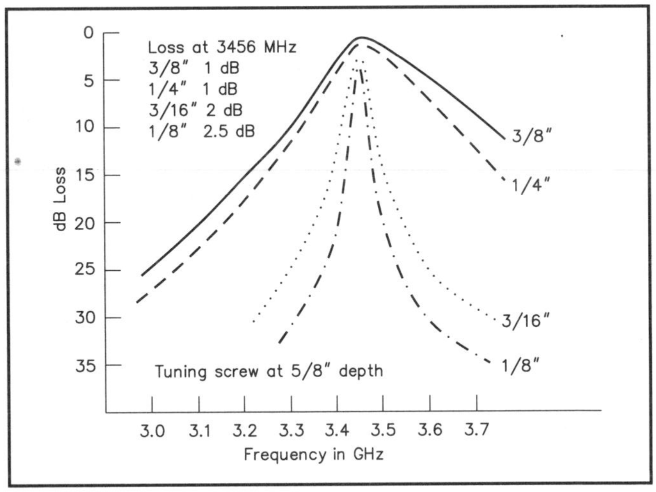

Fig 3—Graph shows insertion loss and bandwidth of 3456-MHz filter with different probe lengths.

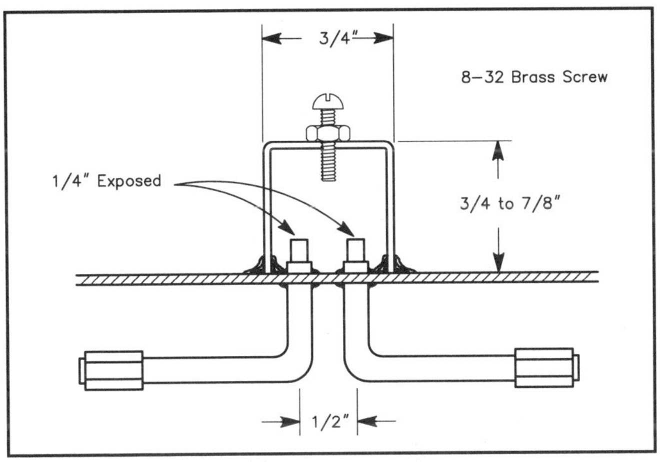

Fig 4—Mechanical details of the 5760-MHz filter using ¾-in. cap.

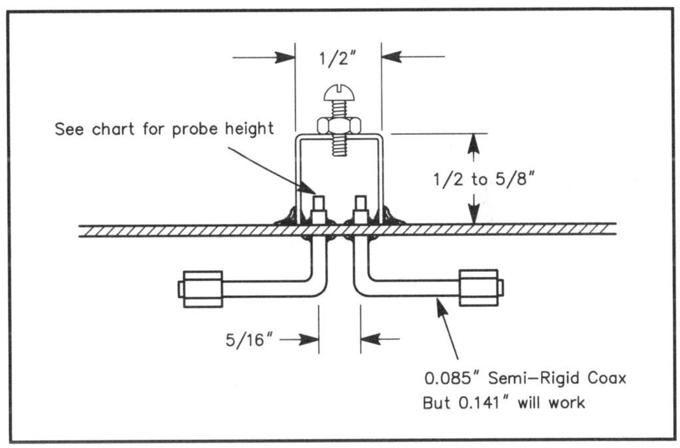

Fig 5—Mechanical details of the 10-GHz filter. Probe lengths of 1/16 to 1/8 in. gave good compromise between loss and bandwidth.

10-GHz Filter

When I first started building these probe-coupled filters, I used them at 10.368 GHz. If you can dig up some 0.085-in. semi-rigid coax, it's much easier to use than the larger 0.141 in., but both sizes work. The filter resonates somewhere between 11.5 and 12 GHz, so while digging through the bins at your hardware store, look for the longer end caps. The longer ones will have slightly lower loss, but it's not worth driving around town. I did try replacing the brass tuning screw with a steel screw on the filter with 1/8-in. probe, but loss went from 1.4 to 2.0 dB (Fig 5).

Waveguide Bandpass Filters

By James D. Green, K5JG

Waveguide filters usually fall in two categories: direct-coupled and quarter-wave-coupled cavities. This is a quarter-wave coupled filter, because requirements for precision are somewhat less than for the direct-coupled filter.[1] For those who are interested in the area inside the guide, the space between a set of posts acts like a resonant section or cavity. The distance between sets of posts is the coupling line. A set of posts may be viewed as a simple parallel-tuned circuit, with the tuning screw located halfway between the post set as the variable capacitance of the tuned circuit. Penetration of the screw into the cavity lowers the resonant frequency and effectively increases the cavity length or, in this case, the distance between the posts. Cavities are usually constructed somewhat shorter than required. The capacitive screw allows for frequency variation, as with any other filter. Since screw penetration lowers the cavity Q, avoid excessive screw penetration.

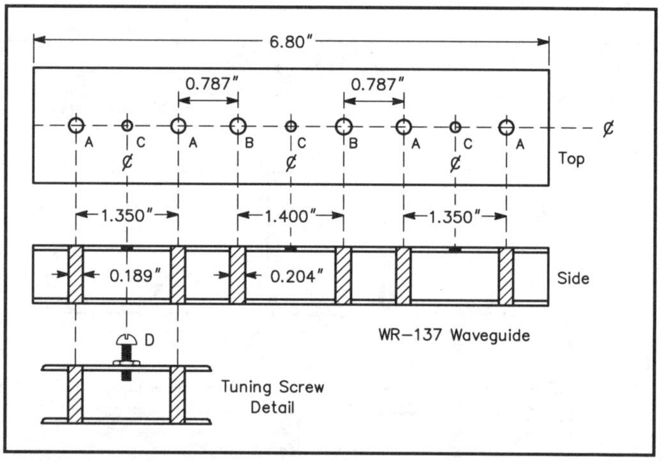

Fig 1—5.7-GHz Waveguide Bandpass Filter. Locations of holes and components are A: No. 11 drill for 0.189-in round brass rod; B: No. 6 drill for 0.204-in. round brass rod; C: No. 36 drill for 6-32 screw; D: 6-32 × ⅝-in. screw with 5/16-in. nut (3 required).

A Word About Plating

All screws should be silver plated, to minimize filter losses.[2] Contrary to popular belief though, plating the interior walls and posts of a waveguide filter of this type is not worth the time, trouble or expense. A poor plating job, which is more likely than not, is worse than no plating at all.[3]

Construction

Filter construction is simple and straightforward. Measurement accuracy is important. All measurements should be made with accurate calipers. Use a very sharp metal scribe to indicate all lines and locations. Remember that small scribing errors can amount to many MHz deviations in the finished product. Cut a section of waveguide to the proper length as shown in Fig 1. Measure and scribe a center line down the middle of the guide. Scribe the location of the first post hole and with the calipers measure and scribe each succeeding post and screw hole. Mark all holes to be drilled with a very sharp center punch. Do not use an automatic punch since there is a danger of bowing or bending the waveguide wall. Lightly punch all hole locations. Secure the waveguide section in a small vise while drilling holes. Do not attempt drilling with a hand drill. All holes may now be drilled to the proper size. Remember that holes for the posts are drilled through the top and bottom of the guide, screw holes only through one face. Holes for the 6-32 screws should now be tapped. Burrs should be removed from the inside of the guide using a very fine file, taking care not to leave any deep scratches on the interior surface. The brass posts should be cut to length and inserted in their proper holes. Soft solder can be used to solder posts in the guide; however, silver solder is highly recommended. Using silver solder eliminates the possibility of posts dropping out when you are attaching flanges. Soft solder also tends to run inside the guide and form a dome around the post ends. It also runs down the outside of the posts, leaving a glob which tends to drastically change the cavity frequency. Silver solder reduces these problems. Immediately after soldering, dunk the entire assembly in a 10% solution of sulfuric acid for a few seconds. Then rinse thoroughly. This will remove all flux and leave the assembly clean, with a nice pink color. The filter is now complete and only flanges need be added. Either soft solder or silver solder may be used on the flanges. After the flanges are attached it would be wise to run the assembly through another dunk-rinse cycle. After drying, the filter inte-

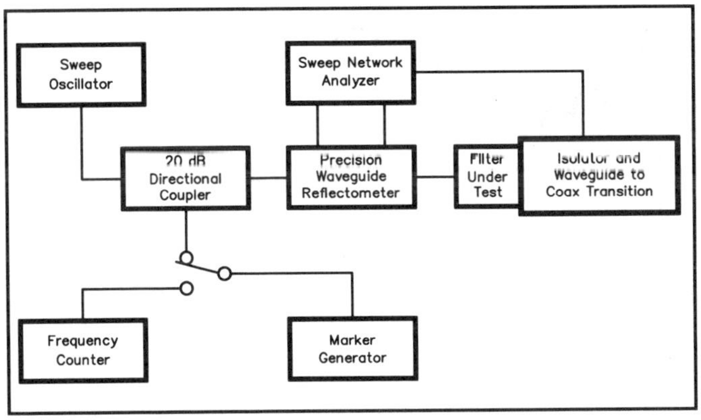

Fig 2—Ideal filter test setup.

rior should be checked for any metal shavings or other obstructions that might have lodged in the guide. Needless to say, protective goggles should be worn during the entire process.

Alignment

Tests and alignment of this filter were conducted using the test set up shown in Fig 2. This is a fairly common arrangement in the commercial field. It is the most convenient way of rapidly determining filter parameters, including return loss, response curve and insertion loss. The average ham usually is not lucky enough to have access to such equipment. With this fact in mind, numerous tests were conducted using the simple set up shown in Fig 3. The filter was aligned first on the band edges and then the band center. After alignment on each frequency, I used a network analyzer to determine if the band pass characteristics and SWR would be satisfactory for amateur use. Although the simple alignment method is not perfect, I found it satisfactory for most uses. The signal source should have at least 10-dBm output and the power meter should be capable of reading down to at least –25 dbm. Although you can't measure SWR, when I used the simple method of alignment, SWR never exceeded 1.4:1. The response of a perfectly aligned filter is shown in Fig 4.

Before starting alignment, place the filter in a small table vise or anchor it in some way. Nothing should move but the screws you will be adjusting. Thread oversize (5/16-in.) nuts on the silver-plated screws, all the way up to the screw head. With the signal source adjusted to the desired frequency, measure and record the power output of the source, by connecting the two coax-to-waveguide transitions face to face. Reconnect the transitions to the filter. It makes no difference which end of the filter goes to the source or load. Common practice for UG-344 or UG-343 flanges is that the cover flange faces the source and the choke flange faces the load. Set the power meter for maximum sensitivity. Insert only the screw for the middle cavity, and slowly turn it inward while watching the power meter for a very slight upward movement. This will depend on the power output of the your source. Turn slowly and watch carefully. Once this small peak is found the battle is half won. Gently run the nut down to the waveguide surface and finger tighten, being sure not to lose your peak. The input and output cavity screws may now be inserted. Alternately adjust the input and output cavity screws for maximum power meter reading. If things are going well, you should have had to change power-meter scales

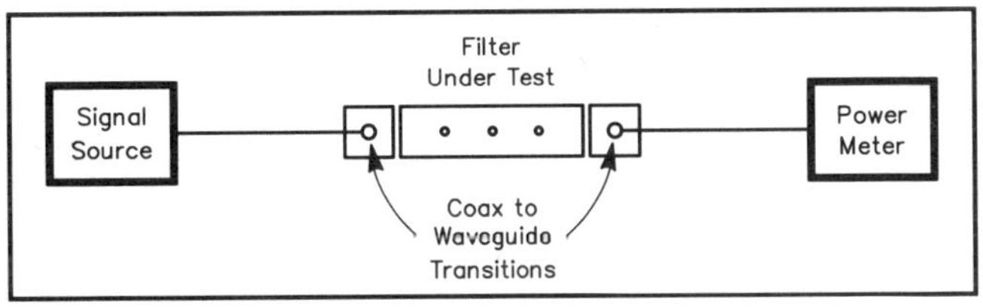

Fig 3—Filter test setup adequate for amateur applications.

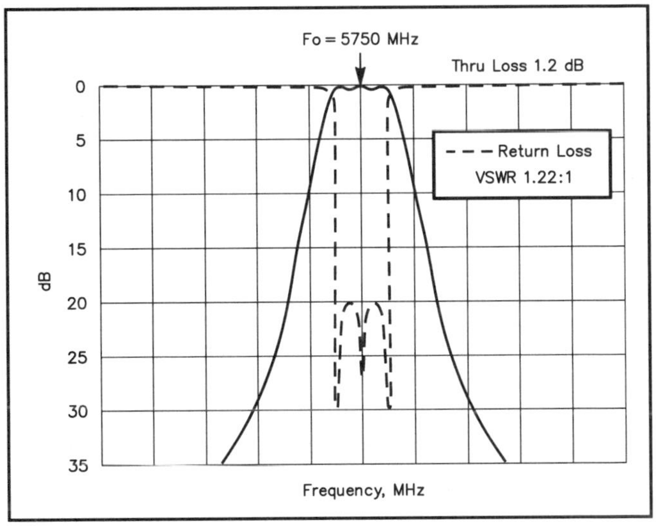

Fig 4—Filter response curve.

a couple of times. You will find that just finger tightening the screws during this portion of the alignment saves a great deal of time and is much easier on the soft copper threads in the waveguide. After you are satisfied that there is no more to be gained from the input and output screws, readjust the middle cavity screw for a peak power reading. Then, alternately adjust input, middle and output in that order for maximum power. You'll have to do this a number of times, as the tuning screws interact. To get an idea of the filter loss, compare the power meter reading against that recorded earlier. Final adjustments should be made with a small wrench on the nuts. Be careful not to overtighten. If you do have an accident and strip the waveguide threads, all is not lost. Rethread the hole for an 8-32 screw and proceed. This filter has been used with screws up to 8-32. When the alignment is complete, check the source power again, to accurately determine the filter passband loss. In most cases, if the pass-band loss is no greater than 1.3 dB, the skirts, stop-band attenuation and SWR should be satisfactory for most amateur work. If you need to paint the filter, be sure that all the screws and nuts are masked, so you may can shift frequency later.

Notes

[1]R. M. Fano, "Microwave Filters Using Quarter-Wave Couplings," *Proceedings of the IRE*, Nov 1947, pp 1318-1323.

[2]E. Tahan, "Microwave Filter Design Techniques," *Microwave Journal*, Mar 1962, pp 111-116.

[3]G. L. Ragan, *Microwave Transmission Circuits*, pp 126-127 (Vol 9 of the MIT Radiation Laboratories Series, New York: McGraw-Hill, 1946).

A Waveguide Band-Stop Filter for 5760 MHz

By James D. Green, K5JG

Where a bandpass does not provide sufficient discrimination against interference, use a band-stop filter.[1] Band-stop filters have the opposite characteristics of a bandpass filter. Signals within a certain finite band of frequencies are suppressed in a band-stop filter, while frequencies above and below that band are passed. Most amateurs have neither the time nor inclination to go through the ordeal of designing such a filter, and their high cost usually precludes purchase. The band-stop filter described here is simple to construct, easy to tune and has sufficient attenuation to clean up practically any interference problem.

Fig 1—5.7-GHz band-stop waveguide filter.

Description

The band-stop filter shown in Fig 1 contains three similar resonators or cavities, spaced along the main waveguide arm. These cavities are coupled to the main waveguide through circular irises, and are spaced to mostly cancel the reactive component presented in the main guide. Each iris looks like an inductance in series with the main guide.[2,3] Each cavity is cut somewhat shorter than the highest frequency desired; in this case, just above 5.925 GHz. The cavities are tuned by a screw centered on their broad walls. Penetration of the screw lowers the cavity frequency.[4] As with any screw-tuned waveguide cavity, the Q is reduced as the screw penetrates the cavity, but is acceptable at the low end of the band, where penetration is maximum. Q can be improved by using silver-plated screws.

Construction

As with any other microwave construction project, use the utmost care in all measurements. Refer to Figs 2 and 3 during the following steps. Scribe and saw the main section and the three cavities from a piece of WR-137 waveguide. If possible chuck the waveguide in a lathe or cut it with a band saw. A hacksaw will suffice if certain precautions are taken: Make scribe lines from the dimensions in Fig 2, but make a saw line slightly greater than the correct dimensions. This oversize cut will allow for errors in squareness. After sawing, the pieces can be filed down to the correct length and squareness of cut. The overall length of the main section is not critical, however,

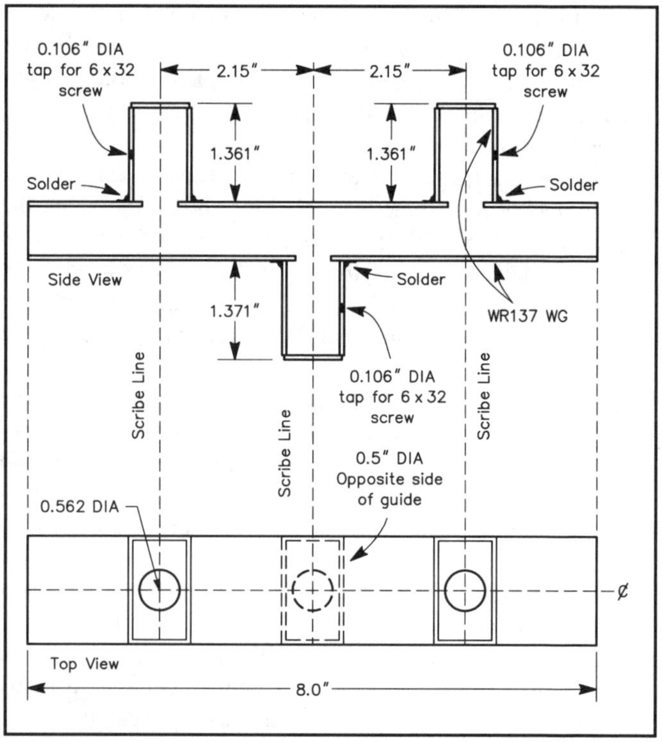

Fig 2—Mechanical details of the 5.7-GHz band-stop filter.

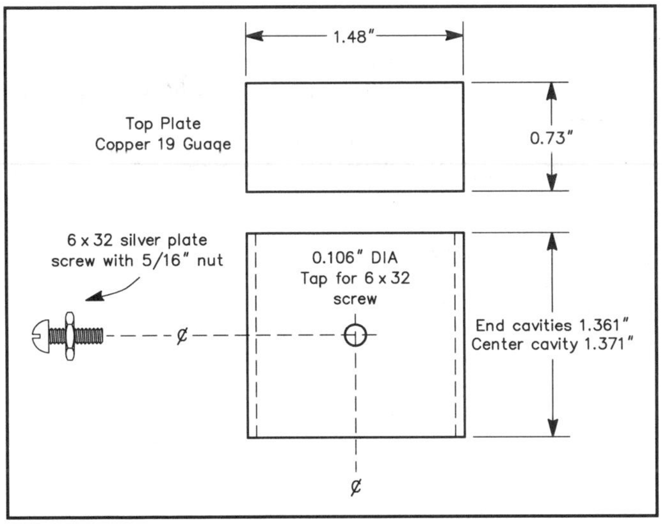

Fig 3—Cavity assembly.

the ends should be square for correct flange fit. The length of each cavity is critical, and each should have perfectly square ends. Smooth all rough edges on the ends with a fine file before proceeding. Determine the center of the broad wall of each cavity, and mark and drill with a No. 36 bit. Take care during drilling not to bend the cavity wall inward. Tap each of the three holes for a 6-32 screw. After tapping, smooth the rough edges of the tapped hole with a fine file. Remember, the inside determines the performance and the outside determines the looks.

Next, locate the center line of one narrow wall of each cavity and scribe a heavy line down that wall. This line will later be used as an alignment marker. Measure and scribe the location of the iris holes in the broad wall of the main waveguide section. Bring the center scribe line for each hole completely across the broad wall and down the narrow wall. (see Fig 2) This line will be used as an alignment marker for centering the cavity over the iris hole. If these holes are to be drilled on a drill press, secure the section of waveguide in a vise, with a C clamp, or both, while drilling. If the waveguide moves, the iris hole won't be round. Mark the initial hole with a small center drill, then use an intermediate-size drill before using the final size. After you've drilled all the iris holes, clean all burrs around each hole, inside and out. This is the area that this filter differs from most commercial versions. Ideally, the thickness of the wall surrounding the iris hole should be as thin as possible. In a commercial filter, this area is milled down across the broad wall of the guide, making the entire wall facing the cavity extremely thin. This refinement is not necessary for general amateur use. You should now be ready to solder the cavities to the main guide. Silver solder is highly recommended. Soft solder can be used, but the joints may melt when flanges and cavity end plates are soldered.

With the main waveguide arm in a vise, place any one of the cavities over the iris hole. Use the center-line scribe marks on the narrow walls to line up the cavity over the center of the iris. The cavity should be held to the main section with a small C clamp or weight. Be sure that only enough pressure is used with the C clamp to hold the cavity securely. It is extremely easy to bend the broad wall of the guide, particularly after it is heated. Use as little solder as possible, to prevent excessive amounts from accumulating inside the cavity. Proceed in a similar fashion with the other cavities. After the last cavity is soldered, dunk the assembly in a 10% solution of sulfuric acid and rinse thoroughly. This will clean the metal surfaces and remove flux residue. Flanges may now be attached to the main guide, using either silver or soft solder. This is the last time you'll be able to see inside the cavities. They should be relatively clean, and free of any obstructions. The copper end plates may now be soldered into position. Notice that the end plates are slightly smaller than the outside end dimensions of the waveguide. This was done purposely. The end plates should be soldered solidly around the waveguide perimeter, with no gaps or holes. The undercut size allows you a better view of how well the solder flows between the end plate and the guide. It will be much easier to soft solder the end plates. Watch this area carefully. It is easy to get a large solder glob inside the cavity if you apply too much solder to the end plates.

Alignment

A filter of this type would normally be aligned with a sweep generator and network analyzer or spectrum analyzer. For amateur use, this precise alignment is not required. Maximum attenuation of the filter stop-band was greater than 60 dB, but it could not be measured accurately. At least 40-dB attenuation should be sufficient to eliminate most interference problems. An alignment setup like the one described in the preceding article should be adequate.

Thread three silver-plated 6-32 screws with oversize 3/16-in. nuts. Insert a screw into the middle cavity, and slowly turn the screw inward with your fingers until the power meter deflects downward. This tuning is extremely sharp. Run the nut down flush with the guide and adjust the screw for half-scale reduction of the original reading. Finger tighten the nut and adjust the other cavities, using the same procedure. Tune each cavity for a half- scale reading, while increasing the power meter sensitivity as required. Tune each cavity a little at a time, until overall maximum attenuation is reached. Do not attempt to tune any one cavity for maximum attenuation. The attenuation in all probability would make further tuning impossible. In other words, one cavity tuning could be lost as far as knowing exactly the location of its resonance. After reaching maximum attenuation or the limits of measuring capability of the test set up, tighten all lock nuts firmly, without changing cavity tuning. A word of warning is in order. Some power-meter detector heads are very easily damaged by overheating. A slip of the fingers or screwdriver during the latter part of the tuning procedure could apply excessive power to the detector mount.

Those fortunate enough to own or have access to proper test equipment allowing return loss measurement over a wide band can make a worthwhile addition to the filter. Since the iris spacings do not represent perfect impedance inverters, the effect of this addition will tend to reduce the Chebyshev ripple in the passband, and also the return loss. This can be accomplished by adding 6-32 screws in the guide broad wall, exactly centered over the iris but on the opposite wall. The screw end should point directly at the center of the iris hole. Add screws for the outside cavities only. Screws should be tuned alternately for minimum passband ripple and return loss in the passband. This addition should only be done if adequate test equipment is available. This modification is for purists: the filter is quite suitable for amateur use as is. After alignment, the filter may be painted if all cavity screws, nuts, and flange faces are masked.

Notes

[1]Mathaei, Young and Jones, *Microwave Filters, Impedance-Matching Networks and Coupling Structures*, (New York: McGraw-Hill, 1948) p 417.

[2]Southworth, "Principles and Applications of Waveguide Transmission," (New York: Van Nostrand, 1950) pp 251-253.

[3]Mathaei et al, pp 434-438.

[4]Montgomery, Dicke and Purcell, *Principles of Waveguide Circuits*, (New York: McGraw-Hill, 1948) p 168.

[5]Mathaei et al, pp 751-755.

24.1-GHz Transition

By Kent Britain, WA5VJB

[From Proceedings of Microwave Update '88]

When I first started some serious work with 24 GHz, one of the first items I needed was a waveguide-to-coax transition. Very few of the commercial K-band transitions are designed for use near our 24-GHz ham band, so I came up with this design optimized for 24.1 GHz (Fig 1).

In constructing this transition, you must use either K or SMA connectors. I highly suggest the 2-hole flange mounts. Avoid stainless-steel SMA connectors, which are impossible to solder. It can be a bit tricky to solder the SMA and the backplate at the same time, so use plenty of clamps or a vise. My thanks to K5SXK for his help with the "Wavelength in Guide" calculations.

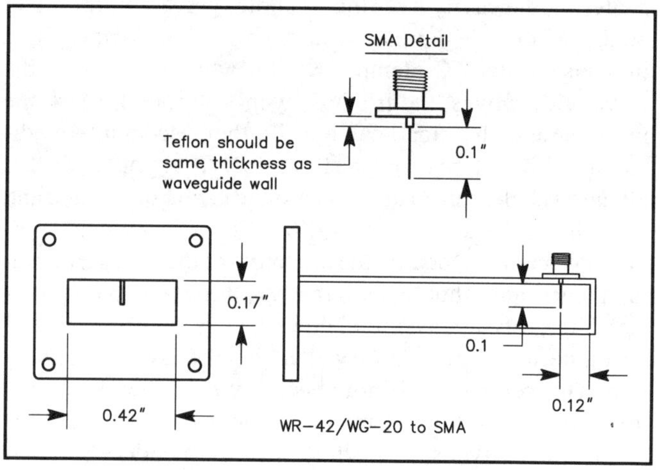

Fig 1—24-GHz coax-waveguide transition.

Multiple-Cavity, Iris-Coupled Waveguide Filters For X Band

By Bob Atkins, KA1GT

(From *QST*, March, 1984)

Waveguide filters have a number of applications in amateur microwave work. Among them are suppression of the image response in a receiver and selection of a desired harmonic from a step-recovery diode multiplier. The type of filter described here is a waveguide cavity, iris-coupled filter. It consists of a number of resonant waveguide cavities, each cavity being coupled to the next by a circular iris.

The characteristics of the filter are determined by three parameters: the number of cavities, the size of the coupling irises and the length of the cavities. The number of cavities primarily determines the skirt response of the filter— attenuation outside its passband.

Typical skirt response curves are given in Fig 1 for 10-GHz filters consisting of 1-4 cavities. The size of the coupling irises predominately affects the response of the filter within the passband. All filters will have some passband ripple, the magnitude of which is a function of the intercavity coupling. Ripple can be diminished by using larger irises, at the expense of lower attenuation outside the passband. Fig 2 shows the passband response of typical filters with 1-4 cavities designed with a 1-dB passband ripple. These are theoretical curves and, in practice, 0-dB insertion loss is never achieved. In general, the more cavities there are, the higher the minimum insertion loss. Typical insertion losses are on the order of 1 dB for a well-constructed filter. If the filters are redesigned for a 0.1-dB passband ripple, the passband pattern remains the same, but the amplitude of the ripple is reduced by a factor of 10, while attenuation outside the passband (see Fig 1) is reduced by about 10 dB (for a 4-cavity filter) at all frequencies.

The length of the cavities determines the resonant frequency of the filter. Resonant frequency can be lowered by using a tuning screw in the broad face of the waveguide (Fig 3). As it is very difficult to construct the filter accurately enough to operate at a given frequency, it is common to design it for a frequency higher than required (for example, 10,500 MHz) and then use tuning screws to tune it to

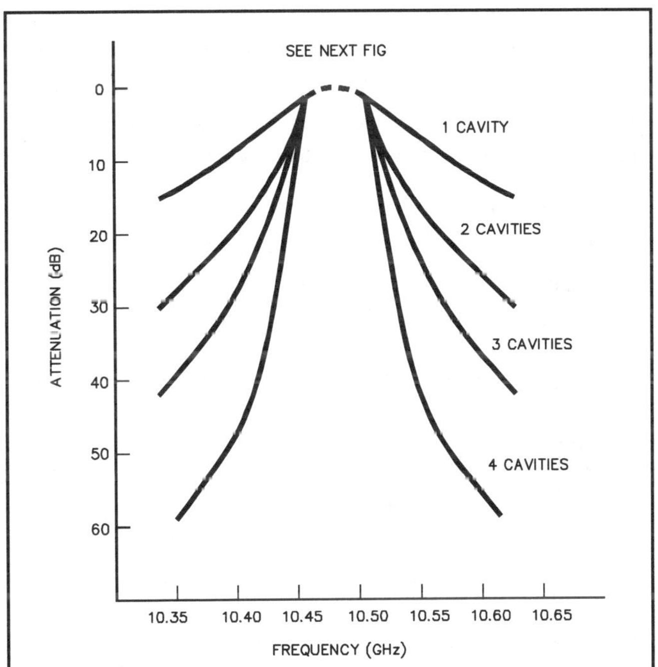

Fig 1—Skirt response of typical waveguide cavity filters with 10.5-GHz center frequency, 60-MHz passband (1-dB bandwidth) and 1-dB passband ripple.

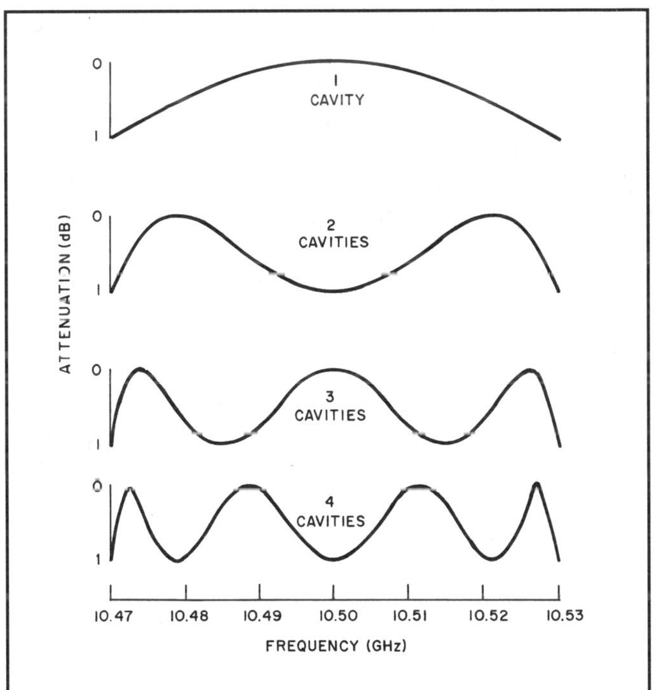

Fig 2—Passband response of waveguide cavity filters described in Fig 1.

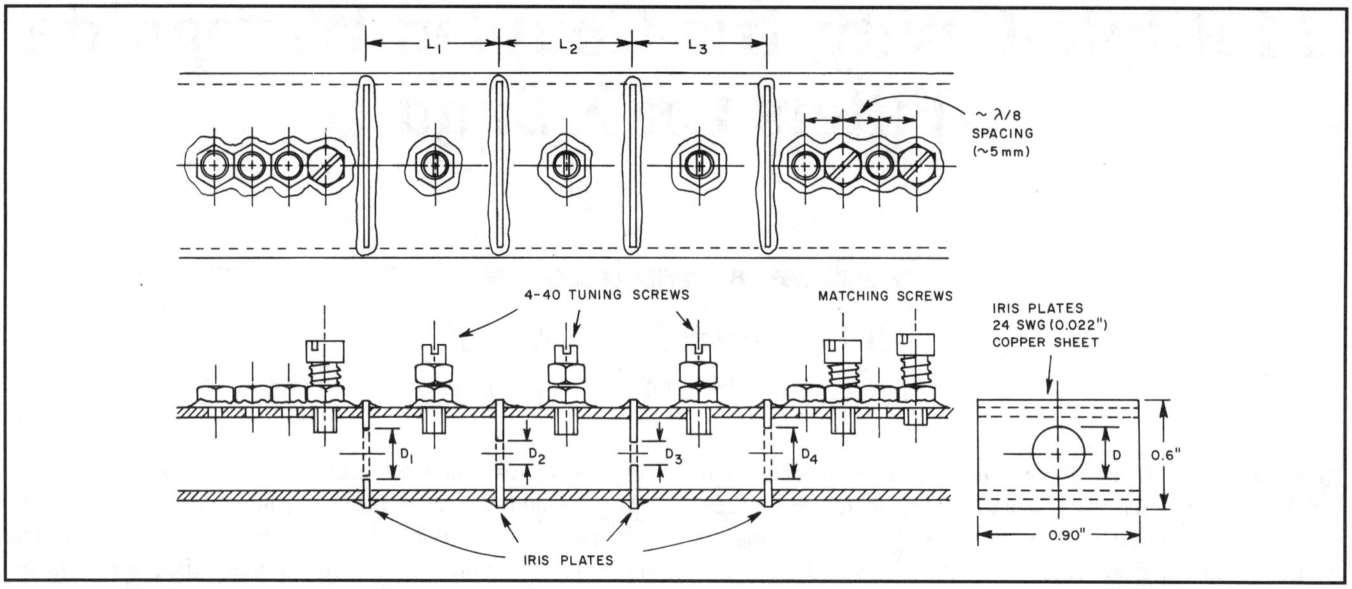

Fig 3—Three-cavity waveguide filter.

the desired frequency. It is also possible to raise the resonant frequency by placing 1/4-20 tuning screws in the narrow face of the waveguide.

Construction Of Waveguide Filters

Figs 3 and 4 show the construction details of a waveguide filter. To minimize insertion loss and maximize the cavity Q of the filters, they should be made of copper, rather than brass waveguide. If you use brass, have it silver plated after construction. It's also important to prevent soldering flux from entering the cavities. Remove flux with solvent after soldering. Tuning screws should also be copper or silver-plated brass.

The first step in construction is to mark out the waveguide as accurately as possible, according to Table 1. Cut slots across the broad faces of the waveguide at the marked positions, using a jeweler's saw or a thin hacksaw blade. The width of the blade cut should be just wide enough (0.022 in.) to allow the iris to be inserted into the waveguide. Cut the iris plates to size from copper sheet and drill the iris holes. Working with thin copper sheet may be easier if the copper is first sandwiched between two thicker sheets of aluminum or brass. The thicker material minimizes distortion while drilling and sawing. Clearance holes for the tuning and matching screws can now be drilled in the waveguide.

At this stage, the inside of the waveguide should be deburred with a file and cleared. Insert the iris plates in the waveguide and solder them, top and bottom, with a large (about 200-watt) soldering iron. Take care not to get solder or flux inside the wave-guide. The iris plates do not need to be soldered to the narrow waveguide faces. Solder the nuts for the tuning screws in place. The nuts may be jigged in place with cadmium-plated screws, which do not take solder well. You can prevent previously soldered areas from melting by apply wet tissues to them.

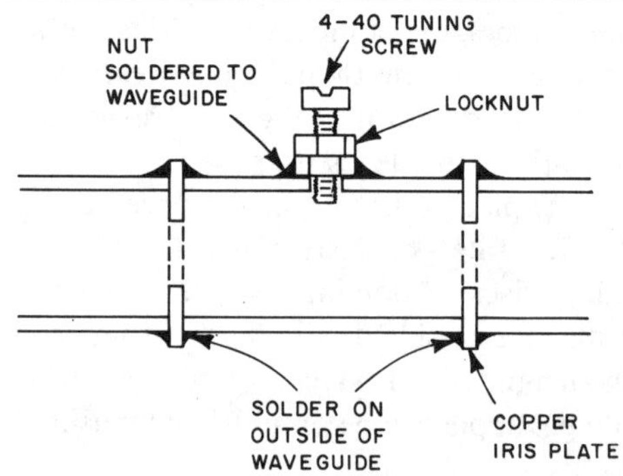

Fig 4—Detail of waveguide cavity.

Alignment and Use

When filters of this type are not tuned to the signal frequency, high insertion loss precludes alignment with a signal

Table 1
Dimensions for Multicavity Filters

No. of Cavities	*D1*	*D2*	*D3*	*D4*	*D5*	*L1*	*L2*	*L3*	*L4*	*Attenuation at 10.530 GHz*
1	7.54	7.54	—	—	—	17.1	—	—	—	1.7 dB
2	6.86	3.67	6.86	—	—	17.8	17.8	—	—	8.4 dB
3	6.74	3.40	3.40	3.40		17.9	18.2	17.9	—	19.2 dB
4	6.70	3.34	3.19	3.34	6.70	17.9	18.2	18.2	17.9	30.6 dB

Note: Dimensions are for a filter with a 10.5-GHz center frequency, a 0.5-dB bandwidth of 30 MHz and 0.5-dB passband ripple. Iris plate thickness is 0.222 in. All dimensions are in millimeters.

source and diode detector. If a sensitive, tunable receiver (Gunnplexer or other) is available, it should be possible to align the filter with a Gunn-diode source. If the equipment is available, connect one end of the filter to a dummy load and the other to an RF power source via a directional coupler. As the filter is tuned to the signal frequency, the reflected power measured by the directional coupler will decrease.

In either case, once the filter is tuned to the signal frequency, matching screws may be inserted as shown in Fig 3. Four positions for matching screws, spaced about ⅛ apart, are available at each end of the filter. These screws should be adjusted to give minimum insertion loss. Most of the time, only one or two screws are required at each end, depending on the degree of mismatch.

7

Low Level Amplifiers

Contents

GaAsFET Preamplifier for 70 cm

By Chip Angle, N6CA

(From *The ARRL Handbook*)

The preamplifier described here and shown in Figs 1 through 3 offers good noise performance and gain, and is suitable for terrestrial, EME and satellite applications at 432 and 435 MHz. Gain is approximately 15 dB and the noise figure is 0.55 dB, measured on an HP8970A noise-figure meter with the HP346A noise source. This preamplifier is easy to build and offers stable operation with little adjustment.

Circuit Details

The schematic diagram of the 70-cm GaAsFET preamplifier is shown in Fig 2. The circuit was originally designed for an NEC NE21889 GaAsFET, but a Mitsubishi MGF1402 or MGF1302 device will work fine. The version shown here uses an MGF1402, but most single-gate GaAsFETs will work.

This preamplifier uses source feedback to bring the input impedance close to 50 ohms. Input and output return loss is typically better than 15 dB. This means that a band-pass filter can be used ahead of the preamplifier without introducing mismatch loss or instability.

The design uses plenty of decoupling capacitors and RF chokes. Good-quality bypass capacitors are a must for stable operation, so use chip capacitors where specified. D1 and D2 are included for reverse and overvoltage protection. Regulator U1 allows virtually any voltage greater than 9-V dc to be used.

Construction

The preamplifier is built on a G-10 glass-epoxy printed-circuit board using surface-mounting techniques. Layout is shown in Fig 3; Fig 4 is a full-scale template. All components mount to the circuit traces. The other side of the board is unetched copper except around the connector center pins, and acts as a ground plane. All grounded pads on the etched side are connected to the ground plane with eyelets or pieces of tinned wire that pass through holes in the board and are soldered to both sides.

The board is mounted to the lid of a die-cast box (Hammond 1590B or Bud CU-124) and is designed to accommodate type-N connectors. Aluminum spacers for the connectors must be fabricated, as shown in Fig 3. These spacers mount between the box lid and the ground plane of the PC board. Q1

Fig 1—The 70-cm GaAsFET preamplifier is built on a PC board that is mounted to the lid of a diecast box. This version uses a Zener diode instead of the 3-terminal regulator shown in Figs 2 and 3. Connector mounting hardware also holds the board in place. Component layout is shown in Fig 3.

has two source leads. Only one is used on this board, and the extra lead may be clipped off. The ferrite bead on the drain lead mounts in a small hole cut in the board. The drain lead passes through this bead and is soldered to a pad.

GaAsFETs are static-sensitive, so handle them accordingly. It's best to assemble the rest of the circuit and install Q1 last. Use a grounded iron if possible, and solder the leads quickly.

Adjustment

The preamplifier shown in Fig 1 was assembled in the ARRL lab and tested before any adjustments were made. For starters, C2 was set at midrange. As assembled, the preamp measured 13 dB gain with a 0.57-dB noise figure—perfectly acceptable performance for a device such as this. After adjusting C2 and bending L3 and L5, the gain increased to 15.25 dB with a 0.54-dB noise figure. This is acceptable for all but perhaps the most demanding EME applications. No amount of spreading or compressing the coils could make the gain fall

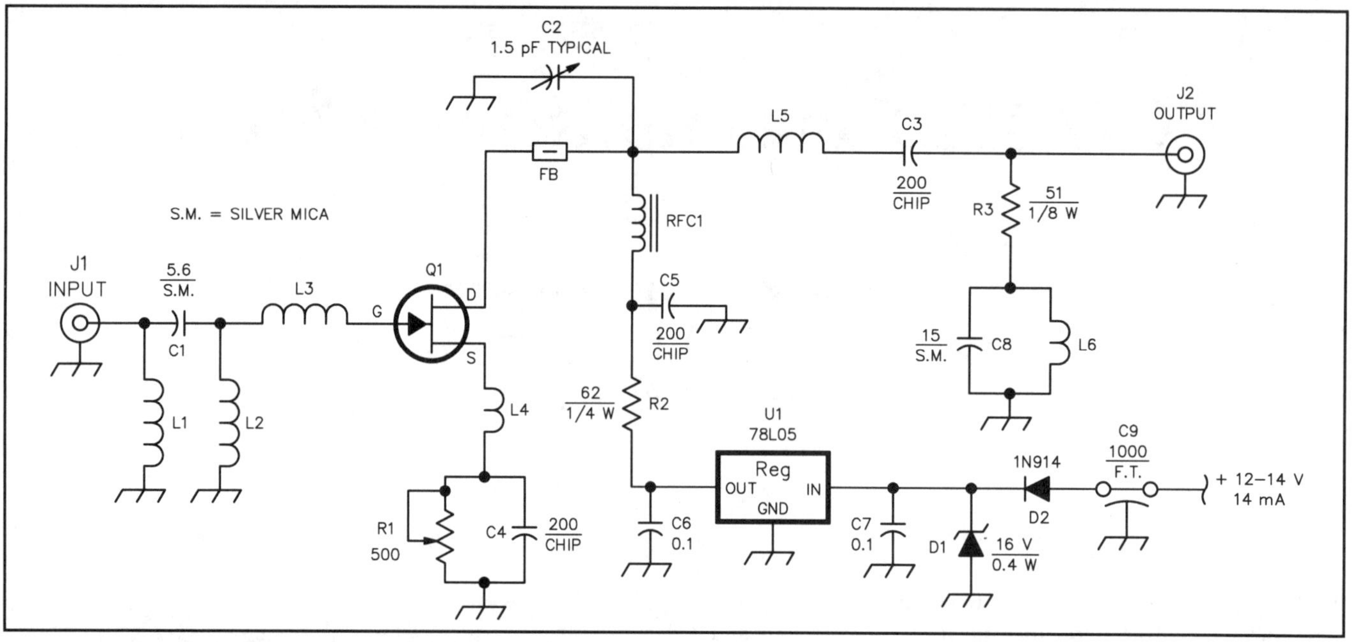

Fig 2—Schematic diagram of the 70-cm GaAsFET preamplifier. Resistors are carbon-composition types. Resistor values are given in ohms; capacitor values are given in pF.

C1—5.6-pF silver-mica capacitor or same as C2.
C2—0.6- to 6-pF ceramic piston trimmer capacitor (Johanson 5700 series or equiv).
C3, C4, C5,—200-pF ceramic chip capacitor.
C6, C7—0.1-µF disc ceramic capacitor, 50 V or greater.
C8—15-pF silver-mica capacitor.
C9—500- to 1000-pF feedthrough capacitor.
D1—16- to 30-V, 500-mW Zener diode (1N966B or equiv).
D2—1N914, 1N4148 or any diode with ratings of at least 25 PIV at 50 mA or greater.
FB—FB-43-101 or FB-64-101 ferrite bead.
J1, J2—Female chassis-mount Type-N connectors, PTFE dielectric (UG-58 or equiv).
L1, L2—3 t no. 24 tinned wire, 0.110-inch ID, spaced 1 wire diam.
L3—5 t no. 24 tinned wire, 3/16-inch ID, spaced 1 wire diam. or closer. Slightly larger diameter (0.010 inch) may be required with some FETs.
L4—1 t no. 24 tinned wire, 1/8-inch ID.
L5—4 t no. 24 tinned wire, 1/8-inch ID, spaced 1 wire dia.
L6—1 t no. 24 tinned wire, 1/8-inch ID.
Q1—Mitsubishi MGF1302 (see text).
R1—200- or 500-Ω Cermet potentiometer set to midrange initially.
R2—62-Ω, 1/4-W resistor.
R3—51-Ω, 1/8-W carbon-composition resistor, 5% tolerance.
RFC1—5t no. 26 enam. wire on a ferrite bead.
U1—5-V, 100-mA 3-terminal regulator (LM78L05 or equiv TO-92 package).

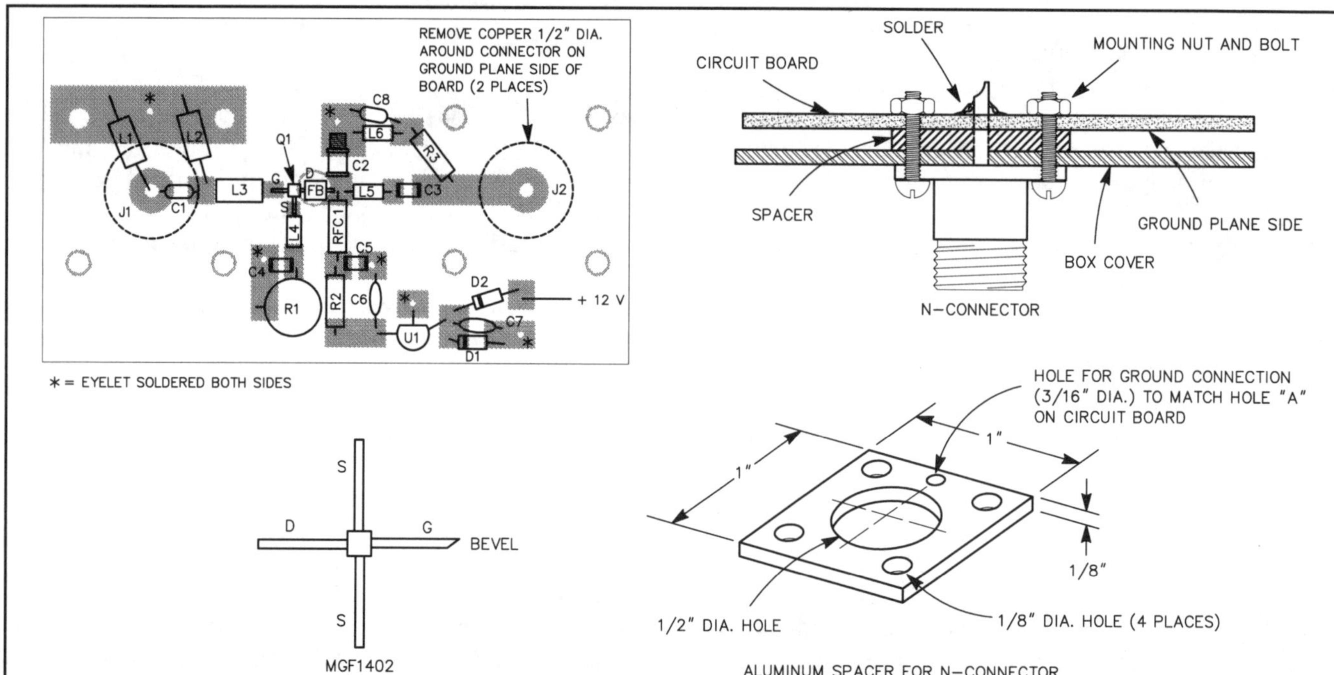

Fig 3—Parts-placement diagram for the 70-cm preamplifier.

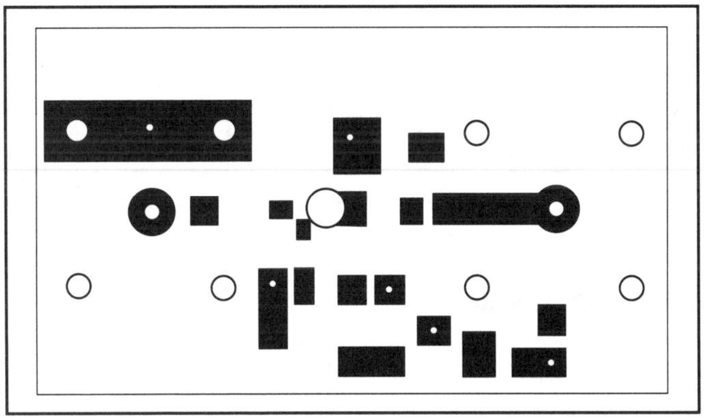

Fig 4—A full-size etching pattern for the preamplifier.

below 12 dB or the noise figure rise above 0.65 dB. At no time did the preamp display signs of instability. In short, this is a reproducible project. You can assemble it without test equipment and be reasonably sure of the performance.

A Half-Watt 903-MHz Amplifier

By Zack Lau, KH6CP

(From May 1988 *QEX*)

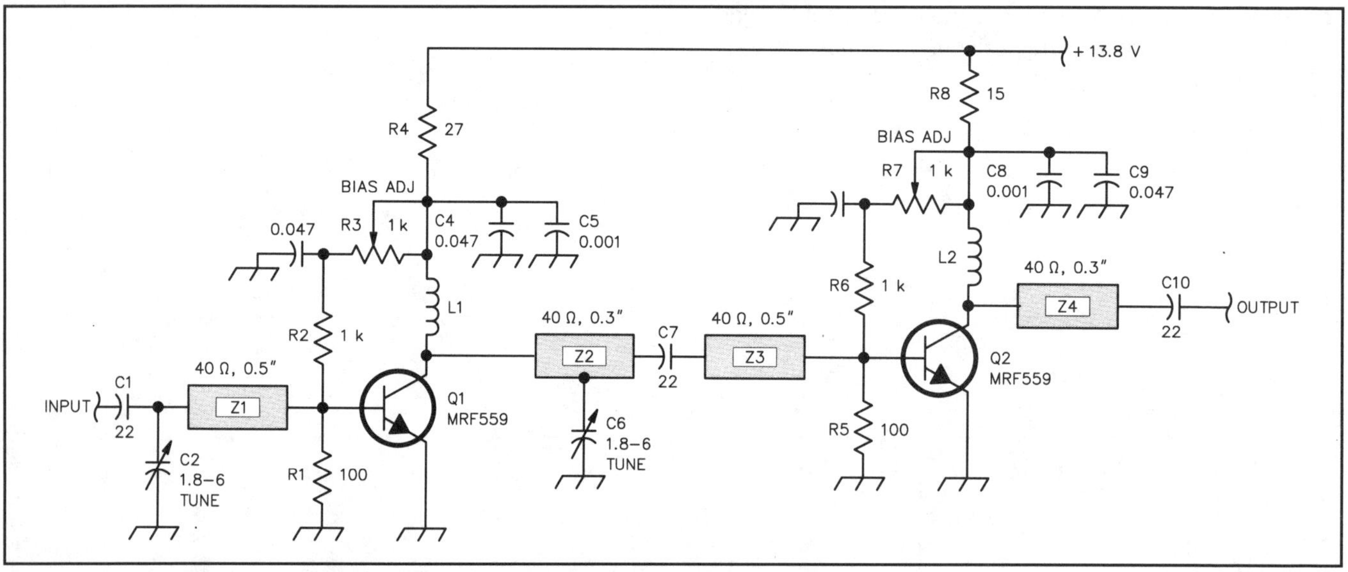

Fig 1—Schematic of an inexpensive 0.5-W 903-MHz amplifier. Resistors are ¼ W. Capacitors are NP0 chip capacitors unless noted, although more expensive porcelain capacitors may work better.

C2, C6—1.8-6 pF trimmer capacitors. Mouser 24AA70.
C3, C4, C9—Tiny ceramic capacitors or chip capacitors.
L1—7 t no. 22 tinned copper wire, space wound; no. 32 drill bit used as temporary form.
L2—5 t no. 22 tinned copper wire, space wound; no. 33 drill bit used as temporary form.
Q1, Q2—MRF559.

The two-transistor amplifier shown in Fig 1 is an inexpensive way to amplify the 10 mW of RF output available from a monolithic microwave integrated circuit (MMIC) to 0.5 W on the 903-MHz band. Q1 and Q2, MRF559s, are inexpensive plastic-cased devices. When driven with an Avantek MSA-0404 MMIC (around 10 mW saturated output), the saturated CW output is 0.5 W as measured with an HP435/8482A power meter and a Bird 10-dB attenuator. A two-tone IMD test indicates that the MMIC saturates before this amplifier does, since the higher order IMD products are down 47 dB. See Fig 2.

The amplifier is built on a 1/16-in., double-sided glass-epoxy PC board (Fig 3). All components mount on the etched side. The other side is left unetched to act as a ground plane. The 40-Ω striplines (Z1-Z4) are made using 0.15-in.-wide traces.

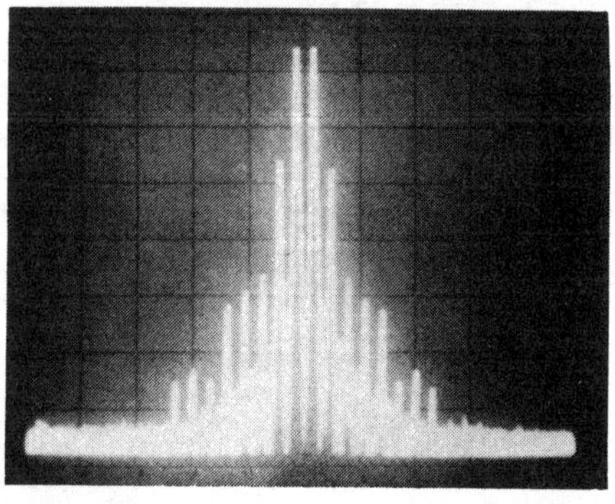

Fig 2—Spectral display of the 0.5-W 903-MHz amplifier during two-tone intermodulation distortion (IMD) testing. Third-order products are approximately 26 dB below PEP output, and fifth-order products are approximately 47 dB down. Vertical divisions are each 10 dB; horizontal divisions are each 10 kHz. The amplifier was being operated at 560-mW PEP output on 903.1 MHz.

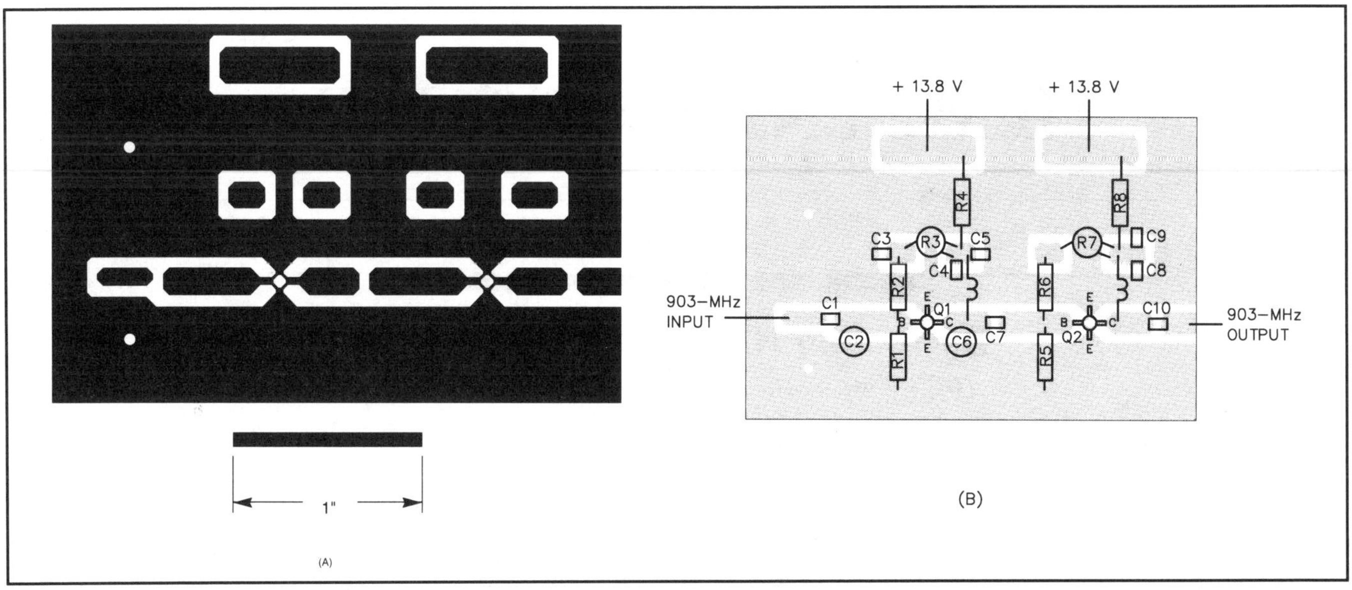

Fig 3—Circuit-board etching pattern (A) and part-placement diagram (B) for the 0.5-W 903-MHz amplifier. The etching pattern is shown full size from the etched side of the board. Black areas represent unetched copper foil. Board material is 1/16-in. G-10, double sided. The other side of the board is unetched to form a ground plane. The shaded area in B represents the copper pattern. All components are mounted on the etched side of the board.

Use good VHF/UHF construction techniques and keep the leads short when mounting components. Drill oversize holes for Q1 and Q2 so the leads lie flush against the board traces. Use copper foil wrapped through the transistor mounting holes to connect the top and bottom ground planes at the emitters of Q1 and Q2. Wrap the board edges with copper foil to connect the top and bottom ground planes.

Use R3 and R7 to adjust bias currents to 22 mA for Q1 and 87 mA for Q2. Note: These are the *total* currents as indicated by the voltage drops across R4 and R8. The bias currents do drift slowly without RF drive, but this is not a problem as long as the supply voltage is switched off during receive. The prototype works fine without a bypass capacitor at the junction of R6 and R7.

Tune-up is simple. Apply drive and adjust C2 and C6 for maximum output. Attempts at matching the output with a tunable network failed to improve gain or output power. Better results may be obtainable by using Teflon board and porcelain chip capacitors.

HEMT LNA for 1296 MHz

By Tommy Henderson, WD5AGO

(From *Proceedings of Microwave Update '91*)

Introduction

EME (Moon Bounce) is a very demanding mode of operation. A tenth of a decibel (dB) gained on receive noise figure (NF) can offer a 0.5- to 1.0-dB system improvement. Most GaAsFET preamps have a difficult time achieving below 0.5 dB NF (35 Kelvin) at room temperature on 1.3 GHz. Cooling the preamp[1] is one way of achieving lower noise figures; but cooling is somewhat difficult to do. Another possibility is to design a preamplifier around super low noise devices called HEMTs (High Electron Mobility Transistors). Data sheets specify HEMTs to have noise figures between 0.2 to 0.35 dB and gain of 17 to 20 dB, at a frequency of 2 GHz. The problem with most HEMTs is maintaining stability at low frequencies, because they are designed to be operated above 4 GHz.

Circuit

The original work to evaluate the HEMTs at 1296 MHz took place on a test fixture, where parts could be installed and removed easily without the use of solder. The first item of concern was to monitor the noise figure while checking for instability (the latter turns out hard to evaluate without the use of computer modeling). Grounding the source leads and not using source bypass capacitors aided in both categories. Negative bias must now be used on the gate and should be turned on before or at the same time as V_{DD}. A battery is the simplest way to apply gate bias; you can also use a power supply.[2,3] Another benefit of grounded source is that the operating perimeters for lowest noise figure are easily found for several different devices through the negative bias adjustment.

HEMTs that were tested included the MGF4303, FHC30LG, NE32184, NE20283, and ATF35176. All devices showed noise figures less than 0.5 dB and gains from 18 to 22 dB! The Avantek ATF35176 is one of the newest line of P-HEMTs and is available from Penstock Inc. Noise figures of 0.39 to 0.55 dB were achieved, while gains ranged from 12 to 18 dB. Best results were obtained with the ATF35176. Because NF results seem to be different from meter to meter, an ATF10135 preamp was used as a standard. Also, the ATF10135 GaAsFET was placed in the test fixture for further comparison.

Stability has been a problem with the HEMTs, and I tried several different methods of taming them. Place conductive foam near the circuit, keep the source inductance less than 0.07 inches, and don't use ferrite beads on the drain lead. Good component layout also helps.

After the original circuit was built and used, an opportunity to have the circuit modeled using the Touchstone computer software design program shed light on the potential instability problems. To correct the low-frequency instability, I removed the input choke to ground; for high frequency instability, a series resistor with a controlled amount of inductance was added. Retuning the circuit yielded a preamp with a stability factor K above one from 100 to 20,000 MHz. The gain dropped 3 dB and noise figure increased only 0.04 dB. It is difficult to obtain the figures given on computer modeling, due to inaccurate S parameter data below 2 GHz, and stray reactances that may enter the real circuit.

Construction Notes

I didn't make a circuit board. Instead, all connections other than grounds were made point to point in air, to reduce losses. You can make a box from unused double-sided circuit board. Solder the source leads of Q1 directly to the box. Alternately, solder the source leads to a ½- × 1-in.-piece of thin brass, and bolt the brass to the box. This method helps prevent overheating the HEMT.

Adjustment

Adjust V_g and V_d for lowest noise figure. Typical values are:

V_g: – 0.40
V_d: 2.5
NF: 0.4 dB
Gain: 17 dB
S11: 9 dB

Conclusion

The circuit in Fig 1 has been tested on HP 8970s with noise figures from 0.32 to 0.45 dB. A two-stage version (the same preamp cascaded and tuned) measured 0.29 dB on NRAO

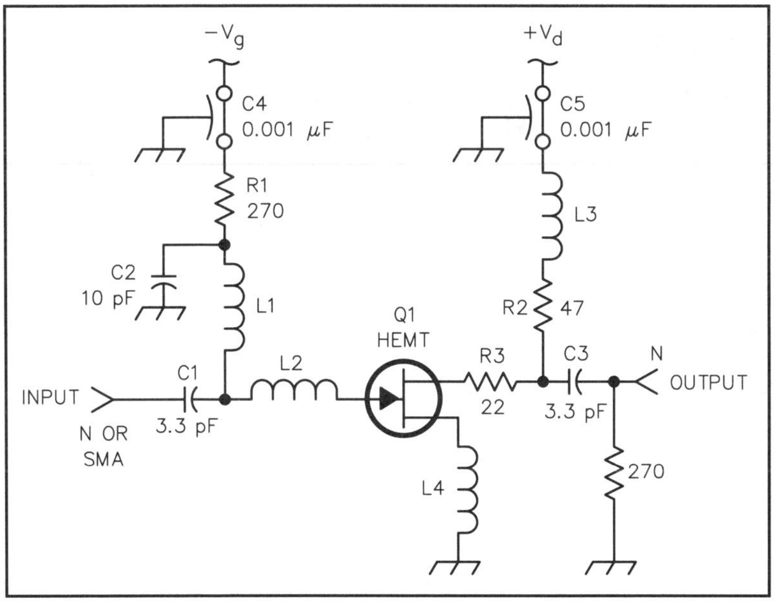

Fig 1—Schematic of the 1296-MHz HEMT preamplifier. Resistors are ¼-W carbon film or composition. Capacitors are 100-mil ATC chips.

C1, C2—3.3 pF, 100-mil ATC chip caps.

L1—4 t no. 30, 0.1-in. ID, 0.25-in. long.

L2—4 t no. 30, 0.06-in. ID, 0.25-in. long. Compress or expand turns for minimum NF.

L3—1 t, 0.07-in. ID (lead of R2, 0.1-in. long on V_d side).

L4—0.06-in. long source leads, 0.05-in. above ground plane.

R3—22 Ω, carbon comp. Make leads 0.1-in. long, and connect 0.2-in. away from Q1.

(National Radio Astronomy Observatory) lab equipment without cooling; the gain was 32 dB. The preamp performs well from 1100 to 1500 MHz with minor adjustment of L1. For over three years, HEMTs uncooled and cooled have been in service in my 1296-MHz EME system with good success. They are worth a look at, if you want to get below the 0.5-dB NF barrier.

Notes

[1]Henderson, T., "Electronically Cooled 1.3 GHz LNA," *Proceedings of the 24th Central States VHF Society Conference* (Newington: ARRL, 1990), pp 1-8.

[2]Ward, A., "Simple Low Noise Microwave Preamplifiers," *Proceedings of Microwave Update '88* (Newington: ARRL, 1988), p 65; also *QST*, May 1989, pp 31-36, 75 (reprinted in this chapter).

Table 1

Device	*Bias*	*NF*	*Gain*	*S11*
ATF10135	1.7 V_{dS} 15 mA	0.55 dB	12 dB	6 dB
ATF10135	2 V_d −0.2 V 20 mA	0.51 dB	13 dB	9 dB
ATF35076	2V_d −0.4 V 18 mA	0.39 dB	18 dB	7 dB

A 2304-MHz Preamp Using the MGF1302/1402

By Jim Davey, WA8NLC

(From *Proceedings of Microwave Update '89*)

Here's a simple preamp for 2304-MHz that is designed around the Mitsubishi MGF1402/1302 GaAsFET. It has been built by several amateurs with good results. If the directions are followed closely, the noise figure will be well under 1 dB. If a noise-figure meter is available, the gate circuit can be optimized for a noise figure of about 0.7 dB. The circuit (Fig 1) is patterned after a 1296 preamp published by W6PO several years ago in the EIMAC *EME Notes*.

The preamp is built on 0.031-in. Teflon board with a dielectric constant of 2.5 (Fig 2). I have not tried it myself, but Duroid 5880 will probably work OK. The board is mounted in a small open top box made of ½-in. wide brass strip stock available at hobby stores. Clearance holes are drilled in the ends for SMA-connector center pins. The connectors can be bolted or soldered to the brass wall. I prefer this type of mounting method to right-angle mounting through the board for both mechanical and electrical reasons.

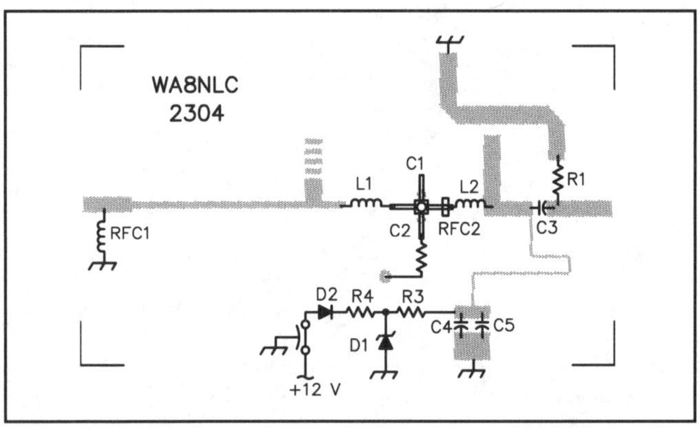

Fig 1—Schematic diagram of the 2304-MHz preamp.
C1-C4—Porcelain chip cap, 10-pF (0.050-in. preferred).
C5—470-pF chip cap.
D1—Zener diode, 4.7-V, 400-mW.
D2—1N914 or equiv.
L1—2 t no. 28 enam, spaced 3 wire diam, 0.055-in. ID.
L2—1.5 t no. 28 enam, spaced 1 wire diam, 0.055-in. ID.
R1—50-Ω, ⅛-W carbon composition or chip resistor.
R2—68-82 Ω, select for I_d = 10 mA.
R3—100 Ω typ., select for V_{ds} = 3.0 V after I_d is set.
R4—270 Ω, ¼ W carbon composition or film.
RFC1—4 t no. 28 enam, closewound, 0.055-in. ID.
RFC2—T12-6 core over drain lead of transistor.

The GaAsFET is supported by a 50-mil size porcelain chip capacitor under each source lead. A small hole just large enough to pass the chip capacitor is drilled at each location and the capacitor inserted so the metallized end is flush with the ground plane on the back of the board. A quick pass with a hot soldering iron and solder will tack the capacitor to ground. If the 50-mil size capacitors are used, the other end will protrude slightly on the circuit side. Center the transistor between the two exposed ends of the capacitors and solder. Leave some lead on one side to attach the bias resistor.

The preamp is adjusted by spreading or compressing the turns on the gate coil and adjusting the length of the stub on the gate for lowest noise figure.

Parts can be purchased from Microwave Components of Michigan, PO Box 1697, Taylor, MI 48180.

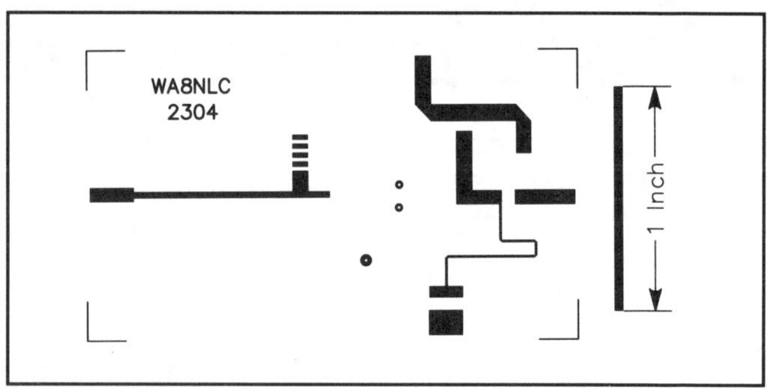

Fig 2—Circuit-board artwork for the preamplifier.

Simple Low-Noise Microwave Preamplifiers for 2.3 Through 10 GHz

By Al Ward, WB5LUA

(From May 1989 *QST*)

One of the hurdles to overcome in building a microwave station is construction of a good low-noise preamplifier. Techniques using lumped constants (capacitors and inductors) that work well at lower frequencies are often difficult to realize above 2304 MHz. Once a design is worked out and the preamplifier built, the unit must then be tuned for minimum noise figure (NF) before any sort of reasonable performance can be expected. Preamplifier tune-up itself presents a hurdle because many hams don't have access to NF test equipment that is accurate in the microwave region.

If these low-noise amplifiers (LNAs) are duplicated exactly from the information presented, no RF adjustments are required. Just bias their active devices properly, and the units are ready to go—with NFs under 1 dB at frequencies up to 5.7 GHz and around 1.5 dB at 10.368 GHz.

The 10-GHz amplifier is designed around the Hewlett-Packard ATF-13136 GaAsFET, and the lower-frequency units are designed around the Hewlett-Packard ATF-10136. Both devices have a nominal gate length of 0.3 micron. The ATF-10136 has a total gate periphery of 500 microns. The ATF-13136 has a gate periphery of 250 microns, making it more appropriate for higher-frequency operation. Best of all, these transistors are not high-priced exotics. Both the ATF-10136 and the ATF-13136 are the short lead versions of the ATF-10135 and ATF-13135 respectively. The long lead versions (35 package) are being phased out of production in favor

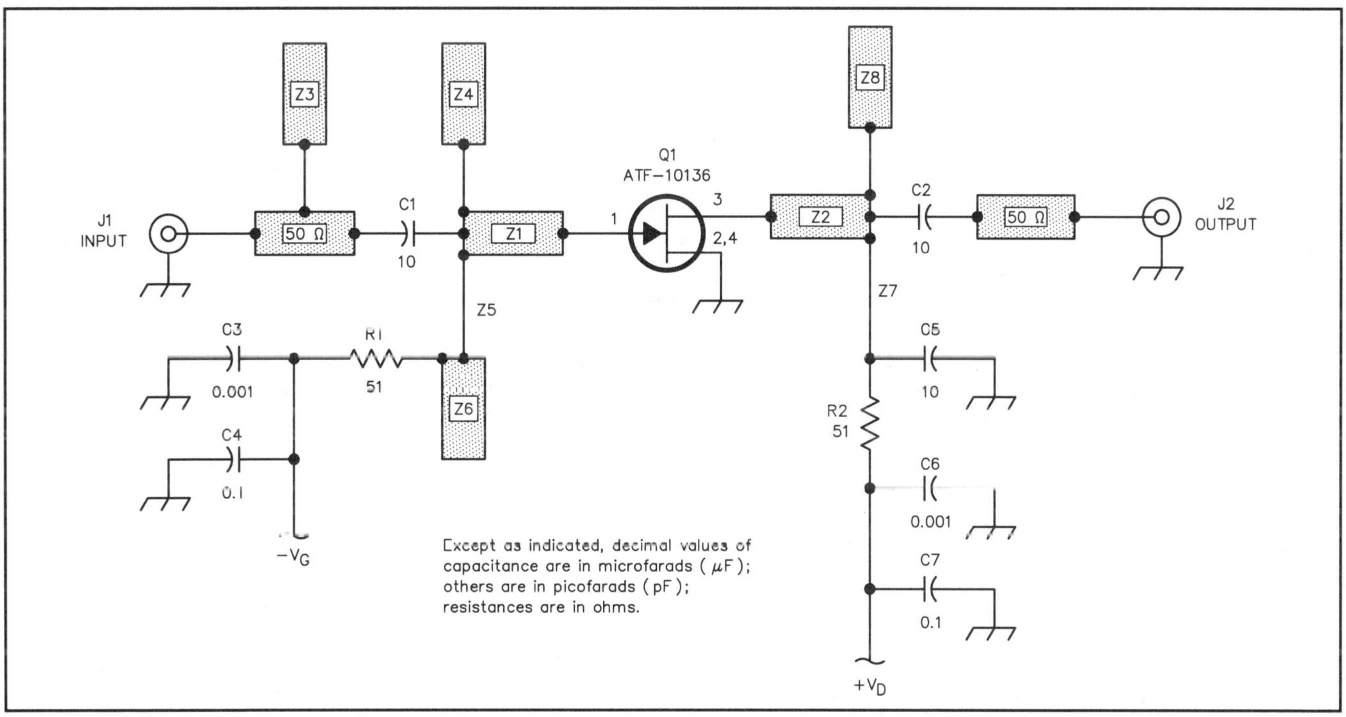

Fig 1—Schematic of the 2.3-GHz preamplifier. Z1 through Z8 are microstriplines etched on the PC board. Shaded rectangles marked "50 Ω" are 50-Ω transmission lines etched on the PC board. All resistors and capacitors are chip types. C1, C2 and C5 can be 0.05- or 0.1-in. square. C4 and C7 enhance "low-frequency" bypassing. J1 and J2 are SMA female connectors; see text.

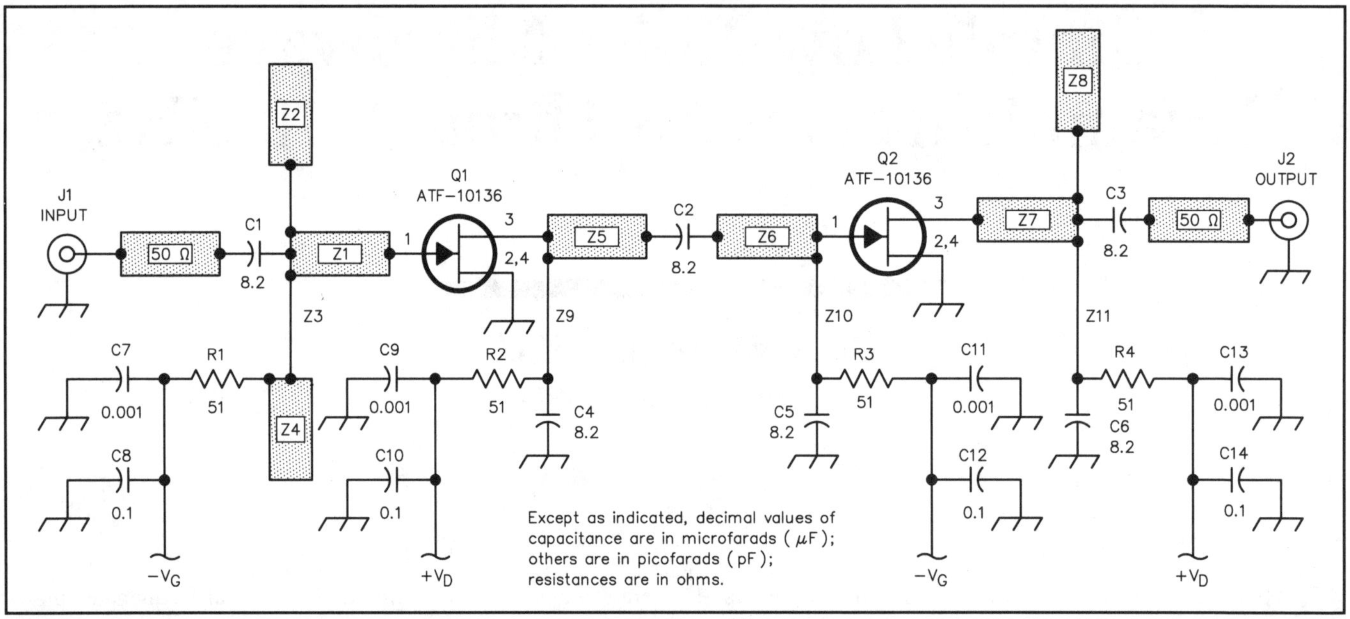

Fig 2—Schematic of the 3.4-GHz preamplifier. Z1 through Z11 are microstriplines etched on the PC board. Shaded rectangles marked "50-Ω" are 50-Ω transmission lines etched on the PC board. All resistors and capacitors are chip types. C1-C6 are 0.05-in. square. C8, C10, C12 and C14 enhance "low-frequency" bypassing. J1 and J2 are SMA female connectors; see text.

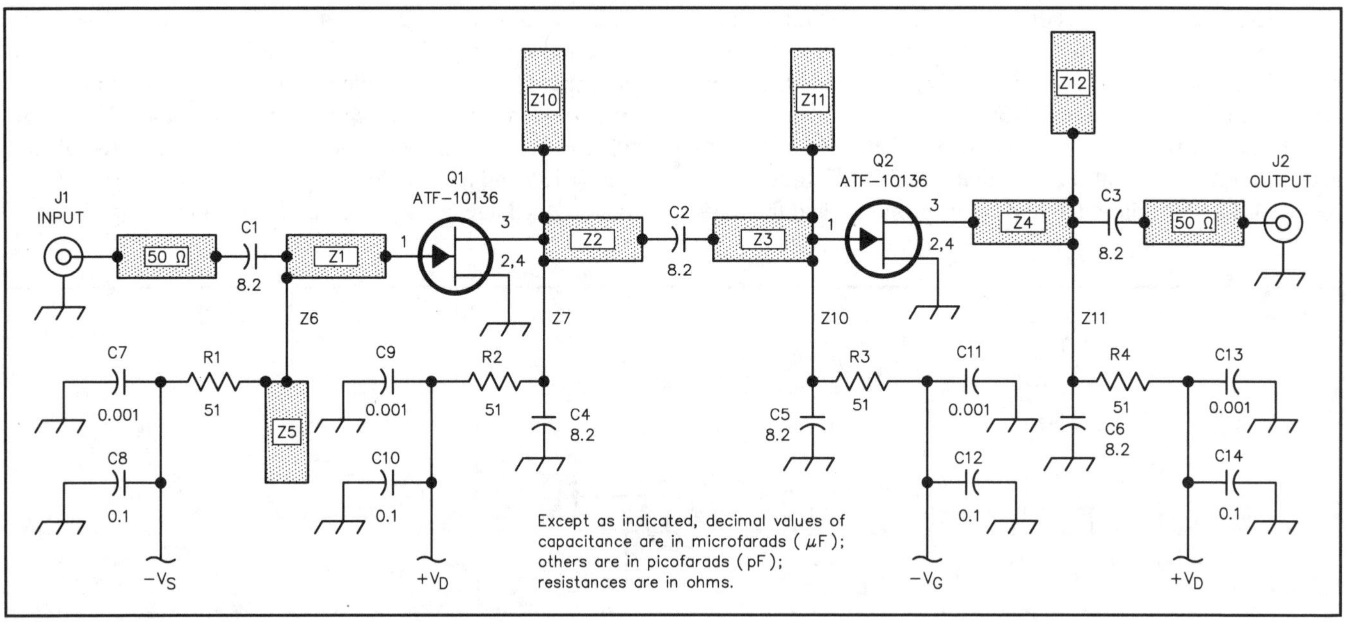

Fig 3—Schematic of the 5.7-GHz preamplifier. Z1 through Z12 are microstriplines etched on the PC board. Shaded rectangles marked "50-Ω" are 50-Ω transmission lines etched on the PC board. All resistors and capacitors are chip types. C1-C6 are 0.05-in. square. C8, C10, C12 and C14 enhance "low-frequency" bypassing. J1 and J2 are SMA female connectors; see text.

of the short lead versions. The short lead versions are required for most commerical "pick and place" high volume assembly lines.

Circuit Description

Schematics for the preamplifiers are shown in Figs 1-5. Fig 1 shows a single-stage 2.3-GHz design. Figs 2 and 3 show 2-stage preamplifiers for 3.4 and 5.7 GHz. Figs 4 and 5 show single- and two-stage 10-GHz LNAs.

The same basic circuit configuration is used for each preamplifier. All impedance matching is accomplished with microstriplines. None of the amplifiers contains adjustable RF components. Fixed-value capacitors and resistors are also used in the gate- and drain-bias decoupling circuitry.

I did the initial design for these preamplifiers with the aid of a Smith Chart and optimized them through computer simulation. LNA design is made considerably easier by computer-aided design software. I was fortunate to have access to Touchstone, an RF design program made by EESOF.

One of the most important parameters that the computer

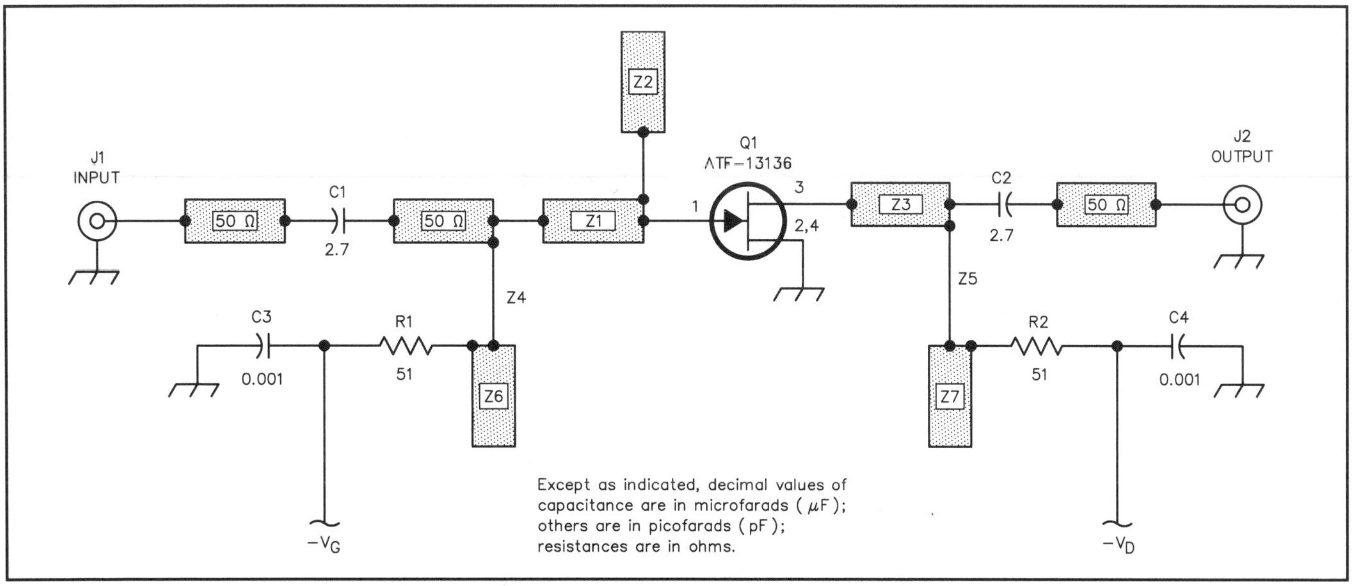

Fig 4—Schematic of the single-stage 10-GHz preamplifier. Z1 through Z7 are microstriplines etched on the PC board. Shaded rectangles marked "50-Ω" are 50-Ω transmission lines etched on the PC board. All resistors and capacitors are chip types. C1 and C2 are 0.05-in. square (ATC Type A capacitors are preferred). J1 and J2 are SMA female connectors; see text.

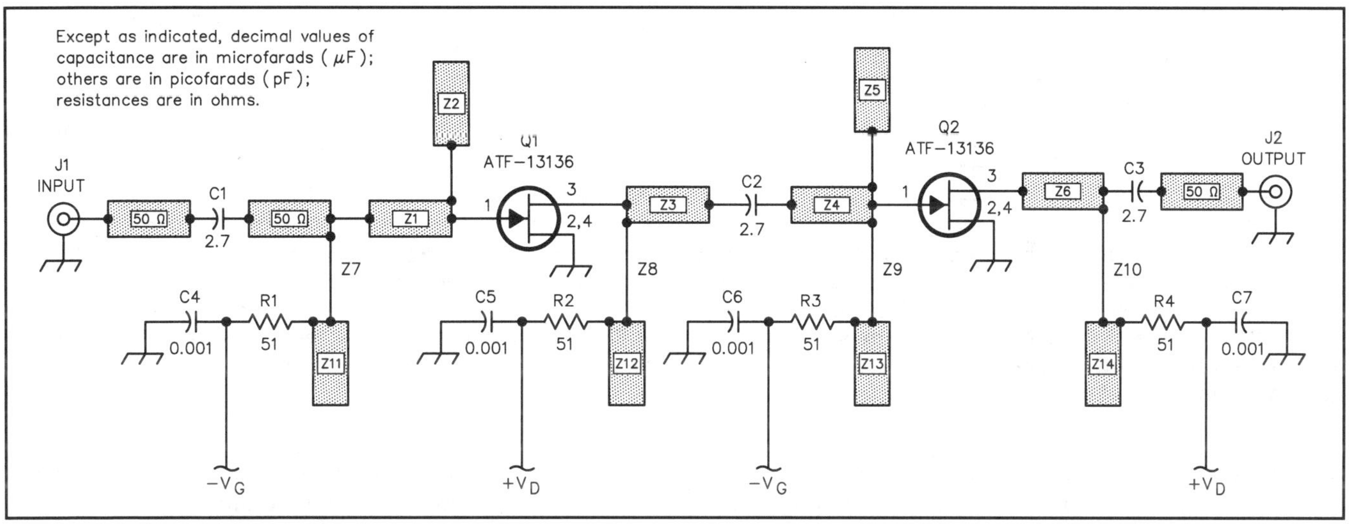

Fig 5—Schematic of the two-stage 10-GHz preamplifier. Z1 through Z14 are microstriplines etched on the PC board. Shaded rectangles marked "50-Ω" are 50-Ω transmission lines etched on the PC board. All resistors and capacitors are chip types. C1-C3 are 0.05-in. square (ATC Type A capacitors are preferred). J1 and J2 are SMA female connectors; see text.

can analyze is stability. Proper choice of components in the bias decoupling networks and the use of source inductance in the form of source-lead length helps to maintain stability. The results are improved input and output SWR, while maintaining low noise figure and moderate gain. In the case of the 3.4, 5.7 and 10-GHz amplifiers, the effect of the "through-the-board" mounting of the source leads is taken into account in the design so that no plated-through holes are necessary to achieve good performance.

Bias Networks

To minimize circuit losses, ground the GaAsFET source leads directly to the ground plane. To operate these devices with their source leads at dc ground, you must bias each device's gate negatively relative to its source. This can be done in a number of ways, so I left bias circuitry off the PC boards.

Three basic bias circuits are shown in Fig 6. Two are passive. The third—and most desirable—is an active bias network that uses a PNP transistor to set the GaAsFET drain voltage and current.

The simplest bias network, shown in Fig 6A, uses a 3.3-V Zener diode to set the drain voltage. A 1.5-V AA cell is used for the bias supply. Bias, applied through R1, sets the gate voltage, which in turn determines the drain current. Generally, the AA cell is connected so that there is always a negative voltage applied to the gate. The preamplifier is then turned on by connecting V_D to a positive voltage source. Because there is a 51-Ω resistor in series with the drain, there is some interaction between drain voltage and drain current: Greater drain

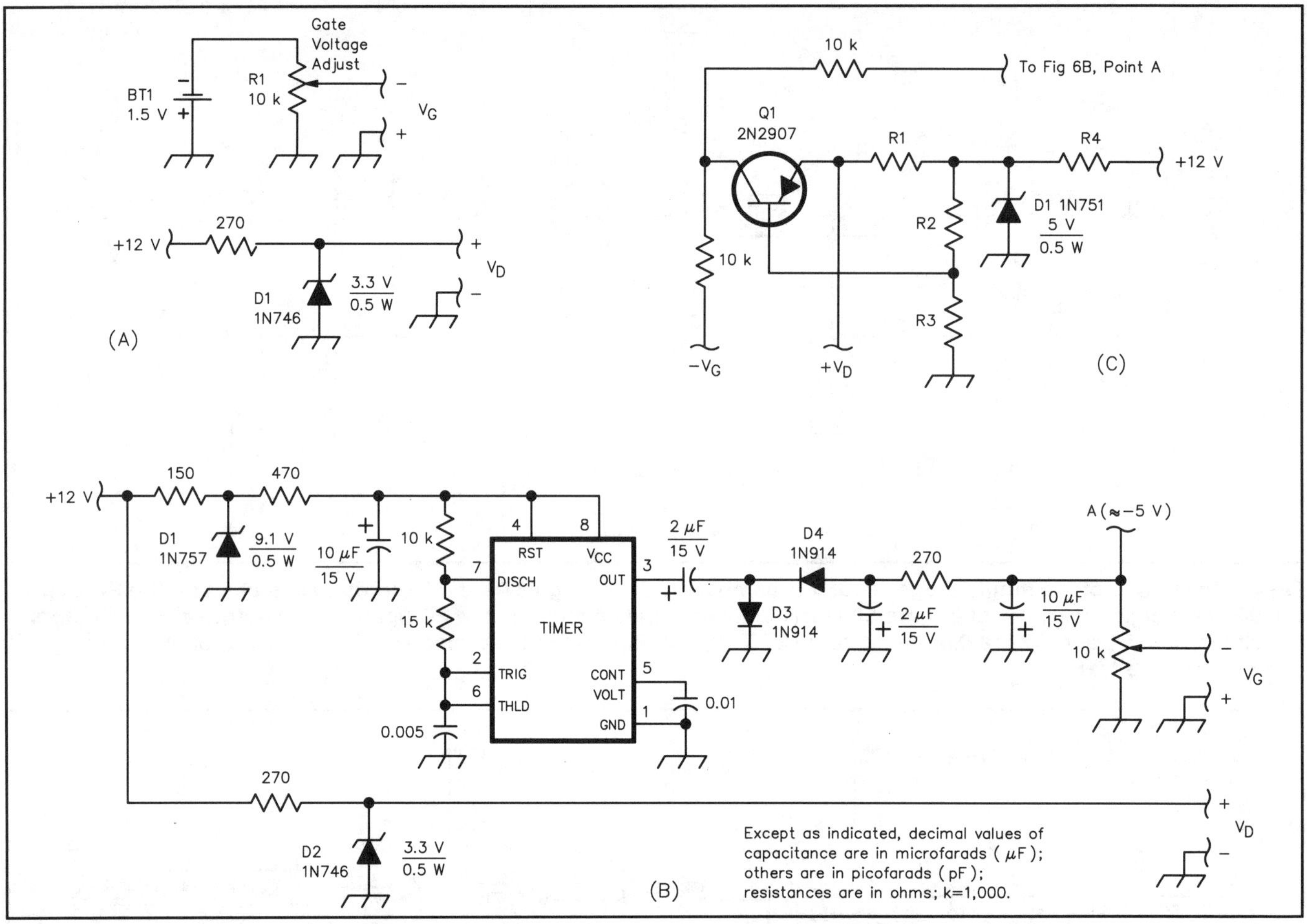

Fig 6—Bias circuits for the preamplifiers. See text for discussion. The passive circuit at A uses a 1.5-V cell for the gate supply and a Zener diode to stabilize the drain supply. The circuit at B, another passive arrangement, uses a 555 timer IC to generate negative gate bias, and a Zener diode to stabilize the drain supply. C shows an active bias circuit. The values of R1, R2 and R3 can be varied for different FET operating conditions; see text and Table 1. The value of R4 should be chosen so that 10-15 mA of Zener current flows when the FET (or FETs, for a two-stage design) is powered at rated bias.

current produces a lower drain voltage. The gate voltage required to properly bias the device varies from unit to unit because of slight variations in pinch-off voltage. (Pinch-off voltage is the gate voltage required to turn the FET off.) A disadvantage of this simple bias circuit is that it lacks compensation for bias changes over temperature variations. Although not optimum, this technique has been used at WB5LUA and WA5VJB with good results. A high-grade, long-life alkaline AA-size cell should last several months before its voltage drops low enough to cause the FET to draw excessive drain current.

An adaptation of the simple passive bias configuration is shown in Fig 6B. Drain voltage is again provided by a 3.3-V Zener diode, but this circuit replaces the AA cell with a positive-to-negative voltage inverter. Several manufacturers make suitable inverter ICs. A less-expensive approach is to use a common 555 timer in the simple inverter circuit shown. With simple battery bias, the negative supply is continuously applied to the FET gate. With the inverter of Fig 6B, gate and drain supplies are simultaneously applied to the FET. This approach has been used by manufacturers of satellite TV receiving equipment for years. Although there can be problems if the drain voltage is applied before the negative gate voltage, the 51-Ω resistor in series with the drain safely limits the maximum drain current—even if the application of gate voltage to the FET is delayed.

An improvement on both passive circuits is the active bias circuit shown in Fig 6C. The negative gate supply uses the inverter of Fig 6B, but the drain supply employs an inexpensive PNP transistor in a circuit that effectively sets both the drain voltage and drain current regardless of device variations. It also offers a more constant bias over temperature than the passive designs. Drain voltage is set by R2 and R3, a voltage divider at the base of Q1. The voltage is then raised by the emitter/base junction voltage of Q1. Drain current is set by R1. The gate voltage required to sustain the drain voltage and current is set automatically by the voltage divider set up by the emitter/collector junction of Q1 and the negative voltage source. About –1 V is supplied to the FET gate.

Table 1 gives resistor values for various bias conditions. I suggest building a separate bias network for each FET stage to properly set each device's bias point. I have, however, used a single active bias supply to power a two-stage amplifier with good success. If the devices are fairly well dc matched (drain

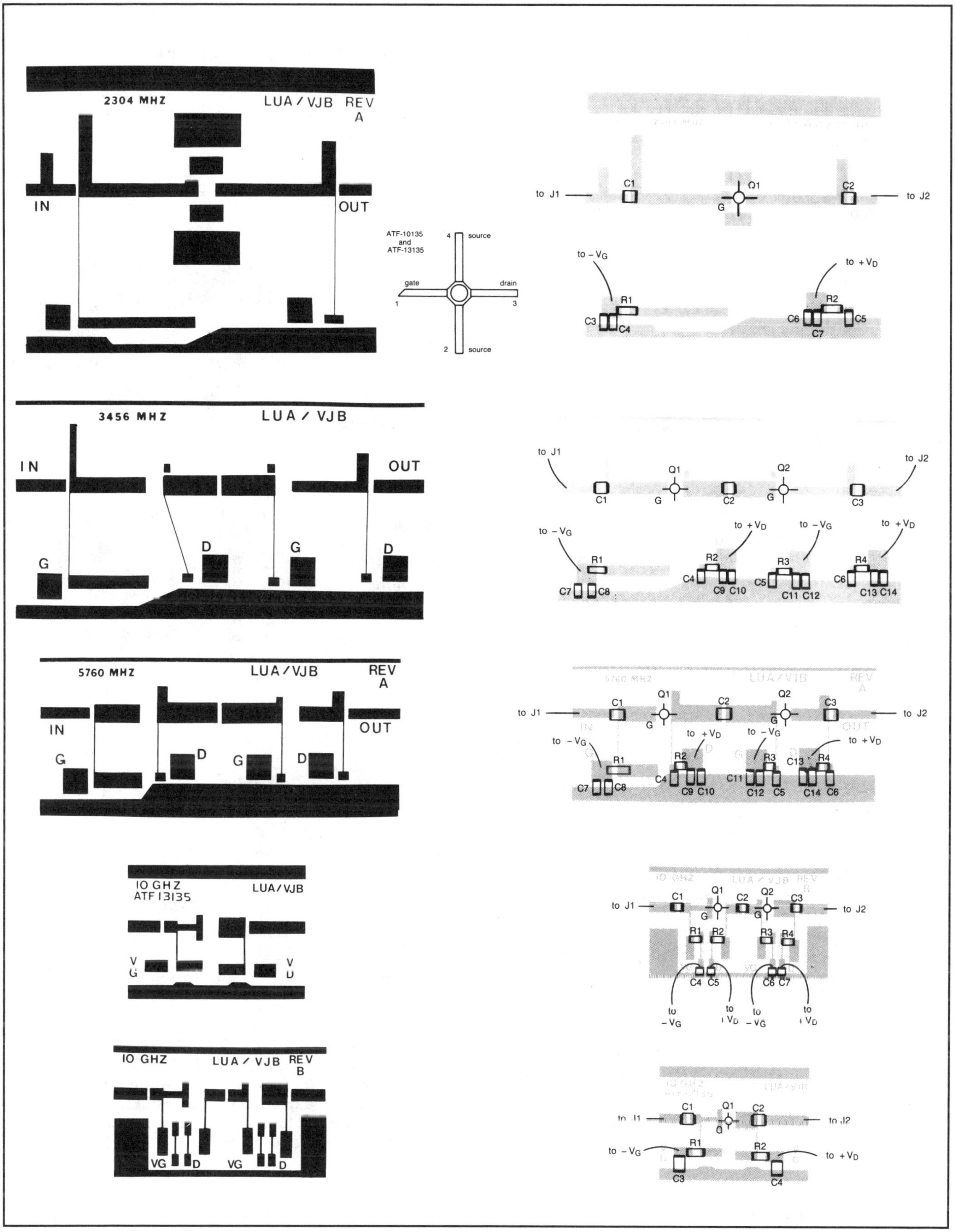

Fig 7—Circuit-etching patterns and parts-placement guides for the preamplifiers. The etching patterns are shown at full size. PC board material is double-sided, 0.031-in-thick Rogers Duroid 5880 or Taconic TLY-5 (dielectric constant, 2.2). Black areas represent unetched copper foil. The back side of the board is left unetched to act as a ground plane. The parts-placement guides are not shown at their actual size. All components mount on the etched side of the board.

current vs gate voltage), this technique will be okay. It will not, however, keep the device drain currents equal if they are not dc matched.

The active bias arrangement can also be used with a battery instead of the voltage inverter. Since the active bias network automatically adjusts gate voltage for a required bias condition, the circuit will adjust the gate voltage as it drops with battery age. The gate requires about –1 V, so the battery can age significantly before the FET bias condition is altered significantly. If this technique is used, it is best to start out with a 5- to 6-volt battery source. The active bias network will compensate for a battery voltage deteriorating to 1-1.5 V. Active bias networks are discussed in greater detail in Hewlett-Packard application note ANA002.

Construction

Construction of all amplifiers is similar. Part-placement guides and etching patterns are shown in Fig 7. All amplifiers are etched on 0.031-in.-thick Duroid 5880 or Taconic TLY-5 PC-board material with a dielectric constant of 2.2. The etched PC board can be installed in a housing such as a die-cast aluminum box. Another method, one that I prefer, is to solder thin (0.02-in.-thick) brass side walls to the PC board to form a shielded enclosure. The brass walls also connect the top and bottom ground planes, which is essential for low-loss "low-frequency" bypassing. Power connections for V_G and V_D can be made via 0.001-µF feedthrough capacitors soldered to the brass walls. See Figs 8 and 9.

SMA-type end-launch connectors are used for J1 and J2 to provide a transition from coaxial cable to the microstripline. Two- or four-hole gold-plated connectors are easily soldered to the PC board or brass side walls, depending on your assembly technique. End-launch connectors are preferred to the right-angle type because of the impedance discontinuity associated with the right-angle transition. Additional amplifier tuning may be required if right-angle connectors are used.

The type and size of the chip capacitors used in these amplifiers becomes increasingly important as frequency increases. For the blocking capacitors, I strongly recommend using good-quality RF-type ceramic chip capacitors, such as those made by ATC. The values specified are common and should not be hard to find. The physical size of the capacitors is especially critical at 10 GHz, where the 0.05-in.-square type *must* be used. Anything larger produces a sizable mismatch on the microstripline.

The value of the "low-frequency" bypass capacitors is less critical. Anything in the 820- to 1500-pF range will work fine. The value of the "high-frequency" bypass capacitors is somewhat more critical, though. Stay within 10% of the values indicated. Again, use good-quality capacitors for "high-frequency" bypassing.

To obtain a low noise figure, the preamp's FET source leads must be properly grounded. In the case of the 3456-MHz and higher-frequency preamplifiers, bend the FET source leads down right at the case and insert them through slots in the PC board. See Fig 10A. This technique works fine for the long leaded devices. For the short leaded devices, I would suggest using 0.050 inch to 0.070 inch wide copper wire straps to bring the bottom ground plane to the top of the printed circuit board where the device can be soldered directly to the "C wire" strap. The "C wire" strap should be positioned close enough to the edge of the package such that the device source lead length is minimal, i.e. less than 0.010 inches of source lead for the 3456 MHz through 10368 MHz preamplifiers. See Fig 10B.

Table 1

Active Bias Circuit Values for Various Drain Currents

V_{DD} (V)	V_{DS} (V)	I_D (mA)	R1 (Ω)	R2 (kΩ)	R3 (kΩ)
3.5	2.5	20	75	2.2	2.8
3.25	2.5	15	117	2.2	2.3
4.0	3.0	20	50	2.2	4.3

Fig 8—This prototype two-stage 3.4-GHz preamplifier was built by the author. The enclosure is made from sheet brass soldered to the PC board (see text).

Fig 9—A completed two-stage 10-GHz preamplifier.

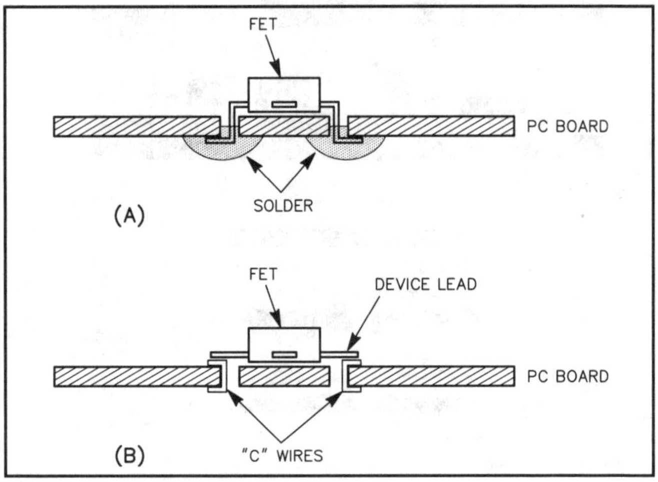

Fig 10—The ATF-10135 and -13135 FET source leads are bent, inserted through slots cut in the PC board, and soldered to the ground plane. The ATF-10136 and -13136 leads are not long enough to pass through the board, so you must add "C wires." See Fig 10B.

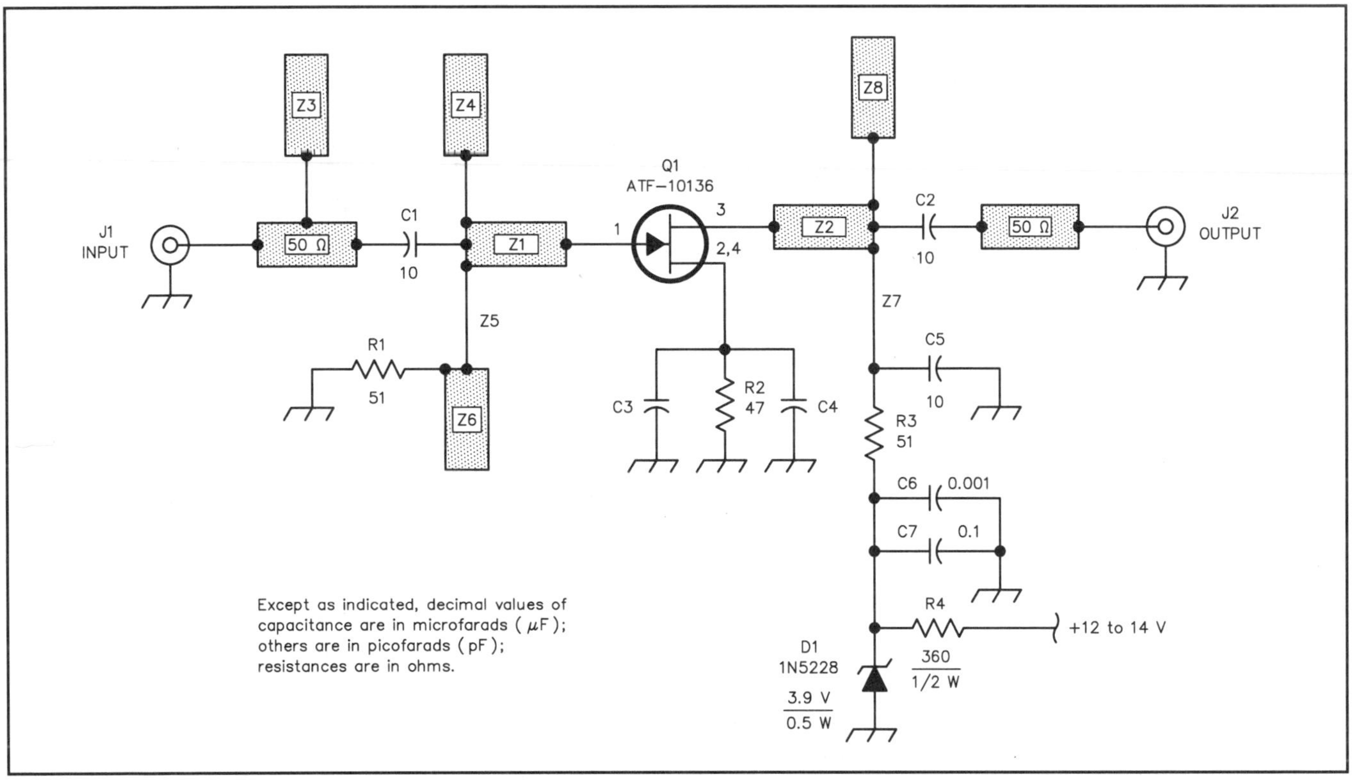

Fig 11—Schematic diagram of the self-biased version of the 2.3-GHz preamplifier. Z1 through Z8 are microstriplines etched on the PC board. Shaded rectangles marked "50-Ω" are 50-Ω transmission lines etched on the PC board. All capacitors are chip types.

C1, C2, C5—0.05- or 0.1-in. square chip capacitor.
C3, C4—470- or 1000-pF leadless, round, disc-ceramic capacitor (see text).
J1, J2—SMA female connectors (see text).
R1, R3—51-Ω chip resistor preferred; ¼-W carbon-film resistor with short leads should be okay.
R2—47-Ω ¼-W carbon-film resistor.

The slots can be made with a sharp hobby knife or something similar. Be sure to clean up the area where the leads will pass through the slots by removing any extra dielectric material or copper. The source leads should be passed through the board, again bent at right angles and laid neatly along the bottom foil. Solder them to the bottom ground plane and try to force the solder to cover the length of the slot if possible.

Device installation is different with the 2304-MHz preamplifier because additional source-lead inductance is required. This inductance is addedby making each FET source lead 0.07 inch long in the case of the ATF-10135, rather than grounding them with the minimum possible lead length. Ground pads have been established on the artwork, and these pads must be properly connected to the bottom ground plane. Cut slots near the edge of the two pads closest to Q1. Insert a 0.1-in.-wide copper strap through the slots and solder top and bottom. Each source lead is then 0.07 in. long when Q1 is centered on the PC board. For the short leaded ATF-10136, I would suggest adding an additional length of 0.020 inch wide copper foil to lengthen each source lead as required. The leads from an "old" transistor will work fine.

Components for these preamplifiers, including chip capacitors, leadless capacitors, chip resistors and SMA connectors are available from Microwave Components of Michigan and Down East Microwave. Etched PC boards are available from Down East Microwave.

Table 2

Actual Preamplifier Performance Versus Computer Simulation and Projected Worst-Case Performance

Device	*Freq (GHz)*	*Bias per device*	*Gain (dB) Type*	*Gain (dB) Worst Case*	*Gain (dB) Simul*	*Noise Figure (dB) Typ*	*Noise Figure (dB) Worst Case*	*Noise Figure (dB) Simul*
ATF-10136*	2.3	2 V @ 20 mA	13.5	12.0	13.9	0.5-0.6	0.8	0.60
ATF-10136**	2.3	2 V @ 20 mA	13.0	12.0	13.0	0.65	0.9	0.70
ATF-10136	3.4	2 V @ 20 mA	23.0	22.0	24.1	0.8-0.9	1.0	0.58
ATF-10136	5.7	2.5 V @ 15 mA	18.0	17.0	20.6	0.9-1.0	1.2	0.85
ATF-13136*	10.4	3 V @ 20 mA	8.5	7.5	10.6	1.25-1.5	1.7	1.25
ATF-13136	10.4	3 V @ 20 mA	18.0	15.0	21.9	1.5-1.7	2.0	1.35

*Single-stage amplifier
**Self-biased, single-stage amplifier

Results

If the preamplifiers are built according to the information given in this article, RF adjustments should not be required. A slight adjustment can be made to the bias point if desired, but performance should be very close to that shown in Table 2 with the bias conditions shown. Based on test results obtained from preamplifiers built by a number of hams using available components, the average builder should be able to meet the worst-case gain and NF specifications shown in Table 2. If slightly different dielectric material or construction techniques are used, the amplifier can be tuned by moving small capacitive tabs along the microstripline while monitoring NF and gain. A properly modified toothpick makes a handy low-loss tool for moving tabs around on the circuitry. Cut the end of a toothpick on a diagonal. Wet the end of the toothpick and use it to move small metal tabs around on the etch.

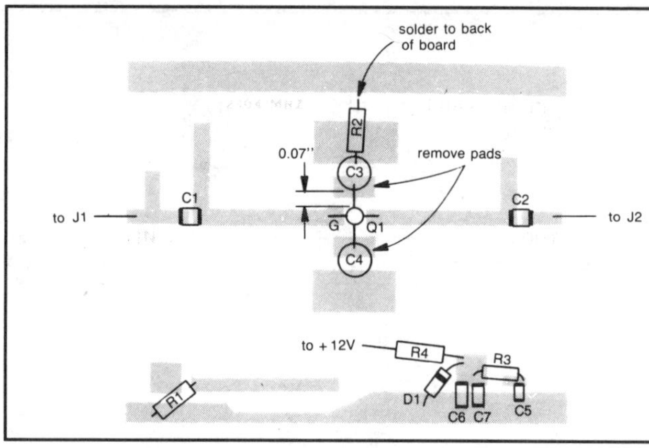

Fig 12—Part-placement guide for the self-biased 2.3-GHz preamplifier. The etching pattern is the same as shown in Fig 7, except two pads are removed and Z3 must be lengthened. See text.

Applications

The 2304-MHz preamplifier provides acceptable receive performance in the 2401-MHz OSCAR band with no modification. The NF at 2401 MHz should be only 0.1 or 0.2 dB greater than that obtainable at 2304 MHz, and LNA gain at both frequencies should agree to within a dB. Using a pair of these preamplifiers at the feed of my 2401-MHz satellite system, I see about 5 dB of sun noise and 10 to 15 dB of signal-to-noise ratio from the Mode S transponder aboard OSCAR 13. My antenna is a 4-foot diameter UHF TV dish with ¼-inch mesh.

A similar two-stage 2304-MHz preamplifier is in use at the feed of my 24-foot home-brew stressed parabolic reflector. The measured NF of this preamp is 0.65 dB. With this system, I have worked W3IWI/8, SK6WM, OE9XXI and W4HHK on 2304-MHz EME.

Variations

In the original design, source leads were grounded directly to obtain the lowest possible noise figure. This necessitates the use of a dual-polarity supply as discussed earlier. The typical approach at VHF is to self-bias the FET by using a resistor connected in series between the source and ground. A capacitor is used to bypass the resistor at RF. As the frequency of operation increases, it becomes increasingly difficult to obtain high-quality, low-impedance bypassing. Although the self-biasing technique is simpler to build and uses a single power supply, some RF performance is sacrificed.

I evaluated self-biased versions with the help of the computer, and several prototypes were built and tested. The artwork was modified to include 0.075-in.-square pads to mount the FET source leads. Chip capacitors were then bridged between these pads and a ground pad. The ground pad was then connected to the back ground plane by using 0.1-in.-wide ribbons through the board. This technique was tried on all preamplifiers except the 10-GHz models. With this configuration, the preamp NF at 2304 and 3456 MHz was 0.2 to 0.25 dB greater than that of the grounded-source design. The gain of the 2304-MHz model decreased approximately 2 dB, while gain of the 3456-MHz model was reduced by slightly more than 3 dB. At 5760 MHz, performance deteriorated even further: NF increased by 0.8 dB, and gain decreased by 6 dB.

A computer simulation predicted similar results. In addition, the computer simulation indicated an oscillation in the 7- to 8-GHz range for all models. Apparently, the inductance associated with the addition of the chip capacitors, associated mounting pads, and ribbon ground returns is great enough to cause instability at higher frequencies.

In looking for a way to avoid this oscillation, I tried using leadless round disc capacitors to bypass the source leads. I soldered the FET leads directly to the capacitors and soldered the capacitors to the bottom side of the PC board. This technique was tried on the 2304-MHz preamplifier; see Fig 12.

The bypassed-source 2304-MHz preamplifier uses the same artwork shown in Fig 7 for the grounded-source version. You'll need to make one change to the artwork, though. The input stub (Z3 of Fig 11) must be made slightly longer (increase its length from 0.22 to 0.38 in.) to help tune out the effect of the source bypass capacitors. Z3 can be extended by soldering a piece of copper foil to the end of the etched line. To mount C3 and C4, the source bypass capacitors, drill holes the diameter of the capacitors through the PC board as shown. See Fig 13, a photograph of the finished unit. Be sure to place the holes so that the FET source leads are 0.070 in. long when Q1 is centered between Z1 and Z2. Solder a brass or copper sheet across the holes on the ground-plane side, and solder one side of C3 and C4 to the sheet. Then solder the Q1 source leads to the tops of C3 and C4.

Fig 13—The prototype self-biased 2.3-GHz preamplifier. The large rectangular pads near C3 and C4 are not needed and were dropped from the final artwork.

According to the computer simulation, this technique yields unconditional stability in the 7- to 8-GHz range. The overall stability of a preamp built in this way is comparable to the grounded-source design. The *measured* performance of this configuration is good too. Compared to the grounded-source model, the gain of the bypassed-source LNA is within 1 dB, and the NF is within 0.1 dB, of that obtained with the grounded-source configuration—and no stability problems are evident. This technique is suggested for the 2304-MHz model only.

Although the preamplifier circuits are designed for minimum noise figure, the ATF-10136 and the ATF-13136 devices are capable of producing moderate power at microwave frequencies. When biased at 4 V and 70 mA, the ATF-10136 is capable of producing 20 dBm (100 mW)—operating at its 1-dB gain-compression point—at 4 GHz. Slightly less power can be expected at 5.7 GHz. At 12 GHz, the ATF-13136 is capable of producing 17 dBm at its 1-dBm gain-compression point when biased at 4 V and 40 mA.

Summary

Any one of these amplifiers can be constructed easily in an evening. They offer low noise figure, acceptable gain, good input and output SWR, and good stability. Several dozen of them are already in use nationwide with good results. Copies of the 10.368-GHz preamplifiers, for instance, were used by WA5VJB and WA7CJO to make the first-ever 10-GHz EME QSO!

Microwave LNA Update

By Al Ward, WB5LUA

(From *Proceedings of Microwave Update '89*)

Hewlett-Packard has a series of low-noise plastic FETs that use the same die as the ATF-13136, a 250-micron Ku-band low-noise FET. Although the S parameters and noise parameters are somewhat different from the ATF-10135 device, I decided to try the ATF-13484 in the 2.3-GHz LNA described in *QST* for May 1989.[1] With a slight amount of rebiasing and a slight extension of the first input stub, I was able to achieve a 0.85-dB noise figure with an associated gain of 15.6 dB at 2.4 GHz. The device should provide nice performance in A0-13 Mode-S operation. A schematic is shown in Fig 1.

Shortly after I published the *QST* article, my company made the decision to supply only the short-leaded ATF-10136 and ATF-13136. The use of automated assembly lines with pick-and-place machines created the need to buy the devices with precut leads, ie, 0.040 in. vs 0.177 in.

Unfortunately, in amateur applications where plated through holes are a luxury item, the shorter lead parts can not be mounted as easily. The original article suggested the use of "C wires" as one means of grounding the source leads on the short-leaded devices. I decided to try using small rivets as an alternative. With the help of Downeast Microwave, I built up two more 10-GHz LNAs using very small tin plated brass rivets (0.050-in. diameter). The center-to-center spacing of the rivets was 0.160 in. I had to be careful that the rivets did not touch the adjacent matching networks. The ATF-13136 device (with 0.040-in. leads) was then centered between the rivets and the input and output matching networks. Noise fig-

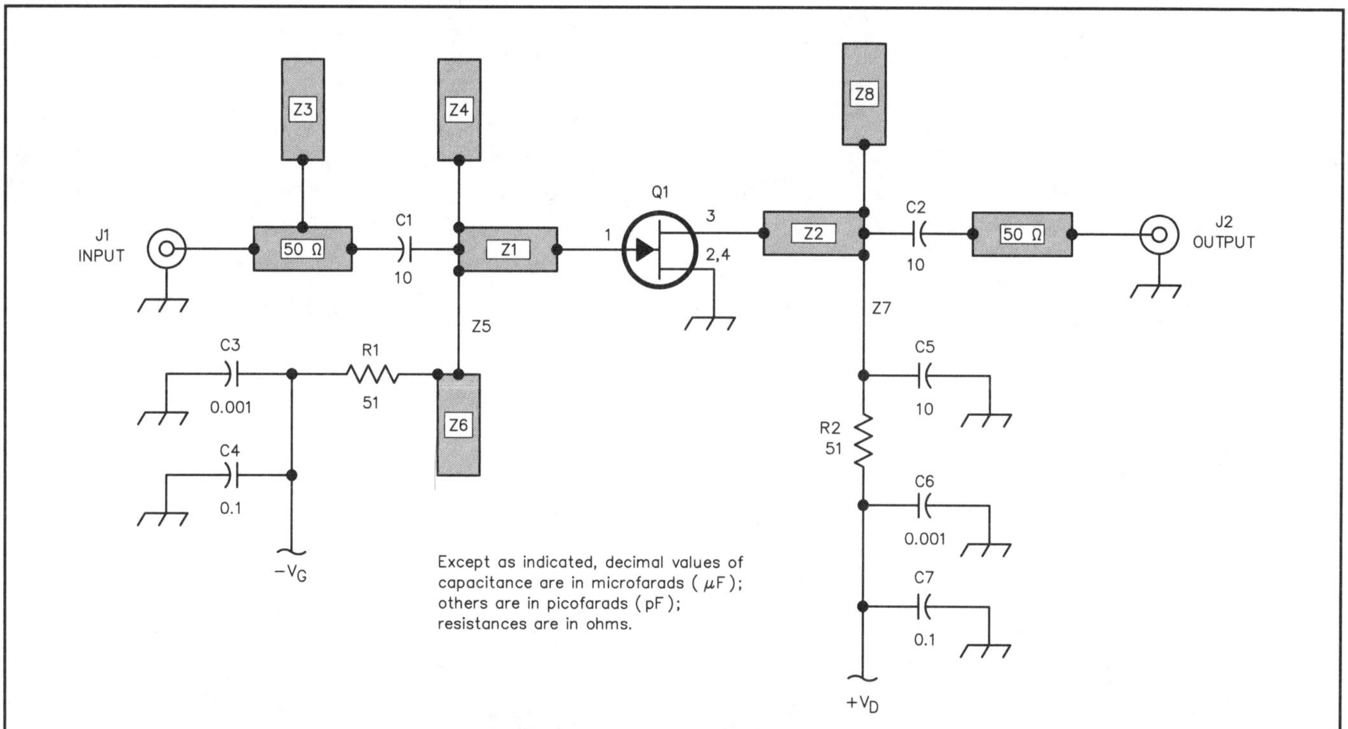

Fig 1—Schematic of the 2.3-GHz preamplifier. See the preceding article for details.

ure and gain performance of both the single and dual stage LNAs was comparable to the typical values shown in the original article.

On the prototype boards, I installed the rivets from the front, and used a small hammer to flatten them out on the back. I then soldered them to the ground plane. As it turned out, the flattened area appeared smaller in diameter than the face of the rivet, so it might have been better to install the rivets from the back. Whatever technique you use, just make sure that they don't touch the etched circuits. Rivets of all sizes are available from International Eyelets, Inc, at 1-800-333-9353.

Note

[1]A. Ward, "Simple Low-Noise Microwave Preamplifiers for 2.3 Through 10 GHz," May 1989 *QST*, pp 31-36, 75 (reprinted in this chapter).

GaAsFET and HEMT Amplifiers for 24 GHz

By Tom Hill, WA3RMX

(From *Proceedings of Microwave Update '91*)

Introduction

Amateur projects using GaAsFETs at 24 GHz are fairly rare.[1,2] There is a fundamental device-technology reason for this. The available (within reasonable cost) packages that the manufacturers put their FET die into, are just not suitable much beyond 18 GHz. There are resonances in the package itself, and worse, even with automated wire-bonding equipment the parts still have so much variation between individual units that they cannot publish specifications at these frequencies. In professional designs, this is overcome by using the naked FET die and ceramic hybrid technology with gold wire-bonding. This technology is still out of reach even for the "above average" microwave experimenter.

This paper describes the steps I went through to build several amplifiers using packaged FETs that are not specified above 18 GHz. While this procedure can be used as a guideline for your efforts, it is not an exact plan for construction.

Project Plan

Active Devices

The goal is to find parts that are as low cost (we *are* on an amateur budget) as possible while having some chance of performance at 24 GHz. I used NE04583, NE32083, and ATF35176. The first is a 0.3-micron surplus MESFET, while the other two are low-cost HEMT parts. All of these are packaged parts with die that have basic performance well above 18 GHz. There are several other parts available today that look quite promising, but I have not yet tried them.

Materials

The first and most far-reaching decision to make, is how to implement the interstage tuning networks, and what board material to use. The problem is that with the impedances unknown in advance, the form of the network is also mostly unknown. Each interstage network will be custom built as the amplifier is tuned up. Moreover, since the individual FETs are so different, an etched board with a copy of the first network discovered to work will likely fail miserably when the next FET is installed. Each amplifier will likely require a large amount of retuning to get the best results. The easiest topology to allow this much tweaking without too much damage is a straight 50-ohm line with a length sufficient to allow both shunt inductive straps and shunt capacitive copper tape "snowflakes" to be moved back and forth along the line until improvement is obtained. If 10-mil thick Duroid material is used, this provides for a 50-Ω line width that is almost the same width as the FET gate and drain leads. This is quite convenient for construction. The Duroid also has fairly low loss at these frequencies. Fortunately we do not need a broadband amplifier in this application, so we can simply tune for maximum at 24.192 GHz, which is relatively easy with this topology.

Bias

The second problem is the bias injection scheme. First I wanted a technique that allows the source of the FET to be directly connected to the ground plane. I also wanted a feed-back bias that will keep current constant as temperature changes. The so-called "PNP Wrap-Around" bias circuit satisfied all these needs. See Fig 1 for the schematic of the 3-stage NE045 amplifier. Fig 2 is the schematic of the 2-stage ATF351 amp.

Next I wanted a scheme that didn't need really good inductors and bypass capacitors that work well at 24 GHz. The solution that I chose is to segment the top-side ground plane into islands that have the gate and source bias on them. These fairly large areas of ground plane are quite effective for bypassing at frequencies above a few GHz. For the lower frequencies, the gaps between these ground planes are bridged with many chip capacitors. I alternate between several values of capacitor to try to avoid a resonance at some particular VHF.

Power Supply

The power supply includes the standard automatic switch that allows the drain voltage to be applied only if the gate bias is present. This keeps the drain current from becoming excessive during turn-on and turn-off transients.

I chose +7 and –5 volts, as these voltages were already required by the 6-GHz IF amplifier in my system. You may change the basic supply voltages if necessary, as long as the actual FET bias conditions are provided for.

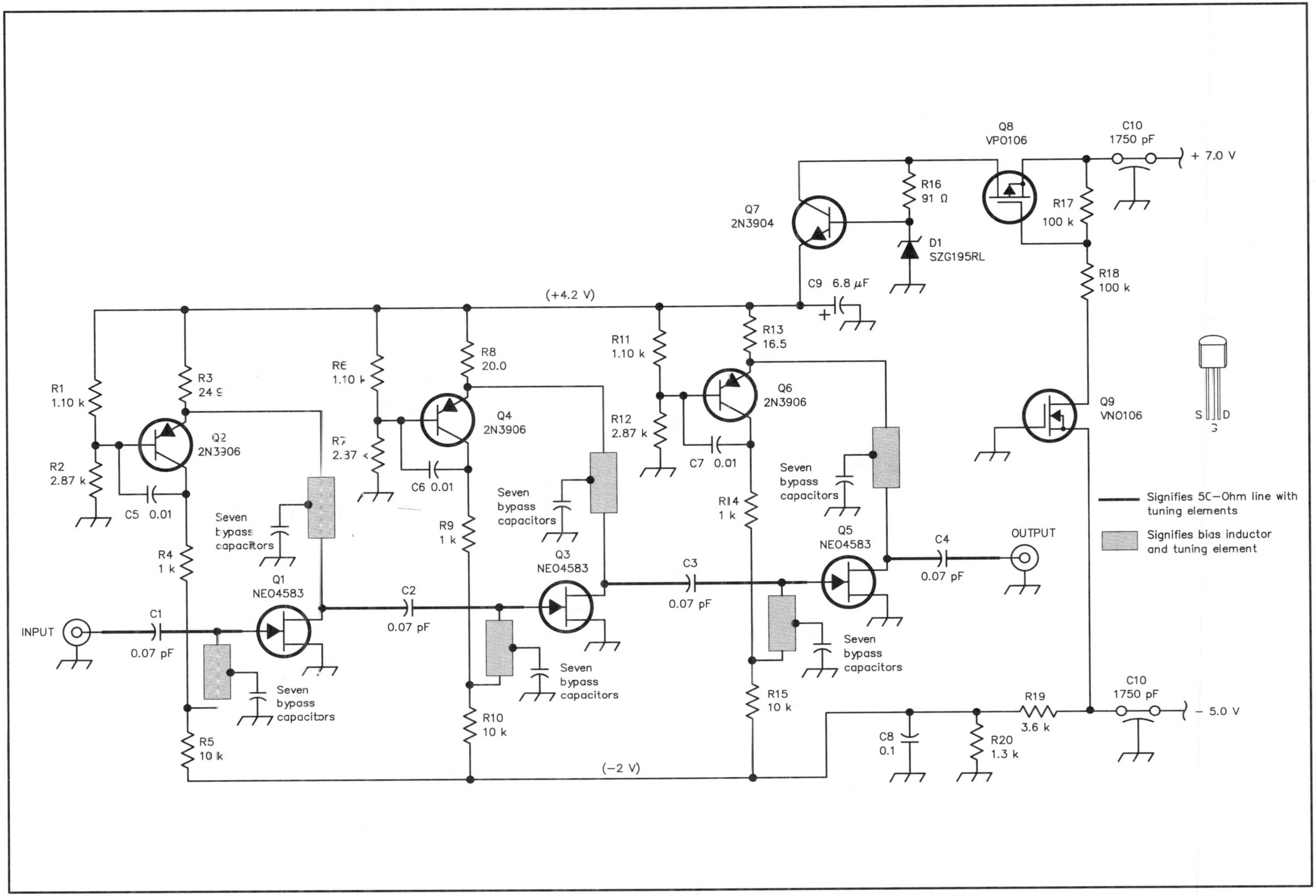

Fig 1—Schematic diagram of the second NE045 amplifier. Resistors are ¼-W carbon-film.

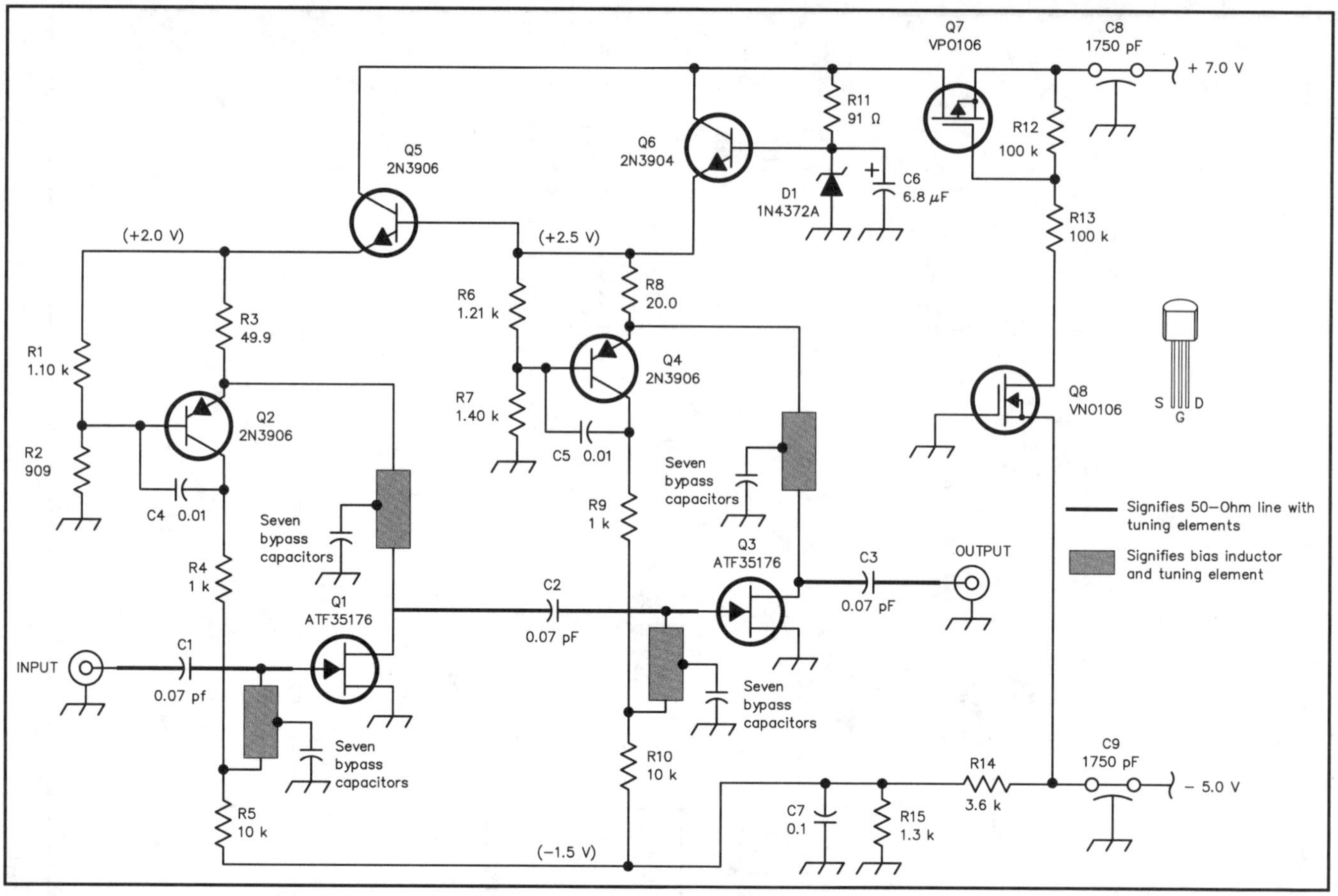

Fig 2—Schematic diagram of the ATF351 amplifier. Resistors are ¼-W carbon-film.

RF Connections

In my transverter[3] I use coax to connect the amplifiers. This adds yet another problem, since the readily available SMA connectors are usually good only up to 12 or 18 GHz. One solution that I have used quite successfully in the past (and what I still used here for the male SMA cables between the amplifiers), is to individually test each connector. I have usually found 20 percent or so of the random connectors obtained surplus will have a severe notch at 22 or 23 GHz with good operation again occurring at 24.2 GHz. For this project I was able to find some surplus female Wiltron "K" connectors. This eliminated one variable from the big list of unknowns, as these connectors are good to 40 GHz, and they mate to standard SMA connectors.

Packaging

For the package I chose a small drawn brass box. I added some microwave absorber material in the top and bottom of the box to eliminate resonances. I don't think that the box is likely very critical to the final outcome of this project.

Construction Procedure

ECB

Since the RF part of the board is really quite simple, I chose to make the boards individually instead of creating filmwork and using photoresist on each board. I simply put plater's tape on the copper where I wanted it to remain. The etchant removes the small areas left. The top-side pattern I used is shown in Fig 3. The bottom ground plane is solid copper.

My calculations determined that a 50-Ω line requires a copper width of approximately 29.7 mils. I rounded this off to 30 mils wide. The gaps in the line that allow for mounting the coupling capacitors are made about 10 mils wide so as to fit as closely as possible the Duroid capacitors' thickness. The packaged FETs have 70 mils between the leads, so that is the width of the space left for them. The line length of 0.3 inches between each adjacent component leaves plenty of room for the tuning elements to be soldered on and moved around.

Capacitors

Very few chip capacitors that can be soldered to circuit boards perform well enough at 24 GHz to serve as coupling caps. I used small pieces of the same Duroid circuit board material as coupling capacitors. A small rectangular piece 0.040 × 0.030 in. (with copper on both sides) is soldered standing up across each of the small gaps in the transmission line in between each of the FETs as well as at the input and output of the amplifier. These form approximately 0.07-pF capacitors, which have excellent performance that I have tested beyond 30 GHz. While soldering these in place, you should ensure that the solder fillet fully wets the capacitor and the transmission line, so that there are no gaps that can create a notch filter in the middle of your amp.

The bias islands are bypassed with alternating 7 pF and

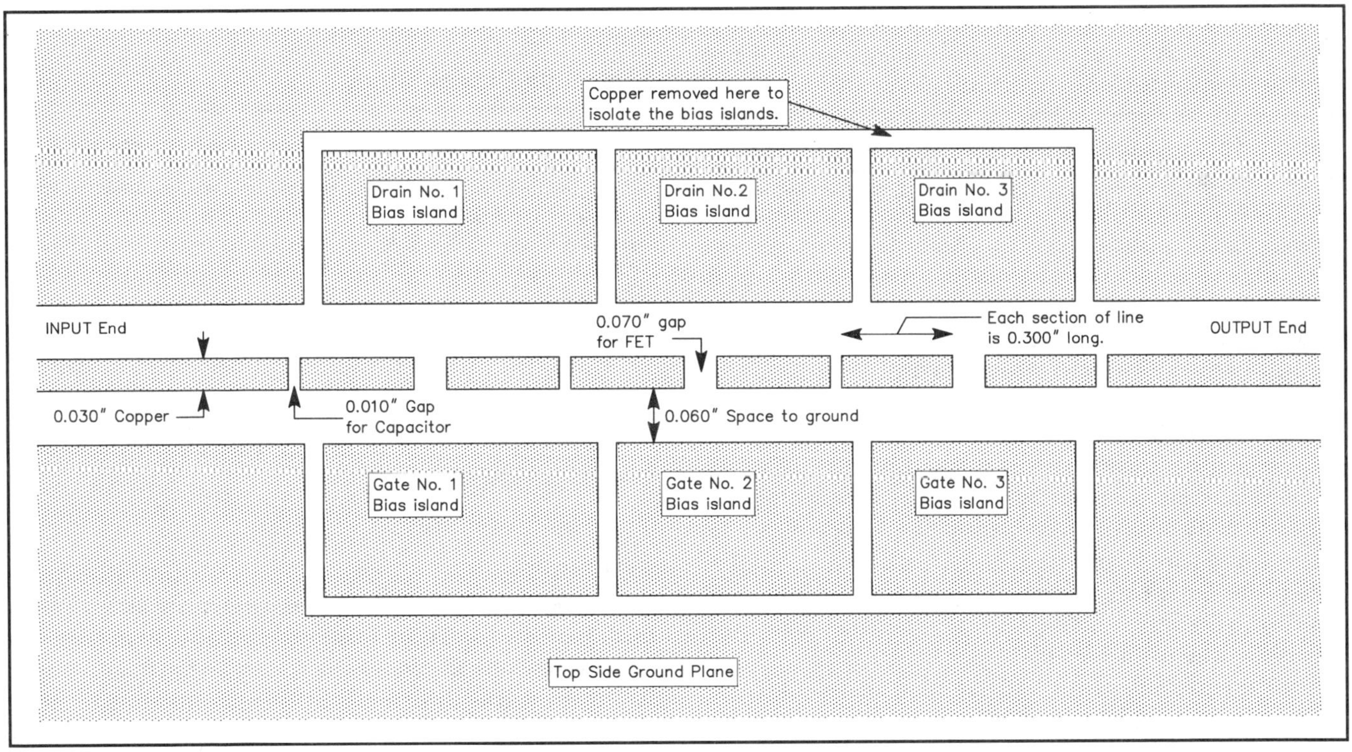

Fig 3—Circuit-board layout for the 3-FET amplifier.

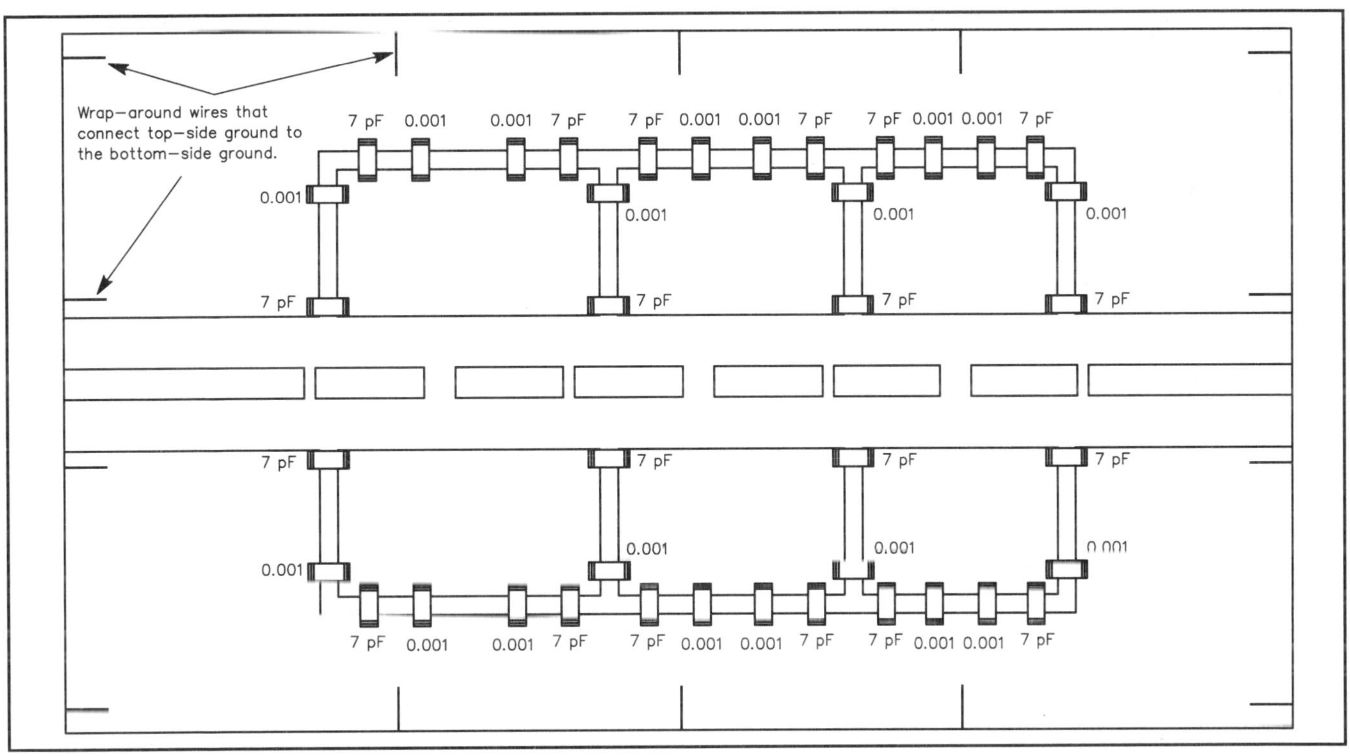

Fig 4—Capacitor locations for the 3-FET amplifier.

1000 pF chip capacitors. For the three-stage amp this is a total of 20 of each value capacitor. See Fig 4 for placement of these.

Grounding

The top-side ground planes must be connected to the bottom-side plane. I put a few straps of small wire around the edges to accomplish this. Particularly important is the edge of the top plane right next to the coax connection (there are four places total like that). These wires are also shown in Fig 4.

FET Testing

I chose to dc test each FET before installation. This is because there is a considerable variation in I_{DSS} between individual FETS of the same type number. I wanted the highest I_{DSS}

parts to be in the output stages. This allows the maximum output power possible. Also this testing allows setting the bias of each stage to be optimized for its particular function. The input stage is biased for low noise, while the output stage is biased for maximum power output. I chose 50% or less of I_{DSS} for the low noise stage, and about 80% of I_{DSS} for the output stage.

This testing also gave me the chance to select the NE045 FETs for the highest I_{DSS} amongst the ones I had available. This will allow for the highest possible output power from the amplifier. See Table 1 for the bias conditions I set for several different amplifiers after I measured the I_{DSS} of the FETs that I planned to use.

Table 1
Prototype Bias Settings

	First FET			*Second FET*			*Third FET*		
	I_{DSS}	I_D	V_D	I_{DSS}	I_D	V_D	I_{DSS}	I_D	V_D
NE045 #1	22	10	3.5	25	20	3.5	29	20	
NE045 #2	32	20	3.5	34	20	3.5	37	30	3.5
NE32 #1	28	20	2.2	37	25	2.2			
NE320 #2	31	20	2.2	32	25	2.2			
ATF351 #1	17	10	1.5	45	30	2.0			

FET Installation

After all of the coupling and bypass capacitors have been installed, it is time to mount the FETs. I cannot stress too hard the desirability of good static precautions. I soldered a 2-in, piece of bare wire-wrap wire to ground, and then tacked it to both of the drain bias islands. This grounds them, and will remain in place until the amp is ready to be tested. Another such grounding wire is soldered to the gate bias islands.

Next I used a no. 48 drill (0.076-in.) to put a hole in the center of each area where a FET will be placed. I did this drilling by hand, with a backup piece of wood, to avoid distorting the board. Then I used a small razor-sharp knife to enlarge the sides of the holes where the source leads will be.

The FETS themselves are installed by bending the source leads down and poking them through the hole in the board. These leads are then bent back straight, which will now slightly distort the board. The 10-mil Duroid is soft enough that the FET leads will lift up the board so that the ground plane in fact rises up to go over the top of the source leads. Now the FETs can be soldered in. Be careful not to get solder bridged to the gate or drain leads while soldering the source leads, as they are just barely visible through the hole in the board.

When the gate and drain leads are soldered, extra care must be used to ensure that solder is wicked up between the leads and the ECB runs. If any space is left here at all, a choke will be formed at some frequency that may well be near where you want this amp to work. Even a few mils can be a disaster.

Bias Circuits

I built the bias circuits by simply soldering the components to each other in the air. This particular circuit lends itself well to this method of construction, especially with the bias islands available as mounting pads.

Connectors

After the bias circuits are done, the last parts needed are the connectors. I waited until last to install these as the junction to the ECB is relatively fragile with my style of mounting. I simply butted the 0.047-in. diameter coax from the connectors directly to the edge of the ECB with the center conductor overlapping the copper run. The shield has a large lump of solder bridging to the ground plane. See Fig 5.

Install Board

The board is installed in the box AFTER initial tune-up. This allows the board to be laid on a solid back-up block so that the board is not bent or otherwise distorted while the tune-up is being done.

Tune-Up Procedure

Equipment

I used a Wiltron 360 network analyzer which is really nice. However, since we really only care about the one frequency, the job can actually be done with the transverter as RF source and a power meter as the indicator. In fact this is how I did the final max-power tune-up. I used an HP 8970A noise-figure meter with my own diode noise head (calibrated to a waveguide noise tube) to test the noise figure. If you are only going for the best you can get, without necessarily knowing just exactly what the number is, you can use a fluorescent light bulb in front of your dish, and a scope on your receiver output for the tuning job.

The worst problem you can really get hung up on is stability. If the amplifier oscillates it can be hard to identify and fix the cause. Certainly, if the power meter reads a bunch out with no input applied, this usually signifies oscillations. Also, an unstable gain or noise figure, particularly when a hand is brought within an inch or so of the board, is another sign of oscillation. When I found these problems, I looked for the part of the RF circuit where I could stop the problems most easily with a light touch of a finger or small piece of microwave absorber material. I then worked on changing the tuning elements in this area until the oscillations went away, before continuing the tune-up procedure.

Adding, Removing, Moving Bias Inductors

Now you may remove those shorting wires and apply dc power. After verifying proper dc operation, and taming any oscillations that might be found, you can start tune-up. With some drive power applied I first found the gain to be about –15 dB or so. That's right! A LOSS of 15 dB.

Don't worry yet; we still have a long road to travel. I first made small (20 mils at a time or less) changes in the positioning of one end at a time of the small wires that connect the bias islands to the gate and drain connections of the FETs. I also sometimes added or removed some of these straps in parallel maybe 100 mils or so away from an existing one. Often these changes might only make 0.1-dB change at a time. I usually got the gain up to maybe +1 to 4 dB with these wires.

Adding Copper "Snowflakes"

I next took a very small sliver of insulator (I used alu-

Fig 5—Installation of coax to the board. See text.

mina) with a small piece of metal (about 20 mils square) on the end to find the places on the transmission line where a bit of capacitance will help. I slid the metal end back and forth on the line until an improvement was noted. Then I would solder on a small bit of thin copper tape in that spot. Now recheck for the same result as the wand. Now continue the process. This would result in another 10 dB or so of added gain. Then another round of fun with the small wires will produce some small improvement. It will probably take three or four cycles between the wires and the copper pieces to get the gain optimized. The first time I did this it took 8 hours or so. I also got over 15 dB of gain as a reward!

Maximum Power Output Procedure

Since these FETs are run on really low current and voltage, they really compress at very low power out. To maximize the power out the tuneup must be redone with as much power out as the amp can manage without more than about 3 or 5 dB of compression. Now tune the output matching section to get best power out. This is only needed if you plan to use this particular amp for transmitting.

Low-Noise Procedure

If this amp is to be used as the input of your receiver, a tune-up for lowest noise will be needed. To do this you will put the amp into your test setup for noise, and tune the input network for best noise figure. You may also need to tune some on the first interstate network, but go easy here. The input stage is the major noise source, and the interstate more affects gain. I got about a 2-dB improvement in noise figure by this procedure. Recheck gain and power output when you are done here.

My first amplifier took about 11 hours total of tune-up time to get to final completion. The last one has taken about 4 hours. So it does get easier with a bit of experience.

My Results

I have built 5 units so far. Two with a 3-stage NE045 amp, two with a 2-stage NE320 amp, and one with a 2-stage ATF351 amp. All of these have showed excellent results. The gain of NE045 amp number 2 is just a bit better than the first one, by about 4 dB. The bandwidth is quite narrow, but this has not proven to be a problem. In fact it may well be able to double as the 24-GHz filter as well. All of the amplifiers are reasonably similar in basic performance after the tune-up marathon.

It would be advisable to be aware that there is lots of gain at some lower frequencies in addition to the intended ones.

The NE045 amp was optimized for max power out. One-dB compression occurred with a drive level of –4.5 dBm. I get a good solid 10 mW out of this unit. The others use weaker parts, and consequently get less power out. The ATF35176 produces about 6.5 mW, and the NE320 manages just 1 mW.

Noise

All of the amplifiers have performed surprisingly well even when tuned solely for maximum gain or maximum power out. The NE045 amps have 7- or 8-dB noise figure. The NE320 (still tuned for max gain) has 5.5-dB noise figure. The ATF351 achieved just under 5 dB, so it is the one that I chose to retune for optimum noise figure. With another hour of tuning time, I got it down to 3.5 dB! This is really quite respectable for 24 GHz.

Summary

My transverter now has two of these amplifiers in cascade. The input is a 2-stage NE320 unit. The second unit is a 3-stage NE045 amplifier. The result is an amp with 30-dB gain, 5.5-dB noise figure, and 8-mW output power. I am also putting the ATF351 amp at the dish to get 3.5-dB noise figure at the dish. The ATF35176 amp, combined with a 2.5-W TWT amp makes a really high-performance SSB/CW station.

Notes

[1]M. Kuhne, DB6NT, "GaAsFET Amplifier," *DUBUS*, Jan 1988.

[2]T. Takamizawa, JE1AAH, "HEMT Amplifier," *DUBUS*, Feb 1991.

[3]T. Hill, "SSB/CW Equipment Concepts for 24 and 47 GHz," *Proceedings of Microwave Update '89*, pp 39-62. This amplifier has now been added to that unit.

A 2.5-Watt Linear Amplifier for 2304 MHz

By Dave Mascaro, WA3JUF

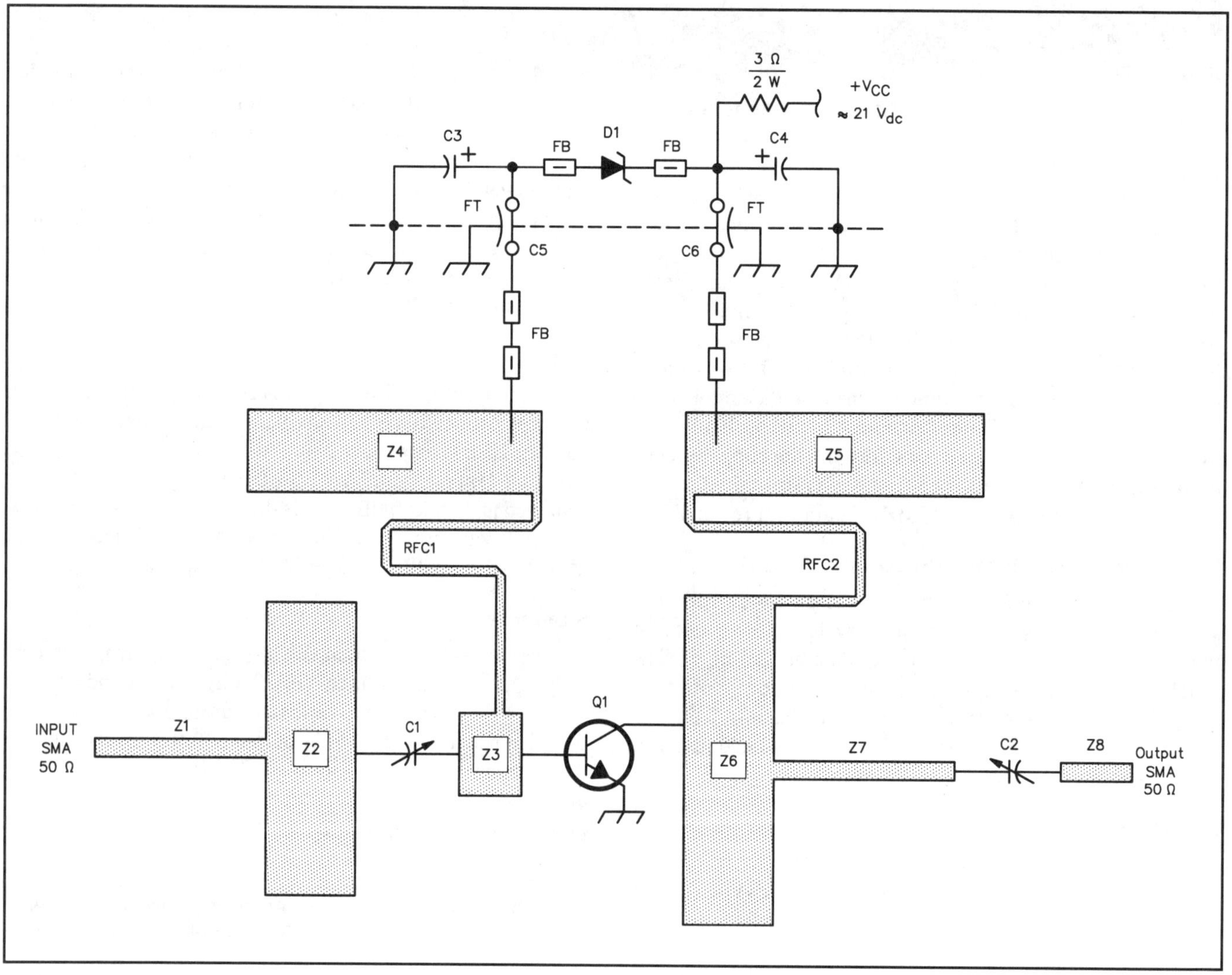

Fig 1—Schematic of the 2.5-W 2304-MHz linear amplifier.

C1, C2—3-pF Johanson Gigi-trim capacitor.
C3—10-µF, 6-V dc electrolytic.
C4—10-µF, 35-V dc electrolytic.
C5, C6—0.001-µF feedthrough capacitor, Spectrum Control SCI 729-303 (series resonant).
D1—20-V, 1-W Zener diode, 1N4747A or equiv.
Q1—Thomson CSF TCC 20 L 25 (SD1855) transistor.
RFC1, RFC2—Microstrip choke, 0.035-in- × 1.10 in.
Z1, Z8—50-Ω microstrip line, 0.080-in. wide.
Z2—0.2 × 0.65-in. microstrip line.
Z3—0.25 × 0.30-in. microstrip line.
Z4, Z5—RF short, 0.275 × 1.0 in.
Z6—0.25- × 0.675-in. microstrip line.
Z7—0.080- × 0.65-in. microstrip line.

This amplifier uses a Thomson TCC 20 L 25 transistor to provide about 2.5 W on the 13-cm band. Drive requirement is about 990 mW. The amplifier operates in Class A. I built the amplifier on 1/32-in. Teflon double-sided PC board. The microstrip circuitry can be etched using Exacto-Edge as I did, or by either negative or positive photo methods.

I milled out a brass carrier and sweat soldered the PC board to the brass. PC-board construction can be used with similar results. You can build a housing from double-sided G-10 PC board, by soldering 4 walls to the microstrip PC board to form a box. Solder the connectors to the ground plane side of the board, with the center pins going through the board and soldered to the 50-Ω input and output lines. Alternately, the connectors could be mounted in the end-launcher method. The connectors are soldered to the side walls of the box, with the center pins soldered to the microstrip input and output lines.

As with all amplifiers, this one needs good RF grounds and power-supply decoupling. You must install ferrite beads where indicated in Fig 1. All excess copper should be etched off the PC board to minimize the effects of RF propagation across the board.

The collector voltage should be about 21-V dc. It is supplied through a 317 three-terminal voltage regulator IC. Adjust the voltage until idling current reaches about 350-400 mA. Idling current value is not critical. You can readjust it for maximum power output after the amplifier is tuned up.

Alignment

Use a razor knife to trim the size of Z6, while simultaneously adjusting C2 for maximum output. Make a straight cut across the foil, but don't remove the excess. If necessary, you can bridge the gap with solder to restore the length. Trim the size of Z2 in the same way, and adjust C1 for best input return loss. If possible, inspect the amplifier with a spectrum analyzer to ensure it is stable.

8

Power Amplifiers

Contents

Power Tubes at 432 MHz

By Steve Powlishen, K1FO

To obtain high power (greater than 500-W output) at 432 MHz within a typical amateur budget, still requires the use of vacuum tubes. 432 MHz is in the gray area between VHF and microwave technology. Discrete component circuits will still function at 432 MHz, albeit with limited performance. More important, 432 MHz is short enough in wavelength that microwave-type circuitry such as cavities and striplines are small enough in size to be effectively used. Likewise, 432 MHz is a frequency where old technology glass insulation tubes will still function (at least some of the lower power tubes will) but with very poor performance. Most of the radial design ceramic/metal transmitter triodes and tetrodes will function at 432 MHz; however, only a limited subset of the available models will provide acceptable performance. True microwave tubes, such as planar triodes work very well at 432 MHz but few of these type tubes exist that will run near the maximum amateur power level.

There are several parameters in a power tube's design that determine if it will function properly at 432 MHz:

- The element spacings must be close enough to minimize transit time problems. Transit time effects can result in poor efficiency or worse make the tube literally non functional.
- The grid inductance (or screen) inductance must be low enough that degenerative feedback from the element inductance doesn't kill the gain of the tube.
- Preferably, we want a tube that has a self-neutralizing frequency above 432 MHz so it can be neutralized in conventional methods without requiring tuned grid or screen circuits.
- The interelectrode capacitances must be low enough that the tube can be effectively resonated in efficient and easy to construct input and output circuits.

For high-power operation (500 W or more output) these requirements are only met by external anode tubes with ceramic insulation. With external anode tetrodes many of the common amateur perceptions about tube operation are either incorrect or of little concern with these type tubes. Assuming that the four criteria listed above are met, that is a given tube design functions properly at 432 MHz, let's examine the tube characteristics that determine how much output power can be run.

CATHODE AREA

The primary limiting factor in how much power can be extracted from a tube is how much cathode current can be drawn. Almost all the designs considered in this article have indirectly heated cathodes. That is, they use a solid surface cathode made of an oxide coating, which will freely emit electrons when heated to a minimum temperature. Although some differences in cathode coatings exist, most are similar. This means that the amount of cathode current that can be drawn is proportional to the cathode area. A general rule of thumb is that, under conservative ratings, each centimeter^2 of cathode area will give 125 mA of continuous-duty CW cathode current. More recently, the widespread use of tubes in Class AB1-AB2 and so called Class-B linear service has allowed more liberal cathode current limits of up to 250 mA per cm^2 of cathode area. Older tube designs, such as the 4CX250B have about 2 cm^2 of cathode area. They are rated at 250 mA of continuous-duty plate current. While newer designs such as the 8874, have only slightly more cathode area, they are rated at 350 mA CW plate current and up to 500 mA plate current in intermittent CW and SSB linear service. Some of the larger tubes such as the 8877, 8938, 4CX1000A and the 7213 have about 10 cm^2 of cathode area. These tubes are conservatively rated at 1000 mA of plate current in continuous carrier service, but they can run 1500 to 2000 mA of cathode current in intermittent service.

GRID INTERCEPTION AND DISSIPATION

Grid interception and dissipation are closely related and not understood very well by most amateurs. Grid interception is the amount of cathode current the grid will grab from the electron stream going from the cathode to the plate. Cathode to grid spacing and alignment along with the plate spacing and physical grid structure size determine how much cathode current will wind up flowing through the grid structure. Older-design tubes such as the 3-1000 (and its ceramic-metal version the 3CX1000A7) have high grid interception, such that their grid current will be 30 to 40% of the plate current. This high grid current means that valuable cathode emission is being lost to the grid and cannot be converted to output power. This high

Table 1

SPECIFICATIONS FOR TRIODE TUBES USABLE AT 432 MHz

	Plate Diss.	*Plate V_{MAX}*	*CCS Plate mA*	*IVS Plate mA*	*Grid Diss. (W)*	*Max Grid mA*	*Freq Max. Ratings*	*Filament Volts*	*Filament Amps*	*Capacitances (pF) C_{in}*	*C_{gp}*	*C_{out}*	*Anode Dia.*
3CX400U7/8961	400	1500	400	500	5	100	1000	6.3	3.0	18.4	.07	6.1	2.10
3CX400A7/8874	400	2200	350	500	5	100	500	6.3	3.0	20.5	.03	6.0	1.63
3CX800A7	800	2250	600	650	4	60	350	13.5	1.5	26.0	.05	6.1	2.50
3CX800U7	800	1800	500	650	4	60	1000	13.5	1.5	23.3	.04	6.2	2.53
Y-730	1000	3000	600	600	1.5	50	1500	5.5	3.2	19.0	.07	6.5	2.79
TH308B	700	2200	600	700	2	70	1000	6.3	5.5	16.0	.13	8.2	2.75
Y-831	1500	3000	600	600	1.5	50	1500	5.7	3.3	21.0	.07	6.9	3.18
Y-846	1500	3000	600	600	1.5	50	1500	5.7	3.3	18.0	.07	7.3	3.18
TH328	750	2200	600	700	2	70	1000	5.5	5.4	19.0	.07	8.2	2.75
TH338	1200	2500	600	700	2	70	1000	6.3	5.5	16.0	.13	8.2	3.20
3CX1500U7/8962	1500	2000	1000	1500	30	200	1000	5.0	11.7	28.0	.04	13.0	3.38
3CX1500A7/8877	1500	4000	1000	1500	25	175	250	5.0	10.5	38.5	.10	10.2	3.38
8938	1500	4000	1000	1500	25	150	500	5.0	10.5	39.0	.14	13.0	3.38

NOTES: 1. All triode tubes are specified for grounded grid, cathode driven operation.
2. Power output levels for all tubes except the 8938 reflect intermittent SSB voice or keyed morse code CW operation.
3. The 8938 is capable of providing 1500 watts output at 432 MHz in continuous duty service.
4. Surplus and used-cost figures are typical and may vary significantly with tube age, condition, warranty terms (if any) and market factors.

grid interception also requires a substantial grid structure that is capable of dissipating the current (ie, power). This large grid structure in turn limits the gain of the tube and can also severely limit the high-frequency capabilities of the tube. If the grid was just made smaller however, the plate current could become uncontrollable.

In the late 1960s a new cathode technology was developed to focus the cathode current between the grid wires. Tubes employing this technology are called focused-cathode or low-grid-interception tubes. By lowering the number of electrons which will be picked up by the grid, the grid can be made smaller, which raises the gain of the tube, makes more cathode current available to be converted to plate output power and when properly designed, both high frequency and linear performance of the tube can be improved. Many long time operators used to tubes with 25-50 W grid dissipations are horrified at new technology tubes, such as the 3CX800A7 with its 4-W grid dissipation. In reality, a properly operating 3CX800A7 will deliver 750-W output power with 25 mA of grid current, whereas a 3-500Z can easily run 150 mA of grid current at that same power level. The only times these low grid dissipations are a problem is if the amplifier is accidentally driven without plate voltage present, or if it is severely mistuned. The danger of harming the tubes can be minimized by the inclusion of proper protection circuits.

PLATE DISSIPATION

Plate dissipation is fairly far down the list of power determining factors for an external anode tube. With glass envelope tubes such as the 3-500Z, the primary means to dissipate plate power is by radiation off the plate. This is readily observed by the red, orange to white color of the plates under modest-to-full power to overload conditions. It doesn't matter how much air is blown by such a tube—there is only so much plate dissipation available. In an external-anode tube, the plate dissipation is directly proportional to how much air is blown through its radiator. In fact, most of the external-anode tube specification sheets will list plate dissipations far greater than their nominally specified value when higher airflows are used. The limiting factor in increasing plate dissipation becomes a combination of the decreasing ability of air to extract heat from the radiator at very high air speeds, and the limitation of the copper anode structure to conduct heat away from the metal-to-ceramic seals at a fast enough rate. In fact, as long as an external-anode tube has reasonable plate efficiency (>40%) a large

Height	432-MHz Linear Operation Useful Power Out (W)	Drive Power (W)	3rd Order IMD (dB)	New	Cost Surplus	Pullouts
2.50	500	25	-35	$420	$200	$100
2.14	605	35	-35	$330	$200	$100
2.52	750	29	-36	$330	$200	$100
2.85	750	25	-36	$450	$220	$100
3.19	800	12	-40+	$1200	rare	$100
3.20	900	25	-40+	$1200	rare	$100
3.19	1000	29	-40+	$1200	rare	$100
3.19	1000	15	-40+	$1200	rare	$100
3.20	1100	12	-40+	$1200	rare	$100
3.20	1100	32	-40+	$1200	rare	$100
3.40	1500	90	-40	$1800	rare	$150
4.02	1500	150	38	$650	$350	$100
3.57	1500+	82	-44	$1800	$600	$200

enough blower will make the available cathode emission the limiting factor in power output.

NEUTRALIZATION

Every tube listed in this information must be neutralized for proper operation at 432 MHz. This includes tetrodes when operated in either grid driven or cathode driven operation. It also includes all of the listed triodes when operated in grounded grid mode. Fortunately many of the listed tubes have self neutralizing frequencies close to 432 MHz and can be neutralized with relatively easy grid or screen inductance tuning. More information is given in the 3CX800A7 amplifier construction project.

TUBE CHARACTERISTICS AT 432 MHz

With this background information about power tube operation in mind, the capabilities and operating characteristics of a number of the available power tubes which will operate at 432 MHz will be described. Tube data sheets don't tell the whole story. Some tubes work better than others. The number of tubes that work well at 432 MHz is small. Some will generate their power with much better intermodulation distortion (IMD) products than others. Some are nearly impossible to neutralize, while others offer completely stable operation. Other traits, not usually spelled out on the data sheets, are power drift stability, power gain in real 432-MHz circuits, allowable amateur intermittent voice and keyed CW power levels and last but not least, availability through surplus channels.

Two tables are included in this article. Table 1 lists the operating characteristics of the external anode tubes which will work at 432 MHz. Models that work well at 432 MHz are included along with ones which are usable, but are not good performers. Key parameters to look for when selecting a tube is its efficiency and drive power. The tables start out by listing specification sheet information for the tubes. Parameters such as plate voltage, plate current, screen voltage, interelectrode capacitances etc. are given. This information is what manufacturers' data sheet provides for an amplifier designer. This table tells you what parameters the tube operates within, but not how it will react at 432 MHz. The tables add to this data sheet information by giving actual operating parameters at 432 MHz. For triodes these operating parameters are given for grounded grid, cathode driven linear operation. Power output levels reflect intermittent amateur SSB voice operation or Morse code keyed CW operation. These ratings reflect long term reliable operation in these intermittent modes.

For the tetrodes listed in Table 2, operating specifications are provided for both linear and Class-C operation. Note that some of the tetrodes are not suitable for Class-C operation. The listed power output levels reflect intermittent SSB voice and keyed CW operation under linear specifications. At higher power output levels the tubes will either be not linear or not perform reliably over a period of time. Class-C operation levels are given for keyed CW operation. For continuous-carrier duty (such as FM or ATV) power levels must be reduced to at or below the tube's CCS ratings. More information on intermittent tube operation which is often also called IVS (Intermittent Voice Service) is given by Bill Orr.[3] Tube prices are also included in Tables 1 and 2. The new list prices are typically for Varian Associates, EIMAC brand tubes. The EIMAC tubes are generally considered a premium grade tube. Many of the more common tubes, such as the 4CX250B, are made by several manufacturers, whose list prices may be different. Note that several of the tubes are proprietary models, that is they are made by only one manufacturer. The list prices are not always hard and fast as some distributors and retailers may discount the manufacturers' suggested retail price. Surplus prices are typical street prices for unused tubes that have become available to amateurs. They will typically not carry a manufacturer's warranty. Pull-out prices are for used tubes. Usually they are from broadcast or military transmitters. There are no set prices or availability for surplus and pull-out tubes. The prices given are sort of an average of what I have seen the

Table 2

Specifications For Tetrode Tubes Usable At 432 MHz

Tube Type	Plate Diss.	Plate V_{max}	CCS Plate mA	IVS Plate mA	Grid Diss. (W)	Screen Diss. (W)	Screen V_{max}	Freq Max. Ratings	Heater Volts	Amps	Capacitances (pF) C_{in}	C_{gp}	C_{out}
4CX250B/7203	250	2000	250	275	2	12	400	500	6.0	2.6	15.7	0.01	4.5
4CX250R/7580W	250	2000	250	325	2	12	500	500	6.0	2.6	17.5	0.04	4.8
8930	350	2400	250	325	2	12	500	500	6.0	2.6	17.5	0.03	4.9
4CX300A/8167	300	2500	250	300	2	12	400	500	6.0	2.9	29.0	0.04	4.0
4CX600J/8809	600	3000	600	700	1	15	300	110	6.0	5.4	50.0	0.13	6.3
4CX600B	600	3000	600	700	3	15	300	500	6.0	4.3	45.0	0.15	5.8
7650	600	2500	500	600	3	25	1200	1215	6.3	7.9	37.5	0.11	5.3
4CX1500BC	1500	3000	900	1250	1	12	350	450	6.0	10.0	81.5	0.02	11.8
7213	1500	3000	1000	1250	15	50	1000	1215	5.5	17.1	42.0	0.17	17.1
YL1050	1600	3000	900	1300	10	30	650	400	3.8	20.5	36.0	—	10.0
GS-23B	1600	3500	1000	1500	5	25	600	1000	6.3	5.7	33.0	0.03	11.5
8792	1800	3500	700	1250	10	50	1000	400	5.5	17.3	38 0	0.03	16.0

Notes:

1. Specified parameters are for grid-driven operation.
2. Specified parameters are for cathode-driven operation.
3. Surplus and used-cost figures are typical and may vary significantly with tube age, condition, warrany conditions (if any) and market factors.

various tubes sell for in recent years.

Let's take a look at the inside story of each of the tubes in the tables.

4X150A/7034

These tubes are not recommended for use at 432 MHz. Through 150 MHz the 7034 is as good as a 4CX250B both in power output and linearity. The 7034 also has a 250-W plate dissipation despite its '150 designation. The difference between the 4CX250B and the 4X150A is that the 7034 uses a glass plate-to-screen insulator. This glass heats up easily at 432 MHz and may develop holes or the seals may break. Do not confuse the 7034/4X150A with the original 4X150. The original 4X150 has both glass base and plate insulators. The 7034 has a ceramic base insulator. The original 4X150 will not even run acceptable power levels at 144 MHz.

4CX250B/7203

The 4CX250B has been a standard tube among amateurs for over 30 years. Twenty years ago the "standard" power amplifier design for 432-MHz DX work was the K2RIW parallel kilowatt.[1] The original amplifier used a pair of 4CX250Bs, and was designed to operate at maximum efficiency in Class-C operation (CW or FM, not SSB linear) at 1000 W of input, which was the legal amateur power limit before 1984. Since the power limit was raised to 1500-W output, tales of incredible power output levels from 4CX250B tubes abound, but in reality the 4CX250B is a mediocre tube in linear service. About 600-W PEP linear output, from a pair of tubes, is all you get before IMD performance seriously deteriorates (2000 V on the plate, 550-mA peak plate current for 2 tubes, zero grid current). In true linear service a K2RIW amplifier will run at about 55% efficiency. When run hard in Class-C service, efficiency will reach 70% with power output levels around 900 W.

Some builders of the K2RIW amplifier have reported instability and thermal drift. These problems are most often due to running the amplifier without neutralization. Some builders have neutralized the K2RIW amplifier by feeding back a sample of the grid voltage (by tapping on the center of the grid line between the tubes) back to the plate. Others have successfully tuned the screen inductance. Most builders report that neutralization is easier with the SK620 or SK630 type sockets, which incorporate screen shielding, than with the SK600 or SK610 sockets used in the original amplifier.

Although some operators have run these tubes at even higher power levels, tube life is compromised. Essentially an oxide-coated cathode has a fixed amount of electrons that it will emit over its life. The choice is yours—do you want to stretch the cathode's life over several years, or do you want to use it up in a few months? If you make the decision to stretch the limits of the 4CX250B, you will be best off running the plate voltage higher than specified (up to 2500 volts under load) and keeping the cathode current within reason (less than 300 mA per tube). Also be sure that the screen current is reasonably low (less than 30 mA per tube). To comfortably reach the present amateur 1500-W output power level, some 432-MHz operators have resorted to using power splitters and combiners to run a pair of K2RIW amplifiers simultaneously. Tim Pettis, KL7WE, described how he built hybrid power splitters and combiners to operate a pair of the K2RIW amplifiers.[2] N9AB has used a similar approach to combine a pair of 8874 amplifiers and a pair of 3CX800A7 amplifiers.

Common surplus availability at reasonable prices makes the 4CX250B attractive. The 4CX250 also comes in some newer flavors, the 4CX250BC/8957, which is a premium-quality high-tolerance version and the 4CX250FG/8621, which has a 26.5-V filament. A water-cooled version of the ubiquitous 4CX250, the 4W300B/8249, is also made. There are even conduction cooled models of the '250, the 8560A and

Max. Dia.	Height	Linear Power Out (W)	432 MHz Linear Drive (W)	Typical Linear Eff. (%)	3rd Order IMD (dBc)	Power Out (W)	Class C Drive Power (W)	Eff. (%)	Notes	New	Cost ($US) Surplus	Used
1.64	2.64	300	10	55	-23	450	25	70	(1)	80	50	10
1.64	2.64	425	12	55	-25	550	30	70	(1)	125	60	15
2.08	2.46	425	12	55	-25	550	30	70	(1)	195	80	25
1.64	2.50	325	10	55	-23	500	25	70	(1)	120	40	15
2.08	2.71	750	20	55	-35	n/a	n/a	n/a	(2)	1000	150	35
2.08	2.45	750	23	55	-35	900	30	64	(2)	1200	rare	50
2.06	2.40	680	30	55	-31	960	38	64	(2)	1100	100	30
3.38	5.13	1500	65	50	-40	n/a	n/a	n/a	(2)	1000	rare	rare
3.75	3.34	1500	70	55	-27	1500	100	65	(2)	2500	250	100
3.74	4.33	1500	75	56					(2)	2000	500	150
3.55	4.25	1500	60	56					(2)		250	rare
3.75	3.34	1500	70	55	-37				(2)	2500	300	100

8560AS. Conduction-cooled tubes are not very practical for a high-power 432-MHz amplifier due to the limited available plate dissipation and because of the additional plate-to-ground capacitance of the tube-to-heat sink junction.

4CX250K/8245

A coaxial-base version of the 4CX250B, the 4CX250K will generate the same power output as a 4CX250B. With a properly designed input circuit a 4CX250K can do it at lower drive levels. The pulse rated 4CPX250K/8590, along with the 26.5-V filament (4CX250M/8246) versions are also usable. I have often seen the 26.5-V filament tubes in the 4CX250 family languish on flea-market tables. Filament transformers with 28-V secondaries are readily available, and it is a simple matter to convert an amplifier to 26.5-V filament operation. So, don't pass up the 26.5-V filament tubes when the price is right.

4CX350A/8321

The 4CX350A is a ruggedized, high-linearity version of the 4CX250B. Its spec sheet looks great—full ratings to 110 MHz, 2500 V at 300 mA and better linearity than the '250B. Several amateurs who have tried the 4CX350A, however, have reported poor efficiency and instability. This "poor" operation could be a result of just substituting the '350As for 4CX250Bs without properly adjusting the amplifier for them. The 4CX350A uses a "stretched" cathode giving it more cathode surface area and, hence higher cathode-current capability. This design change requires retuning if not modifications to the amplifier's input and output circuits, due to the substantially higher interelectrode capacitances of the 4CX350A. The different internal construction of the '350A also requires substantially lower grid-bias voltage. This difference in the grid structure gives the 4CX350A its better gain and IMD characteristics. but at the expense of zero grid dissipation. This makes the '350 usable only in linear service. An operator who is primarily interested in SSB operation may want to consider adapting a K2RIW amplifier to these tubes, but an EME operator who is interested in maximum CW power would be better off using 4CX250Rs. One may also wonder what special magic gives the 4CX350 100 W more plate dissipation, since it is physically the same size. The sleight of hand used is to higher specified air flow. Amateurs get very hung up on plate dissipation. On an external-anode tube you should consider the rated dissipation as a nominal value. It is completely dependent upon temperature altitude and air flow. Cathode emission is the more critical element in limiting tube power! As with the 4CX250B, 26.5-V filament versions exist (4CX350F/8322 and 4CX350FJ/8904). The 4CX350FJ also has improved IMD characteristics over the other 4CX350 models.

4CX250R/7580W, 8930

These are the best models in the 4CX250 family of tubes and the only ones for which I would spend top dollar. Although the '250R is nominally rated the same as the B models, its better internal construction allows for a manufacturer's specification of 325 mA of peak plate current per tube in linear service. The 4CX250R was originally designed to withstand severe shock and vibration, as in airborne equipment. The 8930 is internally an identical tube to the '250R, but has a larger anode radiator than the 4CX250R. This larger radiator gives a nominal plate dissipation rating of 350 W. The 8930 also has a linear service maximum-plate-voltage rating of 2400 V. In amateur service, even the 4CX250R can be comfortably operated with 2400 V on the plate. The better characteristics of the 4CX250R and 8930 allow them to be comfortably run in a K2RIW-type amplifier in intermittent operation at 432 MHz with a plate voltage of 2400 and a plate current for a pair of tubes of 650 to 700 mA. This will provide over

800 W of linear output power and, when gently run in Class C, 1100-W output—only 1.3 dB below the legal limit. If you can locate the sockets at a reasonable price it will be worth investing in new 8930s or 4CX250Rs. If you have to purchase the sockets at new price you may be better off spending the money on 3CX800A7s.

4CX300A/8167

The 4CX300A/8167 is a fairly early external-anode ceramic-metal tube design that dates back to the late 1950s. It is suitable for either Class-C or AB1 linear operation. These tubes have been successfully used in 432-MHz amplifiers. Power capability is slightly better than the 4CX250Bs, however, reports from some users claim bad thermal drift. Other users have not reported any thermal problems, while obtaining significantly better power output performance than from the 4CX250 family. Since I have never used 4CX300As at 432 MHz, I can't comment about the thermal drift. It may be inherent in the tube design, due to an amplifier problem, or simply a case of the amplifier needing neutralization. From looking at tube prices, it should be obvious that you would not want to sink serious money into 4CX300As, but they are certainly usable if the price is right and you can also locate sockets. The 4CX300As use a twist-lock base, like a mini 4CX1000A. They are not directly interchangeable with the 4CX250 family but they are similar enough in plate characteristics to be adaptable to the same circuits as the 4CX250 circuit designs. The input capacitance is almost twice that of a 4CX250B which will necessitate some extra effort to get a good input circuit functional. A push-pull amplifier design for the 4CX300A appeared many years ago in *QST*.[12] At least one builder has reported success in adapting 4CX300As to the venerable K2RIW parallel design. There is also a high-cathode-current version, the 4CX300Y/8561. The 4CX300Y has significantly higher input capacitance, and may require changes to the input circuit if it is substituted for the 4CX300A. The high-current versions are not as common in surplus channels, have a specified limit of 110 MHz for maximum ratings, and are more suitable for use at 144 MHz than at 432 MHz.

8122, 8122W

The 8122 has a bad reputation for thermal drift, poor reliability and high IMD levels. The 8122 has an 11-pin base like the 8874. Sockets are made from 11-pin sockets in combination with separate screen bypass capacitors. I have not used 8122s at 432 MHz so I don't know if the thermal drift can be eliminated. I don't recommend purchasing 8122s for use on 432 MHz. Even if you have some 8122s lying around, you may want to find an HFer with a National NCL-2000 amplifier or NCX-1000 transceiver, sell him the tubes and buy something else.

3CX400U7/8961

The 3CX400U7 is a coaxial-base version of the 8874, designed for use up to 900 MHz. Varian built a large number of 850-MHz cavities, which Motorola used in cellular phone and paging transmitters. As Motorola replaced them with solid-state units, they sold the old tube units to hams for only $50, a great deal for 200 to 300 W at 902. This great deal came to an end when some enterprising hams were caught buying the cavities for $50 and reselling them at a tremendous profit to competing commercial interests. Now, reportedly, Motorola crushes the cavities when they are taken out of service. I wouldn't recommend purchasing new 3CX400U7s, as the money would be better spent on 3CX800A7s or an 8938. If you find a pair of 3CX400U7s surplus they will make a cool, linear and efficient kW of output power at 432, as long as you don't mind making the coaxial sockets.

3CX400A7/8874

This rugged, reliable triode has been around since 1971. It was designed by Varian Associates as a triode version of the 4CX250 family. Since the 8874 was designed a decade after the '250 family, it incorporates within the same physical size better power output and linearity than its tetrode relatives. The 8874 uses an 11-pin base, and early versions looked just like the 8122. In the mid 1970s Varian made some improvements to the 8875 and incorporated the 4CX250B style ceramic in its plate to grid ring insulator along with a '250 style screen ring for its grid ring. The 8874 is easily adaptable to the K2RIW parallel kilowatt design. A pair of tubes in a K2RIW-type amplifier will give 1100 to 1200 W of intermittent service output power in stable and linear format. I also described a single tube 8874 amplifier[4,5] that delivered 550 to 600 W output. The 8874 now costs new, more than the newer and more powerful 3CX800A7, making the '800 the logical choice for one purchasing new tubes. There is a conduction-cooled model of the 8874, the 8873. The 8873 is not suitable for a high-power 432-MHz amplifier for the same reasons described in the 4CX250B section where the conduction cooled 4CX250B is discussed. Another version of the 8874, the 8875 uses cross tube air flow. The 8875 is not a good choice for a 432-MHz amplifier due to its lower plate dissipation and larger size. In addition they are very difficult to find at surplus prices. The 8874 is used in a few commercial transmitters along with a great number of amateur amplifiers. This has made the 8874 widely available in surplus and pull-out channels, however when purchasing them you will be competing with owners of older ETO Alpha amplifiers. Varian also made some 450-470 MHz cavities that used a single 8874. These units occasionally show up surplus and they can be easily converted to 432 MHz. In summary, the 8874 is highly recommended for use at 432 MHz, although if you are purchasing new tubes you will be better off getting 3CX800A7s.

3CX800A7

This is one of the newer tube designs. Introduced in 1984, it is an improved 8874 featuring very high power gain, for a triode, and a large anode radiator. Testing of a single tube 432-MHz amplifier has reliably given long term CW power levels approaching 600 W. An output of 750 W per tube (1500 for a pair) is easily obtainable in intermittent service. Due to the close grid spacing and precise internal construction, running the 3CX800A7s at high power requires grid overload protection circuits. A compact single tube 3CX800A7 amplifier designed and built by the author appears elsewhere in this book as does an amplifier which uses a pair of the 3CX800A7s. Although the tubes are a bit pricey when purchased new, the sockets are low in cost when compared to tetrodes, the operation is stable, efficient, linear and reliable. If you are purchasing new tubes, the 3CX800A7 is probably the most cost-effective solution to high power on 432 MHz when long tube life and reliable operation is considered. At 1500 W output on 432 MHz, the tubes are somewhat stressed, as they are operating at a frequency 23% higher than specified for maximum ratings. At 432 MHz it is advisable to run only SSB and keyed CW, keeping key down tune up times as short as possible when running full power. Several users of 3CX800A7 at 432 MHz have had tubes go soft after a few years. In examining their operating habits it was found that their amplifiers were not neutralized and they did not take care in keeping cathode current within reason. There is also a pulse rated version the 3CXP800A7. For linear amateur service, the pulse version does not present any major advantages over the standard model.

3CX800U7

This is a coaxial-base version of the 3CX800A7. It features about 10% lower input capacitance and 25% less feedthrough capacitance. Its specified CW ratings are valid through 1000 MHz (although these ratings are reduced from those specified for the 3CX800A7). At 432 MHz in intermittent amateur operation, the 3CX800U7 will be capable of similar power output levels as the 3CX800A7 but in a properly designed circuit it will require less drive power and it may have slightly higher output efficiency. The extra cost of the 3CX800U7 is not warranted over the 3CX800A7 for an amateur 432-MHz amplifier. So far the limited volumes in which 3CX800A7 has been produced makes surplus or pull-out availability unlikely.

3CX600U7

This little-known and hard-to find-tube would be perfect for use at 432 MHz. It was designed as a big brother for the 3CX400U7 for use up to 1000 MHz. At 432-MHz, 1500-W output for a pair would be easily obtainable at high efficiency and low drive levels. Unfortunately, the 3CX600U7 was never produced in significant numbers and is now obsolete.

TH308, TH328, TH338

These tubes are members of a large planar tube design by the French company, Thomson-CSF. Their technology has allowed them to develop some very high-power, high-linearity tubes for the UHF frequency ranges. Conceptually, these tubes are built like a giant 7289/2C39-type tube. The cathode area of these tubes is almost 5 times that of a 7289. The TH308 family was developed primarily for use in linear TV translator service through 1200 MHz. Their three-tone IMD performance is typically better than –52 dBc (at 200-W carrier level). These tubes will generate 300 to 600 W at 1296 MHz in intermittent amateur use. In a properly designed 1296-MHz circular-cavity plate circuit, the tubes will run at 30 to 35% efficiency. Pull outs of the TH308 family of tubes (primarily the TH328) are starting to show up regularly in the US, coming from television translator amplifiers. The internal construction of all three tubes is similar, with the TH328 and TH338 having larger anode radiators. At 432 MHz, these tubes should approach 60% efficiency and will give excellent IMD performance. On the other hand, if you are also a 1296-MHz operator and have a limited tube supply, you may want to save the tubes for that band. At 432 MHz, the TH308 should deliver over 900-W output, while the TH328 and TH338 may provide 1000 to 1100 W. The tubes have very high gain for a triode, close to 20 dB for the '308 and '338 and over 15 dB for the '328. These high-gain designs result in very low grid currents (less than 70 mA) and very low grid dissipations of about 2 W. Because of the low grid dissipation, be very careful in operating the tubes—grid protection circuitry is a must. In addition, these tubes are not the right choice for an operator who likes to push tubes. The high gain of the tubes mean the amplifier must be neutralized. One unique aspect of these tubes is that the anode air radiators sit on a tapered base and are easily removable. This makes them adaptable to water cooling, which is very popular at 1296 MHz.

Y730-YU129, Y831, Y846

These tubes are members of a series of Varian and EIMAC large planar triodes. They are patterned after the Thomson TH308 family and are also primarily intended for use in UHF-TV translator applications. These EIMAC models are rated at higher plate voltages, however they have a smaller cathode area and use a different (and lower dissipation) grid than the Thomson tubes. For amateur use at 432 MHz, the purchase of these tubes new would be hard to justify as their power output would only be slightly higher than a 3CX800A7 but their cost is 2-½ times as much. The YU129 is an IMD grade-out version of the Y730 that is suitable for amateur operation, and has a much lower price tag than the YU129. These tubes do not perform as well as the Thomson models at 1296 MHz (20-25% efficiency at best), making them available for use at 432 MHz. At 432 MHz, efficiencies for the tubes will be around 60%, with available output power in the 800-W range. The limiting factor in output power will be a combination of cathode current and grid dissipation. The low output capacitance of these planar triodes lets them be used in simple stripline circuits as well as in resonant cavities. The relatively large diameter of the anode radiators used on these tubes (2.8 in. for the Y730 & TH308 and 3.2 in. for the Y831, Y846, TH328 & TH338) makes the use of a pair of tubes in a single 432-MHz amplifier a difficult (but not impossible) proposition. Therefore, these tubes are best suited to an operator who desires low drive requirement, excellent linearity and is satisfied with about 800 W output. These tube models are now obsolete and have been replaced by a new series of tubes (described below). The information is included because these older tube series can often be found surplus or as pull-outs and are suitable for 432-MHz operation.

YU328, YU338, YU339

These tubes are a set of improved Varian UHF planar triodes. The YU328 replaces the Y730, the YU338 replaces the Y831 and the YU339 replaces the Y846. The new tubes feature improved internal construction (primarily in the grid) and have significantly better performance characteristics, particularly at higher frequencies. Although these improvements make them a better tube for 432 MHz, operation of the tubes is greatly improved at 1296 MHz, which means those trying to locate surplus of pull-out tubes will face competition from 1296-MHz operators. Any of these tubes will be capable of over 1000 W output at 432 MHz. The new price of these tubes is fairly high and as with any new tube it will take a few years before surplus and pull-outs become readily available.

7650, 8791

RCA stopped marketing their products to hams around 1970, so its tubes aren't very well known by most amateurs. After General Electric acquired RCA it sold off the RCA tube business. These tubes and many other RCA designs are still made by Burle Industries, whose main focus is tubes for VHF and UHF television transmitters. The 7650 has been around since around 1960. It is a mid-size tetrode that has over twice the power-output capability as a single 4CX250B. Over 1000 W output can be obtained running Class C in intermittent keyed CW service, but it is not very linear above 750 W output. A pair of tubes will easily run the 1500-W output limit. The 8791 is an updated version of the 7650 that has drastically improved IMD performance. The 8791 is used as a driver stage in some VHF television amplifiers and pulls can occasionally be located at reasonable prices. The 7650 and 8791 are essentially interchangeable in a stripline amplifier with a little adjustment to the tuning. K2UYH has been running a single tube 7650 amplifier for years (one was documented in the *Lunar Letter*).[7] Recently, DL7APV built a pair of 7650s and reports 1500-W out is very easy to obtain. The 7650 is often available surplus, but is usually passed up because few amateurs know what they are or what to do with them. Be careful when getting broadcast pull-outs as some stations use up the entire life of the tubes before replacing them. You should be sure that several tubes are available before building an amplifier. I have also seen a couple of different single-tube 7650 cavities that looked like they would tune to 432 MHz. The 7650 and 8791 are very good tubes to use at 432 MHz if you can find them at a reasonable price, you don't mind the complexities of tetrode operation and if you are capable of fabricating your own coaxial sockets.

4CX600J/8809-Y584, 4CX600JA/8921, 4CX600JB

These very exotic and very expensive tubes were designed for demanding military airborne applications. The 4CX600 family was essentially Varian's answer to the RCA 7650-8791. The 4CX600J and its large anode brother the 4CX600JA/8921 have found their way into various military amplifiers. They can be made to run very linearly and reliably at 750 W output per tube. The base is physically the same as a 4CX250, however, the elec-trical pin out is different, requiring modifications to existing sockets or the construction of your own. The anode of the 4CX600J is the same diameter as an 8930 (2-in. nominal) while the 4CX600JA has a larger 2.5-in. diameter anode. The 4CX600JB is an economy version of the '600J that uses a smooth, non-focused cathode. This results in higher drive requirements, but, due to this low gain the 'JB is more suitable for amateur use. The most-common surplus and pull-out version of the 4CX600 family is probably the Y584. The Y584 is similar to the 4CX600J and is used by Rockwell in an airborne transmitter that runs a pair of tubes in the final. Although the specified highest frequency for maximum ratings is 110 MHz, it will work well in intermittent amateur service through 432 MHz. A pair would make a formidable 432-MHz amplifier if you could overcome the inherent design complexities of a very high-gain tetrode at UHF frequencies. The 4CX600J uses a very fine grid structure to obtain its high performance. Because of this, the 4CX600J requires very tightly regulated grid and screen supplies. Its grid dissipation is only 1 W, so only linear operation is permitted and protection circuits are required. In addition, effec-tive neutralization is mandatory for proper operation. Finally, running them cathode driven would be the way to go. At new tube prices there are much better tube choices for use at 432 MHz. I have often seen large quantities of 4CX600J and Y584 pull-outs at bargain prices. The builder with limited financial resources but lots of time and ambition may want to give the tube a try. A single tube 432-MHz cavity was built and described by Varian.[6]

4CX600B, 4CW800B

These are wide-band UHF versions of the 4CX600J. They use an unusual screw-stud base with a bolt-in-place screen ring. Joe Reisert, W1JR, used a 4CW800B (water cooled) tube in a rectangular cavity for many years. He reports 900 W out is obtainable, but he also seems to go through a tube every few years. The design of the cavity was described in an EIMAC application note.[6] Water cooling has become quite common in amateur circles on 1296 MHz, but is rarely used on the lower bands. As with most tubes aimed at the military market there are 26.5-V filament versions, the 4CX600F and 4CW800F. Again, don't invest money in them but if you find them cheap, they're worth considering.

4CX1000A/8168, 4CX1000K/8352

The 4CX1000A is an old external-anode tetrode designed by Varian in the late 1950s. It is known to amateurs primarily because it was used in the Collins 30S-1 HF linear amplifier. The 4CX1000A has also been used in a number of military amplifiers and is often found surplus. These tubes are not recommended for use at 432. Reportedly, K2CBA has made them work, but they will require a bit of effort to get set up properly. The only difference between them is the ring for the screen contact (instead of 3 tabs) on the K. If used in a proper socket, the screen ring can give higher input-to-output isolation (stability) at VHF. When these tubes are used in their standard sockets (SK810, etc) they will have a self-neutralizing frequency below 432 MHz. If you are going to give these tubes a try the best way to get them to work is to run the screen at ground (use a socket without a screen bypass capacitor) and float the cathode. Bypassed screen operation may require tuning the screen to neutralize it. The very high input capacitance in grounded-cathode operation (84 pF) also makes the cathode-driven configuration a must at 432 MHz. This is the procedure used in almost all large commercial and military UHF tetrode amplifiers. The screen at dc ground method of operation was thoroughly described in a 144-MHz grid-driven amplifier.[6,8] You would probably be better off using a 4CX1000A on 144 MHz or 220 MHz, and using something else on 432.

4CX1500B/8660

This is simply a newer, more linear version of the 4CX1000A. The same caveats as with the 4CX1000A apply—

use it on 144! EIMAC prototyped a coaxial-base version, the 4CX1500BC, but this version of the tube is not readily available.

3CX1000A7/8283

This is a ceramic metal version of the long-lived 3-1000Z that uses a 4CX1000A-type twist lock base. A 432-MHz amplifier was described using the 3CX1000A7.[9] It gave substantial power out, but used lots of drive power (it had about 7-dB gain) and had poor efficiency (less than 40%). The only reason why it gave reasonable apparent efficiency was due to all the drive power it required. The 3CX1000A7 is not a good performer at 432 MHz, due to its high grid inductance and a too-wide cathode-to-grid spacing. The tube is better off in an amplifier for 222 MHz or lower. EIMAC now makes another ceramic-metal version of the 3-1000Z, the 3CX1200A7. This tube is designed to use the same socket as the 3-1000Z, and is not suitable for use above 100 MHz. Still another version of the tube is now made, the 3CX1200Z7/YU181. This tube is a 3CX1200A7 that adds a low-inductance grid-contact ring. It is not suitable for use at 432 MHz, but the 3CX1200Z7 should be investigated by serious 144-MHz amplifier builders.

8877/3CX1500A7

The Varian 8877 is an amateur standard through 222 MHz. It is now commonly available in surplus, due to good acceptance in the commercial market. The 8877 is a great tube below 250 MHz, but it's marginal on 432. Due to element spacings which are too wide for 432 MHz, its efficiency is poor (35 to 40%). In addition, its pin socket arrangement makes the cathode inductance high enough that an effective input circuit is hard to design. A combination of both of these effects limits the 8877's power gain to a relatively low 10 dB. If you do run one, use lots of plate voltage (over 3000 V under load) and lots of air. You can get 1500-W out, but only under intermittent operating conditions. W8YIO, K5AZU and K8WW are among the amateurs with experience running the 8877 at 432 MHz. K5AZU described an input circuit for the 8877 at 432 MHz.[10] It has similar plate characteristics to the 8938 and can be used in 8938-type plate circuits. Henry Radio introduced the 3004A 432-MHz amplifier with an 8877, however, they quickly redesigned the model to use the 8938. A pulse version of the 8877, the 3CPX1500A7 is also made. At least one amateur (WB0QMN) reported slightly better results with the pulse version at 432 MHz. If you have 8877s available at a reasonable price and you have lots of drive power, you may want to consider building an 8877 amplifier and hope that you can find an 8938 to replace it with at a later date.

3CX1500U7/8962

This is a higher frequency version of the 8938, designed for use at up to 900 MHz. It will work very well at 432 MHz, its very close internal spacings limit plate voltage to 2000 V. It will be very easy to obtain 1200-W output at 432 MHz with the 8962, and it could be pushed to as high as 1500 W. The 8962 has been discontinued by the manufacturer. The tube is included in this information because I have seen them at flea markets.

7213, 7214

These are a couple more old work horse RCA tubes. Both are tetrodes with 1800 W of plate dissipation and are CCS rated at 3500 volts and 700 mA of plate current. This is the first single tube on the list which will comfortably generate 1500 W of power at 432 MHz. The 7214 is a pulse-rated version of the 7213 and is also usable. Several surplus 7213/7214 cavity designs exist that operate in the 432-MHz range. A number of amateurs have tried to make the 7213 work in λ/2 stripline circuits, but none that I am aware of were ever successful in getting the designs to work properly. The primary problem with their designs appeared to be the use of ineffective screen bypassing. One solution could be to run the screen at dc ground. Screen inductance, capacitance and coupling between the input and output circuits are all critical parameters, which must be managed before the tube will operate successfully at 432 MHz. One military cavity I saw used a tuned screen cavity for neutralization. Another potential problem with using the 7213 in a λ/2 stripline is that the output capacitance of the 7213 is high enough that the first quarter wave of the line almost disappears within the tube. This can cause problems getting the line to work properly and be able to couple power out of it. K2CBA built a 7213 amplifier which uses a λ/2 line arranged in a coaxial form.[11] This design is reported to work very well, although it is physically fairly large. Another arrangement which should work properly is to use a λ/2 stripline that has the tube located in the center. This circuit will require plate blocking capacitors, as either end of the stripline has to be at RF ground. However, this arrangement forces more balanced RF circulating currents. The same is true for the K2CBA design, which is probably key to its success. Another detail in the K2CBA design worth noting was his placement of the screen capacitor plate below the RF deck chassis, on the input side. This arrangement assures much better plate isolation, a critical factor in obtaining amplifier stability. In cathode-driven operation, (the recommended method) power gain will be over 13 dB. The 7213s are commonly available surplus at reasonable prices. The time required to make an amateur design work right will be time well spent. The 4661 is a similar tube to the 7213. It has a different anode radiator and a taller screen to plate insulator

8792/V1

This is a modern, linear version of the 7213. There is also a wide-band, high-gain version, the 8972/V2. What pertains to the 7213 pertains to the 8792, except that they are much harder to find surplus and they will provide a much cleaner SSB signal. They are designed to work cathode driven (as are the 7213, 7214 and most other UHF tetrodes). The 8792 is used in several high band VHF-TV transmitters and in some FM-broadcast transmitters, making local broadcast stations a possible source for these tubes.

YL1050 Family, GS23B

The YL1050 is a relatively common European UHF tetrode. It is part of a series of UHF tetrodes made by Siemens. It will comfortably run 1500 W output in a proper circuit and is used by a fair number of European 432-MHZ-EME operators. Other members of the YL1050 family are the YL1052, YL1055, YL1056, YL1057 and YL1058. All are similar in dimensions and power capability, with the YL1058 having the highest ratings. The YL1050 through YL1057 are capable of working at over 1000 MHz. There is also a Russian-manufactured version, the GS23B. The GS23B is similar in size and capability to the YL1050 but it is not an identical replacement for it. With the opening of world markets, tubes such as the YL1050 and GS23B may become readily available in the United States. At the current exchange rates with the Russian ruble the GS23B may be the best bargain of all. At the Thorn UHF conference in 1992, brand-new GS23Bs were available for under US$200!

GS35B

The GS35B is a Russian VHF triode that is somewhat similar to an 8877. Its internal construction is different enough that it works well at 432 MHz. A few of the Russian and Ukrainian EME operators are using the GS35B and they report a stable 1500-W output is obtainable at 432 MHz. In addition, tube life is excellent at that power level.

TH331 - TH347 Family

The TH331 is part of a Thomson-CSF family of UHF TV tetrodes. Thomson-CSF developed a technology to make the grid structures out of a special type of graphite. These graphite grids solved many of the problems associated with secondary emission and dimensional changes in the grid structure due to heating from the cathode and RF power, while still allowing them to use fine enough wires in the grid. These dimensional changes can lead to poor and unreliable operation. The TH331 is capable of 1500-W output in intermittent amateur service at 432 MHz. The different models in the TH331 family move up in power level such that the TH347 is capable of 5000-W output or more at 432 MHz. There are also some even-larger members of this tube family, the TH563-TH582 which are capable of power levels up to 40,000 W, through most of the UHF TV spectrum! Various tubes in the TH331 family are used by many European 432-MHz operators. They are excellent tubes to use at 432 MHz as long as the builder is capable of handling a large UHF tetrode.

8938

The Varian, EIMAC 8938 is the UHF version of the 8877. Depending upon plate voltage and drive levels, it will run at 52 to 57% efficiency at 432 MHz. A single tube will easily deliver 1500-W output in continuous-duty service, while being stable and linear. The only technical drawback of the 8938 is its relatively low power gain (about 12 dB when stabilized). The chief problem with the 8938 is its cost. The list price for the 8938 is now $1800, and surplus or pull-out availability is very limited. Varian now only makes a handful of 8938s a month, as only two commercial/military transmitters in active service use it. A $\lambda/2$ stripline amplifier I built using the 8938 appears later in this chapter. The 8938 is also quite suitable for operation in a $\lambda/4$ cavity. At one time, Varian built a commercial cavity for the 430-470 MHz range. Although the cavity is no longer made, a few 432-MHz enthusiasts with access to machine shops have kept the design alive by building cavities patterned after the EIMAC design. There is also a water-cooled version of the 8938, the YU-157.

4628

The RCA 4628 is a brute of a tube, designed for UHF-TV transmitter service. It will very easily run 1500-W output. As with most UHF tetrodes it is designed to run in cathode-driven service. Its gain is relatively low, about 11 dB.

CONCLUSION

There are a variety, of tubes which will operate at 432 MHz. All of them have external-anode construction with ceramic insulation. For amateur service, the simplicity and performance of a modern high-gain low-bias triode makes them the preferred choice. If cost were no object, the EIMAC 8938 would be undisputed first choice for a 432-MHz amplifier. When a budget is factored in, a pair of 3CX800A7s is a good choice, if your prefer triodes and are buying new tubes. In addition, the small physical size of the 3CX800A7 allows amplifier designs made from standard 3-in. high aluminum chassis, eliminating the need for special metalwork. The 7213-8792 is my second choice to the 8938 in terms of power capability, but you'll have to live with the additional complexity and critical tuning of a tetrode. The primary advantage of these types of tetrodes is that they are more available surplus at modest prices. The 7650-8791 are also very good tubes if you can find them at the right price. For the budget-conscious, the 4CX250 family, especially the 4CX250R and 8930, will still give respectable performance.

One of the problems in writing an article such as this one, is it lets many more hams in on the secrets of which tubes are most useful. If you come across any of the desirable tubes, please use them or pass them on at your cost to someone who will. This is only a hobby, so don't mess up the works, as others have, trying to make money off companies and individuals nice enough to sell or give away their surplus equipment to amateurs. And finally, please don't hoard tubes and sockets that you will never use. Unused tubes actually do have a shelf life, as their vacuum seals can start to leak, making that gem on the shelf useless. If you're not going to get on the air, please pass them on to another amateur who will put them to good use.

Notes

[1]Knadle, R, "High Efficiency Parallel Kilowatt Amplifier for 432 MHz," *Part 1*: *QST*, Apr 1972, pp 49-55; *Part 2*: *QST*, May 1972, pp 59-62, 79. Construction information also appears in *Radio Amateur's VHF Manual*, third edition, pp 299-304 (Newington: ARRL, 1972).

[2]T. Pettis, "Hy-Brid Hi-Power," *QEX*, Jan 1990, pp 13-14.

[3]W. Orr, "Intermittent Voice Operation of Power Tubes, *Ham Radio*, Jan 1971, pp 24-30.

[4]S. Powlishen, "A Grounded-Grid Kilowatt Amplifier for 432 MHz," *QST*, Oct 1979, pp 11-14. Construction information also appears in *The ARRL Handbook for Radio Amateurs*, for the years 1981 through 1986.

[5]S. Powlishen, K1FO, "Improving the K1FO 8874 432-MHz Amplifier," *QST*, Jul 1987, pp 20-23.

[6]"An EIMAC Family of 4CX600 Tetrodes," *Application Bulletin Number 14*," (San Carlos, CA: EIMAC).

[7]A. Katz, "K2UYH 432-MHz 7650 PA," *The Lunar Letter,* Mar 1983.

[8]H. Barber et al, "Modern Circuit Design for VHF Transmitters, Part 1: *CQ*, Nov 1965, pp 30-36; Part 2: *CQ* Dec 1965, pp 40-45, 92. Reprinted in EIMAC *Amateur Service Newsletter AS-27*.

[9]J. Chambers, W. Orr, "2-kW PEP Amplifier for 432 MHz," *Ham Radio*, Sep 1968, pp 6-12. Also published as *Amateur Service Newsletter AS-25-1* (San Carlos, CA: EIMAC).

[10]P. May, "8877 432-MHz Amplifier Cathode Circuit," *VHF/UHF and Above*, Jun 1986, pp 25.

[11]J. Snyder, "2-KW Amplifier Using the 7213," *432-MHz And Above EME Newsletter*, Jul 1981. The *432-MHz EME Newsletter* is published by Allen Katz, K2UYH. The Snyder amplifier information also appears in a compilation of the 432-EME newsletter technical articles, published by PA3CSG.

[12]A.T. Margot, W6FZA, "A Practical Kilowatt Amplifier for 432 Mc.," Aug 1964 *QST*.

A 3CX800A7 Amplifier for 432 MHz

By Steve Powlishen, K1FO

In 1979, I described a 432-MHz power amplifier using a single 8874 tube.[1] The popularity of the amplifier far exceeded my expectations. While stable, linear and efficient, the 8874 amplifier had several drawbacks:

1. Construction required considerable metal fabrication, as the RF enclosure dimensions didn't correspond to a standard chassis size.

2. At 432 MHz, the 8874 requires over 35-W drive for full output—more than current 25-W transceivers can provide.

The introduction of the EIMAC 3CX800A7, essentially an improved 8874, provided the inspiration to build a new and improved amplifier. I wanted higher gain and simplified construction. In addition, this amplifier has TR switching voltages and currents compatible with low-power solid-state transceivers, and tube-protection circuitry.

The IMD performance of this amplifier is almost an order of magnitude better than some 432-MHz exciters! While this point may appear insignificant, operators may have to contend with local signals that are 80 to 100 dB above the noise and only a few kHz away. Any serious 432-MHz operator can attest to near loss-less propagation periods when signals 300 miles away are 80 to 90 dB above the noise. Under such conditions, this amplifier will not aggravate other operators.

While expensive when purchased new, the 3CX800A7 may still be a bargain. When you consider that in intermittent service a properly operated 3CX800A7 still may be cooking long after another operator has gone through several sets of 4CX250-class tubes, the initial price appears more reasonable.

The amplifier described in this article will deliver about 620 W with 25-W drive. At maximum ratings, the 3CX800A7 generates 730 W. Efficiency and maximum power output are better than with the 8874 amplifier.

Construction Details

To minimize metal work, the amplifier uses 4 standard aluminum chassis. The RF plate enclosure is made from a 5- × 13- × 3-in. chassis. The cathode circuit is housed in a 4- × 5- × 2-in. chassis. The RF enclosures are attached to the EIA standard 5¼ in. high, 19-in. rack panel by 2 standard 5- × 7- × 2-in. chassis. Using the smaller chassis in this way makes it unnecessary to fabricate mounting brackets for the RF deck, while also providing space to mount the control circuits. Heavy gauge (0.062-in. or thicker) cover plates are preferred on the RF chassis to assure RF sealing and provide mechanical rigidity.[2]

The construction of this amplifier has the RF deck mounted on its side, relative to the construction of most amplifiers. This mounting method has several advantages. The tuning controls can be positive-actuating lead screws, while still providing front-panel access. Alternately, the fish-line tuning arrangement can be used, as I did. Fish-line actuated controls allow the front-panel knobs to be placed for convenience and esthetics. The mounting arrangement used in this amplifier accommodates a convenient control arrangement while minimizing the length and bends in the fish line. As a result, the plate tuning controls operate smoothly and repeatably. An additional benefit to the mounting method is that tube hot air exhaust exits to the rear of the amplifier. Other equipment can be mounted above or below this amplifier, without leaving cooling space. I built matching 3CX800A7 amplifiers for 50, 144, 222 and 432 MHz. All four amplifiers and their 2200-V power supply can be mounted in a single 28-in. high desktop rack.

I took care to make this amplifier easy to duplicate with readily available parts. Complete metal cutting and drilling drawings are provided for the RF sections. If you accurately follow the drawings and use all parts specified in Table 1, the amplifier should go together and tune up like a commercial kit. I don't discuss the layout of the control circuits as they are not critical, and you may wish to tailor them to your station. Some of the specified parts are priced higher than junk-box substitutes. With some ingenuity you may be able to use cheaper parts, but you do so at your own risk. I'm unable to offer advice about finding and using substitute parts.

Plate Circuit Details

The plate circuit is the now-standard half-wave stripline with the tube located at one end. A "flapper" tuning capacitor is mounted at the other end. The stripline (Fig 1) is larger than

Table 1
Parts List for the Single 3CX800A7 Amplifier

Chassis and Hardware Components

RF deck enclosure: 5×13×3-in. chassis, Bud AC-422 or equiv.
Cathode compartment: 4×5×2-in. chassis, Bud AC-1404 or equiv.
Side chassis (2 req.): 5×7×2-in. chassis, Bud AC-402 or equiv.
Rack panel: 5 1/4×19×1/8-in., Bud SFA-1833 or PA-1103 or equiv.
Tube socket: 11-pin EIA, Eimac SK-1900 or Johnson 124-0311-100.
Grid collet: Eimac 720359 assembly (Eimac 882931 can be used).
Anode collet: Eimac 720829.
Grid collet insulator: Eimac 720518.
Chimney: Eimac SK-1906.
Panel bearings: 1/4-in. diam., Millen 10066 (2 req.).
Reduction drives: Jackson 4511/DAF (2 req.).

Components Referenced On Schematic Diagram

C1—1.5-5 pF miniature air variable (butterfly); Cardwell 160-0205.
C2—1.8-8.7 pF miniature air variable; Cardwell 160-0104.
C3-C5—1000-pF, 300-V feedthrough capacitor; Tusonix 327-005-C5UO-102M.
C6—Plate tuning flapper. See text.
C7—Plate loading flapper. See text.
C8—1000-pF, 4000-V feedthrough capacitor; Tusonix 2498-001-X5UO-102M.
C9-C10—1000 mF, 25-V electrolytic.
C11—0.15 mF, 25-V disk or epoxy.
C12-C19—0.01-mF, 50-V monolytic ceramic; Sprague 1C105Z5U103M050B.
D1—5.6-V, 10-W Zener, mounted on RCA SK122/5178A heatsink.
D2—10-A, 400-PIV.
D3-D4—2.5-A, 1000-PIV; R170 or equiv.
D6-D14—1-A, 1000-PIV; 1N4007 or equiv.
F1—2-A, AGC or 3AG fast-blow.
F2—3/4-A, AGC or 3AG fast-blow.
I1—120-V neon, amber; GC Electronics 38-282.
I2—120-V neon, red; GC Electronics 38-280.
J1—Chassis-mount BNC female, UG-1094/U.
J2—Chassis-mount N female, UG-58A/U.
J3—Chassis-mount MHV female, UG-931/U.
J4—6-pin male chassis mount; Cinch P306AB.
J5—6-pin miniature chassis connector; Waldom Molex 03-06-1061.
J6-J8—Phono connector; Switchcraft 3501FR.
K1—180-sec. thermal time-delay relay, 115-V heater, SPST-NO; Amperite 115NO180B
K2—Control relay, DPDT, 24-V dc coil; Potter and Brumfield R10-E1-X2-V700.
K3—Control relay, 4PDT, 24-V dc coil; Potter and Brumfield R10-E1-X4-V700.
K4—Coaxial relay, SPST, BNC connectors, 28-V dc coil.
K5—Coaxial relay, SPST, high-power, N connectors, 28-V dc coil.
L1—2 turns no. 16 copper, 1/4-in. diam., 3/4-in. long.
L2—Brass strip 1/4-in. wide × 1/-3/16-in. long.
M1—Dc milliammeter, 600 or 1000 mA fullscale.
M2—0-1 milliammeter with shunt resistors to give full-scale deflections of 60 mA (grid current); 3 kV (high voltage); 30 V ac (filament voltage).
MOT1—54 cfm blower; Dayton 4C012 or equiv.
Q1-Q2—2N2222A, 2N3903 or equiv.
Q3—2N3053 or equiv.
Q4—2N4037 or equiv.
Q5—2N2904, 2N3905 or equiv.
R1—200-Ω, 25-W wirewound.
R2—1000-Ω, 12-W wirewound.
R3—10-kΩ, 25-W wirewound.
R4—1-Ω, 1-W, 1%.
R5-R10—499-kΩ, 1/2-W, 1% metal-film, type RN-60 preferred.
R11—820-Ω, 1/2-W, metal film. Select value to calibrate HV meter.
R12—1.5-kΩ, 1/2-W metal film. Select value to adjust HV relay trip point.
R13—9-Ω, 1/2-W metal film. Select value to calibrate grid meter.
R14—10-Ω, 2-W, metal film
R15—10-kΩ, 1/4-W miniature trimmer.
R16—1200-Ω, 2-W. Select value to set K1 time delay.
R17—50-Ω, 12- or 25-W adjustable slider wirewound.
R18—1-Ω, 5-W wirewound.
R19—12-kΩ, 1/2-W, film.
R20—2-kΩ, 1/4-W, miniature trimmer.
R21, R22, R25, R27—2.7-k, 1/2 W.
R23, R28—10-kΩ, 1/4 W.
R24—4.7-kΩ, 1/4 W.
R26—2.2-kΩ, 1/4 W.
R29—330 Ω, 1/4 W.
R30—500-Ω, 25-W wirewound.
RFC1—8 turns, no. 18 enameled, 1/4-in. diam., closewound.
RFC2, RFC3—8 turns, no. 16 enameled, 1/4-in. diam., closewound.
RFC4, RFC5—7 turns, no. 18, 3/4-in long, 1/4-in. diam.
S1—DPST toggle.
S2—SPST toggle.
S3—2-pole, 4-position rotary.
S4—SPDT miniature toggle.
T1—Filament transformer: 14-V, 2-A secondary, 120-V primary; Stancor P8556 or equiv.
T2—Control transformer: 10-V, 2-A secondary, 120-V primary; Stancor P8653 or equiv.
W1—Plate stripline.
W2—Cathode stripline.

the one used in the 8874 amplifier, both to fill the larger plate compartment and to accommodate the larger (2.5-in.) diameter of the 3CX800A7 anode radiator. The larger stripline allows for better placement of the tuning controls. The preferred material for the stripline is 1⁄16-in. brass, which is silver plated after the collet is soldered in place. Copper is also suitable, but

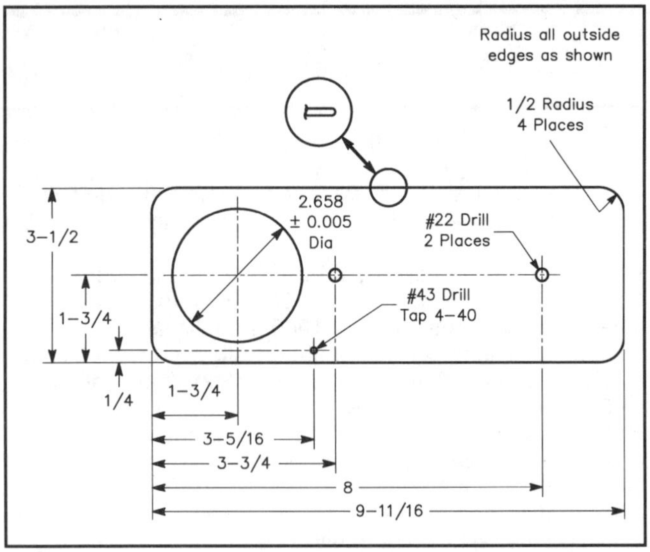

Fig 1—Plate line of the single 3CX800A7 amplifier. Except as noted on the drawing, dimensions are ± 1/32 in. Material is 1/16-in. copper or brass. Anode hole is sized for the EIMAC 720829 collet. Dull-finish (nickle-less) silver plating is recommended after the collet is soldered in place.

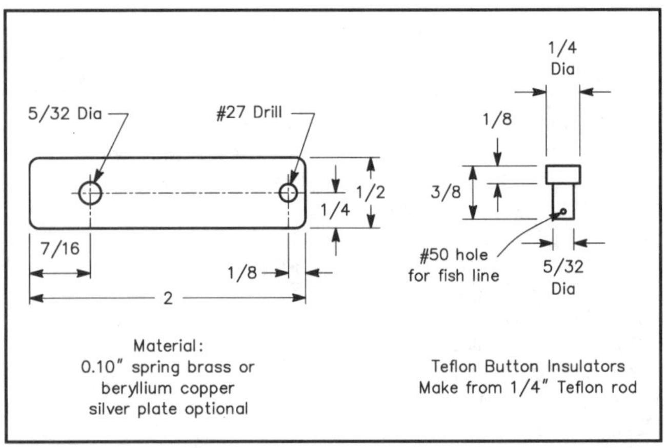

Fig 3—Plate loading capacitor C7 is made from 0.010-in. spring brass or beryllium copper. The button insulator is made of 1/4-in. Teflon rod.

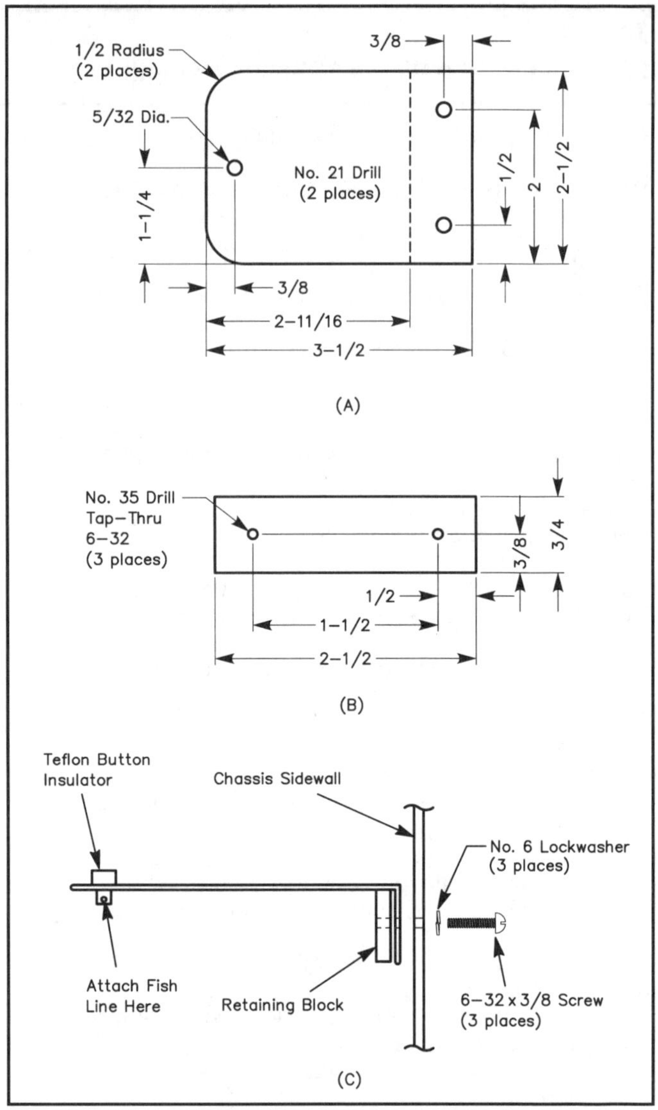

Fig 2—Plate tuning capacitor C6 (A); retaining block (B); mounting of the capacitor and retaining block (C). The capacitor is made from 0.010-in. hard brass or beryllium copper. Silver plating is optional. The mounting block is made from 1/8-in. aluminum.

harder to machine. If you have your line professionally plated, specify a "dull" finish; that is, without nickel content. A plater experienced in RF work will know what you mean. He will copper flash the line before plating it. Specify a minimum silver thickness of 0.001 in. In 1986, the cost to silver plate all parts (W1, W2, L2, C6 and C7) was $63, of which $60 was for setup and only $3 was for material. For minimal additional cost, you and other hams can collect a wide assortment of RF parts and have them all plated at once.

Silver plating is *not* necessary for proper operation. The difference in efficiency between a clean, polished but unplated stripline and a silver-plated line was nearly impossible to measure. Experience with the 8874 432-MHz amplifier has shown that an unplated line begins to tarnish after a few years. When the oxidation is heavy enough, the amplifier tuning drifts as it heats up. Polishing the stripline returns the amplifier to like-new operation with no thermal drift.

The plate line is held in place with two 1½ in. high × ⅜-in. diameter ceramic standoff insulators. Homemade Teflon insulators are even better. They can be made ⅜ or ½ in. diameter. Tap the ends for 6-32 screws. Use brass screws to hold the stripline in place. Steel (and especially stainless steel) can cause unwanted tuning effects. Also use a brass screw to attach the plate RF choke (RFC4). The screws used to attach the standoff insulators to the chassis bottom can be steel. The hole in the plate line for the anode is sized for an EIMAC 720089 collet. Finger stock may be used if you prefer. If you use finger stock significantly larger than the EIMAC collet, the position of the plate-tuning capacitor will be different. Try to obtain finger stock with contacts that are rolled over 180°, sized about ⅛ to 3/16 in. high, with the fingers rolled over about ⅛ in. Be careful to size the anode hold in the plate line correctly.

The tuning and loading capacitors are made from 0.010-in. thick brass shim stock. This material has a spring temper

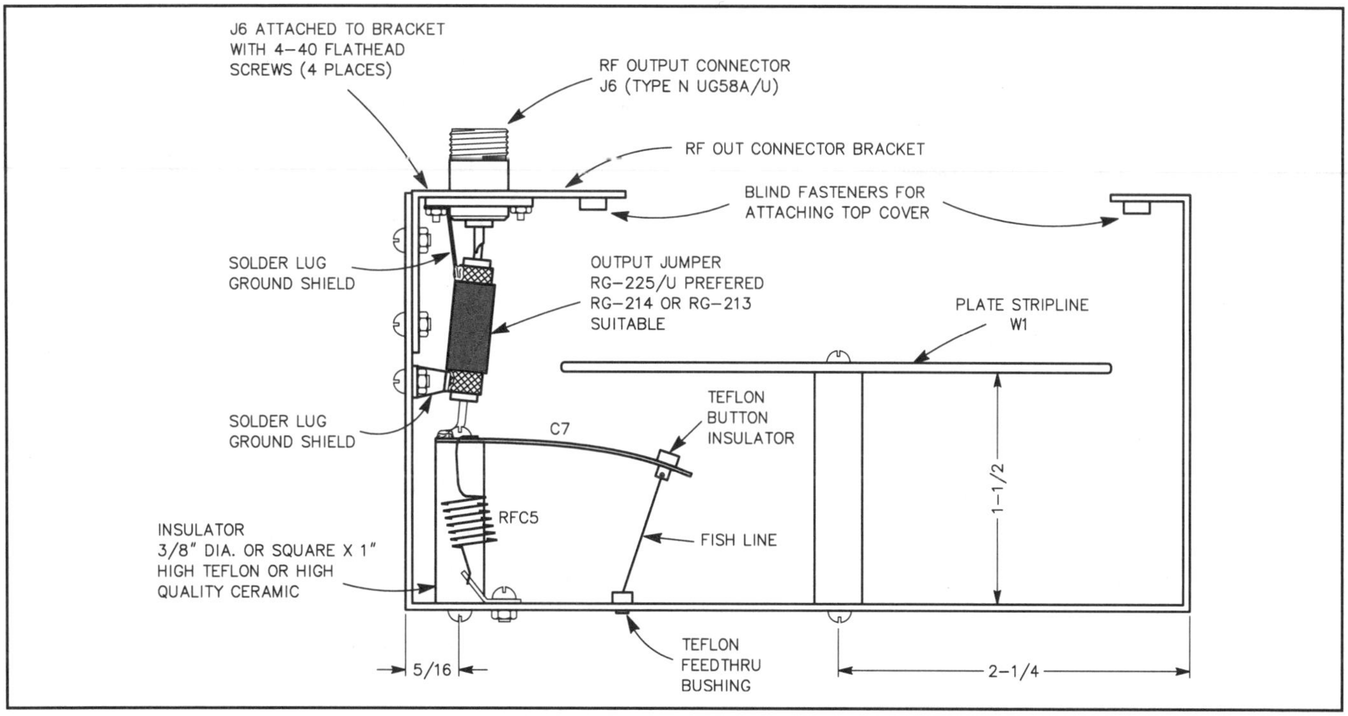

Fig 4—RF output compartment details. The plate line is not centered in the enclosure. See Fig 5 for construction details of the RF output jumper.

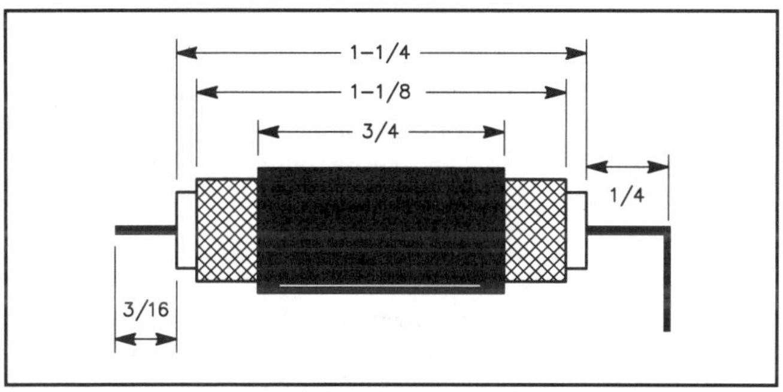

Fig 5—RF output jumper details. RG-225 Teflon-dielectric cable is preferred, but RG-214 or RG-213 cable may be used. Alternately semi rigid solid copper sheathed cable such as UT-141 or UT-250 may be used.

and works well. You can also use beryllium copper. Don't use material thicker than 0.010 in. Thicker stock is more likely to take a set and not spring back to its original position. In addition, thicker material may place too great a load on the fish line and tuning controls. The strain may stretch the fish line or move the tuning control positions, affecting tuning.

The plate flapper (Fig 2) is held in place by an aluminum block ¾-in. wide × 2½ in. long × ⅛-in. thick. Tap the holes for 6-32 thread, for easy installation and removal. The loading flapper (Fig 3) is mounted on a 1-in. tall, ⅜-in. dia. Teflon or ceramic insulator. Both flappers use small Teflon button insulators to secure the fish line. These homemade insulators serve the dual purpose of providing an attachment point for the fish line and preventing the flappers from accidentally contacting the plate stripline.

The preferred fish line is braided Dacron fly line. I used Specialist 18-lb Fly Line Backing, made by Berkley. Be sure

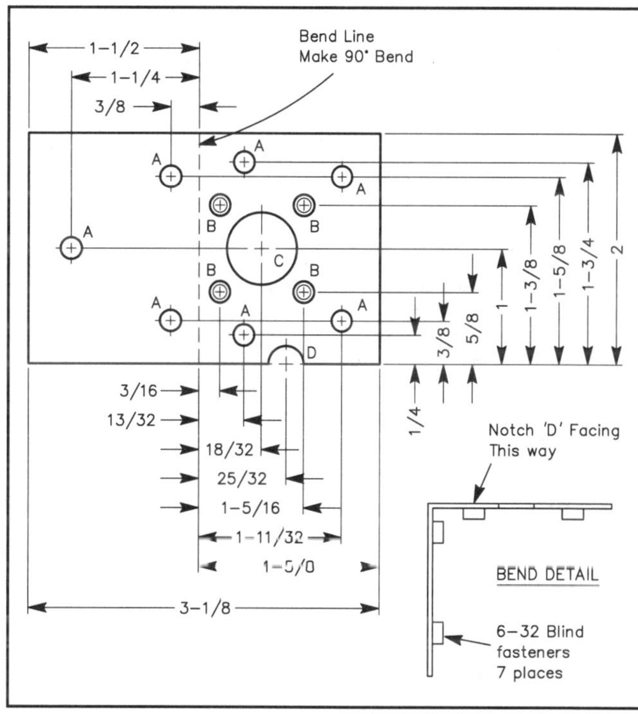

Fig 6—RF-output connector bracket is made from 0.062.-in. aluminum. Holes marked A are 3/16-in. diam., for 6-32 blind fasteners. Asterisks (*) indicate holes that should be located using the top cover as a template when the bracket is mounted. Holes marked B are No. 30, countersunk for flush mount with 4-40 screws. The connector mounting holes may also need to be countersunk. Hole C is 5/8 in. diam. Hole D is a 5/16-in. diam. notch to clear the blower mounting plate.

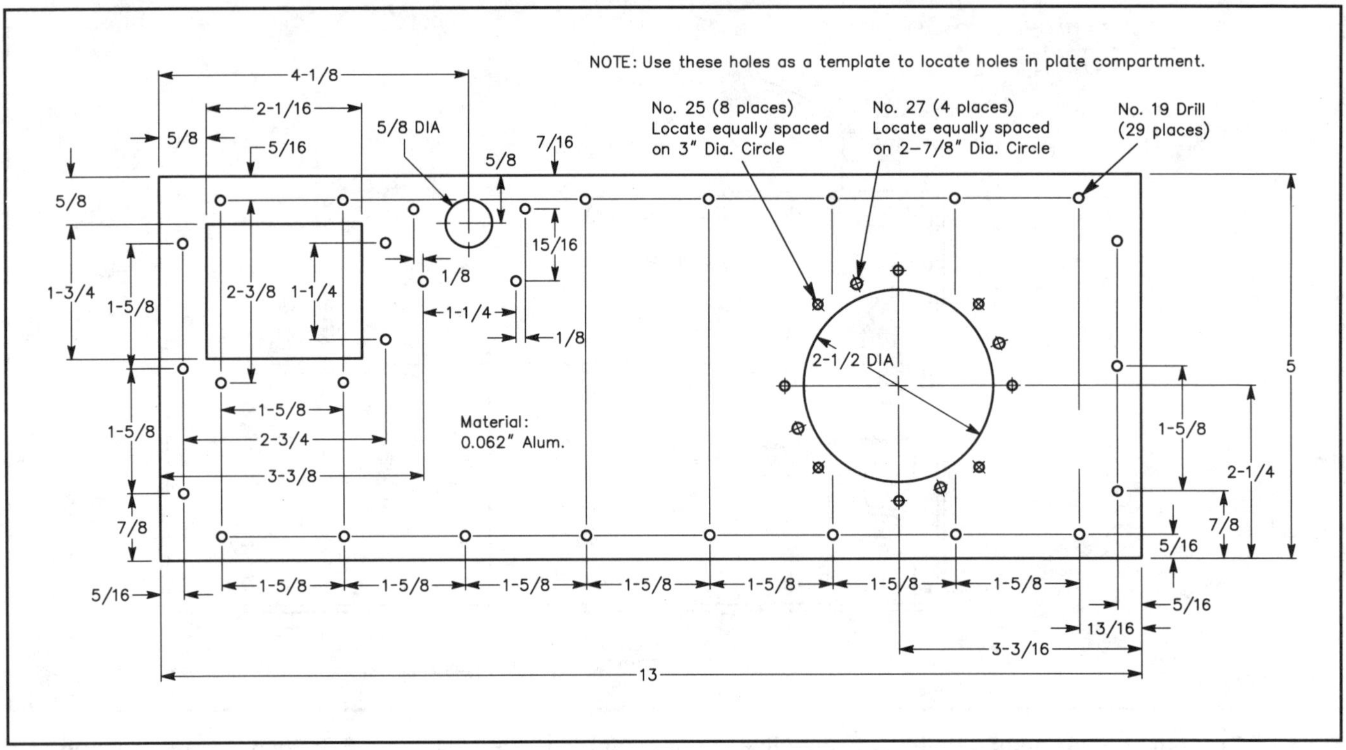

Fig 7—Top cover of the amplifier is made from 0.062-in. aluminum. Use the cover as a template to mark holes in the chassis.

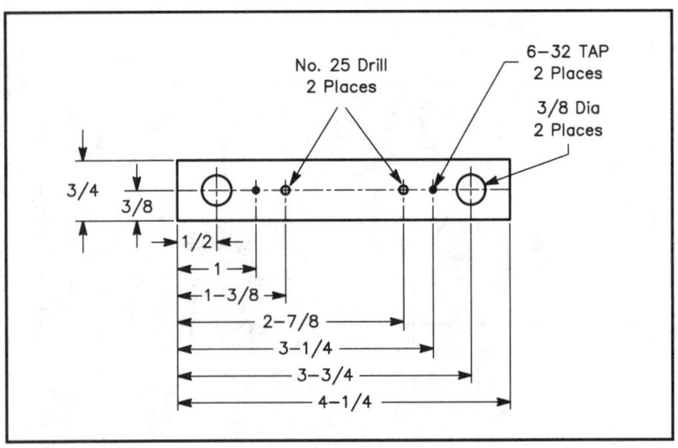

Fig 8—Plate tuning control bracket is made from 0.062-in. aluminum.

all edges on the flappers are smooth and free from burrs. Also make sure the flappers are mounted parallel to the stripline. These precautions assure RF or dc doesn't arc from the plate line to the flappers.

A small piece of RG-225 coax connects the loading capacitor to the RF-output Type-N connector (J2). Fig 4 details assembly of the loading capacitor (C7) and Fig 5 gives dimensions for the coax jumper. RG-225 is similar to RG-214, but uses Teflon dielectric and jacket, along with double silver-plated shields. You can also use RG-213 or RG-214 if you're careful not to melt the dielectric while soldering it in place. Both ends of the shield are grounded through solder lugs. Be sure that the jumper is mounted close to the chassis wall and away from the stripline. The output connector is mounted on a small bracket to allow for rear-panel RF-output connection, while not requiring any disassembly when the plate-compartment cover is removed. Fig 6 gives the layout of the output connector bracket.

Be very careful when notching the plate chassis for the RF- output-connector bracket. Once the lip is cut out it is very easy to bend the chassis if the bracket is not secured in place. The top cover (Fig 7) should be drilled first. Then you can use it as a template to locate the holes in the chassis that will hold the top cover in place. Next, install the blind fasteners (PEM nuts). Once the fasteners are in place, you can notch the chassis for the output-connector bracket. Cut the bracket to size and bend it 90°, as shown in Fig 6. Drill the 3 holes for the bracket mounting screws and install the bracket. Now use the top cover again as a template to locate the output-connector hole and the 4 top-cover mounting holes. You can use the UG-58 output connector as a template to locate its mounting holes.

The plate tuning and loading fish lines are brought through the chassis through small homemade Teflon bushings. The lines run over small pieces of ¼-in. Teflon rod, giving nearly friction-less 90° transition of the tuning controls. Brass shafts, which run through panel bearings that are mounted on a support bracket (Fig 8), control the tuning lines. This bracket is supported by standard 1¼ in. long steel standoffs. To impart a slow, smooth feel to the plate tuning and loading controls, 6:1 ball-reduction drives are used. The drives I used are Jackson Bros. 4511/DAF, and is available from Radiokit or Surplus Sales of Nebraska.[3] The ball drives are

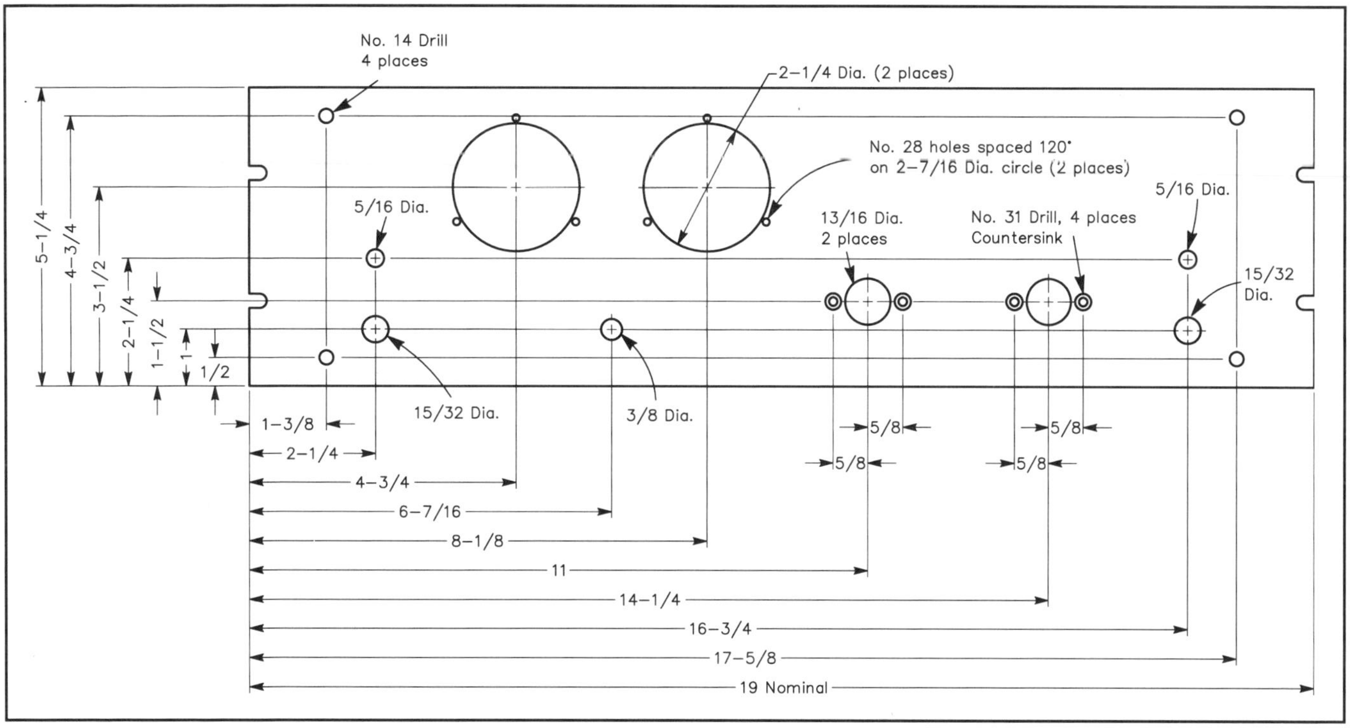

Fig 9—Front-panel layout. The panel is made from a standard 5-1/4 in. aluminum EIA rack panel.

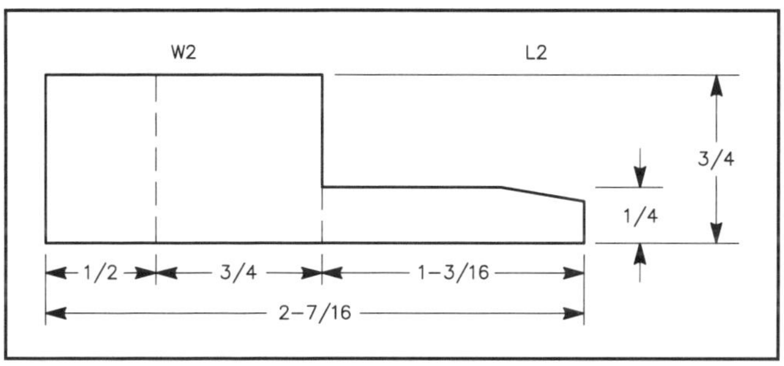

Fig 10—Cathode stripline is made from 0.015-in. brass. Silver plating is optional.

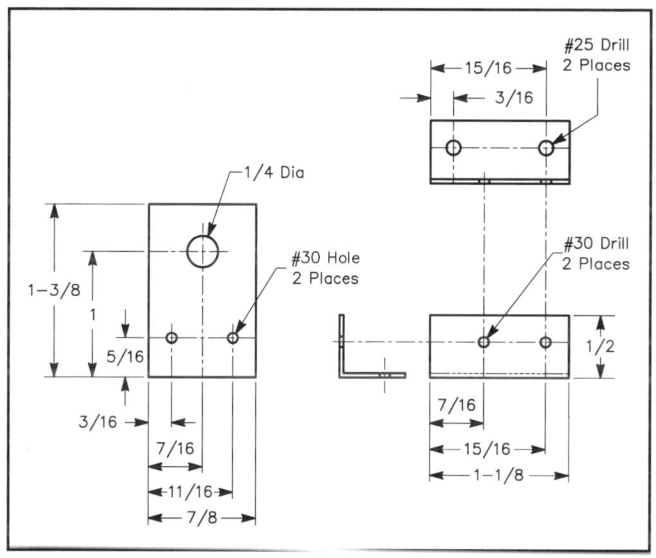

Fig 11—C1 mounting bracket is made from 0.062 X 1/2-in. aluminum angle or 0.062-in. aluminum sheet. The mounting insulator is made from 1/16-in. G10 circuit-board material from which the copper has been removed.

mounted to the front panel with 4-40 flat-head screws. Fig 9 shows the front rack-panel hole sizes and locations for the ball drives and other holes. Use 4-40 nuts to secure the screws, so they act as mounting studs for the ball drives. The nuts also serve as spacers to position the ball drives nearly flush with the front panel. Homemade pointers attached to the ball drives serve as control position indicators.

The specified high-voltage feedthrough capacitor may be hard to find. It may be available from Microwave Components of Michigan.[4] You can also construct a bypass capacitor from 1/16-in. thick brass plates and 0.005-in. Teflon or Kapton sheets. A two-plate capacitor with one plate inside the RF enclosure and another plate on the outside is recommended. Plates sized approximately 2½ × 3¼ in. are suitable. The plates should be flat, polished so they are smooth and free from any burrs, and have rounded edges. If you use Teflon, spread silicone grease on the Teflon to fill in any imperfections. You don't have to coat Kapton, as it doesn't have the porosity or cold-flow characteristics of Teflon.[5]

Cathode Circuit Assembly

The cathode circuit consists of a quasi-half-wave line, similar to that used in the original 8874 amplifier. Changes have been made, due to the higher input capacitance of the 3CX800A7 and to make the circuit more repeatable when duplicated. Fig 10 gives the layout of W2 and L2. Most of the first quarter wave of the half-wave input line is actually inside the tube and socket. W2 forms the rest of the first quarter wave and part of the second quarter wave. L2 completes the second

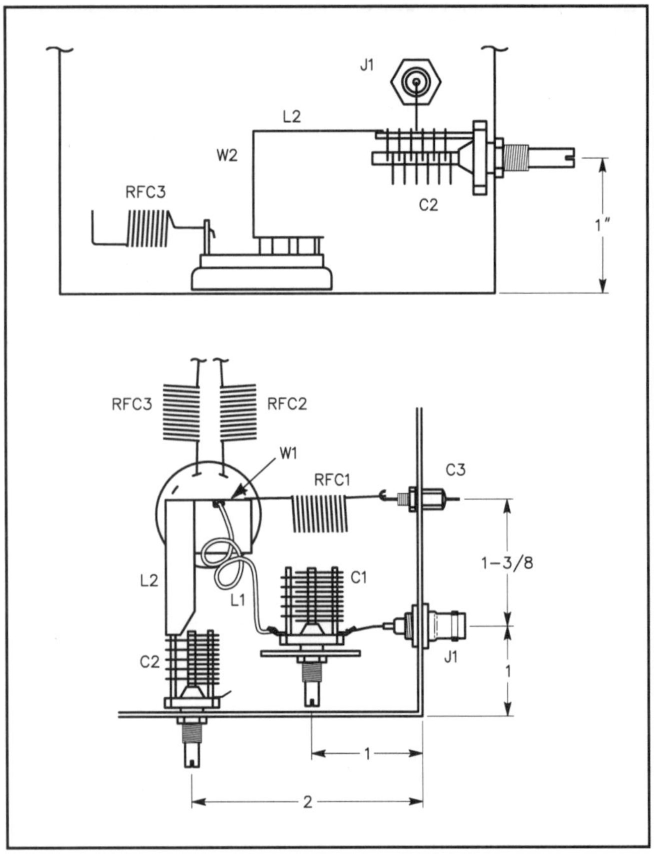

Fig 12—Side and bottom views of the cathode circuit.

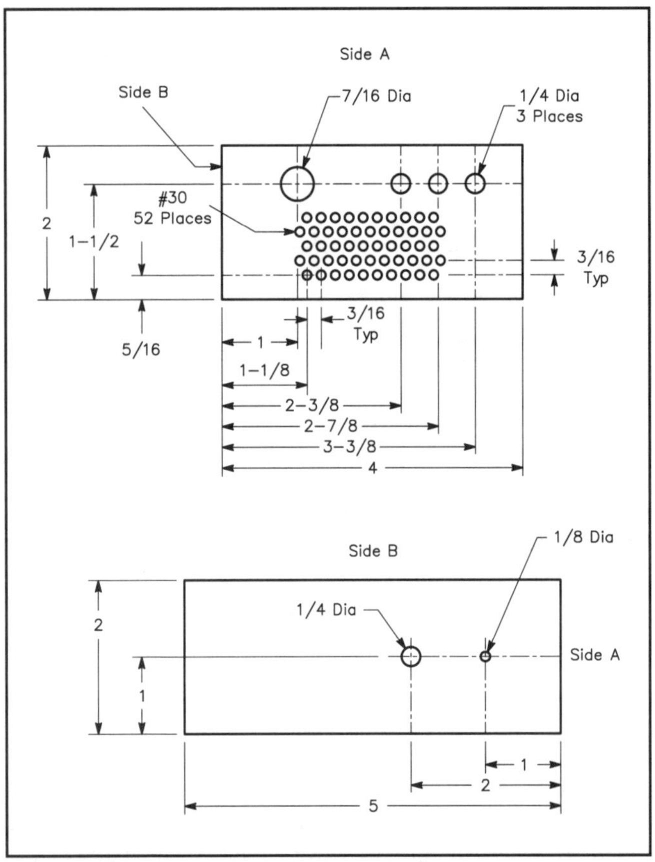

Fig 13—Cathode compartment details. The compartment is made from a 5 X 4 X 2-in. chassis.

half of the line. C2 is simply used to tune the line to resonance. L1 is a series inductance to match the 50-Ω input impedance to the input impedance of the tube. L2 has slightly too much inductance so the series capacitor C1 is used to adjust the total reactance for a proper match. Note that the rotor of C1 must be insulated from ground. A small piece of G-10 epoxy PC board material serves as a mounting insulator. The circuit board is held in place by a small aluminum bracket (Fig 11).

To assemble the cathode circuit, first remove the unused grid pins (4, 7, 11) from the tube socket. Next bend the 6 cathode pins (1, 2, 3, 8, 9, 10) in toward the center of the socket, forming a 90° angle. The bend is made just above the dimple that keeps the pins in place in the socket. The socket is then mounted in place on the grid collet. Note how the 2 filament pins point to the top of the amplifier and the 6 cathode pins point toward the bottom. Mount the rest of the cathode components, C1, C2 and J1. If the layout has been followed, when W2 is soldered to the 6 cathode pins, L2 should line up right on the stator wires of C2. Fig 12 gives 2 views of the cathode-circuit layout. Fig 13 covers the hole-drilling patterns in the cathode compartment. If you follow the layout, the input circuit should tune up easily and you'll obtain an input SWR less than 1.2:1.

Tube Socket Mounting

The recommended EIMAC 720359 grid-collet assembly simplifies construction. It consists of an EIMAC 882931 collet that has been soldered to a 1/16 in.-thick brass mounting ring. The mounting ring has three 4-40 studs that are positioned to match the mounting holes of the 11-pin tube socket. An alternate method of construction is to use the 882931 collet and attach it to a homemade mounting ring. Although the 720359 collet assembly is more expensive, it will save you considerable construction time.

To cut the hole for the tube socket, first punch a 1¼ in. diameter hole in the chassis bottom, located as shown in Fig 14. Next use the tube socket to locate 3 no. 28 holes, orienting the socket per the hole layout in Fig 14. Put the EIMAC 720359 collet in place and use it as a template to drill its 4 mounting holes in the chassis with a no. 27. Remove the collet and socket. Next, drill out the 3 #27 holes with a 9/32-in. drill. Then file out the holes until they form one hole that matches the shape shown in Fig 14. The socket should be able to pass through the hole. Drill a series of 6 no. 43 holes in the collet, matching the pattern shown in Fig 22. Again use the collet as a template to drill 6 no. 42 holes in the chassis. These holes are for bleeding air into the cathode compartment for cooling the tube base. The tube socket may then be mounted to the collet, using 4-40 nuts and lock washers. The collet is then mounted to the chassis with 4 6-32 screws. With everything properly assembled the tube socket will mount flush with the collet.

Additional air is let into the cathode compartment through a series of cooling holes. A metal plate (Fig 15) is used to make an RF filter to prevent leakage from the plate compartment into the cathode compartment. The cathode box is perforated on the side away from the cooling holes. In this way, the air passing into the cathode box is forced across the tube base, to maximize its cooling effect.

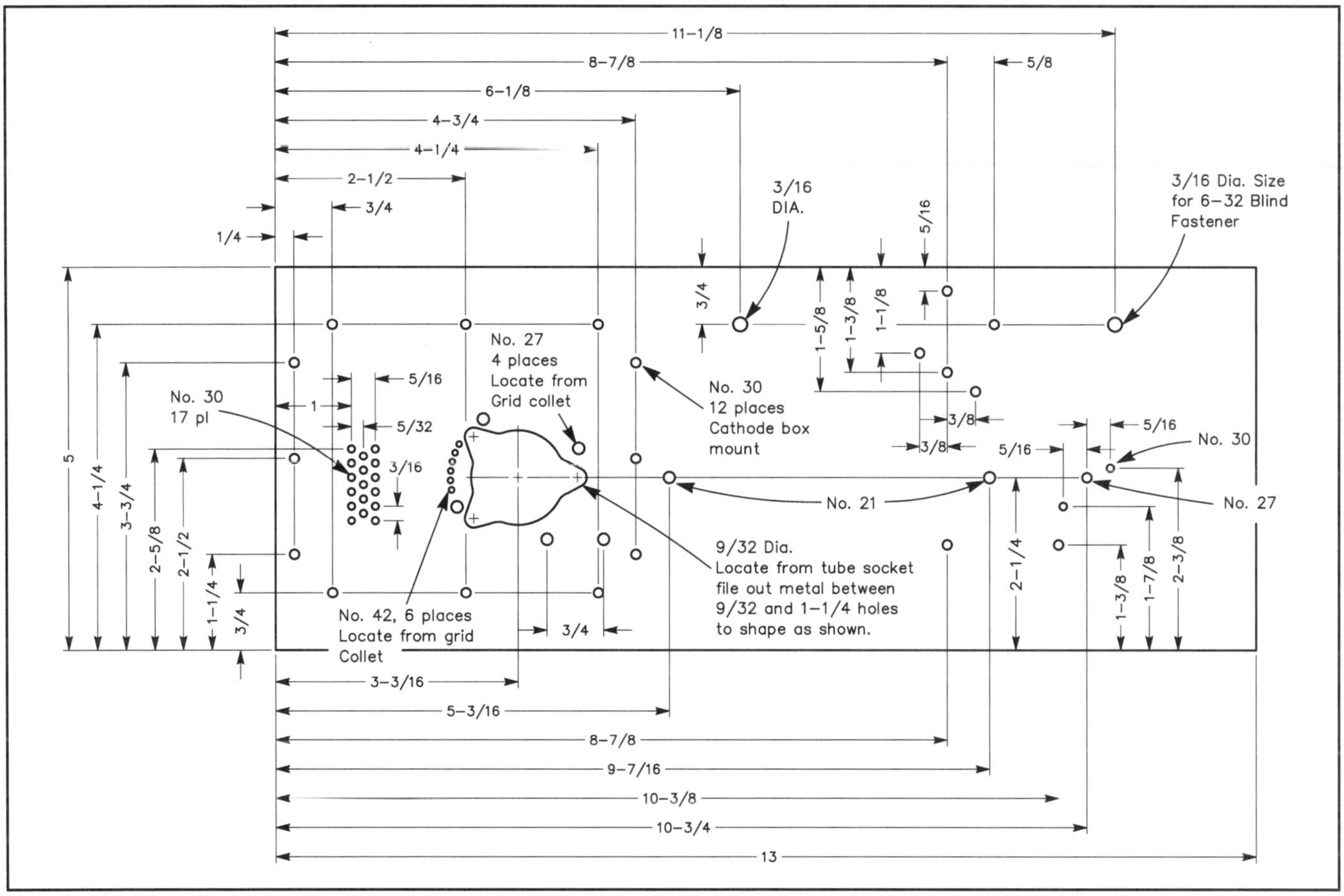

Fig 14—Bottom cover of the plate RF enclosure is made from 0.062-in. aluminum.

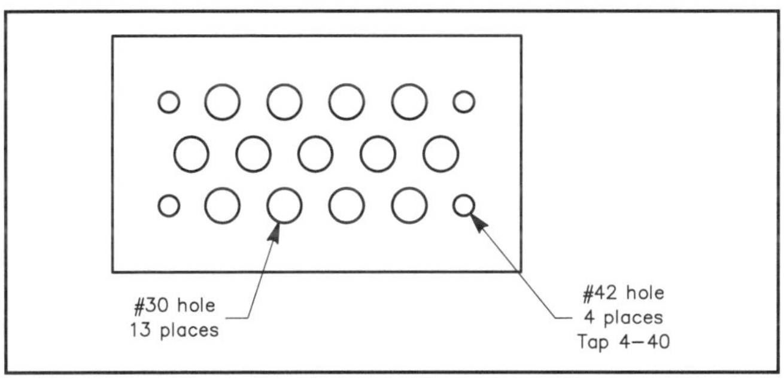

Fig 15—Cathode air vent waveguide is made from 3/16-1/4 in. aluminum.

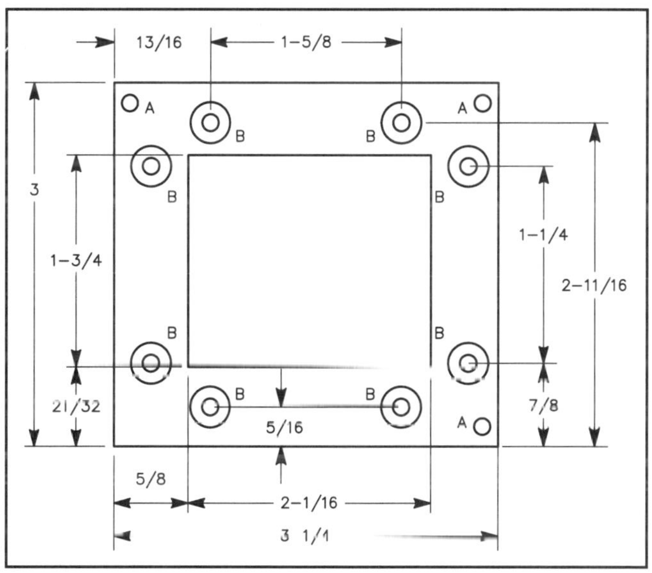

Fig 16—Air inlet retaining plate is made from 3/16-1/4 in. aluminum. Hard alloys, such as 6061-T6 or 2024-T6 are preferred. Holes marked A are No. 7, tapped 1/4-20. Use the blower itself as a template to locate the holes. Holes marked B are No. 25, countersunk for 6-32 flathead screws. Use the top cover to locate these holes.

Cooling Considerations

The air inlet and outlet are RF shielded by means of aluminum cooling screen, sandwiched between the RF deck top cover and ¼-in.-thick retaining plates. Matching hole patterns are cut in the top cover and the retaining plates. For best alignment, use the top cover as a template for drilling the plates. Although these plates may look complicated to make, they were fabricated with hand tools. The large air inlet and outlet holes were first scribed. Then, a series of small holes was drilled along the inside of the marks. The holes are then drilled out with a larger drill so the hole slug can be knocked out. The holes are then simply filed to shape. Fig 16 gives the inlet plate layout and Fig 17 shows details of the outlet plate.

The air-outlet plate is mounted on the inside of the chassis sandwiching the screen between it and the cover plate. The plate also serves to space the chimney down to the plate line. The EIMAC SK-1906 chimney is held in place with 4, ½-in.-

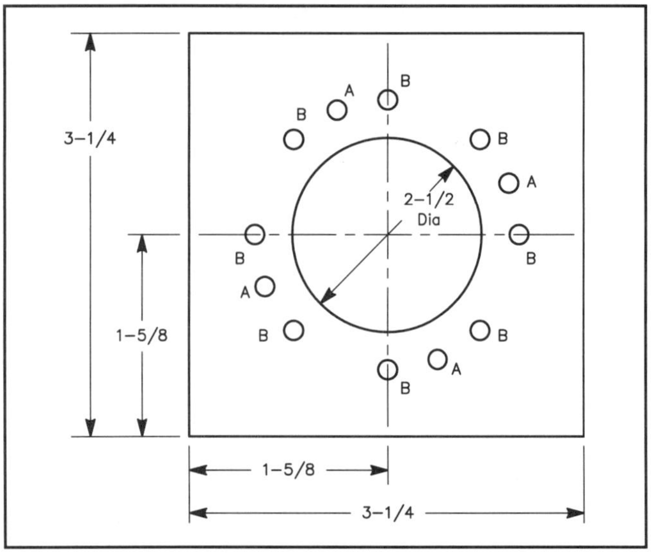

Fig 17—Air outlet retaining plate is made of 1/4-5/16 in. aluminum. Hard alloys, such as 6061-T6 or 2024-T6 are preferred. Holes marked A are No. 27. Locate the 4 holes equally spaced on a 2-7/8 in. diam. circle. Holes marked B are No. 35, tapped 6-32. Locate the holes equally spaced on a 3-in. diam. circle.

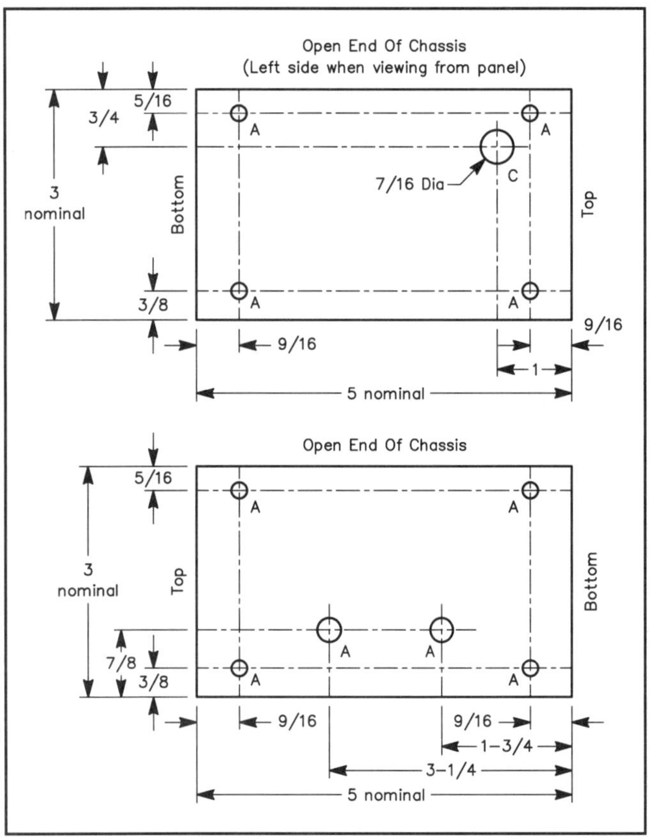

Fig 18—RF deck. At top, tube end; at bottom, flapper end. Holes marked A are chassis mounting holes, No. 23.

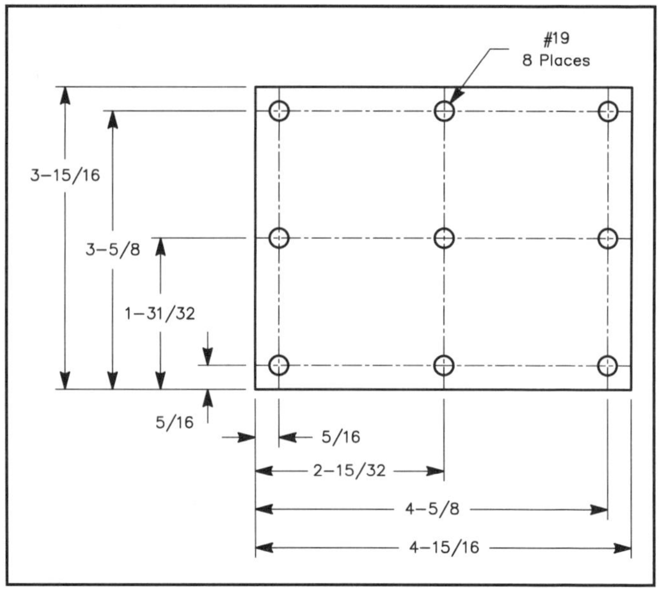

Fig 19—Cathode compartment cover plate is made from 0.062-in. aluminum. Use the finished cover as a template to locate holes in the cathode compartment.

long 4-40 screws, which pass through the cover plate and retaining plate. The threaded metal inserts in the chimney should be removed from the plate line side of the chimney. Alternately, you can use a homemade Teflon chimney. The air-inlet plate and screen are mounted on the outside of the top cover. Tap three ¼-20 blower-mounting holes in the plate. Note that 4 of the inlet plate hold-down screws also secure the top cover. Since the blower covers all 8 of the mounting screws, their holes are countersunk and flat head screws are used. The blower must be removed to take the plate compartment top cover off.

To ensure that adequate RF shielding is maintained, the air inlet retaining plate and the top cover must be fastened tightly. The 5- × 13- × 3-in. chassis is made from 0.040-in.-thick soft aluminum. Sheet-metal screws will easily strip out the thin aluminum after being removed and replaced several times; it's essential to use blind fasteners (PEM nuts or Rivnuts). They are readily available from Small Parts Inc. I used 6-32 fasteners. When installing the fasteners, be very careful not to distort the chassis lip. It is easy to make ripples in the metal, which will cause RF leakage.

The amplifier operates quietly with the specified blower, in combination with effective shock mounting. A ⅛-in.-thick gasket made from high-density foam rubber is placed between the blower and its mounting plate. The blower is attached with three ¼-20 nylon screws, ½-in. long. Rubber grommets are placed on the screws before they are installed. This arrangement assures that there are no solid mechanical contacts to transmit blower noise and vibration.

Metal Finishing

The professional finish on the amplifier is gold-irridite applied by a plating company. Irridite provides a hard and conductive finish. Don't confuse irridite, which is sometimes known as chromate finish, with anodizing. Anodize is not a conductive finish and may contribute to RF leakage and improper operation of the amplifier. All aluminum was wet sanded with 320-grit wet-sanding paper. Sand in one direction in a straight motion. You are sanding the aluminum to remove

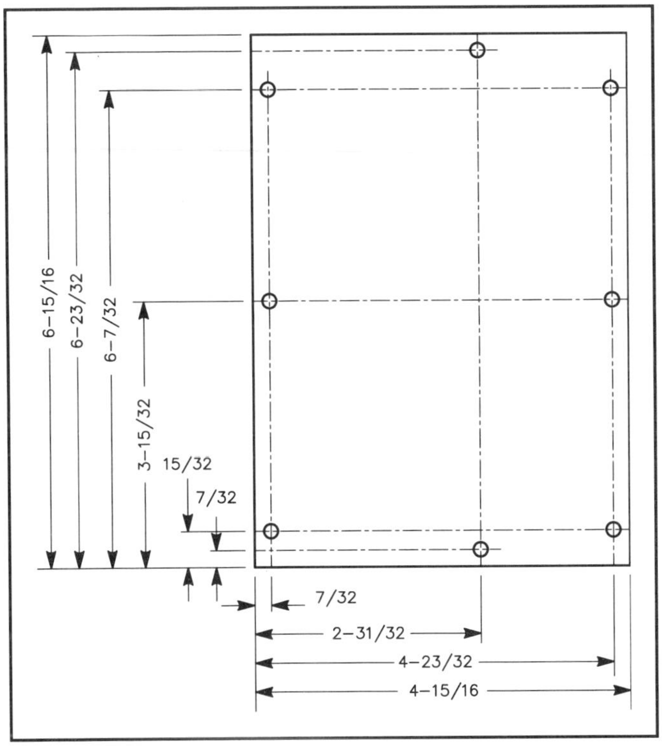

Fig 20—Side-chassis cover plates (2 required). Holes are No. 19. Use cover plates as templates to drill side chassis.

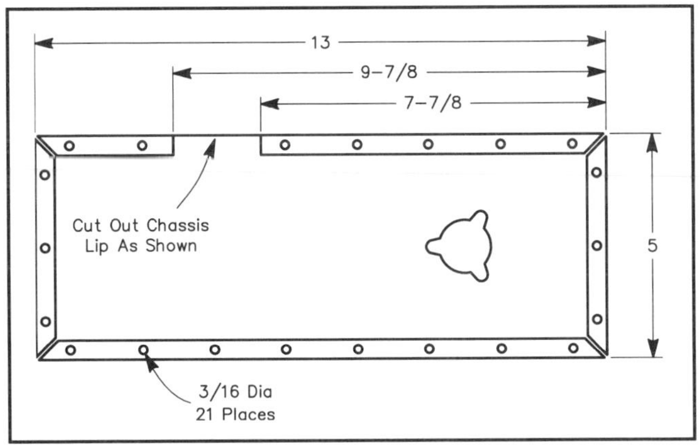

Fig 21—RF plate enclosure. Install blind fasteners in the 21 3/16-diam. holes.

scratches and burrs made during construction. At the plater's specify a deep etch, which will remove any remaining scratches, unless they are very deep. While you can apply irridite at home, I don't recommend you try, though. The aluminum must be absolutely free of contaminants, and the chemicals necessary to ensure it is are not readily available and are dangerous to use. I paid only $32 to have all the amplifier parts irridite plated, in 1986.

The drill work done to make the various air-vent holes (cathode compartment and side chassis) may look as if it was done by magic, to achieve their near-perfect patterns. In truth, these vent holes are simple to create. A piece of standard perforated steel was used as a drilling template, so the many holes do not even have to be marked! After drilling, the burrs are sanded smooth to create the clean, sharp holes. Figs 18 to 21 give additional metalwork drilling and cutting information.

Protection Circuits

The specified filament warm-up time for the 3CX800A7 is 3 minutes. Follow this specification to obtain maximum tube life. As the cathode is heated, the areas directly in front of the filament wires get hot first, and become the first areas to emit cathode current. If you operate the amplifier in this condition, the cathode hot spots will emit all the cathode current, and may vaporize.

An Amperite thermal time-delay relay is used to prevent operation of the amplifier for 3 minutes. A solid-state timing circuit was considered, but the thermal time-delay was chosen because it is simple and heats just like the tube filament. There's no need to wait another three minutes if filament power is only interrupted momentarily. Brief interruptions may be caused by power failures or HV arcing. This kind of flexibility is hard to design into a solid-state circuit.

A 120-second time-delay relay was used because it is the longest delay available in the 9-pin miniature size. I extended its turn-on time to 180 seconds by placing a resistor in series with the heater element. Even if you obtain a 180-second unit, you should check the turn-on time. Most units are specified for use on a nominal 115-V ac circuit. With US line voltage now typically 118 to 120 V, some timers will warm up 15 to 20% faster than the specified time. If necessary, use a series resistor to adjust the warm-up time. Since I built this amplifier, the Amperite 115NO120T (9-pin) and 115NO180 (octal) have become hard to get and very expensive. An inexpensive sealed-plastic-case model (115NO180B) can be used instead.

A small but significant point is the use of a 14-V transformer for the filament. Since the filament voltage is backed off to 13.0 V for operation, you may be tempted to use a commonly available oversized 12.6-V transformer. Doing so will be detrimental to the life of the tube. A well-cared-for tube used in intermittent service may first fail with an open filament. A cold filament has very low resistance (1.5 Ω for the 3CX800A7). Unless some protection is used, the turn-on surge may be as high as 10 A. By using a higher-voltage transformer and series resistors, automatic surge limiting is easily achieved. I used a 1-Ω resistor in series with the secondary of the transformer. A 50-Ω, 12-W variable resistor is used in the primary circuit to set operating filament voltage.. This combination limits turn-on surge to just over 2 A.

To achieve high gain and efficiency, the 3CX800A7 uses a grid structure constructed of small wires. This construction limits grid dissipation to 4 W. It is common practice for HF amplifier designers to include a circuit to sense excessive grid current. Such sensing circuits are not always useful at UHF, as transit-time effects may cause negative grid currents vastly different from real grid absorption and secondary emission. In testing the 3CX800A7 at 432 MHz, the grid current was found to behave predictably, that is positive current which corresponded directly to drive levels was observed. Because of this desirable behavior, a grid-current-sensing circuit was included in this amplifier. R14, in combination with grid shunt R4 develops a voltage sufficient to switch on the grid-trip relay K3, through switching transistor Q2. Contacts on K3 lock in the over-current relay and turn off the OPERATE pilot lamp. No reset switch is used. The circuit is reset by setting STANDBY/OPERATE switch (S2) to STANDBY, then back to OPERATE.

An additional protection circuit senses high voltage. It consists of Q1, which senses the presence of high voltage through the HV metering circuit. Q1 allows the T/R relays to switch on transmit. This circuit prevents the amplifier from keying should plate voltage not be present. While just the grid-trip circuit may seem adequate, the absence of plate voltage will cause the tube to draw excessive grid current when drive is applied. The HV sensing circuit protects against such an occurrence.

Many high-performance 432-MHz stations use a single T/R relay at the array, in combination with an antenna-mounted receive preamplifier. In these stations, in-out RF relays are not used at the amplifier. This could cause a problem if the amplifier ac fuses blow. Since the grid-trip circuit requires that power be present to operate, ac-power loss to the amplifier but not the exciter can result in damage to the tube, unless a driver-to-amplifier interlock is provided. The HV-sensing circuit will protect the tube. Simply unshorting the cathode-bias resistor (R3) will provide sufficient protection for the tube from drive power, but it is better to remove all drive power when plate voltage isn't present.

Another feature of the switching circuitry is provided by S4. This switch, located on the rear panel, allows the operator to determine whether the RF relays actuate every time he transmits, or lock in when the amplifier is switched from standby to operate. This feature can be used if the station includes a tower-mounted preamp with a separate feed line. It eliminates the noise generated by switching the relays in the shack. If an in-out tower preamp is used (with a single transmission line) or a transceiver without an external preamp, this locking feature is not usable. Note though, even if the RF relays are locked in, either an excessive grid-current trip or loss of plate voltage will still drop out the RF relays.

Metering Circuits

The metering circuit has one anomaly: The resistor string is used both for the high-voltage meter and for the high-voltage sensing circuit. Since the HV sensing circuit is referenced to ground, the current from the circuit will return to the HV supply B– lead through the grid-meter shunt. The grid meter will read this current. At idle conditions, the grid-current meter will read negative by about 0.8 mA. If you find this to be a problem, use separate resistor strings for the HV meter and HV sensing. The HV meter would then be connected to the B– line, not to ground.

Note also that the HV meter also reads plate voltage relative to ground. Since the actual power input is determined by plate-to-cathode voltage, the bias voltage must be subtracted to obtain the true plate voltage and power input. As the bias voltage on the 3CX800A7 is only 5.6 V, so the correction is only 0.2%. To obtain a true plate-to-cathode voltage reading, you must reference the HV meter circuit to the tube side of the Zener bias diode. This would again require separate HV meter and sensing resistor strings. Arranging the meter circuit in this way also causes the plate meter to indicate the metering string resistor current.

Many operators will want to build 3X800A7 amplifiers for other bands, and share a common HV supply. This would place the grid shunts in parallel. Grid-current meters on the unused amplifiers will indicate a portion of the operating amplifier's plate and grid currents, uncalibrating all the metering circuits. There are two methods to reduce this problem. One is to place a resistor in series with the grid-sensing resistor (R21). Alternately, you can install a resistor in series with the B– line. T/R relay K2 would short out these resistors on transmit. Use the lowest-value resistors which eliminate the problem. This value depends on the number of amplifiers connected to the same power supply, but a value of 500 Ω is typical. A word of warning: These resistors increase the possibility of damage to the metering circuits when a HV arc occurs. Also note that the 11-Ω effective grid-shunt value (due to the grid over-current sensing circuit) causes a portion of the grid current to flow through the B– safety resistors (R1 on the amplifier and any similar resistors at the HV power supply). R13 can be adjusted to make the grid meter read the correct value when connected to the power supply.

Keying Circuits

A transistorized relay-switching scheme is used. The circuit can be switched by either grounding J6 or by supplying +12 V to J8. Current sinking or supplying capability is only 15 mA, which should be compatible with current 432-MHz exciters. If you are using conventional switching circuits, make sure your relays will reliably contact at only 15 mA of current. As described in the protection-circuit section, a loss of high voltage will prevent the relays from switching, or drop them if they have already switched.

The coaxial relays you use should have dc coils. Dc coils switch faster, have less contact bounce and run cooler than ac types. The coils don't have to be 28 V, as are most available military surplus relays. Several types of suitable coax relays with 12-V coils are available. The switching circuitry can be made to work with 12-V coils but you'll have to change T2. If you have relays with 120-V ac coils, don't despair. Many 120-V ac relays work better on 12 to 28 V dc, so try them on a dc supply.

Amplifier in-out switching relays don't need high isolation. If you use separate input and output relays, the input relay can be a low-power type.

Power Supply Considerations

The high performance of the 3CX800A7 at VHF and UHF is partially a result of the close internal spacings of the tube elements. These close spacings increase the possibility of a plate-to-grid arc over. To protect the tube, power supply, metering circuits and the operator, a 50-Ω 50-W resistor is connected in series with the HV power lead. This resistor substantially reduces the energy dissipated in the tube during an arc over, yet only reduces the full-load plate voltage by 30 V. EIMAC specifies that idle plate voltage should not exceed 2500 V.

I built a modern capacitor-input power supply, which uses a solid-state bridge rectifier in combination with a low-resistance plate transformer. My supply uses a Peter Dahl Hypersil transformer (1800 V at 700 mA, CCS). The electrolytic-capacitor string has a total capacitance of 22 µF. An oil-filled capacitor is preferred. Adequate regulation and ripple filtering can be obtained with less than 25 µF of filter capacitance. No-load voltage is approximately 2510 V. At 600 mA, the plate voltage is 2250 V. Out of the 259-V drop, over 100 V is due to line-voltage sag and the protection resistor. In no case should you use more than 50 µF of filter capacitance. Higher values will not significantly improve regulation, but will greatly increase the possibility of damage during an arc over.

Avoid choke-input filters. I have seen many improperly designed amateur supplies that were made from surplus components haphazardly cobbled together. These supplies can have bad transient-voltage-spike problems, induced from the chokes. These transients will blow rectifier diodes and cause tube arcs. They can also severely reduce the IMD performance of the amplifier. Finally, be sure to use MOV transient suppressors on the ac-input leads.

I highly recommend MHV-type HV connectors. The most dangerous condition to the operator is when high voltage is present and the power-supply-to-amplifier ground connection open. If this happens and you touch the amplifier and power supply (or any other grounded object if the power supply was grounded), you become the ground return. MHV connectors prevent this possibility as the shield makes contact along with the center conductor. RG-59 cable can comfortably handle the 2500-V power that this amplifier uses.

Amplifier Stabilization

Many amateurs mistakenly believe that neutralization of a grounded-grid amplifier is not required. Although operation of this amplifier is possible without neutralization, the simple neutralization procedure used provides for improved tuning, more stable operation and greater apparent efficiency. In addition, tube life may be increased due to the elimination of unwanted higher frequency circulating currents. The amplifier was neutralized by adjusting the grid inductance until the plate current dip matched the maximum output point. The amplifier operates so well that it can be tuned up without a power-output indicator. By dipping the plate current and adjusting the loading for proper grid current one can obtain output power that will be within a few watts of tuning it up for maximum output on a directional wattmeter!

Since the 3CX800A7 is operating below its self neutralized frequency, neutralization is easily accomplished by breaking off grid collet fingers. The so-called self-neutralizing frequency of a tube is simply the frequency at which the combination of tube and socket have the right combination of feedback capacitance and inductance as to create the maximum input-to-output isolation. It should be noted that not all tube and socket combinations have any frequency where their isolation will be acceptable.

By breaking off some of the grid-collet fingers, the effective grid inductance is raised and the maximum reverse isolation point is lowered to 432 MHz. Fig 22 shows the proper pattern for breaking off the grid-collet fingers. Tube characteristics appear to be close enough among different 3CX800A7 tubes such that neutralization for a specific tube is not required. If you want to be safe, however, break off every other

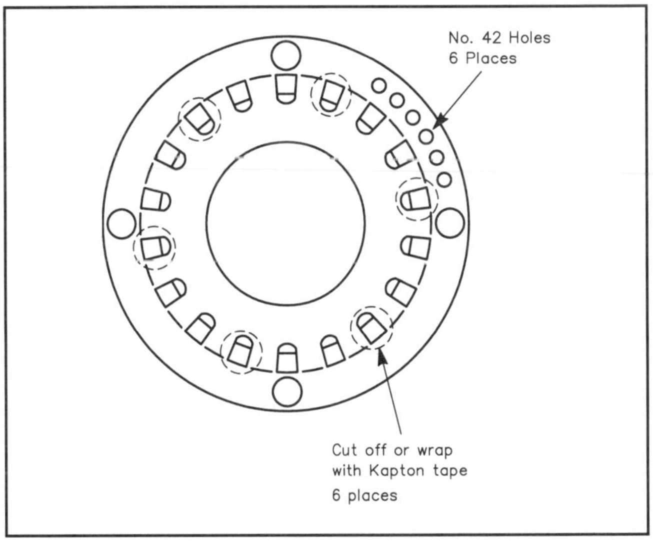

Fig 22—Grid collet modification. The EIMAC 720359 collet is supplied with 36 fingers. Break off every other finger. Wrap 6 of the remaining fingers (as shown) with Kapton tape or break them off. Removing the fingers adjusts grid-circuit capacitance and helps neutralize the amplifier.

finger on the grid collet. There will be 18 remaining fingers. Tape over 6 of them with Kapton tape, spacing the taped fingers equally around the tube. These taped fingers will then allow future tuning adjustments without replacing the grid collet.

An even more scientific approach to neutralization is to look at the reverse isolation of the amplifier. This is done by connecting the output of a signal generator or network analyzer to RF-output connector and measuring the fed-through power with a power meter, spectrum analyzer or network analyzer. Ideally, this test is performed after tuning up the amplifier with all voltages in place. If isolators are not available, there is a danger that the amplifier can oscillate during the test procedure and put out enough power to destroy your signal generator.

The reverse isolation of the amplifier should be a minimum of 20 dB greater than its power gain. Since the 3CX800A7 amplifier has a gain of 14.2 dB, you should see over 34-dB reverse isolation when it is neutralized. This amplifier had less than 20-dB reverse isolation before neutralization and over 30 dB after the collet was modified.

After neutralization, power drift became almost nonexistent. Power output will slowly rise by only 20 to 30 W from a cold start to full operating temperature. This represents a less than 0.2 dB of power drift, certainly an acceptable amount.

Initial Tuning Adjustments

Tune up of the amplifier is quite straight forward. If the dimensions and layout of the amplifier have been closely followed, the following initial settings will place the amplifier very close to optimum tuning at 432 MHz and full power output.

C1: Plates approximately 50% meshed

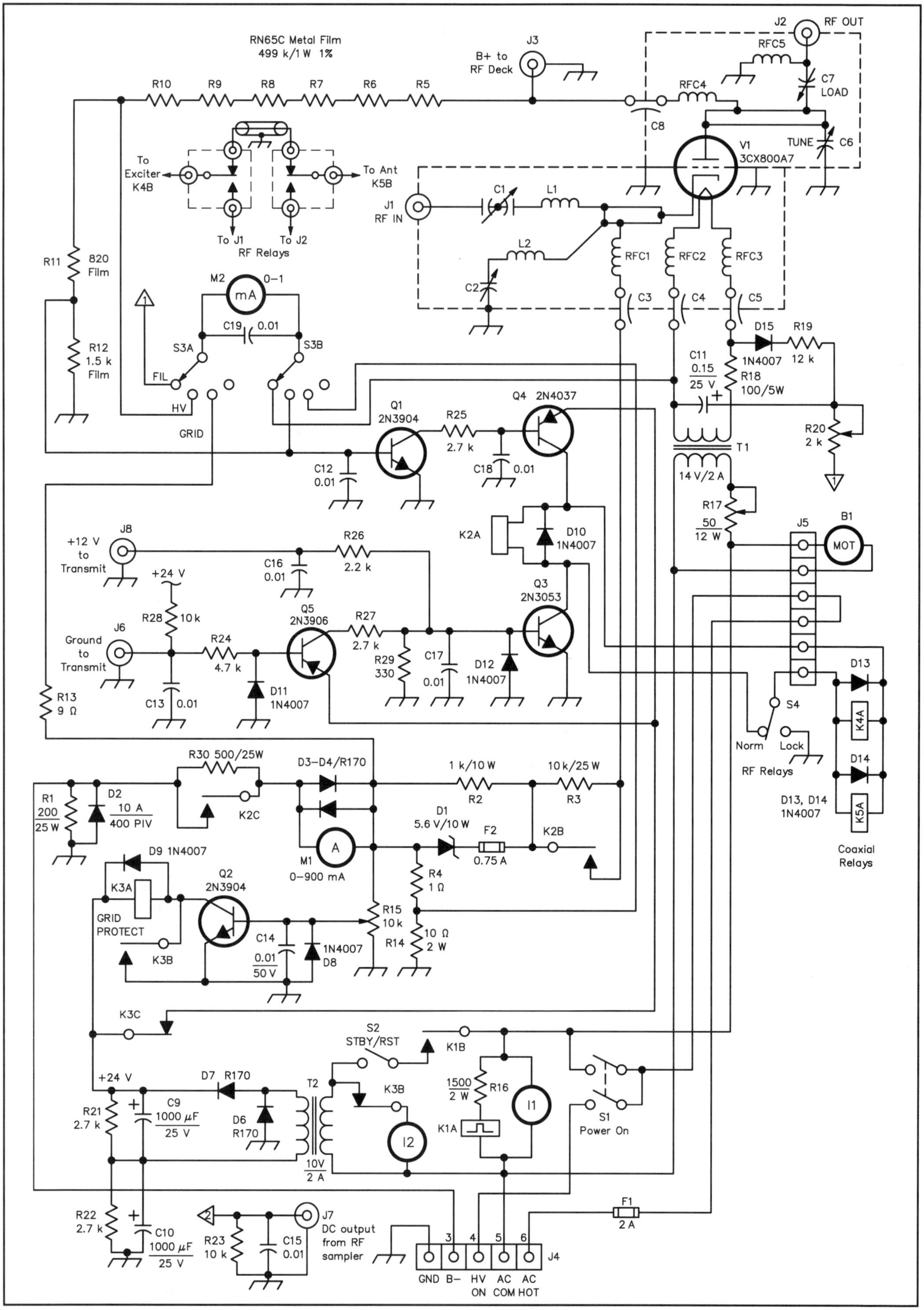

RN65C Metal Film
499 k/1 W 1%
R10
R9
R8
R7
R6
R5
J3
B+ to
RF Deck
J2
RF OUT
RFC5
C7
LOAD
RFC4
C8
TUNE
C6
V1
3CX800A7
To
Exciter
K4B
To Ant
K5B
To J1
To J2
RF Relays
J1
RF IN
C1
L1
L2
C2
RFC1
RFC2
RFC3
C3
C4
C5
R11
820
Film
M2
mA
0–1
C19
0.01
S3A
S3B
R12
1.5 k
Film
FIL
HV
GRID
D15
R19
1N4007
12 k
C11
0.15
25 V
R18
100/5W
R20
2 k
Q1
2N3904
Q4
2N4037
R25
2.7 k
C18
0.01
C12
0.01
T1
14 V/2 A
R17
50
12 W
J5
B1
MOT
K2A
D10
1N4007
J8
+12 V
to
Transmit
R26
2.2 k
C16
0.01
+24 V
R28
10 k
Q3
2N3053
Q5
2N3906
R27
2.7 k
J6
Ground
to
Transmit
R24
4.7 k
R29
330
C17
0.01
D12
1N4007
D11
1N4007
C13
0.01
R13
9 Ω
D13
S4
Norm
Lock
RF Relays
K4A
D14
D13, D14
1N4007
K5A
Coaxial
Relays
R30 500/25W
D3–D4/R170
1 k/10 W
10 k/25 W
R1
200
25 W
D2
10 A
400 PIV
K2C
R2
R3
D1
5.6 V/10 W
F2
0.75 A
K2B
M1
A
0–900 mA
D9 1N4007
K3A
Q2
2N3904
R4
1 Ω
GRID
PROTECT
K3B
C14
0.01
50 V
1N4007
D8
R15
10 k
R14
10 Ω
2 W
K3C
S2
STBY/RST
K1B
+24 V
D7
R170
T2
K3B
1500
2 W
R16
R21
2.7 k
C9
1000 µF
25 V
D6
R170
I2
K1A
I1
S1
Power On
10V
2 A
F1
2 A
R22
2.7 k
C10
1000 µF
25 V
J7
DC output
from RF
sampler
R23
10 k
C15
0.01
3
4
5
6
J4
GND
B–
HV
ON
AC
COM
AC
HOT

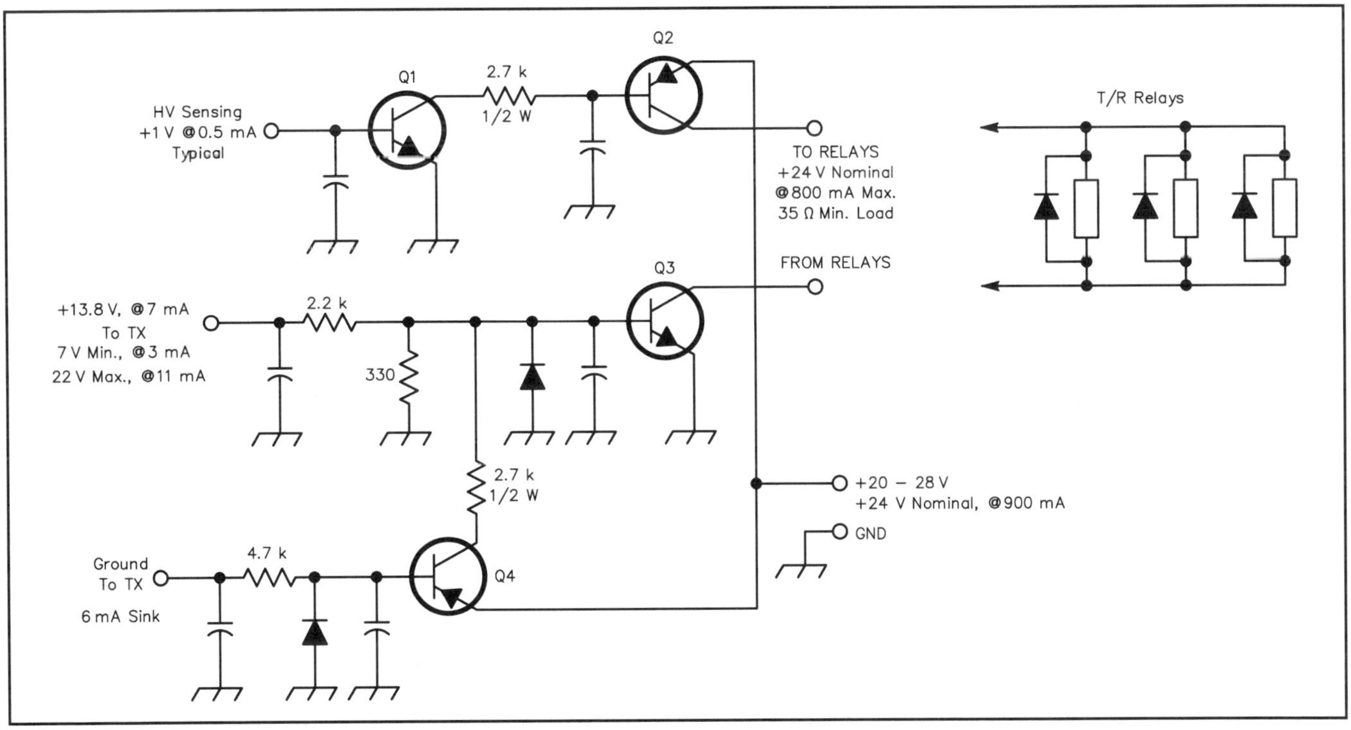

Fig 24—TR switching circuits for use with 24-Vdc relays. These components are mounted on the PC board shown in Fig 25.

←

Fig 23—Schematic diagram of the amplifier. Control circuits protect the tube and amplifier from potentially catastrophic events, such as overdrive, loss of high voltage or loss of heater voltage.

C2: Plates approximately 25% meshed
C6: Fixed end of flapper 1¼ in. above bottom of chassis
Moving end of flapper ¾ in. above bottom of chassis
C7: Fixed end of flapper 1 in. above bottom of chassis
Moving end of flapper ⅞ in. above bottom of chassis

After all wiring has been checked, set R17 (filament adjust) to maximum resistance. Attach an accurate RMS voltmeter to the filament feed-through capacitors. After the filament has warmed up for 3 minutes, adjust R17 until the filament voltage is 13.0 V. Next, adjust R29 such that the filament meter indicates 13.0 V. If the meter cannot be calibrated, change R19.

Apply high voltage and verify the tube draws proper idling current when J6 is shorted to ground (approximately 55 mA at 2400 V). Remove high voltage. Short out cathode resistor R3. Apply a very small amount of drive, increasing drive until the grid current reads 60 mA. Adjust R15 so the grid overcurrent trips at 60 mA. If you are unable to obtain 60 mA of grid current, adjust the cathode tuning controls (C1 and C2) for maximum grid current. After adjusting the grid-trip, reduce drive power and peak C1 and C2 for maximum grid current, being sure not to exceed 60 mA.

Remove the shorting jumper from R3 and turn the high voltage back on. Apply drive power so the plate meter reads 200 to 300 mA. Start adjusting C6 and C7 (plate tuning and loading) maximum power output. As the power output comes up you can increase drive until you are operating at the desired power level. If the amplifier cannot be driven to full power, C1 and C2 may need to be adjusted.

Final adjustments to the cathode circuit must be made at full power. With an in-line power meter connected as close to J1 as possible, alternately adjust C1 and C2 for minimum SWR. You may find that there are combinations of the cathode capacitors that allow the amplifier to be driven but the input SWR is poor. There will be a unique combination of the capacitors that will adjust the amplifier such that the lowest input SWR (less than 1.2:1) corresponds to maximum plate current (lowest drive-power requirement). To obtain this condition will require alternately adjusting C1 and C2 several times. If you cannot obtain a good input SWR, check to see if C1, C2 or both are at minimum or maximum capacitance. If C1 is at minimum capacitance, shorten L1. Lengthen L1 if C1 is at maximum capacitance. If C2 is at minimum or maximum capacitance, it is most likely because you didn't follow the cathode-circuit layout. Using a Bird 43 wattmeter with a 50D element (50 W, 200 to 500 MHz), I can completely null out the reflected power, so that with 30-W drive the wattmeter indicates no reflected power. (This only indicates that reflected power is lower than the directivity of the element, not that I have obtained a perfect match.)

With the input circuit properly adjusted, final adjustments to the plate tuning and loading controls may be made. There will be many settings of C6 and C7 that will deliver lots of power output. There will be a unique combination that will deliver that power output at maximum efficiency. Grid current is an excellent indicator of proper tuning. At full power output the grid current will range from 15 to 30 mA (varies from tube to tube). If the amplifier is properly tuned, grid current should increase as you increase drive. The plate-current dip should

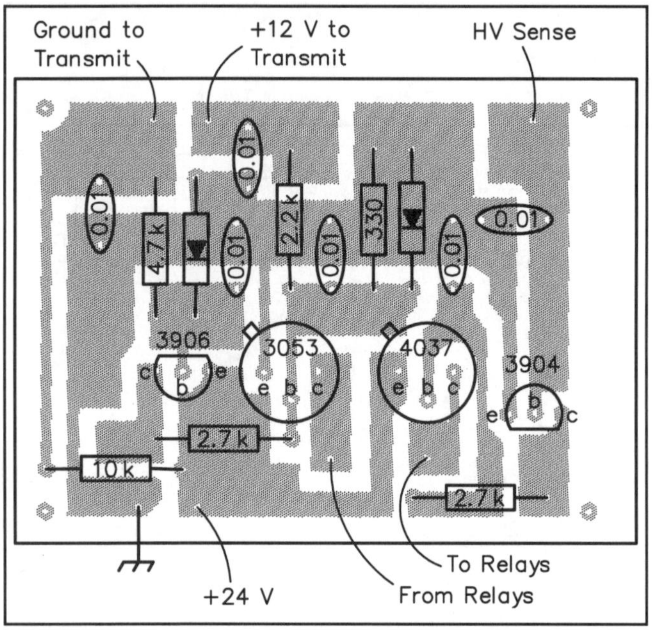

Fig 25—TR switching PC board, component side. While board layout is not especially critical, the board itself must be 1.75 X 3.25 in. or smaller in size.

Table 2
Full Power CW Operating Conditions
Keyed CW and SSB PEP Operation

Plate voltage, full load:	2250 V
Plate current:	600 mA
Dc Input power:	1350 W
RF Output power:	730 W
Apparent efficiency:	54%
Power gain:	14.2 dB
Grid current:	26 mA (will vary from tube to tube)
Drive power:	28.5 W
Input VSWR:	1.16:1
Filament voltage:	13.0 V
Idling current:	55 mA (@ 2400 V; will vary tube to tube)
Bias voltage:	5.6 V

Note: These conditions are for SSB and keyed CW service. Key-down time not to exceed 1 minute. For maximum continuous duty ratings, see text.

also match very closely (within 20 W) of maximum power output. There should also be minimum thermal drift. After letting the fish lines stretch for a few days, power should drift less than 30 W from a cold start.

Amplifier Operation

Efficiency of the amplifier is about 54%. If you are measuring higher efficiency you either have an inaccurate power meter or your high voltage and plate meters aren't calibrated. Use a Bird 1000D element with the Bird 43 wattmeter; 1000E elements typically read about 10% high at 432 MHz. An antenna or dummy load with a high SWR can cause inaccurate power meter readings. This can be verified by placing ¼-wave-length coax sections in series with your feed line, and checking to see if the SWR or indicated power change. Power measurements were made with a directional coupler in combination with a microwave power meter. The directional coupler was measured on a network analyzer to verify the amount of coupling it has at 432 MHz. If your efficiency is below 50%, you may have any of the previously mentioned measurement accuracy problems, a bad tube or just a mistuned amplifier.

At full rated output (about 730 W) the 1-dB bandwidth of the amplifier is 1.3 MHz. The 3-dB bandwidth is 3.9 MHz. The amplifier was designed to operate between 430 and 440 MHz. With the specified plate line and tuning capacitor sizes, the amplifier will operate efficiently at output levels from 300 to 730 W over that frequency range. If operation is desired at higher or lower frequencies, you'll have to adjust the length of W2 or C6.

Maximum CCS ratings of the 3CX800A7 are 2250 V at 600 mA. The specified highest frequency for maximum ratings is 350 MHz. No problems have been experienced running the tube at those ratings in keyed CW and SSB service at 432 MHz. I recommend, however, limiting CW key-down tuneup time to under 1 minute. If the amplifier is to be used in continuous-duty service, such as FM repeater or ATV, limit maximum plate current to 500 mA and reduce full-load plate voltage to 2000 V or less. In continuous service the filament voltage should be reduced to 12.2 V during transmit. A switching circuit should be added to raise the filament voltage to 13.5 V during warmup and standby periods. Maximum CCS output will be 500 W at a drive level of 22 W and 50% efficiency (2000 V at 500 mA). Continuous-duty service at elevations above 2000 feet may also require a larger-capacity blower. When shutting off the amplifier, be sure to let the tube completely cool down. The amplifier should be left powered up (blower running) for at least 5 minutes after the last transmission. The cathode compartment can still be warm after the anode air exhaust is blowing cool air.

Note that in addition to the grid dissipation value of 4 W, the maximum grid current is 60 mA. Under linear service the grid dissipation will be far under the rated value (less than 0.5 W). The grid-current restriction therefore becomes a matter of a total limit of cathode current, which is normally 600 mA of plate current and 60 mA of grid current. At 432 MHz, transit time effects, secondary emission by the grid and back bombardment of the cathode start to occur. The amplifier can be loaded so that negative grid current will be indicated. These conditions require that you be even more careful about grid and cathode current. In general, the amplifier should never be operated with an indicated grid current over 40 mA. An indication of these back-heating effects is grid-current drift. If you are seeing the grid current drift upward at a constant drive and plate-current level, you are running the amplifier too hard (or you may have a bad connector or cable in the system).

You should also be very careful not to overdrive the amplifier. Cathode-driven triodes such as the 3CX800A7 will not exhibit the gain-compression phenomena shown by solid-state amplifiers. That is, the power gain of the tube is just as high at full power as it is at low power. The station should have an exciter that has stable output power. In addition, make provisions to limit the available drive power to 35 W or less.

Table 2 gives full power operating conditions. For SSB operation the conditions represent the PEP point. There is little

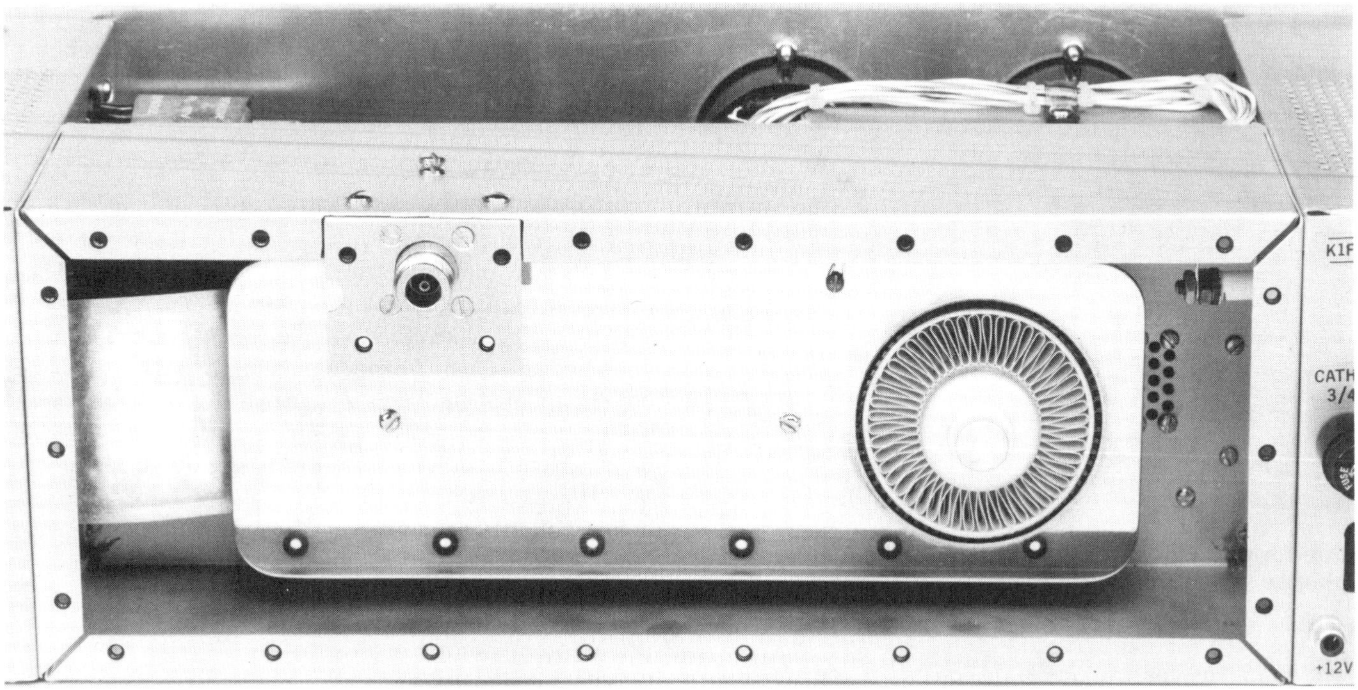

Plate compartment of the single 3CX800 amplifier. Plate tuning capacitor is shown to the left. The RF-output connector is at top.

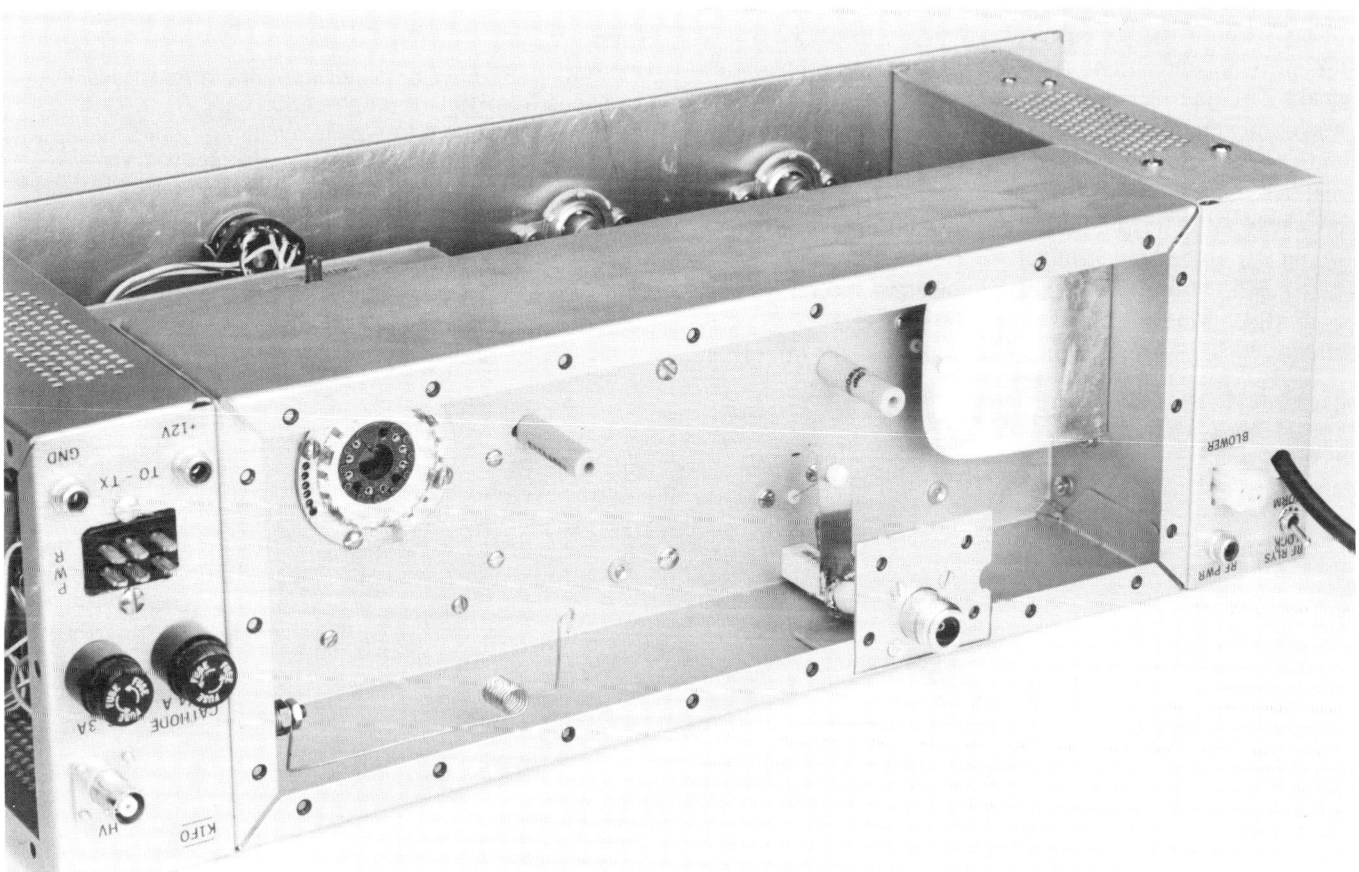

This inverted view of the plate compartment shows the output loading capacitor, which is adjusted by means of a non-conducting cord fastened to its lower edge.

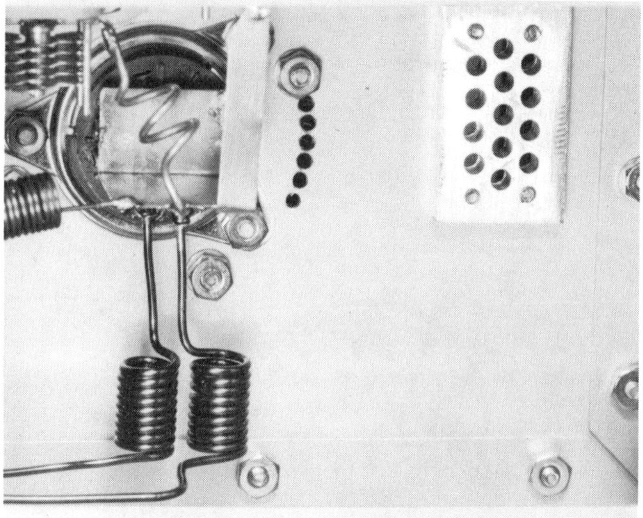

Interior of the cathode compartment. The filament RF chokes are at bottom.

sense in investing in a tube of this quality if you are going to overdrive or mistune the amplifier. During proper SSB operation, without speech processing, indicated plate current on *voice peaks* should be less than 250 mA. Power output as indicated on a slow-responding average-power meter (such as the Bird 43) should be less than 200 W on peaks. If an effective RF peak clipper is used (10 dB of compression and clipping), indicated plate current on voice peaks can approach 400 mA. Average-power watt meters may indicate up to 325 W on peaks. While driving the amplifier harder may give you the satisfaction of seeing higher meter readings, the additional power will be transmitted primarily as distortion products, and won't make your on-frequency signal any stronger.

Pay attention to the cable, connectors and relays you use. All high-power connections should use Type-N connectors. Assemble each connector properly so the shield has good contact and the center pin is aligned and at the proper depth. This much power at 432 MHz will destroy poorly assembled connectors. Use ½-in. or larger Hardline for the antenna feed line. For flexible jumpers, use RG-225 coax.

Using An 8874

This amplifier will also work with an 8874 tube. To use an 8874, set the filament voltage to 6.0 V RMS. Substitute a 6.3-V, 3-A transformer for T1 and eliminate R18. Make the hole in the plate line (W1) 1.75-in. diameter. An EIMAC 008294 collet makes for a simple connection between the plate line and the 8874 anode radiator. Make the air-outlet hole 1⅝ in. diameter, You'll have to make a chimney (sheet Teflon). In the input circuit, L1 and L2 may need to be made longer. Neutralization of an 8874 is slightly different.

The maximum ratings of an 8874 are 2200 V at 500 mA, in intermittent amateur service (keyed CW and SSB). Peak power output will be 570 W. Drive power for that output level will be 35 to 38 W. In continuous duty, limit the 8874 plate current to 350 mA.

Conclusion

This amplifier is easy to build and requires minimal special metal fabrication. Performance is excellent. The investment in parts and careful assembly time will pay off with years of trouble-free operation.

Notes

[1]The 8874 432-MHz amplifier appears in ARRL *Handbook* editions from 1981 through 1986.

[2]Aluminum sheet cut to size is available from Chassis Kit. Charles Byers, K3IWK, 5120, Harmony Grove Road, Dover, PA 17315. Tel 717-292-4901.

[3]Surplus Sales of Nebraska, 1502 Jones St., Omaha, NE 68102. Tel. 402-346-4750.

[4]Microwave Components of Michigan, P.O. Box 1697, Taylor MI 48180. Tel 313-753-4581.

[5]S. Powlishen, "Improving the K1FO 8874 432-MHz Amplifier," *QST,* Jun 1987, pp 20-23.

A Parallel 3CX800A7 Amplifier for 432 MHz

By Steve Powlishen, K1FO

Many 432-MHz DXers, especially those contemplating EME operation, desire to run the US 1500-W output-power limit. This amplifier, which uses a pair of Varian 3CX800A7 ceramic-metal triodes, is capable of reaching that power level in intermittent amateur service (SSB voice and keyed CW). Its design is very similar to the single-tube 3CX800A7 amplifier previously described. The plate circuit is a half-wave stripline with the tubes located at one end of the line. The metering and control circuits are identical to those used in the single-tube amplifier, except that meter calibrations and filament power are set for two tubes instead of one.

Selecting a Design

An amplifier design process involves a number of factors. A pair of 3CX800A7s were chosen for the following reasons:

- They can reach the 1500-W power level.
- They have the simplicity and stability of triode operation. (ie simple zener diode bias and no screen supplies or grid and screen bypassing is needed).
- They have relatively low drive requirements as their gain is very high for a grounded grid triode.
- They have very good IMD performance, making them an excellent choice for SSB voice operation.
- They are small enough to allow an amplifier to be built using standard commercial aluminum chassis.
- Their price point is within most amateur budgets
- They are a modern tube design, readily available today and for the foreseeable future.
- Their acceptance in commercial amplifiers have made them available at reduced cost via pull-out and surplus units.

Few items in our world are perfect and the 3CX800A7 is no exception. Some of the problems encountered in using the 3CX800A7 at 432 MHz are:

- Their anode radiator is large enough in diameter such that making an amplifier with a pair of tubes presents some design challenges.
- Their frequency for maximum ratings is 350 MHz.
- Their efficiency of the 3CX800A7 at 432 MHz is lower than desired.
- 1500-W output power operation at 432 MHz is strictly intermittent if long tube life is to be expected.
- While the price of a pair of tubes is not totally exorbitant, they are priced high enough that they are certainly not disposable items.

When all the pros and cons of the 3CX800A7 are considered and compared to the pros and cons of other tubes that are usable at 432 MHz, however, the 3CX800A7 becomes a good choice for a 432-MHz amplifier.

With the tube model selected, the next task is to design input and output circuits that will deliver, with minimum losses, the RF power that 3CX800A7s are capable of generating. The first design considered was to operate the tubes in parallel with a half-wave stripline, with the tubes located at one end of the line. This is the same arrangement that K2RIW used with 4CX250B tubes. Schematically this type of circuit is shown in Fig 1. The primary advantage of the parallel half-wave circuit is, it is the simplest circuit for an amateur to build. There are no critical or hard-to fabricate blocking capacitors, nor are there any complicated sliding contact surfaces. The primary disadvantage of the half-wave plate line is that circulating currents are not evenly distributed around the tubes seals. This could potentially shorten tube life.

The second disadvantage is that the large diameter of the

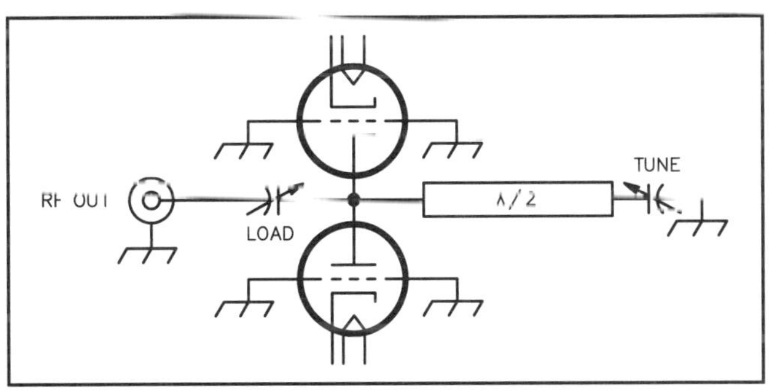

Fig 1—Schematic of a half-wave stripline amplifier using parallel triodes. While easy to build, the large diameter of the 3CX800A7 tubes results in a plate line that is wider than optimal. See text.

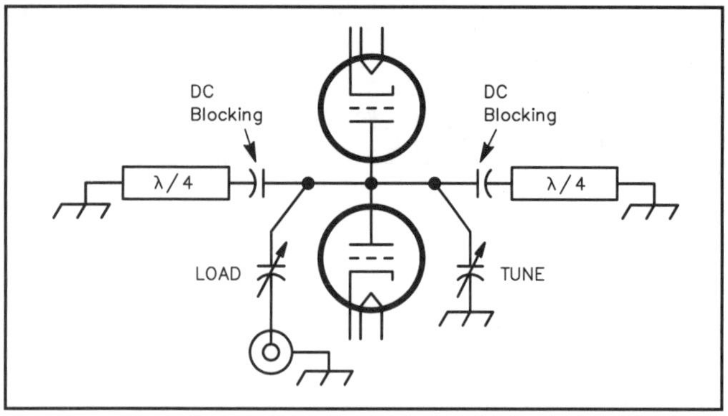

Fig 2—Schematic of a half-wave stripline amplifier in which the tubes are mounted at the center of the line. This circuit has all of the disadvantages of the circuit shown in Fig 1, and is more difficult to build.

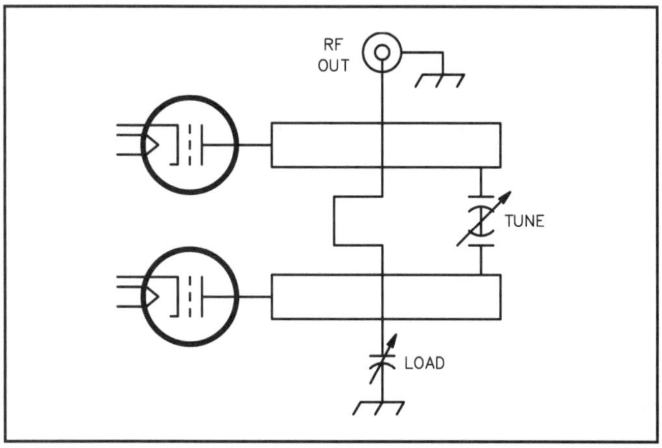

Fig 3—Schematic of a push-pull amplifier. This circuit is also difficult to construct. Balanced loading of the tubes is more difficult to obtain than with an amplifier using parallel tubes.

3CX800A7 anode radiator requires a plate line that is very wide, which yields a plate-line impedance considerably lower than desired. Theoretically, a cavity or stripline resonator will have its best loss and bandwidth characteristics when it has a characteristic impedance of around 75 Ω. An additional design problem with the 3CX800A7 in a parallel stripline design is the large diameter of the tubes necessitates using a relatively wide spacing between them. The wider the tube spacing the more difficult it can be to keep the tubes operating in parallel mode (ie, push-pull resonances must be placed far from the operating frequency for best operation).

Other designs were considered. A λ/2 stripline with the tubes located in the center of the line was a possibility. This arrangement would improve the circulating current problem by distributing the current paths to two sides of the tubes. However, all of the other previously covered stripline design problems would remain, that is lower than optimum line impedance and potential push-pull modes. In addition, locating the tubes in the center of a half wave line makes the inclusion of a critical plate-blocking capacitor necessary. This blocking capacitor adds another potential failure point and makes construction more complicated. The extra complexity of this circuit type can be seen in Fig 2. Moreover, designing and building plate tuning and loading capacitors which would both not disturb the symmetrical current benefits of this type of plate circuit and couple equal power from both tubes is a difficult task.

Although not often used today in amateur amplifiers, the push-pull amplifier was very popular until a few years ago. When designed properly the push-pull design offers some significant benefits. The primary advantage is that tube output capacitances appear in series instead of adding together, as in a parallel design. This effectively raises the output capacitive reactance of the tubes, giving the designer more flexibility in designing the plate resonator. Another advantage of a push-pull UHF amplifier is that, since the tube anodes are not connected together, low-loss plate lines near the optimum 75-Ω impedance are relatively easy to incorporate.

In addition, the push-pull arrangement could be made in a fashion similar to the old "plumber's-delight" 144-MHz amplifiers. This type of design can have better current distribution than a stripline. A proper push-pull design would also be less sensitive to the spacing between the tubes as parallel responses near 432 MHz would not be likely. A push-pull schematic representation is in Fig 3. The down side of the push-pull design is that the plate circuit fabrication is more complex than a K2RIW-style amplifier since standard commercial chassis could not be used. At 432 MHz it becomes very difficult to obtain balanced loading from the two tubes which creates another design problem and potential adjustment headache. Still another problem area is the construction and adjustment of a push-pull input circuit which will maintain proper tube balance can be much more difficult than in a parallel amplifier.

A final design option considered was to simply incorporate two single-tube amplifiers on a single chassis and use a hybrid power splitter to feed input power to the amplifiers and a similar hybrid combiner to couple the amplifiers' output together. Although it would be much easier to obtain near optimum plate-line impedance, it was quickly decided that the losses in the hybrids would be greater than predicted circuit losses in a parallel design. Combining two amplifiers requires a high-power dump load, tuning procedures become tricky and metering circuitry is more complex if both amplifiers are operated off the same B+ supply. In addition, construction complexity is increased due to the need to make additional combining components. All of these factors made this approach less desirable than the alternatives.

Parallel Plate Circuit Design

After analyzing all design options it was concluded that building a half-wave stripline, K2RIW-style, plate circuit was the best compromise. The calculated line impedance, loaded and unloaded Q gives an expected line loss of less than 0.05 dB which is under 17 W at full power, or about a loss of 1% in plate efficiency. These losses were considered acceptable given the simple construction. The unbalanced current design of the half wave stripline and its plate-circuit losses may not be considered acceptable for a continuous-duty amplifier. However, intermittent amateur operation of the single-tube 3CX800A7 half-wave stripline, 432-MHz amplifier at 750-W output has resulted in an immeasurable loss in output power and efficiency after 6 years of use. This validates the suitability of this type of circuit for intermittent amateur use. The final decision in selecting the half-wave plate line design is that it is absolutely the easiest design for a typical amateur to duplicate and get operational.

A tube center-to-center spacing of 2.75 in. was selected, as it is the closest possible spacing that could be used with the 2.50-in.-diameter 3CX800A7s, and still leave room for the plate-line-to-tube-contact collets. The closest possible tube spacing is desired, as it will move push-pull responses higher in frequency, and far enough away as to not cause multimoding problems. Swept-frequency-response measurements were made on the plate line. Resistive loading was used in place of a resonant input circuit, to assure that only plate-circuit resonances would affect the results. With the plate circuit tuned to 432 MHz, a significant response was seen at 657 MHz. At first it was assumed that this was a push-pull mode resonance. Some further testing revealed a similar response in the single-tube 3CX800A7 amplifier. The two-tube amplifier exhibited the 657-MHz resonance with the plate line removed, and whether the two tubes' anodes were shorted together or separated. I finally decided that the 657-MHz response was a self resonance in the 3CX800A7. In the neutralization section I describe how I handled this resonance. Another resonance, dependent upon the plate line was found at 878 MHz. This is possibly the 1/4-wave push pull resonance. Fortunately the close tube spacing kept this response to 2 times the operating frequency. After the amplifier was neutralized no significant resonances including the 878-MHz mode, were observed (Naturally with this type of design harmonic resonances will be evident such as a third harmonic resonance observed in this amplifier).

The close tube spacing also lets the plate line be as narrow as possible, so its impedance will be as high as possible. To simplify construction, it was intended to use standard 3-in. high chassis thus mandating that the plate line be as narrow as possible. The only other solution to the plate-line impedance problem would be to use a taller RF enclosure. This would, however, require custom metal work and defeat the simple-to-construct objective. With the chosen 2.75-in. tube spacing, 6 in. became the minimum width of the plate line. The characteristic impedance of the line is 37 Ω. With the selected 8 15/16 in. length and tuning capacitor size, the plate circuit will resonate from about 350 MHz to 445 MHz. This line length was selected to have the amplifier operate optimally at 432 MHz. If you desire to use the amplifier across the entire amateur band (420 to 450 MHz), shorten the plate line to 8 13/16 in. The exact construction dimensions for the plate stripline (W1) are given in Fig 4. Note that the corners of the line are rounded, as are the edges, to minimize the possibility of the plate line arcing to ground or the tuning capacitors.

The hardest part of constructing the amplifier will be cutting the two 2.660-in. diameter holes in the plate line. These size holes are required for the finger stock used to contact the tubes to the plate lines. This leaves only 0.090 in. of material in the plate line between the two holes. If you are going to use a fly cutter to make the holes, be sure to cut the holes very slowly to avoid distorting or ripping the material in the center

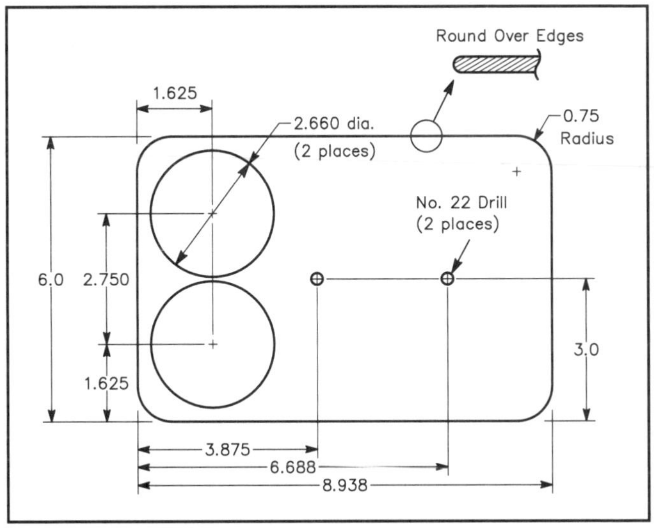

Fig 4—Plate stripline for the parallel 3CX800A7 amplifier is made of 1/16-in. copper or brass. Nickel-less (dull-finish) silver plating is recommended after the plate collets are soldered in place.

or at the edges. If a small machine shop with either a CNC vertical mill or simply a rotary table on a standard vertical mill are available, you could contract them to cut the holes. Cutting the holes with a small diameter end mill (about 1/8 in.) will cut clean holes with square edges and a minimum of material distortion. The plate-line material used was 1/16-in. thick brass sheet 6-in. wide by 8 15/16 in. long, which was silver plated. If you don't plan on silver plating the plate line, you may want to use copper instead, although copper is a harder material to machine. Any material thickness between 1/32 and 1/8 in. can be used, as long as the Teflon spacer insulators are adjusted in length.

Since the tube spacing of 2.75 in. is also the same spacing used on the original K2RIW 2 × 4CX250B amplifier, it allows owners of K2RIW-style amplifiers to convert their units to 3CX800A7s. The K2RIW amplifier used different spacing of the tube centers to the RF enclosure sidewall (2 1/8 in.) which could complicate a conversion. These different dimensions may require that the loading capacitor be retained to the side of the plate line as in the original K2RIW amplifier. Also note that many builders of the K2RIW amplifier who used SK-620 or SK-630 sockets (including most of the ARCOS amplifiers) used 2.50-in. tube spacings which would also hinder such a conversion.

The finger stock used to contact the tube anodes to the plate stripline was selected because I found it at a hamfest flea market. It is approximately 1/4 in. high and is rolled over such that it becomes 0.100 in. thick. Looking in the Instrument Specialties catalog the finger stock looks similar to their part number 97-360. Instrument Specialties 97-380 also would be suitable. It is similar to 97-380, but is only 3/16-in. high. If finger stock cannot be located, EIMAC 720829 plate collet for the 3CX800A7 can be used. I prefer the selected finger stock as it is a bit more flexible than the EIMAC collet which makes aligning the socket positions with the holes in the plate line a bit less critical. The EIMAC collet looks as if it is made from Instrument Specialties 97-251 finger stock which also could

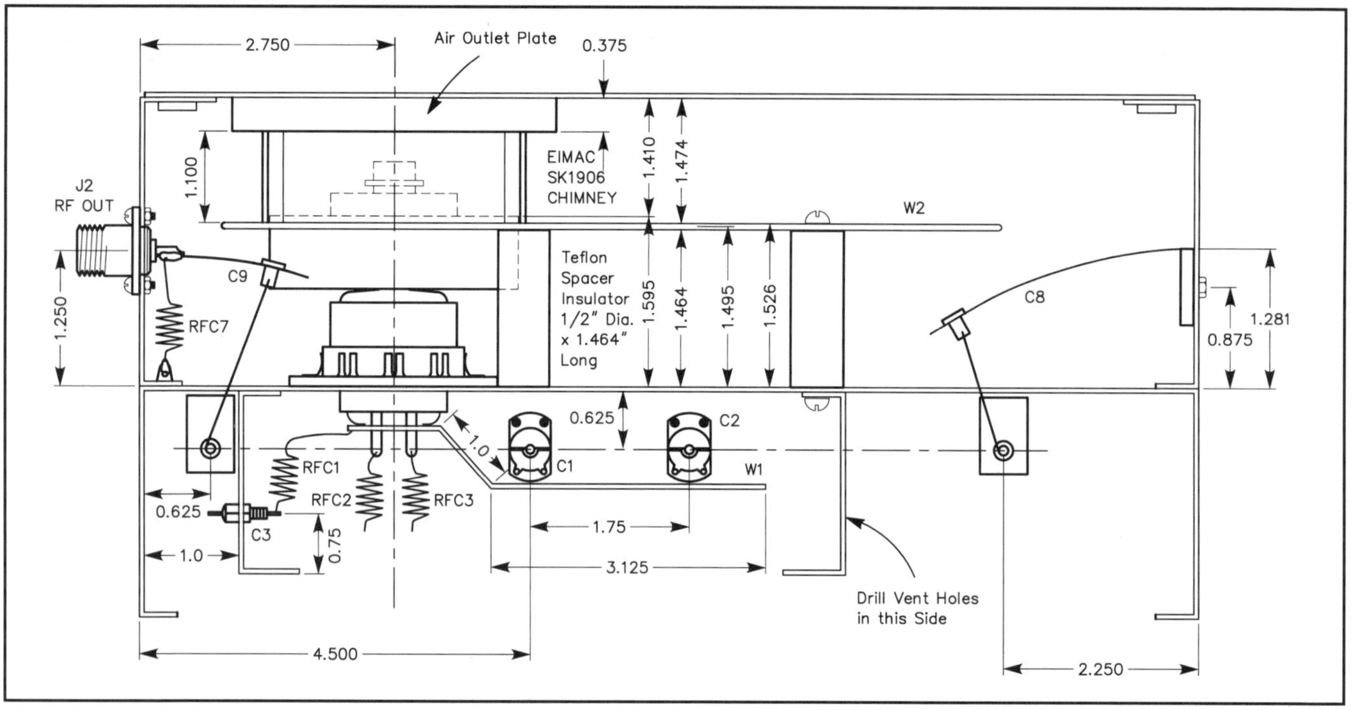

Fig 5—Rear view of the plate and cathode compartments. This view shows alternate mounting of the RF output jack (J9) on the side of the RF output enclosure.

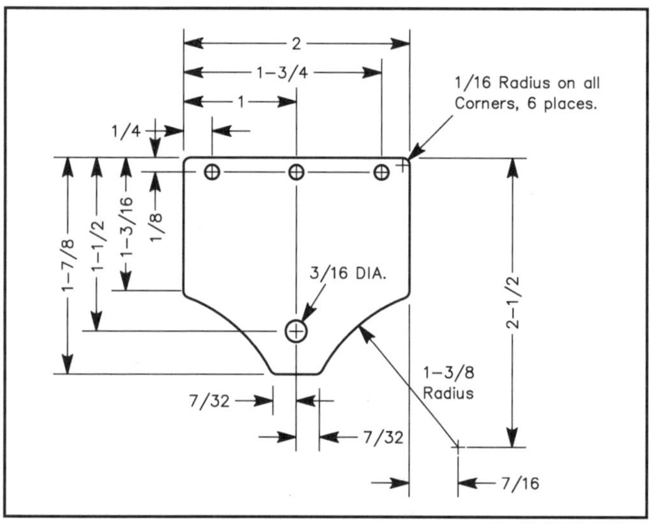

Fig 7—Plate loading capacitor C9 is made of 0.010-0.015 in. hard brass shim stock or beryllium copper. Have the capacitor silver plated to 0.0001-in. minimum thickness. Specify nickel-less plating.

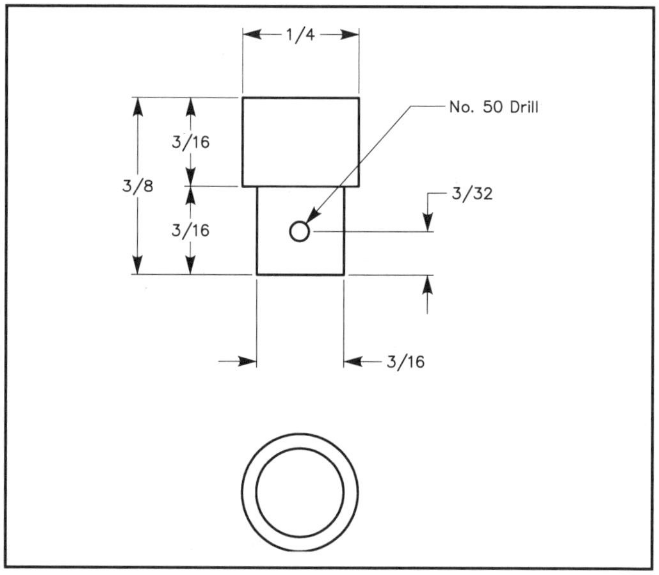

Fig 8—Insulators for C8 and C9 are made from 1/4-in. diam. virgin (electrical grade) Teflon rod.

be used. Most any reasonably sized finger stock can be used. Variations in the inductance of the different finger stock can be compensated for in the positions of the tuning and loading capacitors.

One difference in the 2 × 3CX800A7 amplifier from the original K2RIW design was to locate the output loading capacitor flapper on the tube end of the plate line between the two tubes. Theoretically, this arrangement will result in more balanced power levels delivered from the two tubes. I feel that this arrangement is more important on this amplifier than the original K2RIW amplifier due to the wider plate line. Locating the loading capacitor away from the plate-tuning capacitor also minimizes interaction between the controls. Another benefit of this loading arrangement is that it will help to more evenly distribute the RF current on the tubes and stripline. The amplifier originally was built with the output Type-N connector mounted on the sidewall of the chassis. This arrangement is shown in Fig 5. After initial testing, a concession to esthetics was made and the output connector (J2) was moved to the rear of the amplifier. This necessitated the use of a short coax jumper from the output loading capacitor (C9) to the connector. RG-225, which is Teflon dielectric 50-Ω cable was used

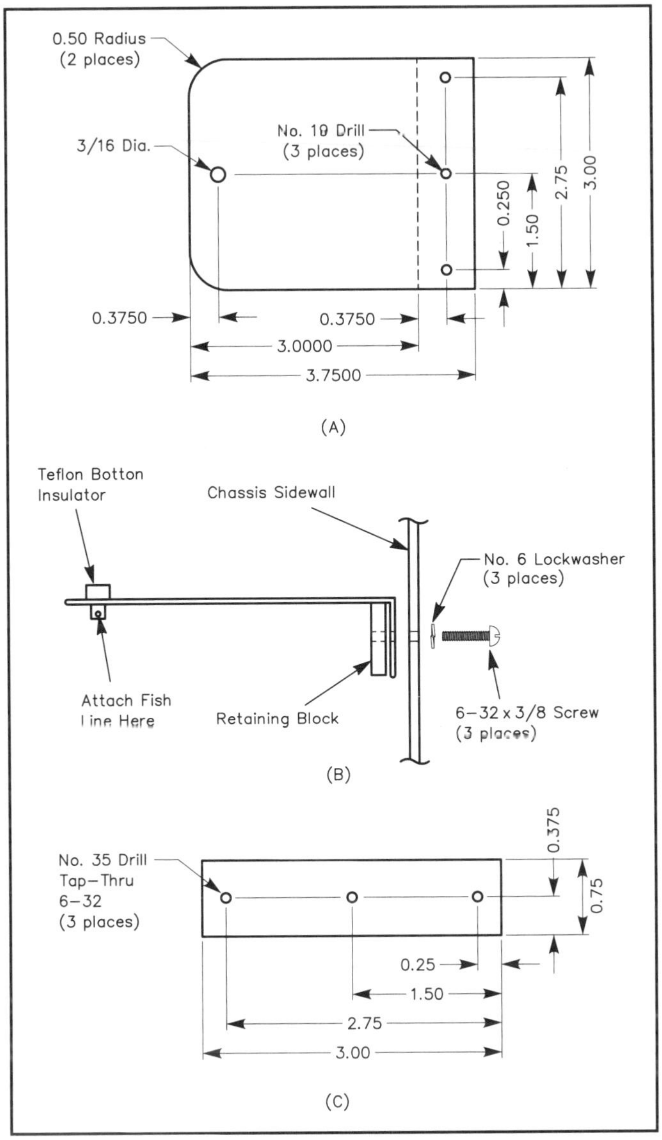

Fig 6—At A, plate tuning capacitor C8 is made of 0.010-in. hard brass or beryllium copper. Silver plating is optional. B shows the capacitor retaining blocks made of 1/8- × 3/4-in. aluminum bar. C shows how the capacitor and retaining blocks are mounted.

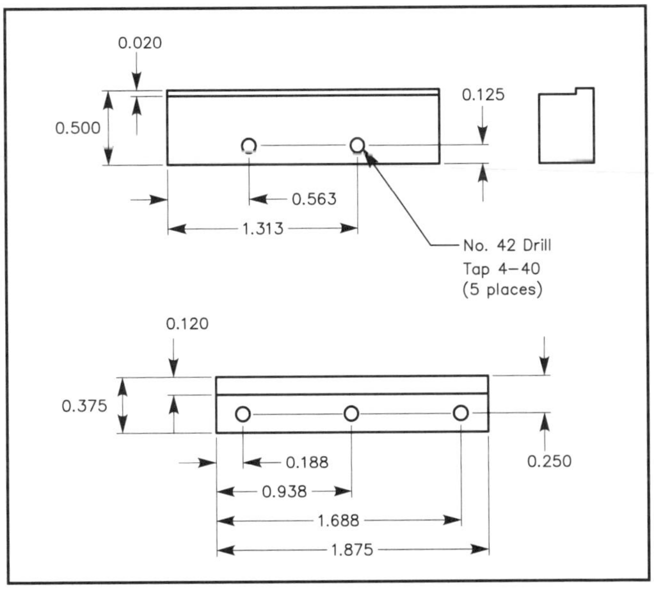

Fig 9—C10 mounting insulator is made from 1/2-in. thick virgin (electrical grade) Teflon rod.

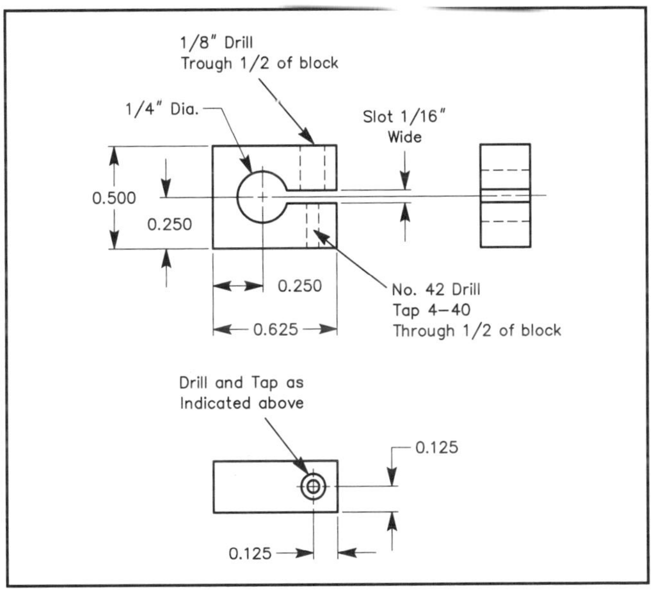

Fig 10—Adjustable stops for the plate tuning and loading controls are made from 1/4-in. aluminum.

for the jumper. RG-225 is similar to RG-214 but with Teflon dielectric and a Teflon jacket. RG-213 or similar polyethylene cables (including foam types and 9913 types) are not usable as the 1500 W from the amplifier will quickly melt such cables. If you cannot locate any high power cable for this jumper it is suggested that the output connector be mounted on the side and the jumper cable be eliminated. The flapper tuning and loading capacitors (C8 and C9) are made from 0.010-in. thick beryllium copper. Hard brass shim stock will work if beryllium copper cannot be located. C8 is shown in Fig 6 and the layout of C9 is covered in Fig 7. Be sure that the edges of the flappers are sanded smooth with fine emery cloth after they are cut to size. This will minimize the possibility of arcs from the plate line to the flappers. The 3/16-in. holes in the flappers are to accommodate Teflon button insulators (Fig 8) which serve to both prevent the flappers from contacting the plate line and they provide a convenient way to attach the fish line to the flappers. C8 is held securely to the chassis by means of a tapped brass or aluminum retaining plate (Fig 6). The loading capacitor is held in place by means of a Teflon insulator which was made from some rectangular Teflon bar stock (Fig 9). The small relief was cut to offer extra protection from arcing to the chassis sidewall. It is not necessary to cut the relief in the Teflon block. The tuning shafts were made from brass as they give a smoother feel when they are turned than aluminum shafts will. The plate tuning capacitor control uses a 6:1 vernier ball drive to impart a slow, smooth feel to the control. A vernier drive was not used on the loading control as its feel is good without any rate reduction. The end of the loading control shaft where the fish line wraps around was turned down to

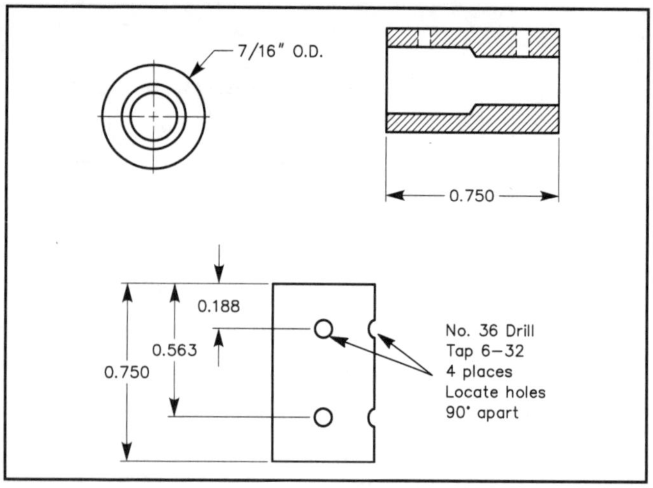

Fig 11—Coupling to adapt 3/16- to 1/4-in. shaft is made from 7/16-in. diam. aluminum rod.

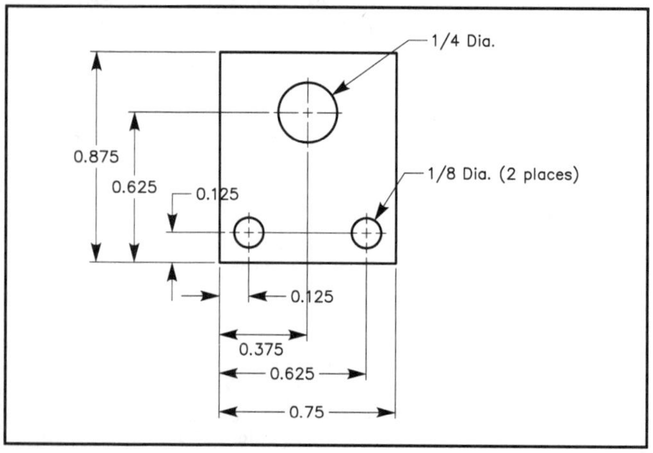

Fig 12—Input circuit capacitor C1-C2 mounting insulators are made from 1/16-in. thick G-10 circuit board material, from which the copper has been removed.

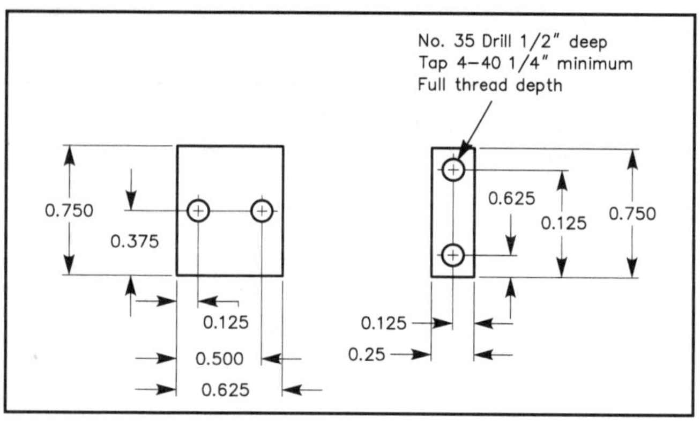

Fig 13—Mounting brackets for the insulators shown in Fig 12 are made from 1/4-in. aluminum or brass bar (aluminum is preferred).

3⁄16 in. to give a slightly slower tuning rate and to limit the loading capacitor adjustment range. Adjustable shaft stops (Fig 10) were made from brass bar stock. They slip over the tuning and loading control shafts and allow the tuning limits to be set without worrying about how long the dial cord is. Screws mounted on the chassis provide the stop points for the clamps. These stops both control the amplifier tuning and loading range by limiting the tuning shaft rotation to just under a full turn. The stops also prevent accidental breakage of the fish line by keeping the controls from turning too far. The panel bearings were intentionally misaligned to keep the tuning controls in place against the force of the flapper capacitors. Shaft locks as are used on potentiometers can also be used to prevent unwanted tuning shaft slippage.

Input Circuit

The tube spacing is actually more critical in obtaining proper operation of the cathode circuit. This is due to the much higher input versus output capacitance that the 3CX800A7s have. The 2.75-in. tube center-to-center spacing also allows use of a standard 5- × 7- × 2-in. chassis for the input circuit compartment, again simplifying construction. The cathode circuit layout was intentionally made as similar as possible to the original K2RIW amplifier, so the owner of a 4CX250 amplifier could convert his amplifier to use 3CX800A7s. To facilitate such a conversion, the position of the input tuning and coupling capacitors to the tube center line is the same as was used in the K2RIW amplifier.

The actual cathode stripline is different in size and shape from the K2RIW grid circuit due to both the different configuration of the 3CX800A7 sockets and due to the higher input capacitance of the tubes. The resultant input circuit works amazingly well given its simplicity and compromise to past mechanical layout decisions. It will tune from approximately 370 MHz to 510 MHz. When tuned to 432 MHz there are no significant resonances, besides the desired 432 MHz resonance and third harmonic responses in the 1050- to 1200-MHz range.

The cathode circuit is tuned by miniature butterfly air variable capacitors. The sections are connected in series to lower the capacitance, raise the voltage breakdown and to eliminate the effects of the sliding metal contact between their shafts and mounting bushing. If the butterfly capacitors cannot be located standard air variables could be used in place of the butterfly capacitors. The miniature capacitors have 3⁄16-in. shafts which require the fabrication of 3⁄16- to 1⁄4-in. diameter shaft couplings. The couplings are detailed in Fig 11. Garolite G-10 grade insulated shafts are used from the couplings to the front panel knobs. Garolite G-10 is very similar to standard G-10 epoxy printed-circuit substrate. If Garolite G-10 rod cannot be located, commonly available linen bakelite or Delrin rod can be used. The input capacitors are mounted on a small piece of 1⁄16-in. thick G-10 epoxy PC board (the copper foil is removed). These mounting plates are dimensioned in Fig 12. These epoxy plates are held to the chassis via brackets made from brass or aluminum bar stock (Fig 13). A full layout drawing of the cathode compartment and cathode stripline is given in Fig 14. A side view of the input stripline and tuning capacitors is also shown in the amplifier section view (Fig 5).

Socket Mounting

The sockets are mounted in identical fashion to the

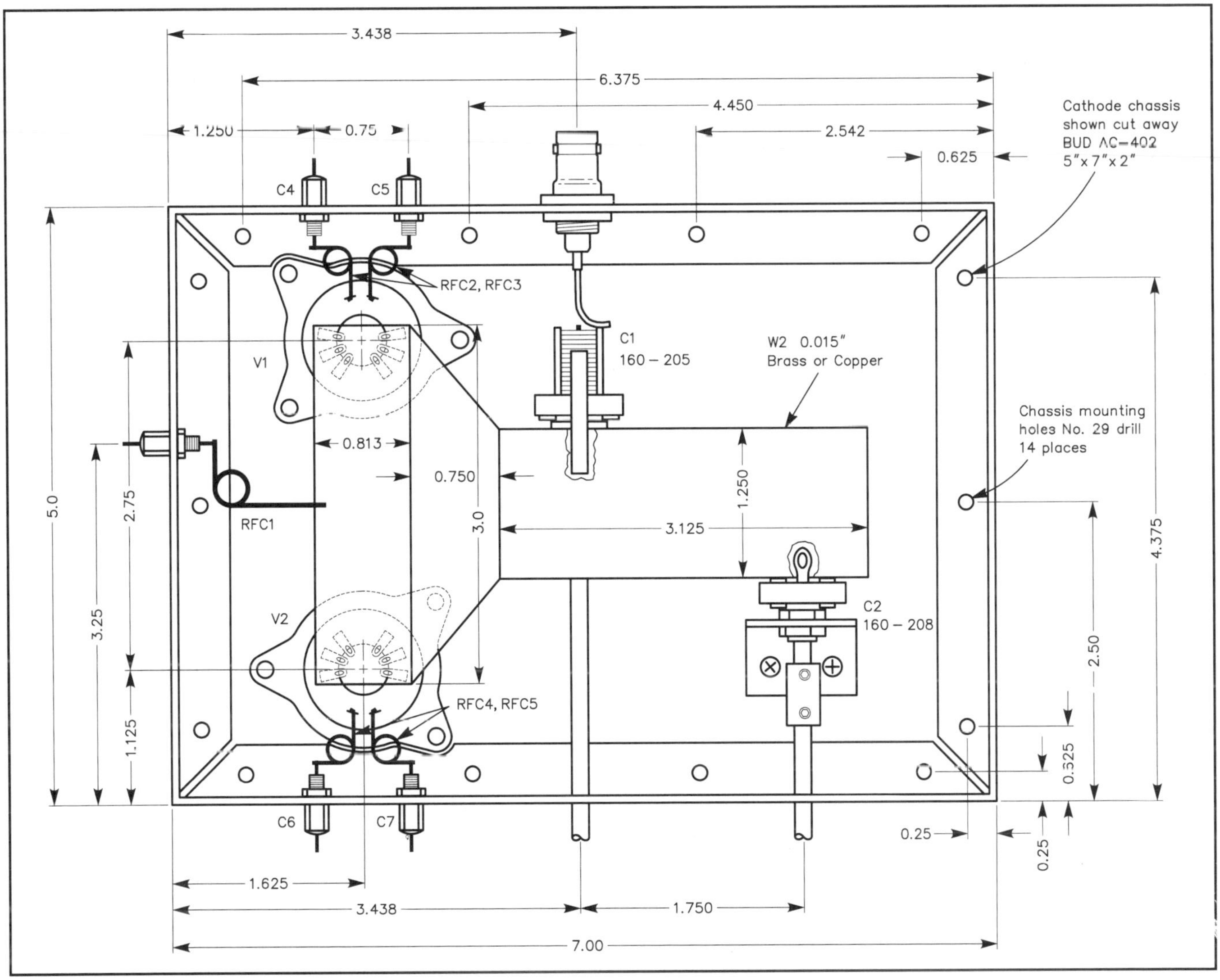

Fig 14—Cathode circuit layout. Enclosure is a 5 × 7 × 2-in. aluminum chassis (Bud AC-402 or equiv.) cut away as shown.

method described for the single tube 3CX800A7 amplifier, so that section should be carefully read. Note that the sockets are mounted mirror imaged that is one socket is turned 180° with respect to the other. This was done to place the cathode pins as close together as possible. This minimizes the width of the cathode line and again the possibility of undesirable push-pull modes present. Some other builders of parallel 3CX800A7 amplifiers have used considerably wider spacings between the tubes. While their amplifiers have worked acceptably, the required drive power was substantially higher than expected leading to the speculation that the wider tubes spacings were causing some form of a push-pull mode response.

Varian part no. 720359 grid-collet assemblies were used for grid grounding. These assemblies use a no. 882931 collet, which is soldered to a 1⁄16-in. brass mounting ring that also has three 4-40 studs for socket mounting. Unfortunately, since the parts for this amplifier were obtained, the price of the 720359 assembly has become quite high. Those interested in saving time may decide that the collet assemblies are worth the expense. Other builders with fewer funds but more time can either obtain the 882931 grid collets and attach them to their own brass mounting plate or they can use finger stock for grounding the grids. The same finger stock which was used on the plate line could also be used for the grid collets. In either case, the simplest construction method may be to use a single 1⁄16-in. thick brass plate that is approximately 5 ½ × 3 in. An alternate grid collet assembly is described in Fig 15.

Air Inlets and Outlets

The air inlets and outlets used on the 2-tube amplifier are different than the other amplifiers. Since a fairly thick plate (3⁄8 in.) was needed to space the EIMAC SK1906 chimneys down from the top cover, the idea of using this thick metal plate as a waveguide-beyond-cutoff RF filter occurred. The attenuation of a signal that leaks through a hole in an RF shield is proportional to the depth of the hole. For holes substantially smaller than a wavelength (less than $\lambda/10$) they will have adequate attenuation if the depth of the hole is twice its diameter. When multiple holes are needed, each individual hole should have sufficient attenuation that the total leakage from all holes in the RF compartment maintains leakage below an acceptable level.

The 3⁄8-in.-thick air chimney spacer plus the 1⁄16-in.-thick cover plate gives a rather substantial depth of 7⁄16 inch. A per-

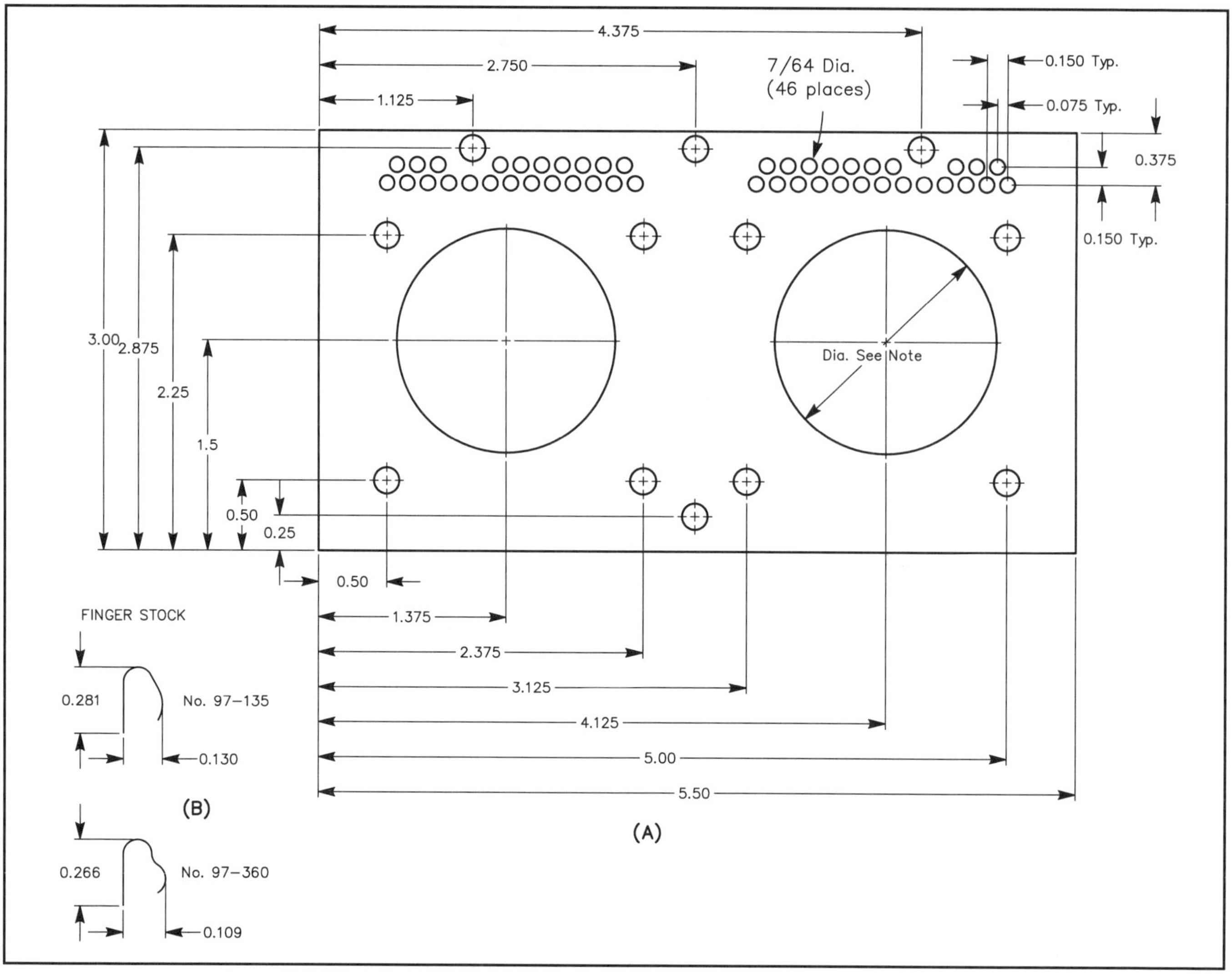

Fig 15—Alternate grid-collet mounting plate for use with finger stock (A); dimensions of the two usable Instrument Specialties finger stocks. The mounting plate is made from 0.062-in. brass or copper, silver plated if desired. The diameter of the large holes depends on the grid-contact method used. For type 97-135 stock, the diameter is 1.600 in.; for 97-360 stock, 1.540 in.; for EIMAC 882931 collet, 1.500 in.

forated sheet of steel with 1/8-in. diameter holes spaced on 3/16-in. centers was available to use as a drilling template, to make a symmetrical pattern of exhaust air holes. After the initial 1/8-in. holes were drilled the plate was bolted to the top cover. The holes were then drilled out with a no. 20 drill (0.1660 dia.) in order to have the holes in the plate and top cover match perfectly. These fancy looking air inlets and outlets are actually easy for the home builder to duplicate. The 3/8-in. plate can be obtained from small volume specialty metal distributors (such as the Dillsburg Aeroplane Works).[1] They can be cut with a hacksaw and filed to shape. Although it will be time consuming to drill the large number of small holes in these air outlets, these parts are easier to make at home (and much safer!) than attempting to cut a 2½-in. diameter hole in a 3/8-in. plate with a fly cutter, which you'd have to do to use window screening. The air-inlet plate is described in Fig 16 and the outlet is covered in Fig 17.

The 0.1660-in. diameter holes would act as an effective waveguide at frequencies around 50 GHz. At frequencies around half the waveguide wavelength (25 GHz) an individual hole 0.166 in. diameter and 0.4375-in. deep has about 84-dB attenuation. At 432 MHz (less than 1/100 the waveguide frequency) the estimated attenuation for all 302 holes in combination is over 90 dB! This means that when the amplifier is operated at the 1500-W level, approximately 1.5 μW will leak through the air outlets. There will be about as much RF power leaking through the cracks at the corners of the chassis joints as through the air outlets. To test the effectiveness of the shielding, I probed the amplifier with a microwave power meter having 0.3-μW sensitivity (HP 435A/8481A). Using a short probe connected to the power meter's sensor, less than 6 μW was detectable when the probe was placed up against the air outlets. About 5 μW was read when the probe was held next to the chassis joints. This is substantially better than the window screening used on the air outlet of the single 3CX800A7 amplifier. On the single-tube amplifier, an indication of about 75 μW was read with the probe against the window screening air outlet while it was run at the 750-W output power level. To put these readings in perspective, a commercial 100-W "brick" read over 200 μW near the 12-V power leads, and over

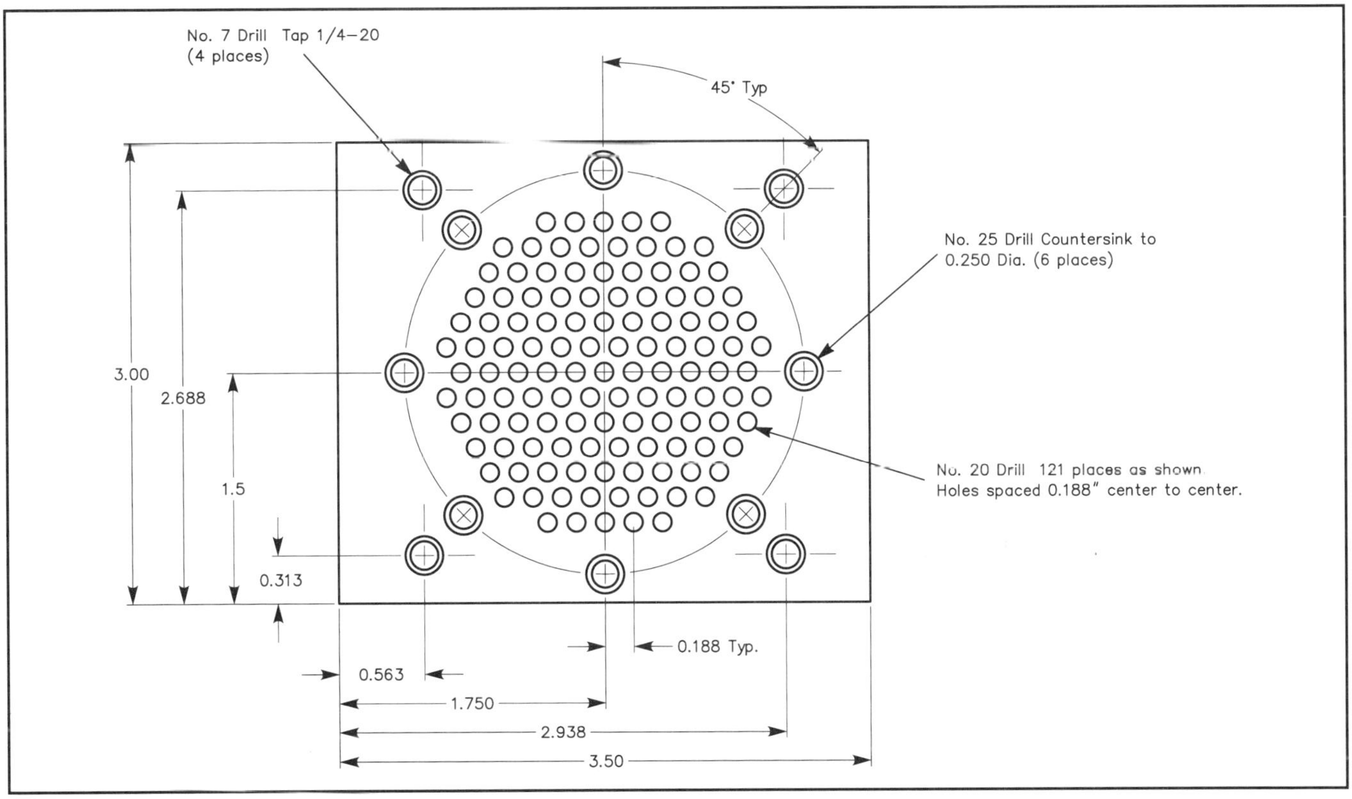

Fig 16 Air-inlet plate is made from 3/8-in. 6061-T6 aluminum plate. The hole pattern is obtained by using a piece of perforated sheet steel as a template. See text for details.

100 mW around its joints and indicator LED holes. After bypassing the power leads the brick leakage was reduced to about 50 μW on its power leads. Another reference point: When the probe was held up to the RG-142 (double-braided shielded cable) from the driver to amplifier, about 30 μW was measured. As a final note on RF field levels, these field strengths are several orders of magnitude lower (more than 1000 times lower) than what emanates from a 5-W hand held. In addition, in normal operation the amplifier will typically be located 2 to 3 feet away from the operator, which further reduces the strength of the RF fields near the operator. In other words, you can feel safe building the amplifier with either the window-screen air outlets or especially the drilled-plate air outlets. As a final note on RF leakage, this amplifier uses an Eric high voltage EMI feedthrough filter instead of a normal feedthrough or bypass capacitor, to bring the B+ into the plate compartment. In addition, the plate stripline was probed to confirm the electrical center point of the line, to minimize RF leakage through the B+ lead. RF leakage on the high-voltage cable was less than 10 μW, compared to the 100 μW or more measured on other 432-MHz amplifiers, which used conventional HV-bypass capacitors.

Mechanical Construction

The overriding consideration in all design decisions for the amplifier was to make it as easy to duplicate as possible. This mandated the use of standard commercial aluminum chassis. An 8- × 12- × 3-in. chassis was used for the plate compartment, and a 5- × 7- × 2-in. chassis was used for the cathode compartment. An 8- × 12- × 2½ chassis forms a base for the amplifier to be built on. The amplifier was built primarily for portable use on EME DXpeditions. This determined the table-top configuration, rather than a rack-mount assembly. The "cabinet" is formed simply by using oversize top and bottom covers (10¾ × 12 × 0.062-in. thick). Side covers are made from 5½ × 10¾ in. plates, also 0.062-in. thick. The front panel is 5⅝ × 12⅛ in. aluminum, again 0.062-in. thick. The front panel is held to the cabinet by ½-in. aluminum angle stock which was purchased in a local hardware store. The 0.062-in. aluminum sheet can be purchased cut to size from Chassis Kit.[2] Although the plate enclosure is the same size as the original K2RIW parallel tetrode amplifier, this amplifier is constructed in mirror image to it. I did this primarily because of the configuration of the Dayton 4C443 blower I used. This layout of the plate circuit is shown in Fig 18. Drilling and punching drawings for most of the metalwork are given in Figs 19 to 22. These drawings allow you to easily duplicate the amplifier without wasting time laying out the components on the chassis.

Blind fasteners (PEM nuts or similar) secure the top and bottom covers securely to the chassis, without danger of stripping the threads. Fasteners threaded 6-32 were used in most places. Be careful not to distort the chassis when installing the fasteners. If the chassis lip is bent or mushroomed, RF leakage can occur between the cover-to-chassis joints. As discussed in the air outlet section, the symmetrical vent hole patterns were easily made by using perforated steel as a template. Note that there are vent holes in the side of the cathode compartment and in the amplifier cabinet bottom cover. These holes are very important to allow cooling air to circulate by the socket bases,

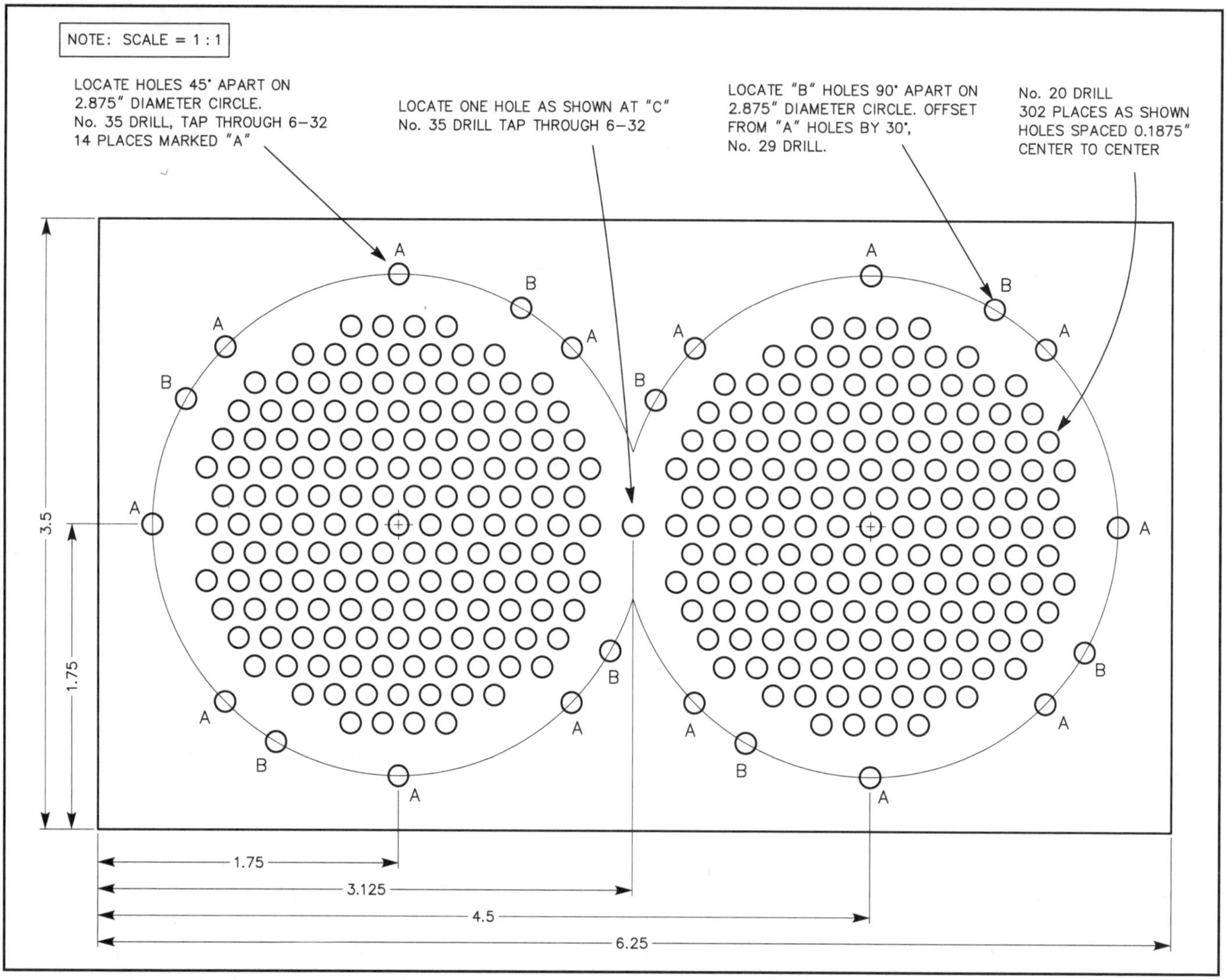

Fig 17—Air-outlet plate is made from 3/8-in. 6061-T6 aluminum plate. See the text and the caption for Fig 16 for details on drilling the hole pattern.

to keep the tube base seals cool. The amplifier must be set off the operating table by feet at least ½ in. to allow this cathode cooling air to escape. If the amplifier must be directly mounted to anything, vent holes should be drilled in the left side panel instead of the bottom.

The amplifier could also be mounted easily on a 7-in. rack panel. The RF deck could be mounted directly to the panel eliminating the need for the side brackets used in the desktop model. Top and bottom covers would be made 8 × 12 in. The meters and filament transformer would then be mounted on the side of the RF deck.

Neutralization

The 2-× 3 CX800A7 432-MHz amplifier required neutralization (as does virtually any UHF tube amplifier) for proper operation. The amplifier was swept from 1 to 1350 MHz to look for any undesired responses. As mentioned previously, an apparent self resonant tube response was evident at 657 MHz. Without neutralization this response was only 3 dB down from the desired parallel 432-MHz tuning. Another resonance in the plate circuit was indicated at 878 MHz which was about 20 dB down from the primary resonance. Additional responses in the 1050 to 1160 MHz range that were 25 to 30 dB down were also observed. Most likely these are third harmonic resonances that are moved lower in frequency by stray capacitance and inductance effects. As with the single tube 3CX800A7 432-MHz amplifier the 2-tube amplifier without neutralization exhibited signs of thermal instability and its maximum power output occurred near a plate current peak not at a dip. Reverse isolation was measured at 17 dB which is only 3.5 dB higher than the gain of the amplifier.

The amplifier was then neutralized by adjusting the grid inductance. Best operation was obtained with the maximum reverse isolation around 420 MHz (38 dB) and 32-dB isolation at 432 MHz. This grid collet fingers were broken off until proper isolation was obtained. A somewhat different pattern was used in removing the grid collet fingers compared to what was done on the single tube amplifier. The proper grid collet finger pattern is shown in Fig 23. Under this condition, maximum power output coincides with minimum plate current and maximum grid current. Power drift was also minimized when

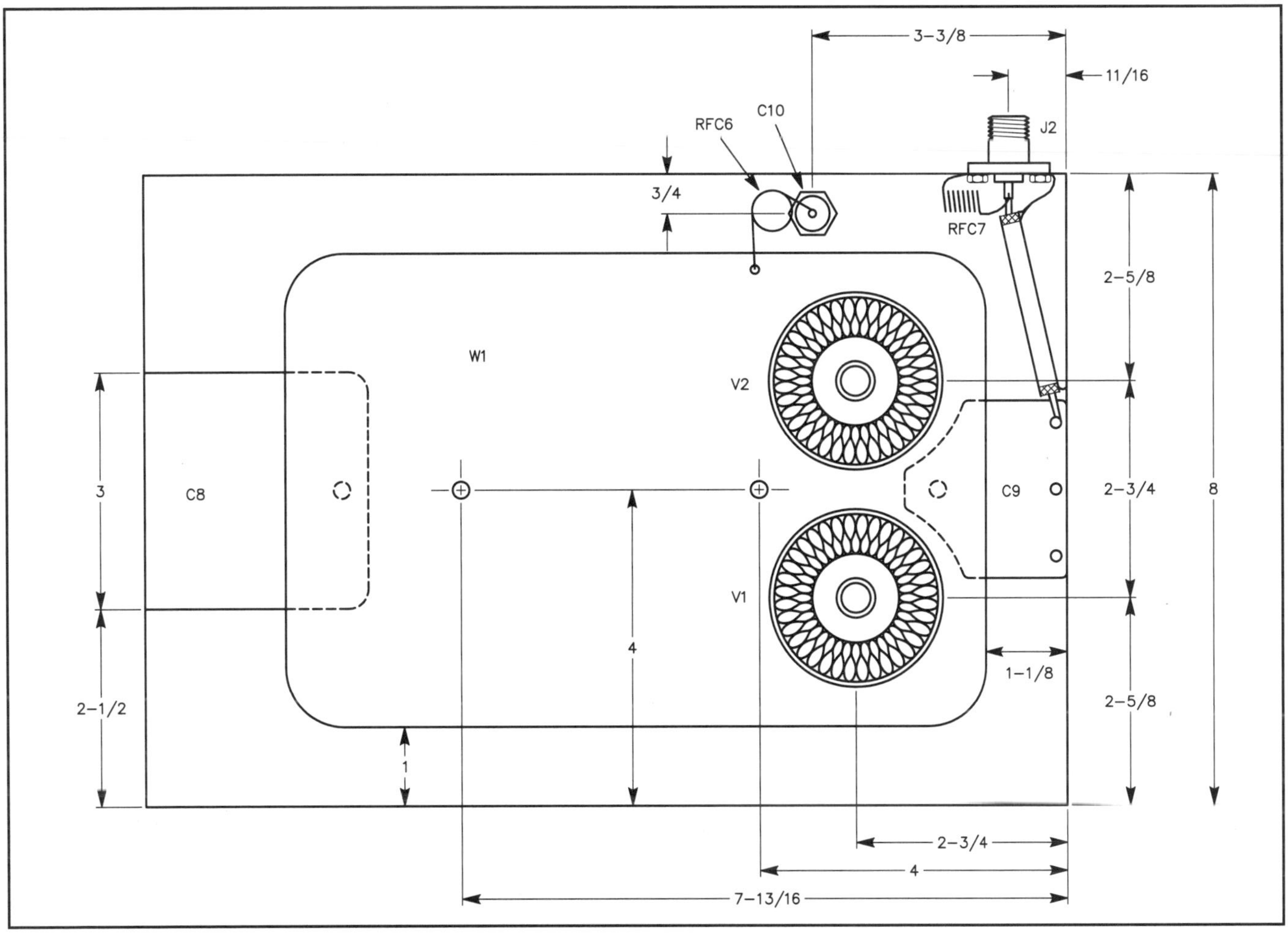

Fig 18—Plate circuit layout. See text for details.

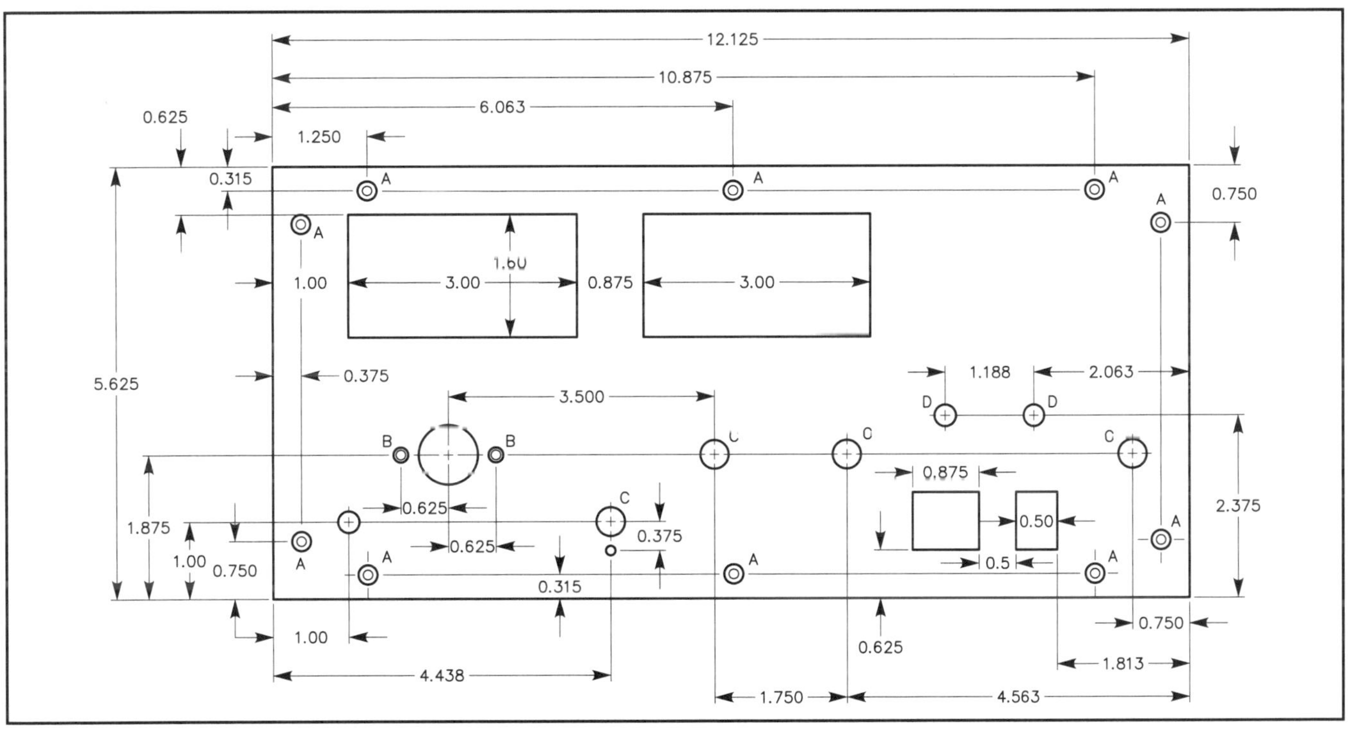

Fig 19—Front-panel layout. See text for details. Hole data: A—0.156 diam, countersink 0.280, 100°; B—0.125 diam, countersink 0.220, 100°; C—0.375 diam; D—0.3125 diam.

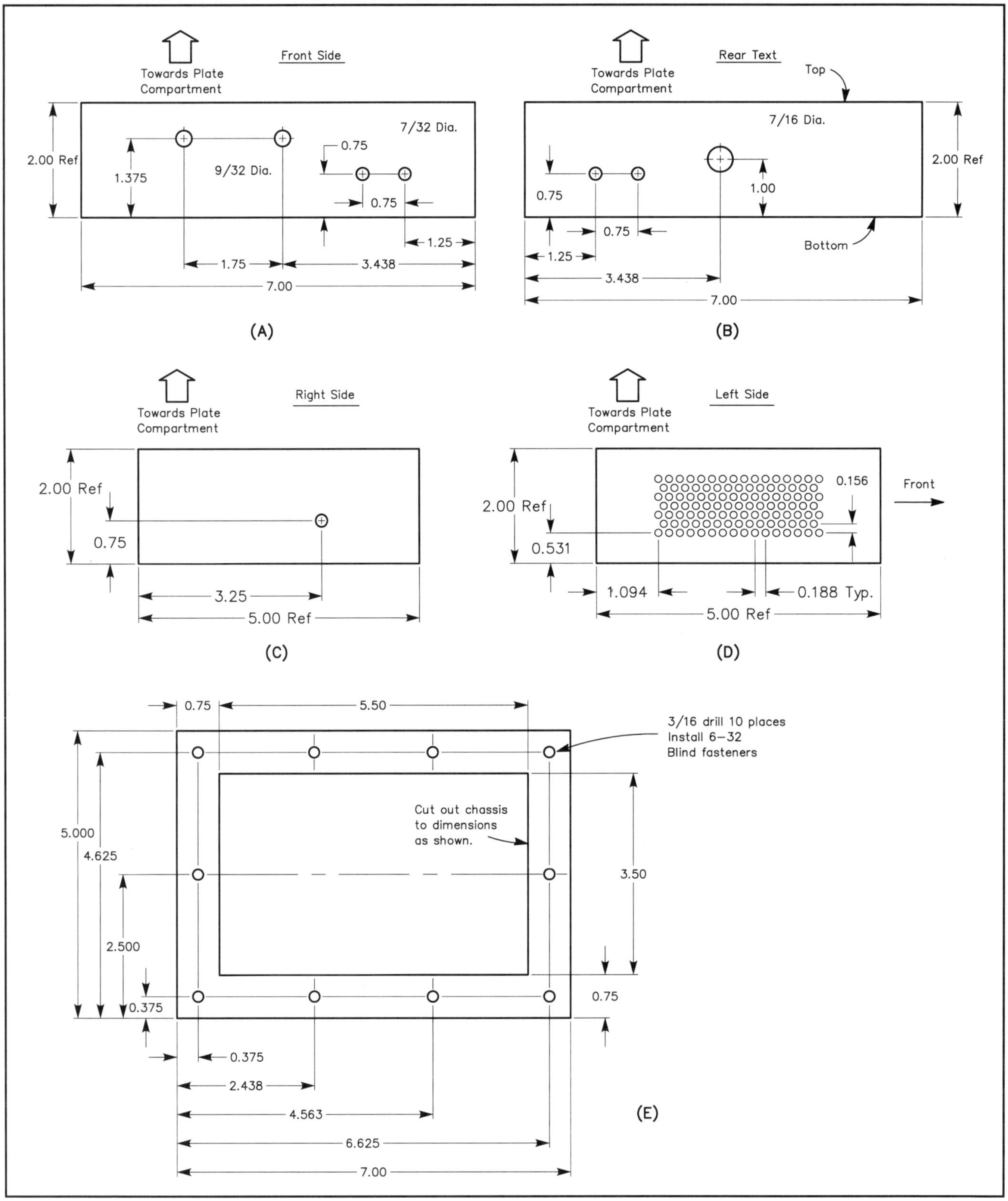

Fig 20—Cathode compartment layout. At A, the front side; at B, the rear side. At C, the left side; at D the right side. See the text and the caption for Fig 16 for details on drilling the hole pattern.

the amplifier was neutralized. A swept response of the amplifier was then measured again. The 657-MHz tube/socket self resonance was reduced by about 20 dB with the amplifier neutralized. The 878-MHz response virtually disappeared. The apparent third-harmonic responses were not affected in magnitude by neutralization. They are high enough in frequency, however, that it is unlikely the 3CX800A7 will exhibit any significant gain there. The magnitude of these responses is also more than 35 dB below the amplifier's forward gain at 432 MHz.

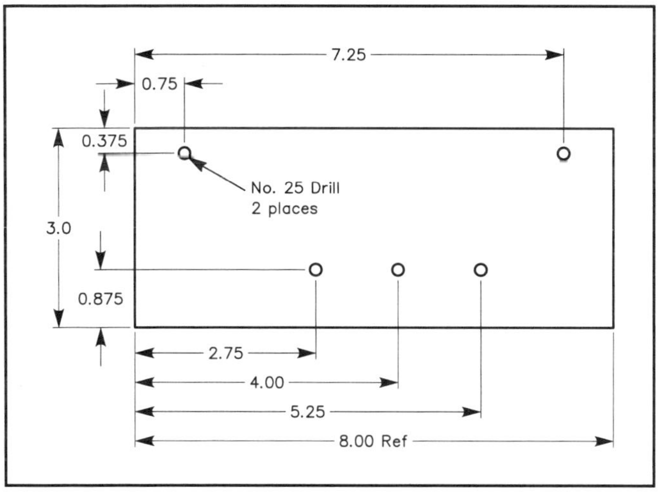

Fig 21—RF enclosure, left side drilling guide.

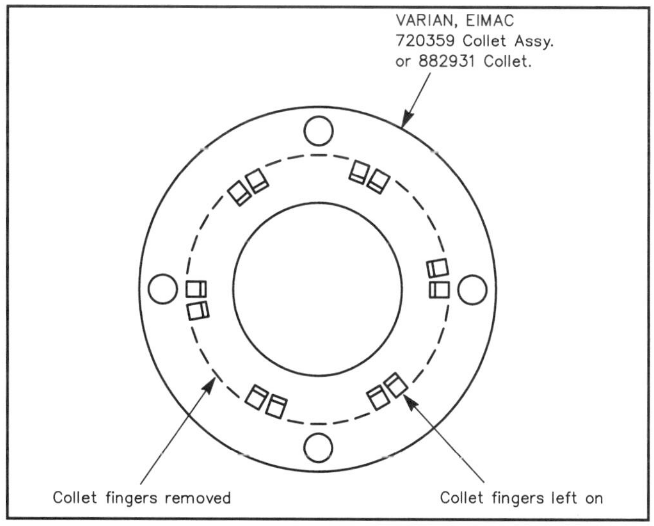

Fig 23—Grid collet modifications. Of every 6 fingers, break off four consecutive fingers as shown. This procedure is necessary to provide the proper grid-circuit capacitance to neutralize the amplifier.

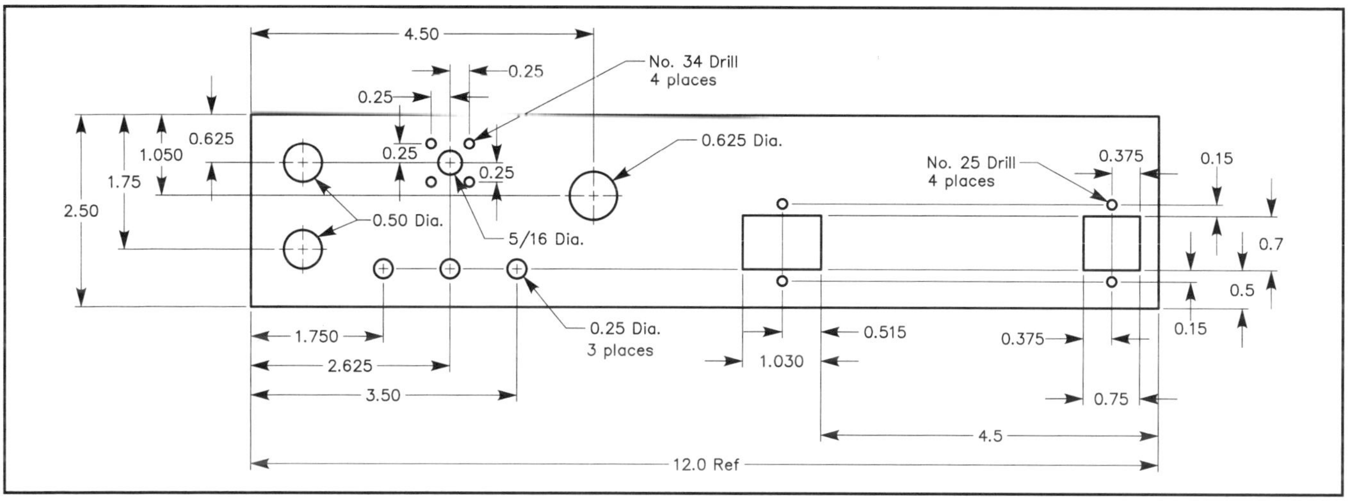

Fig 22—Lower chassis, rear, layout.

Protection Circuits

The 3CX800A7 is a rugged, reliable tube when properly operated. To obtain its high performance (low IMD products and high gain), the grid structure had to be made very fine. This fine grid structure has relatively low dissipation. While not a problem in normal use, should B+ be lost or the antenna be disconnected, enough grid current can be drawn to destroy the tubes. To protect the tubes, both B+ sensing and grid over-current protection is included.

The B+ sensing circuit consists of the B+ meter string (R5-R10) and Q1. These devices sense the presence of B+. The exact pull-in voltage is set by R11 and R12. Q4 is used to switch the T/R relays in combination of a logic circuit (Q3 and Q5) which allows either a low current to ground or a low current +12V to transmit enable the amplifier. If high voltage is not present this circuit prevents the amplifier from switching to transmit mode. If high voltage is lost during transmit, the amplifier is switched to standby.

Q2 is used in conjunction with R4, R14 and R15 to sense grid current. R15 sets the exact amount of grid current at which Q2 will switch K3, the over-current relay. In normal operation should the operator mistune the amplifier, R3 may only chatter, thus limiting the overload condition. If a major failure occurred, such as a missing antenna connection or defective T/R relay,grid current will rise fast enough to lock in the over-current relay, which will drop the T/R switching circuit. To reset the grid current trip, the Operate/Standby-Reset switch (S2) is flipped to Standby-Reset for a few seconds (to allow the 24-V supply to discharge). Switching back to Operate places the amplifier back in the ready-to-operate mode. Although the cathode-bias resistor (R3) will protect the tubes, the most protection is available when an RF relay is used on

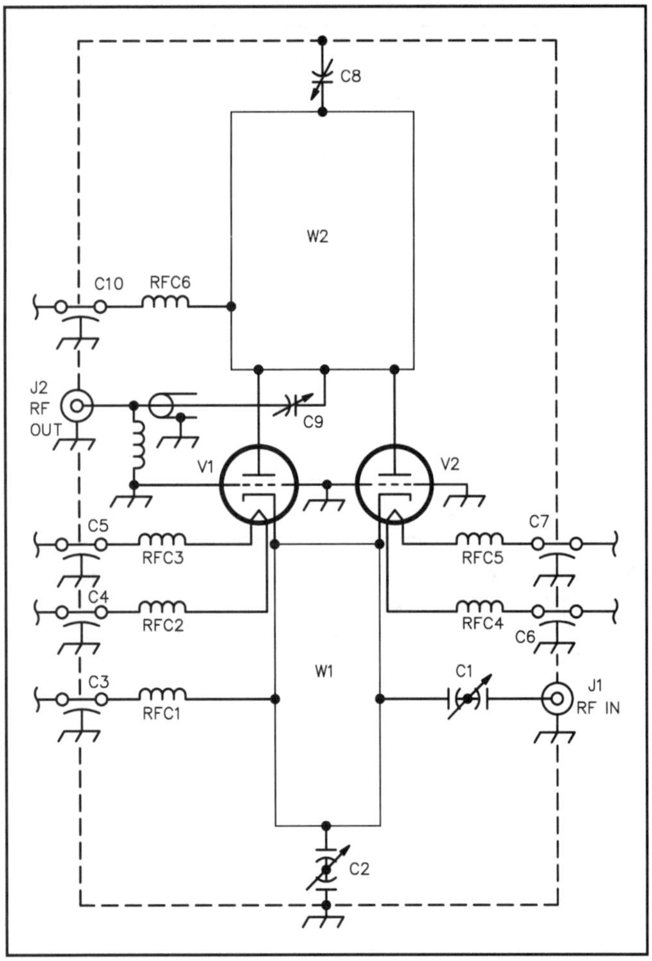

Fig 24—Schematic diagram of the amplifier.

the input circuit. This relay will drop drive power from the amplifier in case of a B+ failure or grid over-current trip.

A thermal time-delay relay (3 minutes) is used to assure the tube cathodes are up to full temperature before drive is applied. The time delay prevents the 24-V control circuitry power supply from turning on until the time delay is complete.

A 50-Ω 100-W resistor should be connected in series with the B+ supply. This resistor will serve to limit the current from the power supply in case of an arc. A fuse in the HV line is another means to protect the tubes from arcs. The HV supply should have as good regulation as possible. I recommend you use a bridge-rectifier circuit with a transformer that has winding resistance as low as possible. The use of a capacitive-input filter is also preferred. Do not use too much filter capacitance (50 µF maximum) as more capacitance will only increase the possibility of damage should a high-voltage arc occur.

Initial Adjustments

If all of the design dimensions and construction methods are closely followed the amplifier will tune up and operate without any "trimming in." For operation at 432 MHz, set the cathode tuning capacitor C2 so its plates are 50% meshed, and the input coupling capacitor C1 so that its plates are 60% meshed. The plate tuning capacitor is preset to have its fixed end 1 9/32 in. from the plate compartment bottom, and the moving end is 19/32 in. from the chassis. The plate loading capacitor should be located so its fixed end is 1 3/16 from the chassis bottom and its moving end is 1 1/16 in. from the chassis. With these settings the amplifier will be closely tuned for running 1500-W output at a plate voltage of 2250 V.

After verifying that all wiring has been correctly completed, prepare for the application of power by setting the filament voltage adjusting resistor (R17) to maximum resistance (minimum filament voltage). Ac power can then be applied to the amplifier, but do not connect the B+ at this time. Connect an accurate ac-RMS voltmeter to the filament power feedthrough capacitors, C4 and C5. Then adjust R17 for 13.0 V RMS. Set the meter switch, S3 to the FIL position, and adjust R20 for a reading of 130 on M2, which corresponds to 13.0 V. Be sure the filaments are fully warm (3 minutes minimum) before making final adjustments to R17 and R20. The filament voltage is set to the low side of the specified filament voltage (13.5 V ± 0.6 V). This is done to both maximize the filament life and to minimize any cathode back-heating effects. If any trouble is encountered in obtaining full plate current, try increasing the filament voltage.

Next, set the grid over-current circuit. Disconnect B+ from the tubes by removing RFC7 from the plate line. Turn on the B+ (needed to activate the HV-sensing circuit) and ground J6 to switch to transmit. Verify that the HV sensing circuit is working and that the T/R relays (K2 and any RF relays which are used) are switching. Apply a small amount of 432-MHz drive power. Since plate voltage has been removed from the amplifier, be extremely cautious, as the tubes will draw a large amount of grid current (retune C1 and C2 if necessary). Increase drive power until the tubes draw 120 mA of total grid current. Quickly adjust R15 for a grid over-current trip. Note that the relay may chatter and not lock in. If this is the case, adjust for consistent chatter at 120 mA. Now turn off the HV supply and, after verifying that the HV has bled down to zero, reconnect plate HV choke RFC7.

The amplifier is now ready for RF power-on testing. Connect power meters to the input and output lines. Turn power back on, including B+. After the filament is warm, the amplifier should key and draw about 90 mA of plate idling current (at around 2550 V. Apply increasing drive until 300 to 400 mA of plate current is drawn. Adjust the plate tuning and loading capacitors until you observe maximum output. Increase drive and retune the amplifier until full power out is observed (1500 W). Adjust the input tuning (C1 and C2) for minimum SWR under full-power conditions. With a 100-W Bird wattmeter element on the input side, you should be able to completely null out the reflected power. If you cannot, verify that your drive signal is free from spurious signals and check the input circuit for any assembly errors. As a final operational check, verify that the plate-current dip corresponds to maximum power output (within 50 W). Grid current should also roughly peak with the plate current dip, however, improper loading will affect this. If these conditions do not occur, check the grid collets for proper neutralization and check that your load or antenna is capable of handling the power and has a reasonable SWR (1.2:1 or better preferred).

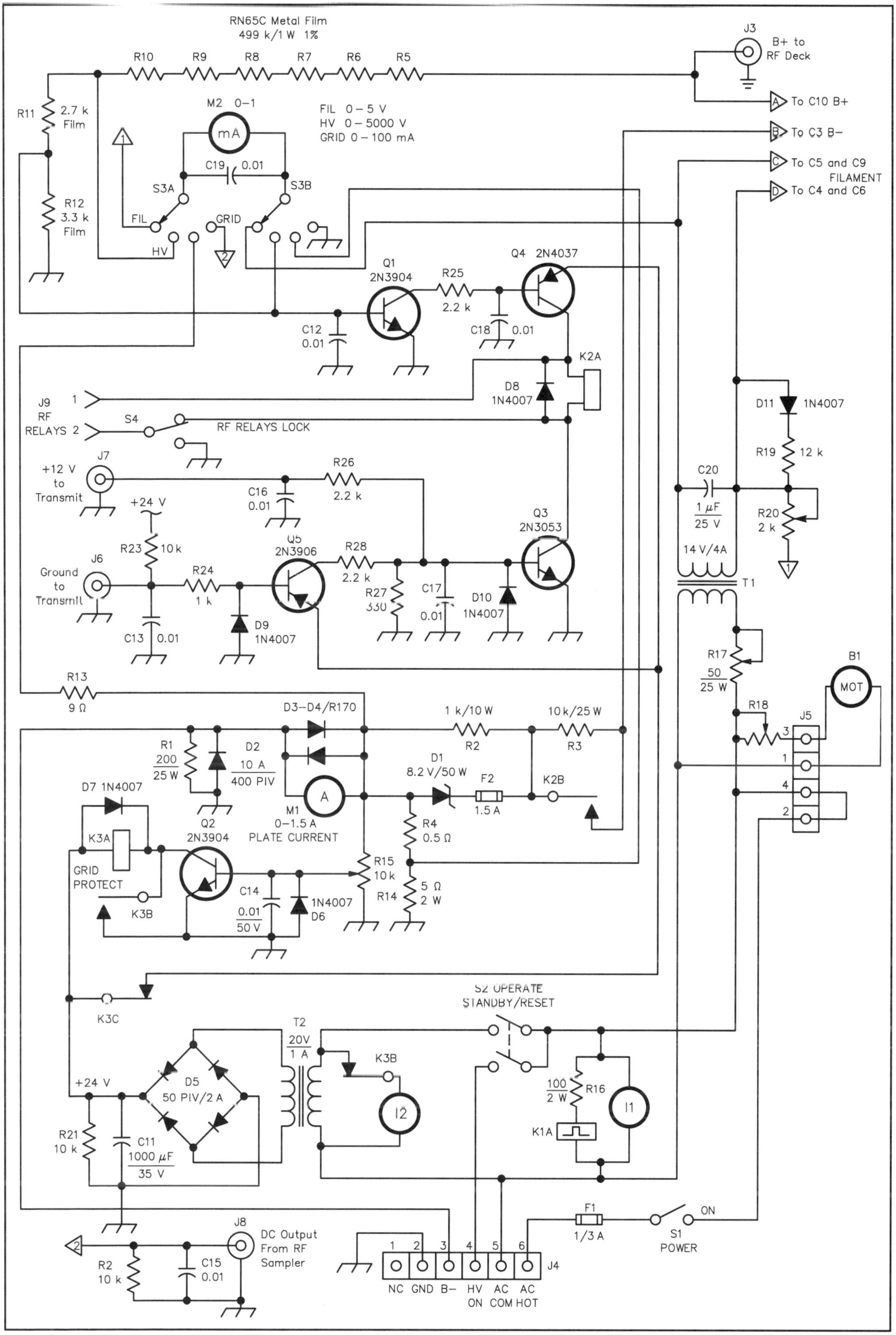

Fig 25—Schematic diagram of the amplifier control circuits.

Table 1
Full Power CW Operating Conditions

Parameter	*Test 1*	*Test 2*
Plate Voltage, Full Load	2230 V	2380 V
Plate Current	1290 mA	1200 mA
Input Power	2877 W	2856 W
Output Power	1500 W	1500 W
Apparent Efficiency	52.1 %	52.5 %
Drive Power	68 W	63 W
Power Gain	13.4 dB	13.8 dB
Grid Current	18 mA	10 mA
Input SWR	1.16:1	1.15:1
Filament Voltage	13.0 V	13.0 V
Idling Current	75 mA @ 2480 V	90 @ 2620 V
Bias Voltage	8.2 V	8.2 V

Notes: 1. Grid current may vary from tube to tube.
2. Plate voltage was measured from the cathode to B+ feedthrough; ie, the bias voltage and voltage drop across the B+ protection resistor are not measured.

Operation

Operation is best (and tube life will be longest) when the amplifier is heavily loaded. If the amplifier is over loaded, grid current is negative and plate current is higher than optimum. When under loaded, grid current rises rapidly as drive power is increased, however output power will not increase substantially. Several different tubes have been tried in both the parallel and single tube amplifiers. All of the tubes gave maximum power output at 432 MHz with very little positive grid current (typical was 7 to 20 mA per tube). The apparent efficiency of the amplifier should be checked (power out/($I_p \times V_p$). If over 50% apparent efficiency cannot be obtained, either the amplifier is not correctly loaded or the plate current and HV meters are not calibrated. If efficiency is over 55%, either the plate meters (voltage and current) or power-output meter are out of calibration. If the antenna has a high SWR, the power meter can indicate high, due to the reflected power traveling up and down the transmission line. Table 1 lists the observed operating conditions for the amplifier. These readings were made after carefully calibrating all metering circuits and verifying power out-

Table 2
Parts List, Parallel 3CX800A7 Amplifier

B1—100-CFM free-air flow blower; Dayton 4C443 or equiv.
C1—Miniature air-variable butterfly. 1.8-5.1 pF per section; Cardwell 160-205.
C2—Miniature air-variable butterfly. 2.2-8.0 pF per section; Cardwell 160-208.
C3-C7—Feedthrough capacitor, 1000 pF, 300 V; Tusonix 327-005-X5UO-102M.
C8—Plate-tuning flapper (see text and Fig 2).
C9—Plate-loading flapper (see text and Figs 3 and 4).
C10—1500-pF, 2500-V, pi-section feedthrough; Murata/Erie 1280-060.
C11—1000-µF, 35-V electrolytic.
C12-C18—0.01-µF/50-V, monolithic ceramic preferred; Sprague 1C10Z5U103M050B.
C19—0.01 µF, 1 kV ceramic
C20—0.5 µF, 25 V.
D1—8.2-V, 50-W Zener diode; IR-Z3307.
D2—100-PIV, 10-A silicon rectifier diode.
D3, D4—2.5-A, 1000-PIV silicon rectifier diode; IR R170 or equiv.
D5—50-V, 2-A silicon bridge rectifier.
D6-D12—1-A, 1000-PIV silicon rectifier diode, 1N4007 or equiv.
F1—2 A, AGC or 3AG fast-blow.
F2—1½ A, AGC or 3AG fast-blow.
I1—120-V pilot lamp, amber.
I2—120-V pilot lamp, red.
J1—Chassis-mount BNC female connector, UG-1094.
J2—Chassis-mount Type-N female connector, UG-58A.
J3—Chassis-mount MHV female connector, UG-931.
J4—6-pin, male chassis-mount connector, Cinch P306AB.
J5—4-pin female chassis-mount connector, Cinch S304AB.
J6-J8—RCA-type phono receptacle; Switchcraft 3501FR.
J9—2-pin female chassis-mount connector; Cinch S302AB.
K1—Thermal time delay, 115-V heater, SPST, NO; Amperite 115NO180B.
K2, K3—Control relay, 24-V dc coil, 4PDT; PB R10-E1-X4-V700.
K4—SPST low-power coaxial relay with BNC connectors, 28-V dc coil.
K5—SPST high-power coaxial relay, Type-N connectors, 28-V dc coil.
M1—Dc milliammeter, calibrated 1.5-A full scale.
M2—0-1 mA dc milliammeter, with appropriate shunt resistors: (0-150 mA grid, 0-3 kV plate, 0-15 V filament).
Q1, Q2—NPN low-current switching transistor, 2N3904 or equiv.
Q3—PNP medium-current switching transistor, 2N3053 or equiv (TO-5 package).
Q4—NPN medium-current switching transistor, 2N4037 or equiv.
Q5—PNP low-current switching transistor, 2N3906 or equiv.
R1—200 Ω, 25-W wirewound.
R2—1k, 12-W wirewound.
R3—10k, 25-W wirewound.
R4—0.5-Ω, 1-W 1%.
R5-R10—499k, ½-W, 1% metal film, type RN-65C.
R11—2.7k, ½-W metal film, select valve to calibrate HV meter.
R12—3.3k, ½-W metal film, select value to adjust HV relay drop-out point.
R13—9-Ω, ½-W metal film, select value to calibrate grid meter.
R14—7.2-Ω, 2-W, wire wound.
R15—10k ¼-W miniature trimmer.
R16—100-Ω, 2-W, select value to adjust time delay (see text).
R17—50-Ω, 25-W wirewound potentiometer.
R18—2500-Ω /50-W wirewound.
R19—12k, ½-W, film.
R20—2k, ¼-W, miniature trimmer.
R21, R23, R27—10k, ½-W, film.
R24—4.7k, ¼ W, film.
R25, R28—2.7k, ½-W, film.
R26—2.2k, 1/4 W
R28—10k, ¼-W, film.
R29—330-Ω, ¼-W, film.
RFC1-RFC5—8 t no. 18 enam, ¼-in diam close wound.
RFC6, RFC7—6 t no. 18 enam, ¾-in. long, ¼-in. diam.
S1—SPST toggle or rocker switch.
S2—DPST toggle or rocker switch.
S3—2P5T rotary switch, non-shorting.
S4—SPDT miniature toggle switch.
T1—Filament transformer, 14-V ac, 4 A; Stancor P-8557.
T2—Control transformer, 20-V ac, 1 A; Stancor P-8604.
V1, V2—EIMAC 3CX800A7 ceramic-metal triode.
W1—Plate stripline (see text and Fig 1).

put via a calibrated directional coupler.

The loading capacitor has enough range to allow the amplifier to efficiently operate from 700 to 1500 W output when the plate voltage is around 2300 V under load. Enough tuning range is available such that load impedance from about 35 to 70 Ω can be handled. If you desire to operate the amplifier at lower power levels, reduce plate voltage.

From a cold start, the amplifier will put out about 1450 W and drift up to 1500 W (that is if the amplifier has previously been tuned up at full temperature) over the first few minutes of operation. After this warm-up time, the amplifier will repeatedly come right up to full power. If you are observing more power drift, check all connectors and cables in your antenna system. The legal limit (1500 W) at 432 MHz will melt RG-213 type polyethylene cables in a few minutes. High power and temperature cable such as Teflon-dielectric cables like RG-225 can be used for jumpers as well as Andrew ½-in. SuperFlex cable. "Hardline" cables such as Andrew LDF4-50A or larger sizes should be used for the feed line to the antenna.

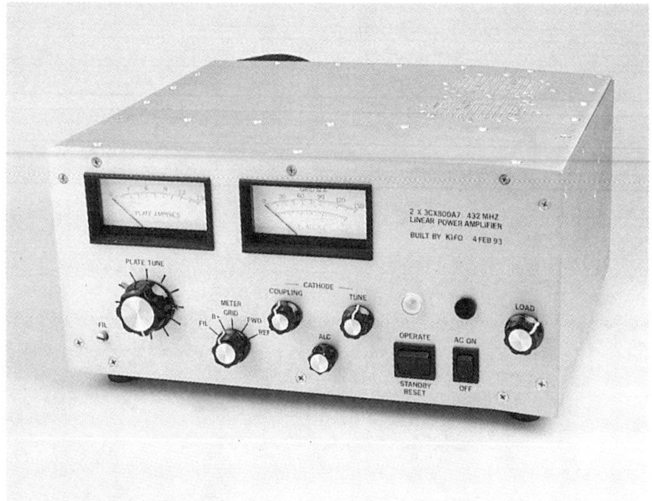

A pair of 3CX800A7 tubes provides a cost-effective means of producing the legal limit on 70 cm.

W2—Cathode stripline (see text and Fig 10).

Tube sockets (2 req)—11-pin EIA, EIMAC SK-1900 or E. F. Johnson 124-0311-100.
Grid collets (2 req)—EIMAC 720359 assembly. (EIMAC 882931 can be used).
Grid insulators (2 req)—EIMAC 720189.
Anode collets (2 req)—EIMAC 720829 or finger stock.
Chimneys (2 req)—EIMAC SK-1906.
Fuse holders (2 req)—Bussman type HKP, 2 required.
Panel bearings (9 req)—¼-in. diam.
Ball drive—Jackson 4511/DAF.

RF deck enclosure—8- × 12- × 3-in. chassis; BUD AC-424 or equivalent.
Cathode compartment—5- × 7- × 2-in. chassis; BUD AC-402 or equivalent.
Bottom chassis—8- × 12- × 2½-in. chassis; BUD AC-1419 or equivalent.
Miscellaneous sheet metal—2, 10¾- × 12- × 0.062-in. aluminum (top and bottom covers); 1, 5- × 7- × 0.062-in. aluminum (cathode compartment cover); 2, 5⅝- × 12⅛- × 0.062-in. aluminum (front panel); 2, 5½- × 10¾- × 0.062-in. aluminum (side panels); 3 ft, ½- × 0.062-in. aluminum angle stock (to hold front panel to cabinet), 1, 6¼- × 3½ × ⅜ in., 6061-T6 aluminum (air-outlet plate); 2, 3½- × 3- × ⅜-in. 6061-T6 (air-inlet plate); 1, 6- × 4- × 0.062-in. perforated steel sheet, ⅛-in. holes on 3/16-in. centers; 1, 6- × 4- × .020 brass or copper sheet (for W2 and C8 tuning pointer); 1, 6- × 9- × 0.062-in. brass or copper sheet (for W1); 1, 3- × 6- × 0.062-in. aluminum (C8, C9 and R17 shaft mounting brackets)
Other metal—2 in., 7/16-in. brass rod (for C1 and C2 shaft couplings); 1, 2- × ⅝- × 0.125-in. aluminum or brass bar (C6 retaining plate).
Other items:
10 ft no. 18 enam wire.
100 ft no. 20 insulated stranded hookup wire.
Terminal strips.
HV-meter pc board.
TR-control pc board.
20 wire ties (small).
Cable clamps.
4-in. RG-225 50-Ω Teflon-dielectric cable.
2, 1.43 in. high × ½-in. diam. Teflon or ceramic standoffs
1 — ½- × ⅜- × 2-in. Teflon rod (to hold loading capacitor).
2 — 1½- × ¼ × ⅜-in. diam. threaded steel standoff.
12-in. ¼-in. dia. Garoite, or Bakelite rod (for C1 and C2 shafts).
½-in. ¼-in. dia. Teflon rod (for C8 and C9 insulators/fishline attachment).
14-in. ¼-in. dia. Brass Rod (for C8 and C9 tuning shafts).
2½- × ⅝- × 1/16-in. G-10 epoxy board (C1 and C2 mounting plates).
3-in. ⅝- × ¼-in. Brass bar stock for C8-9 shaft stops and C1-2 mtg. brkts
3 — ¼-20 × ½-in. long nylon screws.
2' Dacron polyester woven fly line.
Shrink tube.
Assorted grommets.
4, 3-48 × ¼-in. Phillips round-head screws.
8, 4-40 × ¾-in. Phillips flat-head screws.
4, 4-40 × ⅜-in. Phillips flat-head screws.
18, 4-40 × ¼-in. Phillips pan-head screws.
3, 4-40 × ¼-in. brass screws.
2, 4-40 × 1-in. Phillips round-head screws.
46, 6-32 × ¼-in. Phillips flat-head screws.
11, 6-32 × ⅜ in. Phillips flat head screws.
8, 6-32 × ½-in. Phillips flat-head screws.
28, 6-32 × ¼-in. Phillips pan-head screws.
16, 6-32 × ¼-in. Phillips round-head screws.
2, 6-32 × ⅜-in. Phillips round-head screws.
2, 6-32 × ¼-in. brass round-head screws.
2, 6-32 × 1-in. Phillips round-head screws.
8, 6-32 × ¼-in. hex socket set screws.
22, 6-32 hex nuts.
18, 4-40 hex nuts.
20, no. 4 lockwashers.
32, no. 6 lockwashers.
4, no. 6 solder lugs.
62-6-32 self-clinching blind fasteners for 0.050-in. metal.
Knobs—1, 1⅜-in. diam., ¼-in. shaft AlcoSwitch PKES-120B-¼
1, 1-in. diam., ¼-in. shaft, AlcoSwitch PKES-90B-¼
1, ⅞-in. diam., ¼-in. shaft, AlcoSwitch PKES-70B-¼
2, ¾-in. diam., ¼-in. shaft, AlcoSwitch PKES-60B-¼

Inside the plate compartment, with the tubes removed.

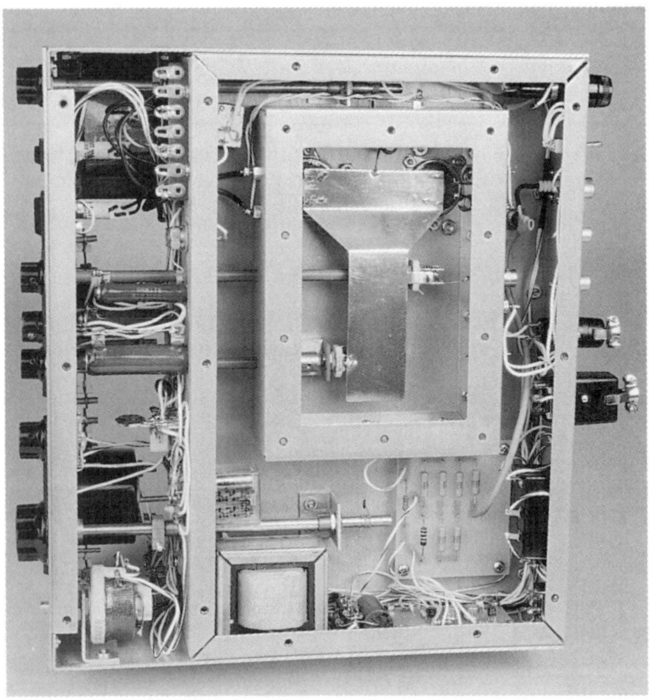

Underside view, showing the cathode compartment and tuning controls.

Since the 3CX800A7 has relatively high gain, you should be certain that drive power level is stable before blaming the amplifier for any power drift. Also be sure that the drive signal is clean. Several different solid-state brick amplifiers were tried as drivers for the parallel 3CX800A7 amplifier. Some handled the amplifier input circuit well, while others weren't so forgiving. One somewhat "flaky" solid-state driver amplifier was cured with the installation of an isolator between the driver and amplifier. Also be sure that the drive signal is free from spurious signals, such as the 404-MHz local oscillator signal if a 28-MHz-IF transverter is used. While testing the

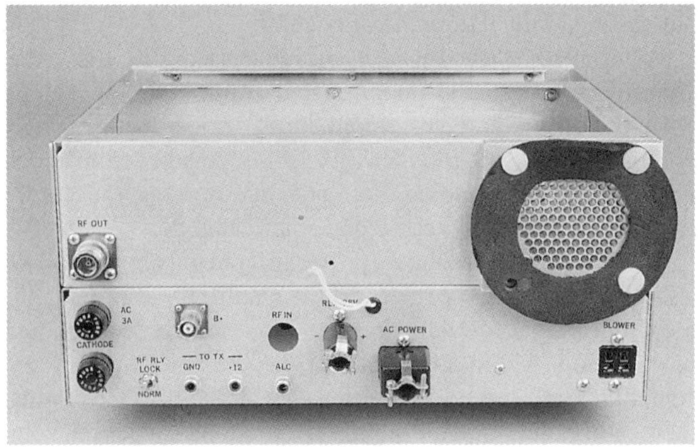

Rear view, with blower removed.

amplifier I found that one of my transverters put out a substantial amount of 404-MHz LO signal, which was then amplified more by a solid-state driver amplifier than was the desired 432-MHz signal. The result was that drive power was substantially higher than expected and amplifier efficiency was poor. In addition, the input circuit could not be tuned for a low SWR. The installation of a $\lambda/2$ filter between the transverter and solid-state driver cured the problem.

I have already discussed the slight stress on a pair of 3CX800A7s when running at 1500-W output at 432 MHz. This is due to operation about 25% higher in frequency than the specified maximum for the tubes (350 MHz). At 432 MHz, the efficiency of the tubes is starting to drop due to transit time effects. Note that in order to obtain 1500-W output power either the maximum plate current or voltage must be exceeded. Table 1 details full operating parameters for the parallel 3CX800A7 432-MHz amplifier. In Test 1, the plate voltage is maintained within specification and the plate current rating is exceeded. In Test 2, the plate voltage was increased until 1500-W output was obtained within the rated 1200 mA of plate current. For this reason I recommend 1500-W operation be limited to SSB voice service and keyed CW (Morse code). For tuning adjustments, key-down times should be limited to 1 minute or less with an equal or greater off time. If you have a well regulated B+ power supply, it is more desirable to run the plate voltage higher than specified and keep the plate current at 1200 mA. By following these precautions a pair of 3CX800A7s will last 10 to 15 years or more in typical heavy amateur EME or tropo operation. If the amplifier is to be used in a continuous mode such as FM or ATV, I recommend that the plate voltage be limited to 2000 V and the plate current be limited to 1000 mA for the pair of tubes. For continuous transmission modes, the cathodes should be checked for back heating. This is done by reducing the filament voltage until power output starts to drop. The filament voltage should be then increased to just above the point where power dropped. The filament voltage should only be reduced during key-down transmit periods. During warm-up and receive periods the filament voltage should be switched to full voltage.

Using 8874 Tubes

The design could be easily adapted to the Varian 8874

tube, if they are available. If you're going to use new tubes, the 3CX800A7 is preferred. To use 8874s set filament voltage to 6.0 V ac-RMS. Use a 6.3-V, 6-A filament transformer for T1. The plate stripline holes should be sized for the 1⅝-in. diameter of the 8874 plate anode (1.785 in. diameter if the same finger stock is used). The air-outlet holes should be reduced to fit within 1⅝-in. diameter circles. Neutralization may be slightly different with the 8874. Ratings for the 8874 are 2200 V at 350 mA continuous and 500 mA in intermittent service (SSB and keyed CW). At full power, required drive power is 80 to 85 W and about 1200 W output power may be obtained.

Conclusion

The parallel 3CX800A7 432-MHz amplifier is an effective means to obtain 1500 W output in amateur SSB and CW service. The performance of the amplifier is only 1.5% lower in efficiency and 0.6 dB lower in gain than a "perfect" design. This slight degradation in performance is an acceptable design trade off, considering the simplicity of its construction compared with a design capable of better performance. Operation is completely stable at 1500 W output. It has been completely reliable and has very good IMD performance, with low drive requirements.

Kit Availability

Chassis kits are available from Charles Byers, K3IWK, 5120 Harmony Grove Rd, Dover, PA 17315, tel 717-292-4901. Complete kits and factory wired units are available from Lunar-Link Systems, 816 Summer Hill Rd, Madison, CT 06443-1604, tel 203-421-3377.

A 1500-Watt Output Amplifier for 432 MHz

By Steve Powlishen, K1FO

The change in regulations which now allow amateurs to run 1500 W PEP output power called for a new amplifier. The 1500-W level is of most significance on EME, but other propagation modes such as tropo-scatter also benefit from the new regulations. The design goal was an amplifier capable of running 1500 W continuous output, be extremely reliable, stable, free from power drift annoyances and have excellent linearity (low distortion products).

I considered using solid-state devices. The best commercially available devices usable at 432 MHz are power MOSFETs, such as the Motorola MRF175GU (28 V) and MRF175LU (56 V). These devices contain two transistors in a single package, and are designed for push-pull operation. Each package can deliver 150 W of linear power output (200 W saturated). Reaching the 1500-W level in reliable linear service would require combining 10 amplifiers! At 432 MHz these devices are 45% efficient at best, which means quite a power supply: 28 V at 120 A or 56 V at 60 A, regulated. Consider also the heat sink required to dissipate almost 1800 W. In addition, when all of the power input splitters and output combiners are considered, the overall system gain would not be much greater than 10 dB. Almost twice the drive power required by a triode amplifier would be needed to excite this solid-state amplifier to full output. Lacking better solid-state devices, I decided a 1500-W solid-state amplifier was impractical for amateur construction. My design process was forced back to familiar vacuum-tube ground.

The list of power tubes that work acceptably above 400 MHz is small, especially when you consider price and performance. A list of available tubes and their performance capabilities was compiled. The 4CX250 family was ruled out first. Although they are readily available for reasonable cost in the surplus market, the use of the 4CX250 family would require more than 2 tubes to reach the desired 1500-W output level within the CCS ratings of the tubes. In addition even when the '250 family is run within their CCS ratings their IMD performance (3rd-order IMD down –23 to 25 dB) is somewhat marginal. The only member of the family designed for high linearity (the 4CX350A) is a poor performer at 432 MHz. The thought of pushing two 4CX250-type tubes to the 1500-W output level, even in intermittent amateur service threatened to be an exercise in frequent tube changes.

The 8874 family (8873, 8874, 8875, 3CX400U7 and 3CX800A7) were examined next and met all requirements except 1500-W output on a continuous basis for 2 tubes. Only the 3CX800A7s appeared capable of generating 1500 W output while only moderately exceeding their ratings.

The 3CX600U7 presented an interesting possibility, but with little or no prospect of surplus tubes, the cost of a pair of them plus the time and effort required to build sockets called for a look at the "big" bottles.

There are many benefits to construction of single-tube amplifiers. Better IMD, freedom from balance worries, ability to use coaxial input or output circuits, ease of use in cavity circuits and simplicity of construction are among the features of single-tube design. The 4CX600 family and 7650 were ruled out quickly as simply not in the 1500-W output class with a single tube. Cost considerations made them a poor choice compared to some larger tubes. If you have good surplus connections, a pair of 7650s would make a formidable 432-MHz amplifier. The 4CX1000 family (4CX1000A, 4CX1000K and 4CX1500B), along with the 8877, were discarded because of their poor efficiency at 432 MHz. The 3CX1000A7 has been used at 432 MHz[1] but suffered the same malady.

This left the three final contenders: the 7213, 4CX1500BC and the 8938. The 7213 is capable of 1500-W output by slightly exceeding its CCS ratings. The 7213 will give acceptable performance, but its significantly higher cost new, and scarcity in surplus scratched it from the finals. The 4CX1500BC was an interesting possibility. It is essentially a coaxial-base UHF version of the 4CX1500B. The 4CX1500B is an improved 4CX1000. The improvements are a higher efficiency anode radiator and improved grid and cathode structure for higher linearity. The excellent IMD, power output and low drive appeared to make the '1500BC the right choice. When the project was started, however, the 4CX1500BC was not in full production. The combination of both uncertain availability and higher cost eliminated any low-drive-tetrode versus simple-triode-circuit arguments.

The winner, therefore, was the EIMAC 8938. The 8938 is essentially a UHF version of the popular 8877. It has been used in both commercial and amateur amplifiers in the 400-

Table 1
Parts List

B1—100 CFM free air flow, 50 CFM @0.7", Dayton, 4C443 Note: wire grate on blower input was cut out to increase air flow.
C1—Min. air var. butterfly. 2.2-8.0 pF/section Cardwell 160-208.
C2—Min. air var. butterfly. 1.8-5.1 pF/section Cardwell 160-205.
C3—Plate loading flapper made from 0.010 brass. See drawing.
C4—Fixed plate tuning capacitor, made from 0.062 aluminum. See drawing.
C5—Variable plate tuning capacitor, made from 0.010" brass. See drawing.
C6—1000 pF/4 kV feedthrough, Tusconix 2498-0C1-X540-102M
C7-9—1000 pF/300 V feedthrough type CK70.
C11—1000 μF/35-V electrolytic.
C12-19—0.01 μF/50 V, monolithic ceramic, preferred Sprague 1C10Z5U103M050B
C20—1 μF/25 V.
D1—25 V/50-W Zener diode, RCA SK560/5265A.
D2—100 PIV/10 A silicon rectifier diode.
D3-D4—3.0 A/1000 PIV silicon rectifier diode, 1N5408 or equivalent.
D5—50 V/2 A silicon bridge rectifier diode.
D6-12—1 A/1000 PIV silicon rectifier diode, 1N4007 or equivalent.
F1—3 A fuse, AGC or 3AG fast blow type.
F2—1-1/2 A fuse, AGC or 3AG fast blow type.
I1—120 V pilot lamp amber.
I2—120 V pilot lamp red.
J1—Chassis mount BNC female connector, UG-1094/U.
J2—Chassis mount N female connector, UG-58A/U.
J3—Chassis mount MHV female connector, UG-931/U.
J4—6 pin male chassis mount connector, Cinch P306AB.
J5—4 pin female chassis connector, Cinch S304AB.
J6-7—RCA type phono receptacle, Switchcraft 3501FR.
J9—2 pin female chassis mount connector, Cinch S302AB.
K1—Thermal time delay, 115 V heater, SPST/NO, Amperite 115N0180B.
K2-3—Control relay 24 Vdc coil 4PDT, P&B R10-E1-X4-V700.
M1—DC mA meter, 1.5 A full scale.
M2—0-1 mA dc meter, with appropriate shunt resistors 0-100 mA grid, 0-5 kV plate, 0.5 V filament.
Q1-2—NPN low current switching transistor, 2N3904.
Q3—PNP medium current switching transistor, 2N3053.
Q4—NPN medium switching transistor, 2N4037
Q5—PNP low current switching transistor, 2N3906.
R1—200 Ω/25-W wirewound.
R2—1000 Ω 12-W wirewound.
R3—10k 25-W wirewound.
R4—0.5 Ω 1-W 1%.
R5-14—499k, 1/2-W, 1% metal film, type RN-65C.
R15—2.7k, 1/2-W metal film, select valve to calibrate HV meter.
R16—3.3k, 1/2-W metal film, select value to adjust HV relay point.
R17—9 Ω, 1/2-W metal film, select value to calibrate grid meter.
R18—5 Ω, 2-W, metal film.
R19—10k, 1/4-W miniature trimmer.
R20—100 Ω, 2-W, select value to adjust time delay.
R21—50 Ω, 25-W wirewound potentiometer.
R23—12k, 1/2-W, film.
R24—2k, 1/4-W, miniature trimmer.
R25-26—10k, 1/4-W.
R28—1k, 1/4-W.
R29-31—2.2k 1/4-W.
R32—330 Ω, 1/4-W.
RFC1-3—7 turns no. 18 enameled wire 1/4" dia. close wound.
RFC5—6 turns no. 18 wire 3/4" long, 1/4" diameter.
S1—SPST toggle or rocker switch.
S2—DPST toggle or rocker switch.
S3—2P5T rotary switch, non-shorting.
S4—SPDT miniature toggle switch.
T1—Filament transformer, 5 V@ 10 A. Stancor P-6135.
T2—Control transformer, 20 V @ 1 A. Stancor P-8604.
W1—Plate stripline. See text and Fig 1. 4.438" × 7.80" × 1/16" brass.
W2—Cathode stripline. See text and Fig 10. 0.020 brass.
V1 Varian, Eimac 8938 ceramic metal triode.
Filament pin contact: Eimac 135310.
Filament collet: Eimac 135307 attached to brass ring.
Cathode collet: Eimac 135306 attached to brass ring.
Grid collet: Eimac 135305 collet attached to 1/16" brass plate.
Anode collet: Make from Instrument Specialties #97-135 finger stock.
Chimney: Varian, Eimac SK-2216.
Fuse holders: Bussman type HKP, 2 required.
Panel bearings: 1/4" dia. 9 required.
Ball drive: Jackson 4511/DAF.
RF deck enclosure: 6.5 × 10.5 × 4", see drawings.
Cathode compartment: 4 × 6 × 2 inch chassis BUD AC-431 or equivalent.
HV meter compartment: 4 × 5 × 2" chassis, BUD AC-1404.
Control circuitry chassis: 4 × 6 × 2" chassis, BUD AC-AC431.
2—10-1/2" × 6-1/2" × 0.062" aluminum top & bottom covers.
1—4" × 6" × .062" aluminum cathode compartment cover.
2—7" high × 19" wide × 0.125" aluminum, rack panel.
2—6" × 10-3/4" × 0.062" aluminum, side panels.
3'—1/2" × 0.062 aluminum angle stock (to hold front panel to cabinet).
1—4-1/2" × 4-1/2" × 1/4"-6061-T6 air outlet plate.
2—3-1/2 × 3" × 3/8"-6061-T6 air inlet plate.
1—6" × 4" × 0.062 perforated steel sheet, 1/8" holes on 3/16" centers.
1—6" × 4" × 0.020 brass or copper sheet (for W2 & C8 tuning pointer).
1—5" × 8" × 0.062 brass or copper sheet (for W1).
1—3" × 6" × 0.062 aluminum, C8, C9 & R17 shaft mounting brackets.
2"—7/16" brass rod (for C1 & C2 shaft couplings).
1—2" × 5/8" × 0.125 aluminum or brass bar for C6 retaining plate.
6'—#18 enameled wire.
100' #20 insulated stranded hookup wire.
Terminal strips.
HV meter PC board.
TR control PC board.
20—Wire ties (small).
2—2.00" high × 1/2" dia. Teflon or ceramic standoffs.
2—1-1/2" × 1/4" - 3/8" dia. threaded steel standoff.
12"—1/4" dia. Garoite or Bakelite rod (for C1 & C2 shafts).
1/2"—1/4" dia. Teflon rod for C8 & C9 insulators / fishline attachment.
14"—1/4" dia. brass rod (for C8 & C9 tuning shafts).
2—1/2" × 5/8" × 1/16" G-10 epoxy board (C1 & C2 mounting plates).
3"—5/8" × 1/4" brass bar stock for C8-9 shaft stops & C1-2 mounting brackets.
3—1/4-20 × 1/2" long nylon screws.
2'—Dacron polyester woven fly line.

to 500-MHz range. It offers excellent linearity (3rd- and 5th-order IMD products down over 40 dB), an ample 1500-W nominal plate dissipation, reasonable filament power (5 V at 10.5 A) and a modest bias voltage (25 V), which allows a simple cathode-bias circuit. The chief drawbacks of the 8938 are its modest efficiency at 432 MHz (50% typical), relatively low gain (less than 13 dB at 432 MHz) and high cost (in amateur terms) of the tube and socket. The cost of the socket can be eliminated by building one. The cost of the tube is relative. When one considers that the 8938 should easily last 10 to 15 years in heavy amateur usage, it begins to look more competitive with pushed 4CX250s, even without considering its superior power capability.

The next decision was to select the plate-circuit design. Determining which type of circuit would be used, as with the selection of which tube would be used, (and virtually all engineering decisions) became a trade off between advantages and disadvantages of any approach. The design of a 432-MHz amplifier is dominated by the characteristics of the tube or tubes used. The capacitive reactance of the 8938 is low enough that a λ/4 resonant circuit virtually disappears within the tube. This limits the choice of tank circuit to that of a cavity or resonant line of λ/2 or longer.

The cavity offers wider bandwidth (provided a λ/4 cavity can be used) and a more equal distribution of circulating currents through the base of the tube. Both of these features should result in reduced power drift. The wider bandwidth makes any tuning drift that does occur of less an effect. The primary cause of power drift at 432 MHz in a mechanically sound amplifier is RF dielectric heating of the ceramic insulators; primarily the one between the grid and anode. (Or, in the case of a tetrode, the screen-to-anode insulator.) In a symmetrical cavity, the circulating currents are more evenly distributed, and cause a more equal and lesser heating effect than a circuit that appears "single ended" and concentrates the currents toward a side of the tube. A properly designed cavity resonator should also be free from stray resonances that can plague single-line resonators.

But, the cavity approach is not without its drawbacks. First and probably foremost for most amateur constructors, is increased construction complexity. For an amateur designing from scratch who has minimal test equipment one must either construct his cavity with sliding walls so the amplifier can be adjusted to account for any calculation discrepancies versus real-world operation or be prepared to make a test amplifier before the real one is constructed. Such sliding-wall construction most likely requires higher tolerances in building the pieces of the amplifier.

The plate-bypass capacitor can be another headache in a cavity amplifier. A mistuned amplifier can arc a bypass capacitor that could safely handle the dc plate voltage. In general, it is harder to determine the proper output coupling for a cavity versus a capacitively coupled stripline. Again, in normal amateur construction where the prototype is the production model, there may be significantly less cut and try with a stripline. If you are like me and like to build compact equipment that can be mounted in a rack with limited access to the rear, a design that lends itself to all-front-panel tuning controls is desired. This can be difficult with cavity amplifiers.

A capacitively loaded and tuned half-wave stripline was selected for the plate circuit of this amplifier as the advantages of a cavity circuit did not justify its increased complexity and time and cost of construction. A commercial cavity available from EIMAC was prohibitively expensive. A half-wave stripline with simple capacitive input coupling was also used for the input circuit. The design is essentially an adaptation of the K2RIW parallel kilowatt for 432 MHz.[2]

Since the actual circuit is dominated by the output reactance of the 8938, the actual impedance and width of the stripline is not very critical. The actual dimensions for the plate enclosure were based on the desire for a compact amplifier. The small size of the plate enclosure also ensures that strange resonance modes are minimized. The height of the 8938 does not allow standard chassis to be used in the construction. The metal work was done by a local sheet-metal shop and their abilities determined that the enclosure was made with separate top and bottom covers, rather than the more conventional single-cover chassis. The custom chassis work also allowed heavy gauge aluminum to be used (5052 alloy, 0.063-in. thick). The lips where the top and bottom covers attach were made ¾-in. wide to give superior shielding where the parts join. If you cannot arrange for the metal work to be done, construction by using flat plates for all sides, held together with aluminum angle stock will work. This type of construction has been described in an 8874 amplifier for 432 MHz.[3] If you deviate from the dimensions I used for the plate compartment your amplifier will most likely work, however, the size and positions of the plate tuning and loading capacitors will change. If you attempt to change the amplifier design dimensions you are on your own. I am unable (and unwilling) to lend advice on such variations.

A close inspection of the plate line reveals that all the corners are rounded. The edge of the line has a smooth radius applied to it. Several users of 8938 stripline amplifier have reported arcing from the plate line. The spacing between the line and the tuning and loading capacitors was several times greater than what the safe distance for a 4000-V dc plate voltage. What was happening was that the RF plate voltage which swings close to twice the dc voltage was arcing from the field concentration points to the tuning or loading capacitor. When this amplifier was being tested and before the radiused edge was put on the plateline, RF arcs could be consistently drawn off it during full power. If the arc was severe enough a dc arc would also be created, tripping the power supply circuit breaker. Once a strong-enough ionized path was created the dc potential would follow the path. Inspection of the plate components would usually reveal carbon tracks at the sharp corners of the plateline. The ability to create smooth edges was one of the reasons that a solid brass stripline was chosen over use of a printed circuit board.

After the plateline edge was smoothed over, the frequency and intensity of the arc diminished significantly, however an occasional arc would still occur. Inspection of the components showed that the arcs were now starting at the relatively sharp edge of the 0.15-in. thick plate tuning capacitor. Adjusting the tuning capacitor size as to increase its spacing from the plateline did not improve the situation. At this point the dual tuning plate arrangement now used in the amplifier was tried. The theory

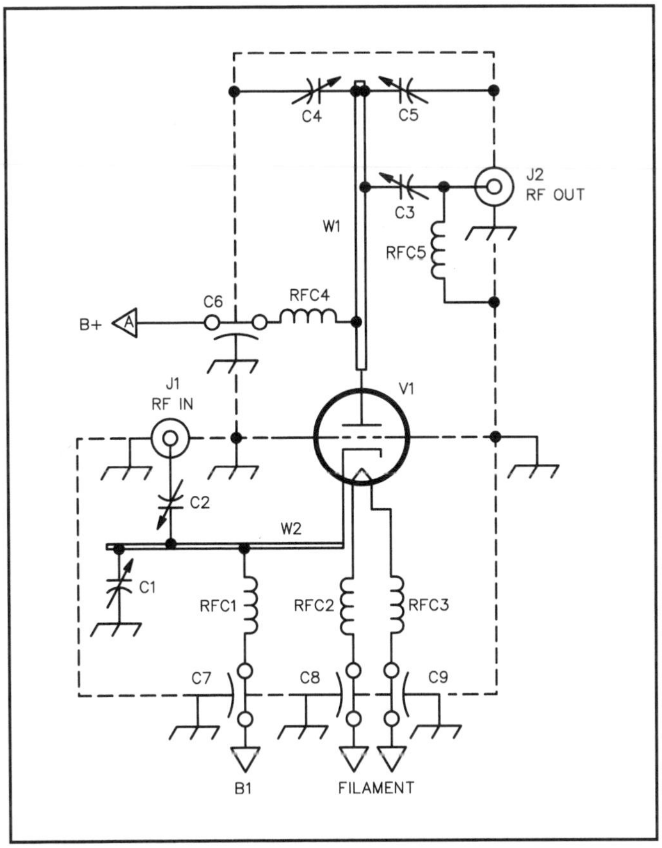

Fig 1—Schematic diagram of the 1500-W linear amplifier for 432 MHz.

behind it was that a capacitor plate on either side of the plate stripline would distribute the field more evenly. This should have the same effect as increasing the spacing of a single flapper to the plateline by four times. The second plate had the desired result and there has not been an arc in the amplifier since it was incorporated. The second plate also offers the ability of being adjusted to set the variable flapper to a desired tuning range. (The mounting holes on the fixed plate are slotted for this purpose.) The fixed plate may be made out of copper, brass or aluminum. You should make it out of material thick enough (0.030 in. or thicker) to allow a smooth radius to be applied around its edge the same as described for the plate stripline.

The plateline is mounted on 2-in. high home made Teflon standoffs. Ceramic standoffs have been used without any problems also. I recommend using brass screws to hold the plateline to the insulators. I also recommend using a brass screw to attach the plate stripline to the plate choke.

The actual choice of the finger stock for the 8938 anode to the plate stripline contact is not critical. I used some stock about 7⁄16-in. high and with folded-over fingers. I got it at a flea market and I do not know who made it. Do not order the EIMAC anode collect (no. 135304), as it is sized to fit over the anode ring on the 8938. It is too small to fit around the anode radiator. The type of finger stock you use determines the size hole in the

→

Fig 2—Schematic diagram of the control circuits for the 1500-W, 432-MHz amplifier.

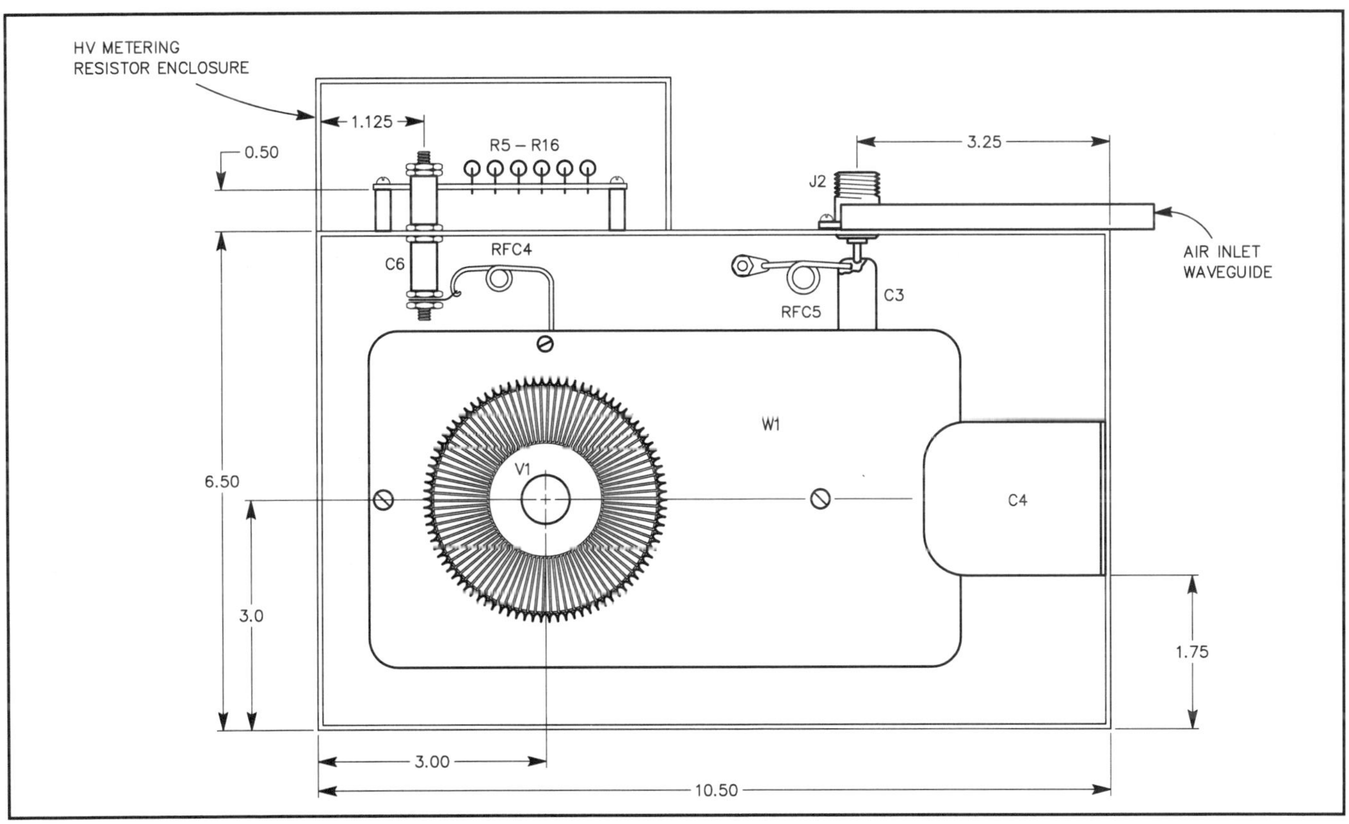

Fig 3—Layout of the plate compartment, showing locations of the plate stripline and the plate tuning flapper capacitor. Also shown is the enclosure for the HV metering resistors.

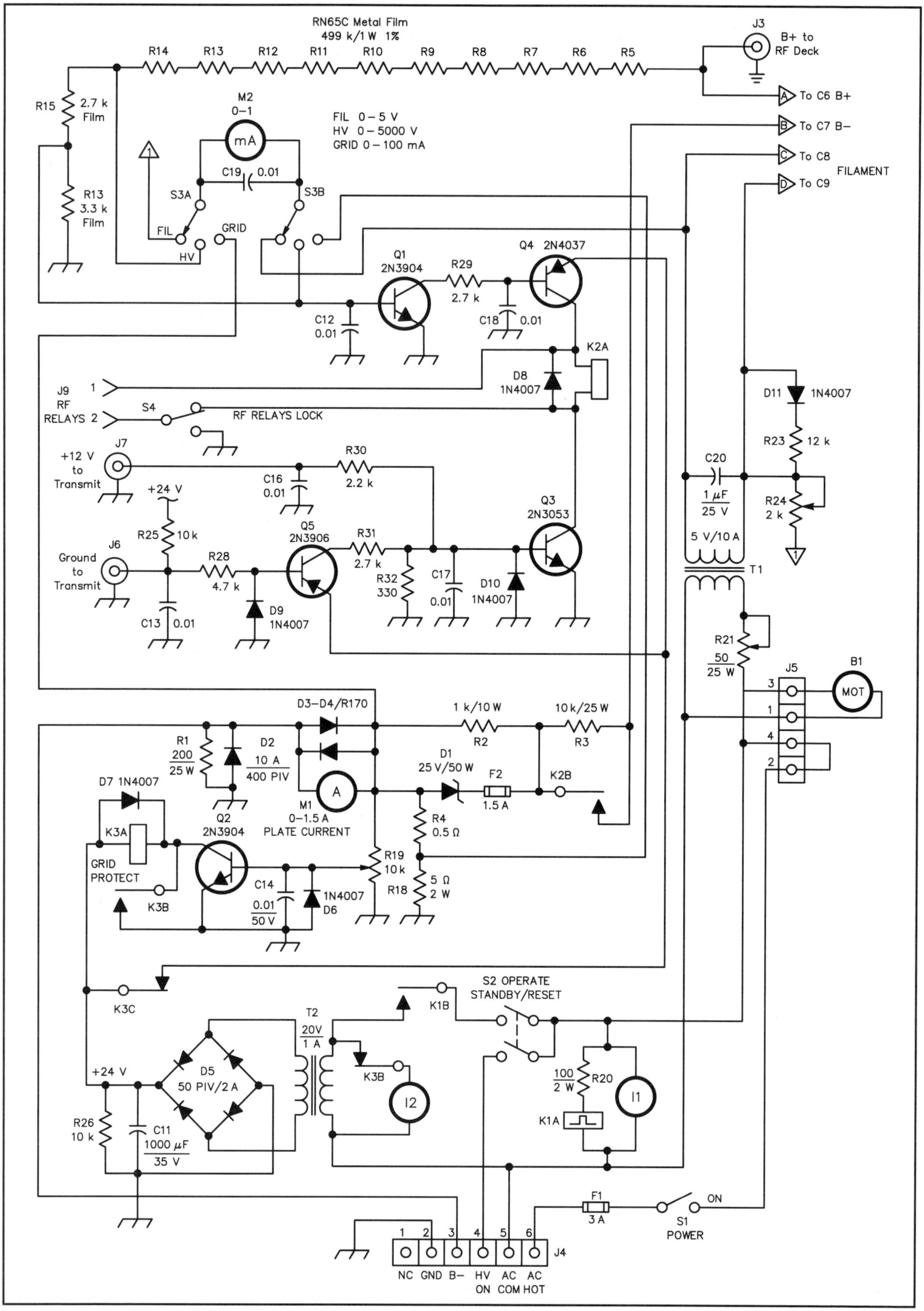
RN65C Metal Film
499 k/1 W 1%
R14 R13 R12 R11 R10 R9 R8 R7 R6 R5
J3
B+ to
RF Deck
To C6 B+
To C7 B−
To C8
To C9
FILAMENT
R15 2.7 k Film
R13 3.3 k Film
M2
0−1
mA
FIL 0 − 5 V
HV 0 − 5000 V
GRID 0 − 100 mA
C19 0.01
S3A
S3B
FIL
GRID
HV
Q1
2N3904
R29
2.7 k
Q4 2N4037
C12
0.01
C18 0.01
K2A
D8
1N4007
J9
RF
RELAYS
S4
RF RELAYS LOCK
D11 1N4007
R23 12 k
J7
+12 V
to
Transmit
R30
2.2 k
C16
0.01
C20
1 µF
25 V
R24
2 k
+24 V
R25 10 k
Q5
2N3906
R31
2.7 k
Q3
2N3053
5 V/10 A
J6
Ground
to
Transmit
R28
4.7 k
R32
330
C17
0.01
D10
1N4007
T1
C13 0.01
D9
1N4007
R21
50
25 W
J5
B1
MOT
D3−D4/R170
1 k/10 W
R2
10 k/25 W
R3
R1
200
25 W
D2
10 A
400 PIV
D1
25 V/50 W
F2
1.5 A
K2B
D7 1N4007
Q2
2N3904
M1
0−1.5 A
PLATE CURRENT
R4
0.5 Ω
K3A
GRID
PROTECT
K3B
C14
0.01
50 V
1N4007
D6
R19
10 k
R18
5 Ω
2 W
K3C
K1B
S2 OPERATE
STANDBY/RESET
T2
20V
1 A
+24 V
D5
50 PIV/2 A
K3B
I2
100
2 W
R20
I1
K1A
R26
10 k
C11
1000 µF
35 V
F1
3 A
S1
POWER
ON
J4
NC GND B− HV ON AC COM AC HOT

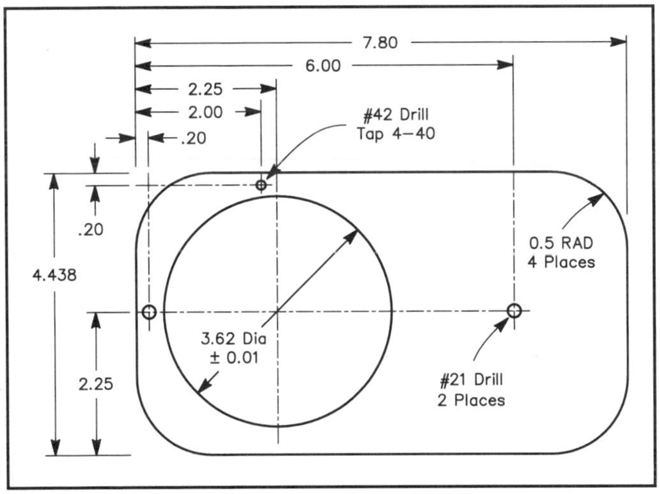

Fig 4—The plate stripline is made from brass sheet, 1/16-in. (0.063-in.) thick. Dimensional tolerance is ± 0.02 in., except the cutout for the anode fingerstock, as noted on the drawing.

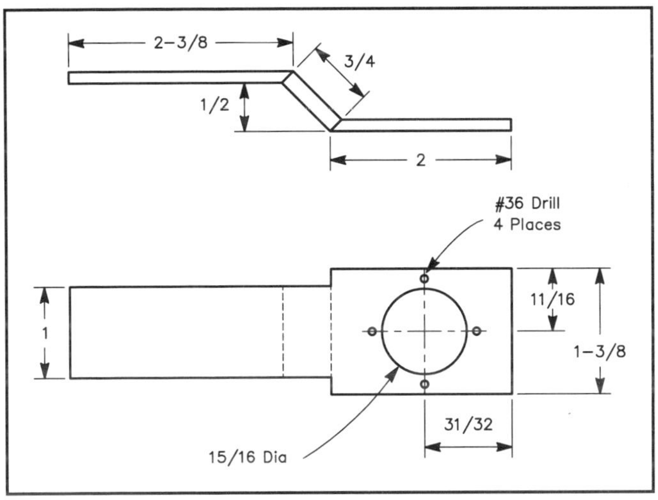

Fig 5—Cathode line is made from 0.020-in. brass, silver plated to a minimum thickness of 0.0001 in.

stripline. A different type of finger stock will slightly change the effective inductance of the stripline, which will alter the amount of tuning capacitance required to resonate the circuit. This is of little concern, especially with the dual-plate tuning capacitor arrangement this amplifier uses.

The choke between the plate-loading capacitor (C3) and ground protects against an arc between the plate line and loading capacitor. It also gives protection against accidental contact between the loading capacitor and the plate stripline. This possibility should be minimized by the Teflon buttons used on both the plate loading and tuning flappers.

The plate and load capacitors are adjusted by the now widely used "fishline" method. The lines are attached to the flappers through home-made Teflon button insulators. They are made from ¼-in. diameter Teflon rod which is put in a drill chuck and turned down with a file. A small hole is drilled through them for tying the tuning line to. This insulation from the flappers makes the selection of the tuning line not critical. Dacron polyester line is preferred if it can be found. Standard monofilament fish line can be used in this arrangement even though it is usually made from a nylon derivative.

The tuning lines are routed to the front-panel controls by Teflon bearings. The bearings are made by putting pieces of ¼-in. Teflon stock in a drill and cutting a grove in it with a hacksaw blade. Teflon bushings are used for feeding the fishline through the chassis. The bushings are made from standoff insulators, by simply pushing the solder pin out of the standoff. The tuning controls use 6:1 Jackson ball drives to impart a slow tuning feel to the controls. The ball drives have provision for attaching pointers to them. The pointers were made out of 0.20-in aluminum sheet and painted red. Tuning-range stops were made by inserting a screw through the ½-in. brass tuning control shaft. A second screw in the tuning control bearing plate limits the plate and load controls to a single turn. This amount of tuning and loading range is more than adequate to tune close to a 20-MHz range for power levels from 500- to 1500-W output. The stops protect the lines and flappers from accidental damage. The pointers on the ball drives allow the tuning controls to be quickly set to different power levels for frequencies of operation.

The grid-grounding ring was made from an EIMAC grid collet (P/N 135505) and a 3½-in. square piece of 1/16-in. brass. The hole for the collet is 2¼ in. in diameter. Twelve mounting holes, drilled to clear 6-32 screws, are equally spaced around a 2¾-in. diameter circle. The outside size and shape of the grid collet plate are not critical. It can be made circular or square. If you do not silver plate your grid collet mounting plate, the side which touches the aluminum chassis should be tinned. This applies to any other copper or brass to aluminum joint. This is necessary to avoid any long term corrosion problems which can be a problem when these metals touch each other. The grid collet contact tabs are broken off in the pattern described in Fig 1 to neutralize the amplifier. Removing the fingers raises the grid inductance which lowers the amplifier's gain at higher frequencies. This is done in order to avoid higher frequency parasitic oscillations. These higher frequency parasitic oscillations can cause dielectric heating of the grid-to-plate insulator, which causes power drift. This condition could also potentially damage the tube by overheating its seals.

If you have not soldered finger stock or collets to a thick brass plate, I suggest you get some extra material and practice first. All the collet assemblies used in this amplifier (plate, grid, cathode and filament) were soldered with a standard propane torch and rosin-core solder. The first step is to thoroughly clean all parts. Steel wool leaves a residue. I use Brillo or SOS pads to clean and polish the parts as their detergents are good grease cutters. Next, wipe down the plate with a low-residue solvent, such as lacquer thinner or acetone, to remove the detergent residue. The finger stock or collets are then fitted (after thoroughly washing and drying your hands!). A shim wedged into the gap in the collet holds it in place. Then a very small amount of liquid resin (the type as used in solder reflow machines) is applied to the joint. The torch is then applied to the brass only. Finger stock is made from very thin tempered beryllium copper.

Excessive heat destroys the temper and will ruin the collet. The reflow resin will change the color or the brass, just as it's getting hot enough to melt the solder. Use the smallest diameter solder you can get (0.20- or 0.30-in. diameter). Apply the solder sparingly to the joint behind the torch flame. Use the torch to sweat the solder into the joint and flow it around the joint. Do not disturb the assembly until you are sure that the solder has set. The excessive resin may be cleaned off with lacquer thinner.

The cathode compartment is built in a standard 4- × 6- × 2-inch chassis. The top of the chassis is cut out to allow a removable cover. The cover gives you access to the cathode compartment. The bottom of the chassis is fastened to the bottom plate of the amplifier compartment. The end of the cathode box away from the tube socket is perforated. A solid bottom cover is used on the cathode box. This is done to force cooling air into the cathode compartment through the grid collet, which passes the socket and cools the filament contacts and seals of the tube.

Figs 2 and 3 detail the rings used to hold the cathode collet (EIMAC no. 135306) and filament collet (no. 135307). You will need a lathe to make these rings. The collets are fastened by screws to a G-10 epoxy circuit board, ⅛ in. or thicker. The second filament contact is made with an EIMAC heater pin contact (no. 135310). To make the socket base, a hole for the heater pin is drilled in the center of the circuit board. The filament pin is inserted. The cathode and filament rings assembled with their collets are placed on the 8938. The circuit board is then put on the 8938. It will be held in place by the filament pin, allowing you to align the collet rings.

The cathode line is made from 0.020-in. brass sheet. Its thickness is not critical. The line is soldered to the cathode ring. Be sure not to get the cathode ring mounting holes filled with solder. The cathode line is tuned and loaded by miniature butterfly capacitors, as in the original K2RIW amplifier. The filament ring uses a fifth screw that bottoms into a tapped hole to connect to the filament transformer. This assures a good contact. If you experience any erratic operation or power drift, be sure to check for good filament contacts. The socket assembly is mounted to the chassis by using ½-in. metal standoffs.

EIMAC recommends reducing the filament voltage to 4.3 V for average power outputs of a kilowatt or greater at 400 MHz. Amateur SSB and CW operation is only 25% duty cycle. This is true even in the thick of an EME contest, which is probably the most grueling use of the amplifier. In addition, linear operation with heavy loading (the tuning condition under which best linearity usually occurs) minimizes cathode back heating. I found that for the intermittent operation that characterizes CW and SSB amateur operation, 4.7 to 4.8 V was the optimum filament voltage. There should be some resistance in either the primary or secondary circuits of the filament transformer, to allow a soft filament turn on. High input surges when the filament is cold are the primary failure mode of a power tube filament. EIMAC does not specify a minimum warm up time for the 8938. For maximum tube life, especially in light of the reduced filament voltage used with this amplifier, I recommend a minimum filament warm-up time of 3 minutes. If this amplifier is to be used in a more continuous mode, such as FM-repeater service or ATV, a lower filament voltage should be used.

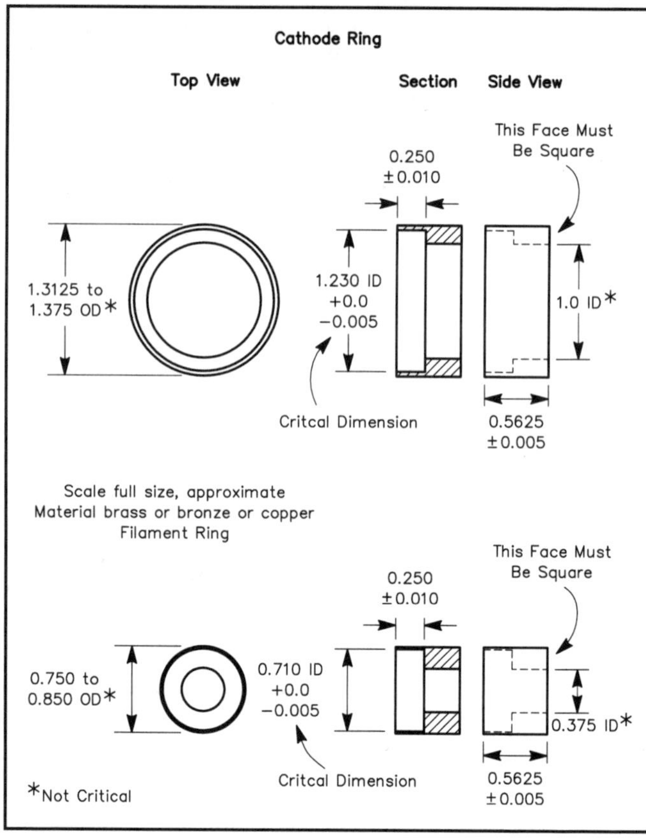

Fig 6—Cathode and Filament rings, made from brass, bronze or copper. Note the critical dimensions given in the drawings.

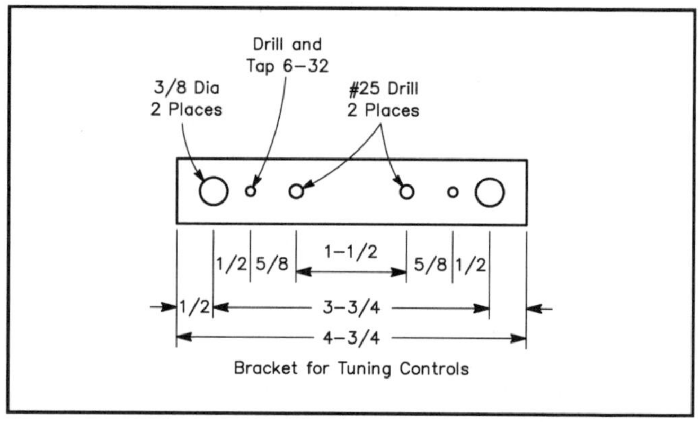

Fig 7—Tuning controls bracket, made from aluminum sheet (thickness not critical).

A filament voltage metering circuit is included in the amplifier. A bridge rectifier and filter capacitor creates a dc voltage which can be read by the 1-mA multimeter. R18 and R19 form a simple dropping network to allow the voltage to be set for a full scale of 5.0 V. Pin jacks are also included on the back panel to allow easy double checks with a high accuracy AC voltmeter.

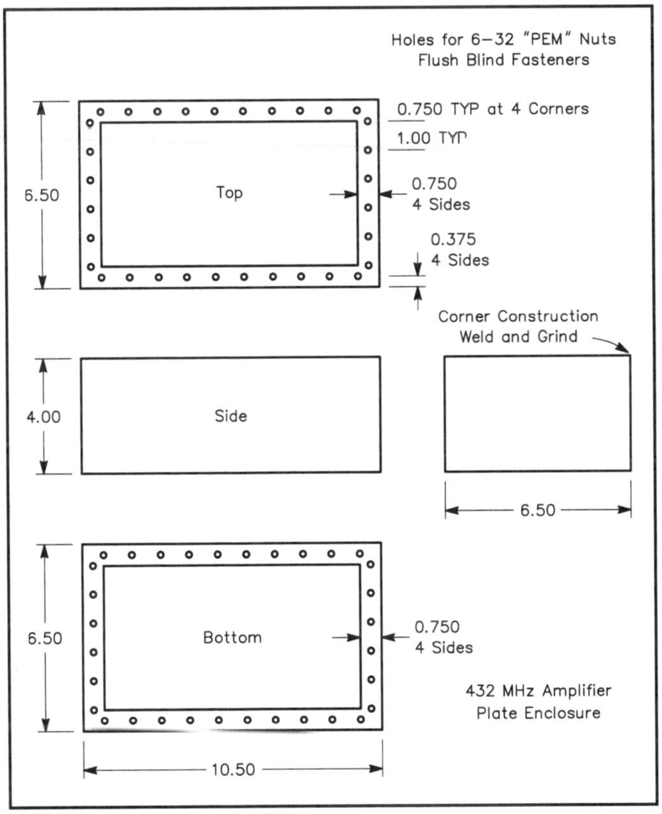

Fig 8—Details of hole drilling and covers for the plate enclosure.

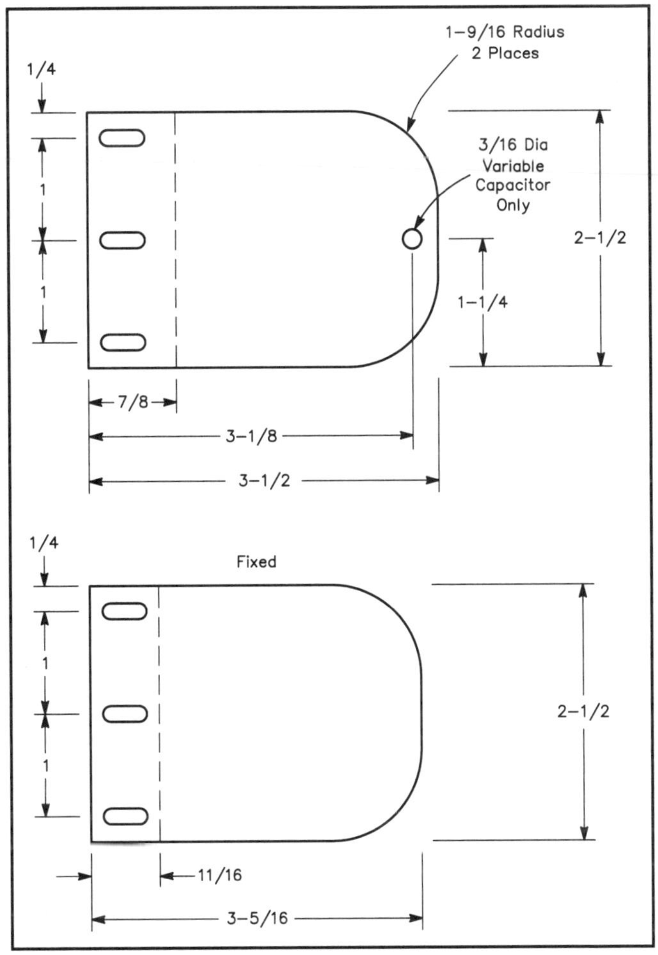

Fig 9—Details of the plate-tuning capacitor. The variable plate (top) is made from spring brass or beryllium copper, 0.010- to 0.015-in. thick, silver plated. The fixed plate (bottom) is made from aluminum or brass, 0.040- to 0.060-in. thick. Do not drill the 3/16-in. mounting hole in the fixed plate.

The high-voltage metering circuit may appear overly complex to some, however, there are a number of reasons why 10 resistors are used in the dropping string. If you're going to measure plate voltage, you might as well build it so it not only reads accurately, but will do so over the life of the amplifier. If you simply want an indication that high voltage is present, a neon lamp will suffice.

Most amateurs do not have an appreciation for the effects of high voltage. I have seen metering circuits where one, two or three resistors were used in the dropping string. This is poor engineering practice for the following reasons: First, all resistors have a maximum allowable voltage drop specification. This is the maximum potential which may be placed across the resistor without ill effects. The ill effects are corona discharge across the resistor. This will show up as noise in a sensitive 432-MHz receiver. A high potential also causes premature aging of the components. Specifically, this accelerates the resistance change all resistors suffer over their life. Second, any excessive voltage across a resistor can lead to catastrophic failure of it and possibly an expensive meter.

When circuit stability is required, good engineering practice calls for a resistor to dissipate less than half its rated value. The resistors in the metering circuit are ½-W units, however, they are only called on to dissipate ¼ W during full plate voltage. The chassis that holds the dropping resistors is perforated on top and bottom, to allow cooling air circulation by convection. Even though the entire resistor string dissipates less than 3 W, that is enough to warm a closed box appreciably. Note also that the dissipation rating of all resistors goes down as the temperature goes up.

The type of resistor selected has a large effect on the long-term accuracy of the circuit. Carbon-composition resistors have the poorest tolerance and greatest resistance change due to both heat and age. Carbon-film types are only somewhat better. Wirewound types are not readily available in the high values required for such a circuit. The only choice was 1%-tolerance military type RN-65 metal-film resistors.

The RN-65 is specified to have ½-W dissipation at 75° C. Most ½-W carbon-composition types are specified at 50° C or lower temperatures. The allowable voltage across an RN-65 resistor is 900 versus 350 for a typical ½-W composition. Proper component selection for something as simple as the plate-voltage meter makes an amplifier much easier to live with during its life.

R15 is used to both keep the B+ metering circuit complete when the multimeter is switched to another position, and to calibrate the meter. The dropping string was purposely selected low (4.99 M versus 5.00 M) to allow for a

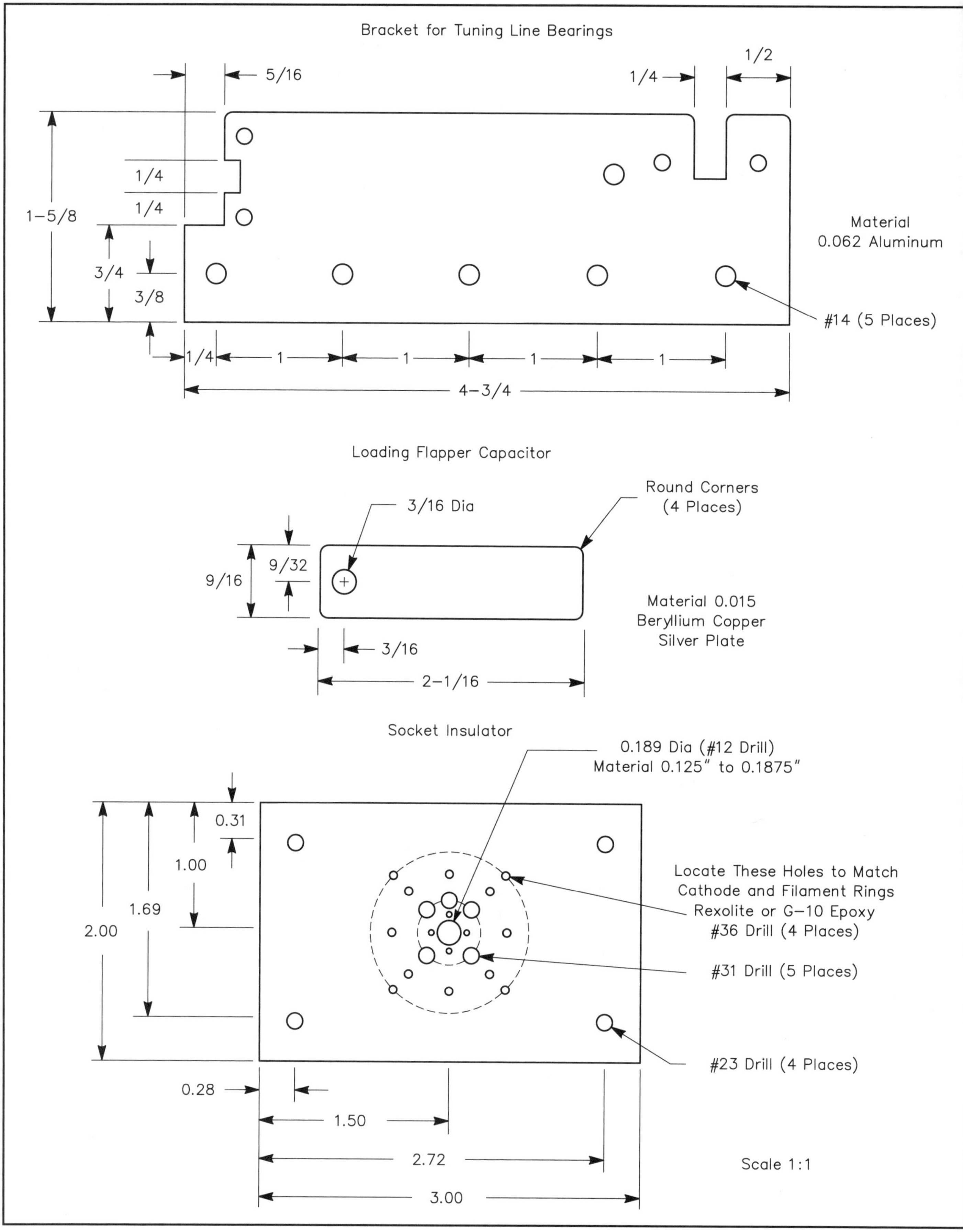

Fig 10—This drawing shows the tuning-line bearings bracket (top), made from 0.062-in. thick aluminum; plate-loading flapper capacitor (center), made from 0.015-in. thick beryllium copper, silver plated; tube socket insulator, made from Rexolite or G-10 epoxy sheet.

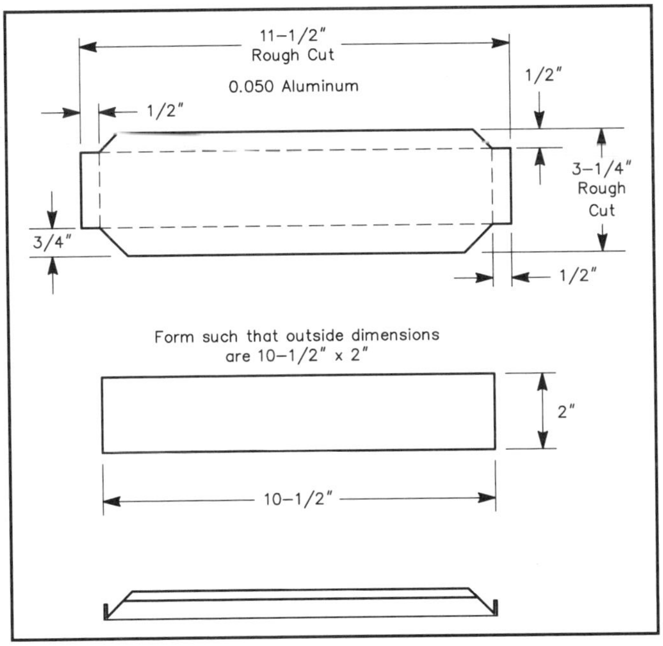

Fig 11—Bottom cover of cathode compartment, made from 0.050-in. aluminum sheet.

calibration adjustment and to generate a voltage that could be used for a high-voltage-present sensing circuit. The 1500-Ω value worked for the multimeter used, which has a 40-Ω internal resistance. Since this resistor is called on to make less than a 5% adjustment to the meter reading, both its initial value and normal aging value changes will have a very small effect on the HV meter accuracy. For example, a 10% change in the resistance of that 1500-Ω calibration resistor will cause a 0.27% change in the meter reading. A value change that small is unreadable on even a very high-quality analog meter movement.

The high-voltage sensing circuit uses a simple switching transistor (Q1) to turn on a relay (K1). The plate-voltage meter dropping string provides the required turn-on voltage to the base of Q1. With the 4.7-kΩ resistor used (R16), the sensing relay pulls in at about 1900 V when high voltage is applied, and drops out at about 1200 V when plate voltage is lost. The voltage at which the relay pulls in can be changed by adjusting R16. To raise the pull-in voltage, increase R16. A quick and sure way to blow a very expensive 8938 is to apply full drive without plate voltage present. The grid will be asked to draw a destructive amount of current during absence of plate voltage. This sensing circuit becomes a cheap piece of insurance. I use this circuit to open a relay in the IF drive to my transverter. With this arrangement I feel secure, even during unattended operation.

Plate voltage is fed from the power supply via RG-59 coax and MHV high-voltage connectors. Coaxial cable is highly recommended in the interest of safety. RG-59 should be capable of withstanding 5500 V dc. The MHV connectors are rated at 5000 V. They provide a ground connection that must be made before the center pin makes contact. One of the worst possible situations is to lose the ground connection between the amplifier and power supply and still have the high voltage connected. The use of MHV connectors prevent this from occurring.

The HV power supply should have a 25-Ω, 50-W resistor in series with the B+ line. This resistor absorbs the energy stored in the filter capacitor in the event of an arc. Chances are pretty good that you will experience arcing when you fire up your 8938. All new tubes go through a process where particles left over from the manufacturing process are burned off or collected to the walls of the tube. A tube such as the 8938 requires relatively close grid-to-plate spacing to make it efficient at 432 MHz. When one of these particles gets between the grid and plate an arc results. In this amplifier, the grid is at ground. The arc has the effect of discharging your 3000-V supply through the grid. The results can be catastrophic. I know of at least one operator who lost his brand new 8938 this way. When a tube suffers any sort of shock, which may happen when the tube or amplifier is moved or shipped, some of the particles can break loose and again induce arcs. Gradually this particle-induced arcing subsides. Even after initial burn in, this resistor protects against RF-induced dc arcs from the plate line, which can easily happen if you forget to connect your antenna. This high-voltage protection resistor has the additional benefits of operator safety and component protection for items such as the meters. In the event of an arc it minimizes the tendency for the B line to go below ground potential, which is both a safety hazard and potentially destructive to components in the metering and cathode-bias circuit.

When firing up the amplifier for the first time, it is a good idea to add a 50-Ω, 100-W resistor in series with the plate-voltage supply. This resistor gives additional protection against arcing. It is also a good practice when firing up a brand new tube to let the amplifier sit with the filament and plate voltage on for a half hour, to collect any stray particles in the tube that may have broken loose in transit, *before* applying drive. If you have closely followed the dimensions given for the amplifier, the following initial tuning capacitor setting will set the amplifier very close to optimum tuning at 1500-W output with 3300 plate volts under load. If you are using a different plate voltage, different loading and tuning capacitor settings will be required.

- Input tuning and loading, plates approximately half meshed.
- Loading capacitor edge away from the output connector about 1⁵⁄₁₆ in. above the bottom of the plate compartment.
- Variable plate-tuning capacitor free edge 1⁵⁄₁₆ in. above the bottom of the plate box.
- Fixed plate tuning capacitor: far edge ¾ in. above the top of the plate line.

Once the filament and plate voltages are verified, short R3 and let the amplifier idle. Idling current should be 100-150 mA if your plate voltage is over 3000 V. If you are unsure of operation it might be wise to initially tune up at reduced voltage. I have a 2000-V tap on my plate transformer for that purpose. Next, apply enough drive to get 200 mA or so of plate current. Quickly adjust the input tuning controls for a peak in plate current, reducing drive if necessary to keep plate current within reason. You don't have to be critical, as the input controls will have to be readjusted under full power. Then start adjusting the plate tuning and loading controls for maximum efficiency, gradually increasing drive power.

When you are near the 1500-W output level, put a directional wattmeter in the input circuit and adjust the input tuning and loading for best SWR at full drive. You should easily be able to adjust the input for under 1.2:1 SWR. For those using Bird 43 wattmeters, at 72-W forward on a 100-W slug you should barely be able to see the needle move off the zero mark in the reflected position. Once the input circuit is adjusted, it should never have to be touched unless the tube is replaced or the amplifier is disturbed. You can now go back and repeak the plate controls for best efficiency. If the grid current is high (above 30 to 40 mA) you have too little loading or plate tuning capacitance. If the grid current is near zero or negative, the amplifier is over loaded. On 2 different amplifiers and 2 different tubes, best efficiency was obtained when the indicated grid current was 10 to 20 mA. You will probably have to spend some time trying slightly different loading and tuning combinations before you find the best efficiency point. This should also correspond very closely to the plate current dip as the plate-tune capacitor is moved through its range. If you are in doubt, remember it is safer both in terms of linearity and stress on the tube (back heating) to over load than under load.

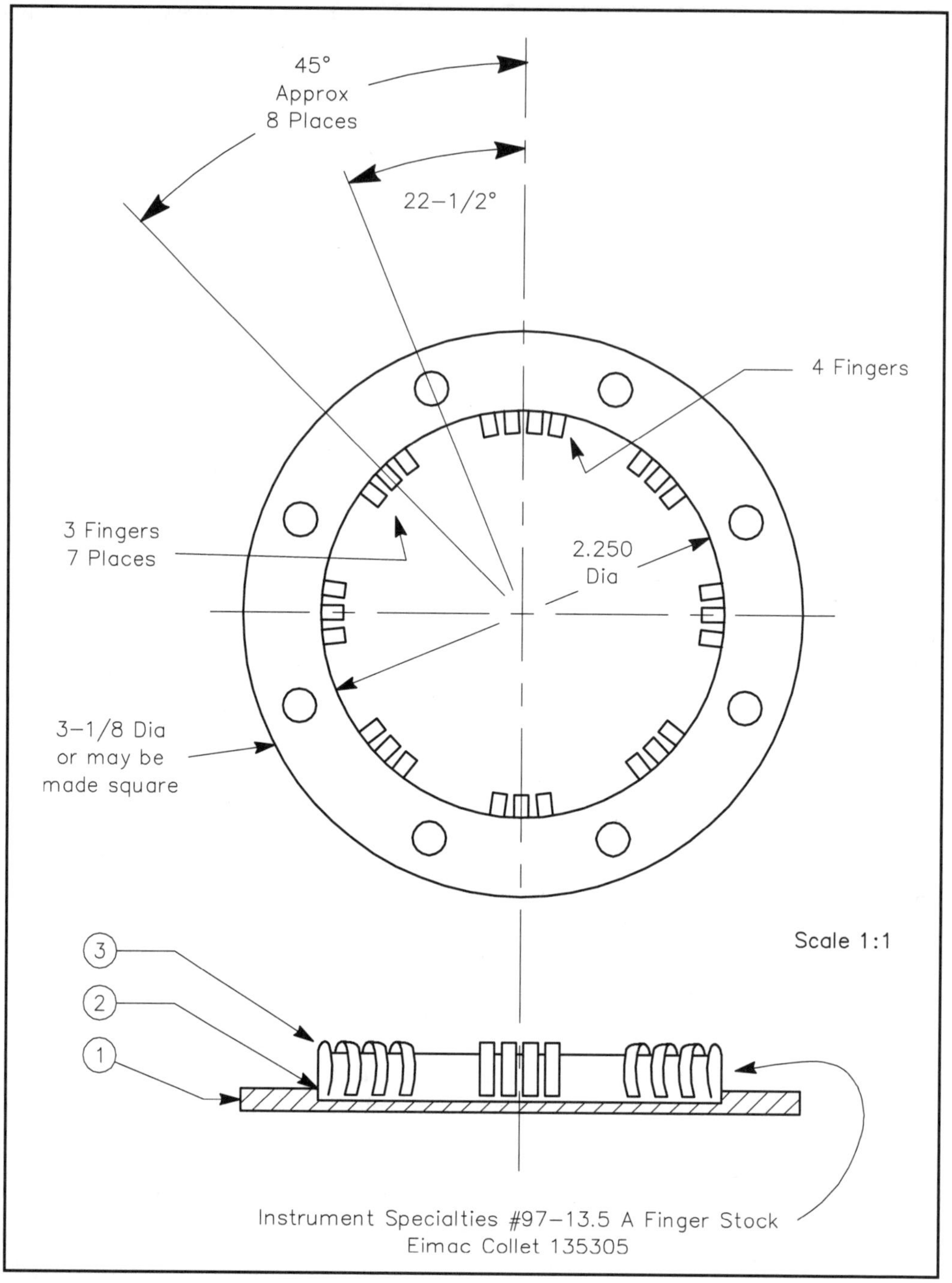

Fig 12—The tube socket finger stock or Eimac collet must be modified before installation. Remove fingers as shown.

Performance of the amplifier has met all expectations. Operation at 432 MHz is completely stable. From a cold start the amplifier, with full drive applied, will come up to 1500 watts output within a few seconds. After this it will hit 1500 watts instantly every time the key is hit. The amplifier has been tested running 1500-W output continuously for over two hours with negligible power drift observed. The amplifier has survived several EME contests (25 hours each of slow-speed CW in 2.5-minute transmissions) and many VHF/UHF contests without a hiccup. The most worrisome part of its operation is how the coax connectors and cable will hold up. RG-213 coax gets very hot and the dielectric will melt with 1500-W power continuously flowing through it. Even RG-331 (½-in. alumifoam) gets warm to the touch during extended CW operation.

There have been many stories floating about telling tales of blown Type-N connectors at high power levels. I have never blown a properly assembled Type-N connector. Type-N connectors were in line during the 2-hour continuous running test also. The connectors are noticeably cooler than the cables which they are connected to during normal operation. I have blown two Type-N connectors on cables I inherited from somewhere. Disassembly of both after the failure revealed poor shield attachment to the connector. My antennas are always tuned for a low SWR (under 1.2:1)

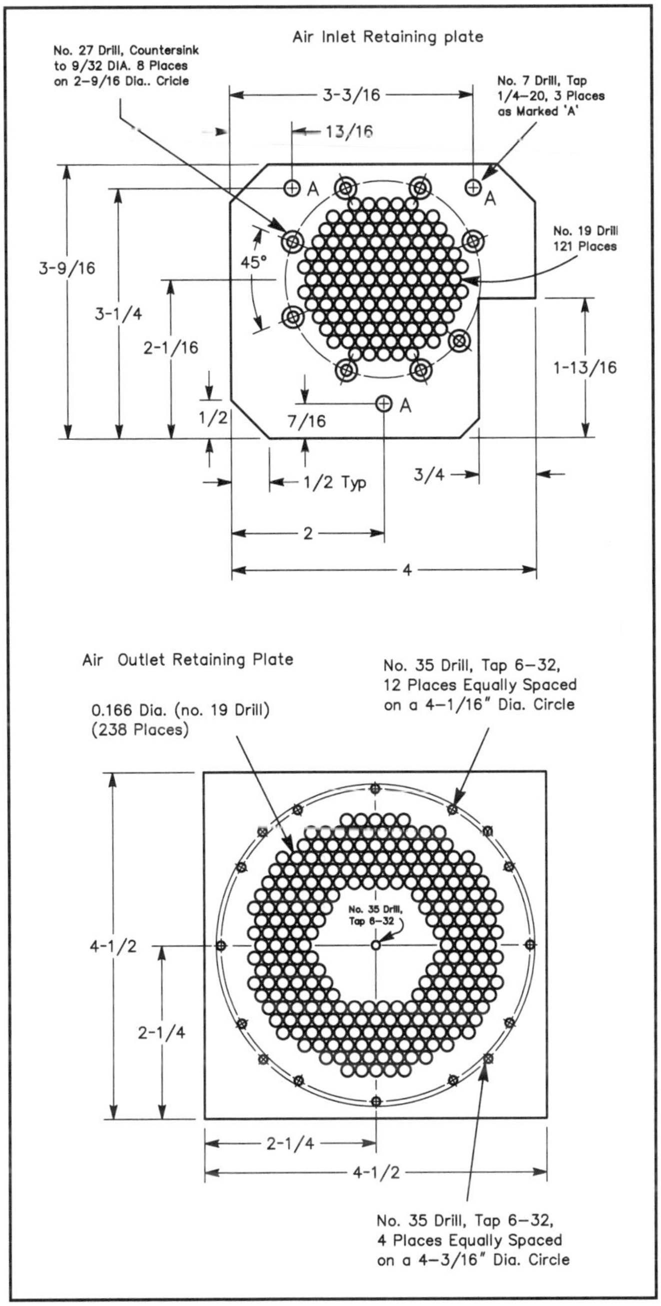

Fig 13—Air inlet retaining plate (top) and air outlet retaining plate (bottom). If you use a blower other than the unit specified in the parts list, you may have to modify the hole locations and sizes shown on the drawing. Make the plates from 0.125- to 0.250-in. aluminum.

and I do not operate the amplifier at full power if the SWR rises above 2:1 for any reason.

Relays become another problem at this power level. I have found that the Transco D type are not suitable for this power level, even intermittently. There is a high-power D type, the DX, but I have not been fortunate enough to obtain any. The Transco Y type has been used without any problems but I have never run continuous power through one for more than a couple of minutes at a time. Dow Key type 60 and 260 relays have been used without any problems, but they do put an impedance bump in the line, as they are not 50-Ω impedance devices. The higher power Amphenol relays (with Type-N connectors) have also been used without problems. There are probably several other relays that can be used. With several companies making copies of some of the various Transco and Amphenol relays, there should be an adequate selection. One of the main causes of relay failure is contact bounce. You may have a system that will work correctly 999 out of 1000 times. With a 1500-W amplifier all you need is that 1 out of 1000 times to put full power to a not-quite seated contact. I highly recommend that you use a delay sequencer that gives the transmit-receive coax relay a minimum or 60 to 80 ms, to be certain it is switched and seated before output power can be obtained.

Specific operating parameters at the 1500-W output level are:

Plate Voltage: 3250 V full load
Plate Current: 880 mA
Idle Current: 75 mA (at 350 W)
Grid Current: 12 mA
Input Power: 2860 W
Power Output: 1500 W
Efficiency: 52.4% (apparent, including drive feedthrough)
Drive Power: 72 W typical, 60 W-85 W depending on tube
Input SWR: 1.14:1
Power Gain: 13.2 dB typical
Bias voltage: 25.2 V

When measuring the performance of an amplifier such as this one, be certain that your measurements are meaningful. I have heard numerous impossible performance claims for various amplifiers. Specifically, the SWR of your test load must be very low. All throughline-type power meters are confused by reflected power. Even Bird wattmeters have limited directivity. When using a load with even moderate SWR (1.5:1), you can obtain several different readings by changing the length of the feed line. Harmonic energy confuses the watt-meter and, in general, makes it read higher than it should. Be sure that your plate-current and plate-voltage meters are really accurate. The approximate efficiency was also checked by measuring the temperature rise of the exhaust air. This measurement indicated that the efficiency of the amplifier may be slightly better than the 52.4% figure that was measured on an inline wattmeter.

If you do not have 3300 V or greater plate voltage available, 1500-W output power can still be obtained from the amplifier. In intermittent keyed CW or SSB service the 8938 can safely handle 1.2-A plate current. At a lower plate voltage, the apparent efficiency of the amplifier is greater, due to the higher drive power required to run 1500-W output. In no case should the 8939 be run at greater than 1.35-A cathode current. Higher cathode currents can cause catastrophic failure of the tube. This failure will occur as the oxide coating on the cathode literally boils off. The cathode current is the sum of the plate and grid currents. The grid current is somewhat unpredictable at 432 MHz, due to transit time and secondary emission effects which may give negative grid current indications if the amplifier is over loaded. To be safe, assume that grid current is 7 to 10% of the plate current no matter what the grid meter indicates.

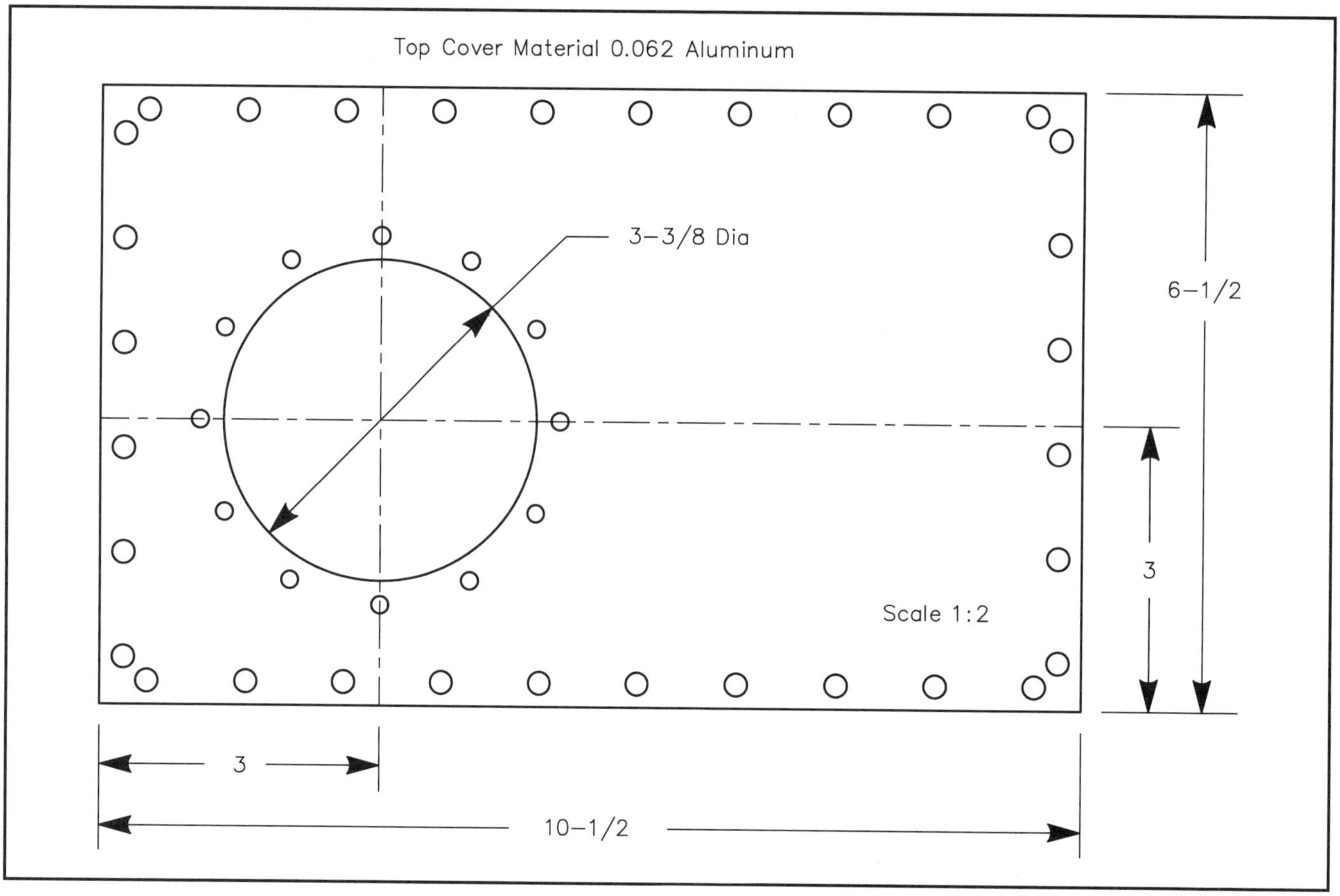

Fig 14—Dimensions of the top cover, made from 0.062-in. aluminum sheet. Note that this drawing is not to scale.

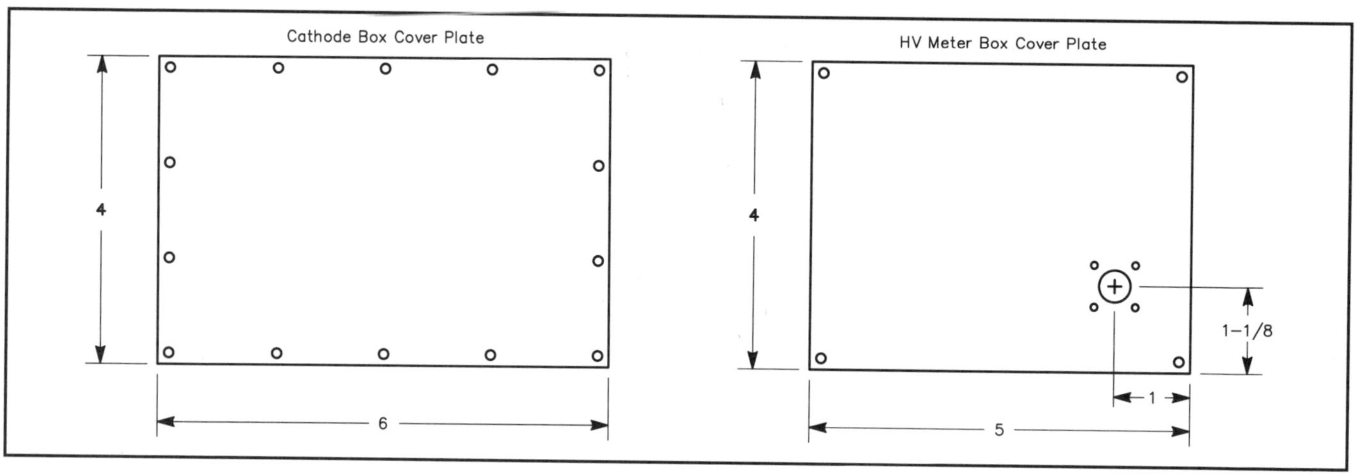

Fig 15—Dimensions and hole-drilling information for the cathode box cover (top) and he HV-meter resistor box cover (bottom).

With a little extra thought in planning and construction, a high-power amplifier can be a pleasant addition to your 432-MHz station. Don't settle for less than an amplifier which is highly linear, easy to tune, stable, reliable and safe, no matter what the power output level.

Notes

[1]J. Chambers, W. Orr, "2-kW PEP Amplifier for 432 MHz," *Ham Radio*, Sep 1968, pp 6-12. Also published as *Amateur Service Newsletter AS-25-1* (San Carlos, CA: EIMAC).

[2]Knadle, R, "High Efficiency Parallel Kilowatt Amplifier for 432 MHz," *Part 1*: *QST*, Apr 1972, pp 49-55; *Part 2*: *QST*, May 1972, pp 59-62, 79. Construction information also appears in *Radio Amateur's VHF Manual*, third edition, pp 299-304 (Newington: ARRL, 1972).

[3]S. Powlishen, "A Grounded-Grid Kilowatt Amplifier for 432 MHz," *QST*, Oct 1979, pp 11-14. Construction information also appears in *The ARRL Handbook for Radio Amateurs*, for the years 1981 through 1986.

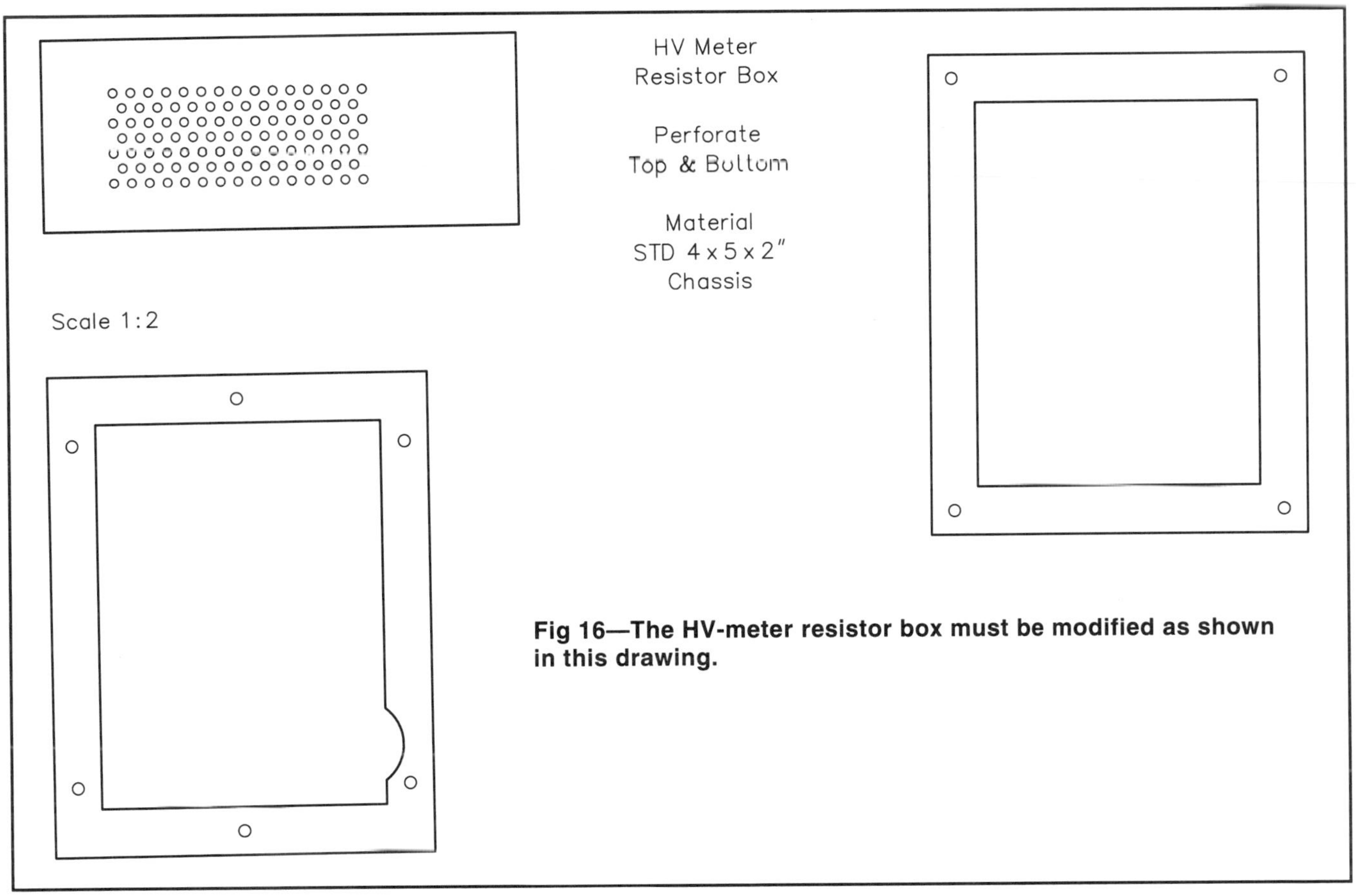

Fig 16—The HV-meter resistor box must be modified as shown in this drawing.

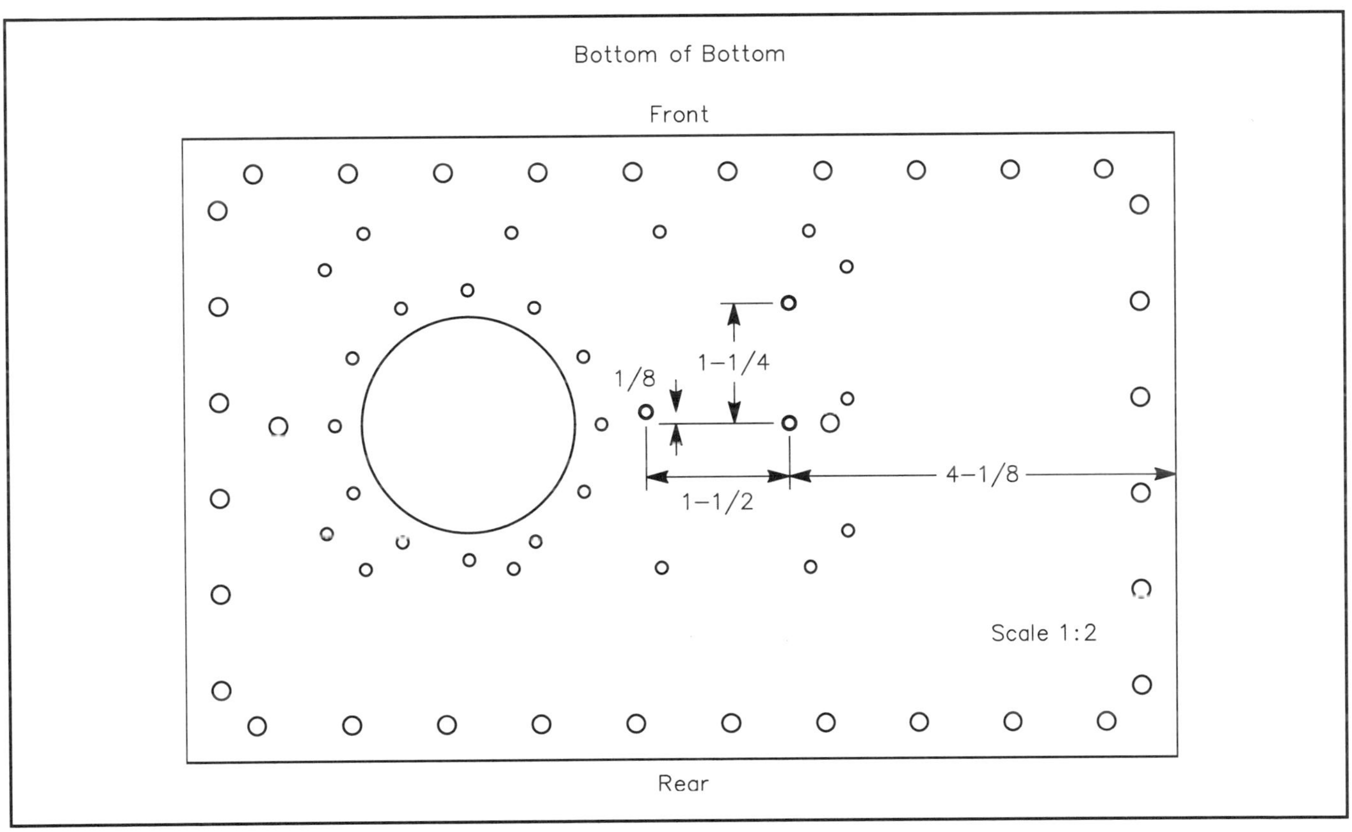

Fig 17—Bottom view of the bottom cover of the amplifier enclosure, made from aluminum sheet (thickness not critical).

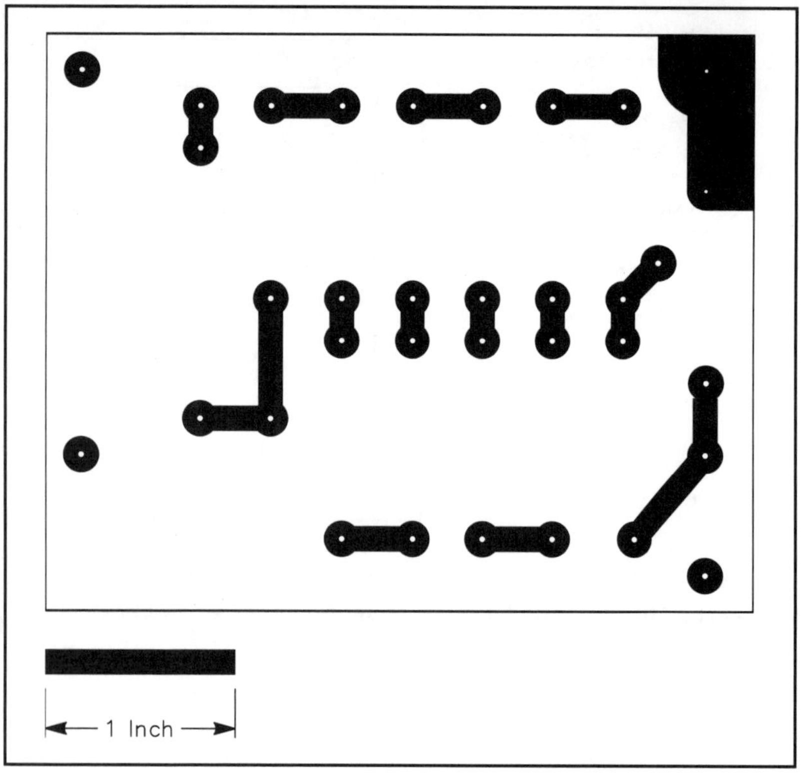

Fig 18—Foil side of the HV-meter resistor circuit board.

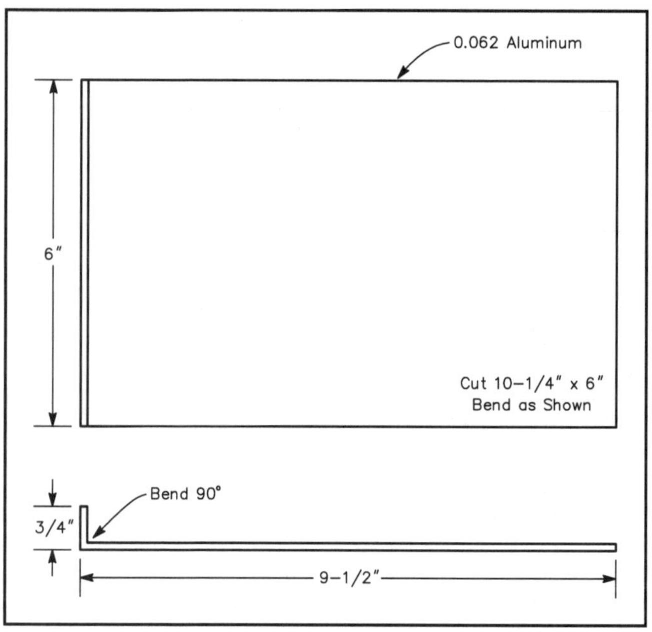

Fig 19—Details of the brackets used to mount the amplifier to the rack panel. Two are required.

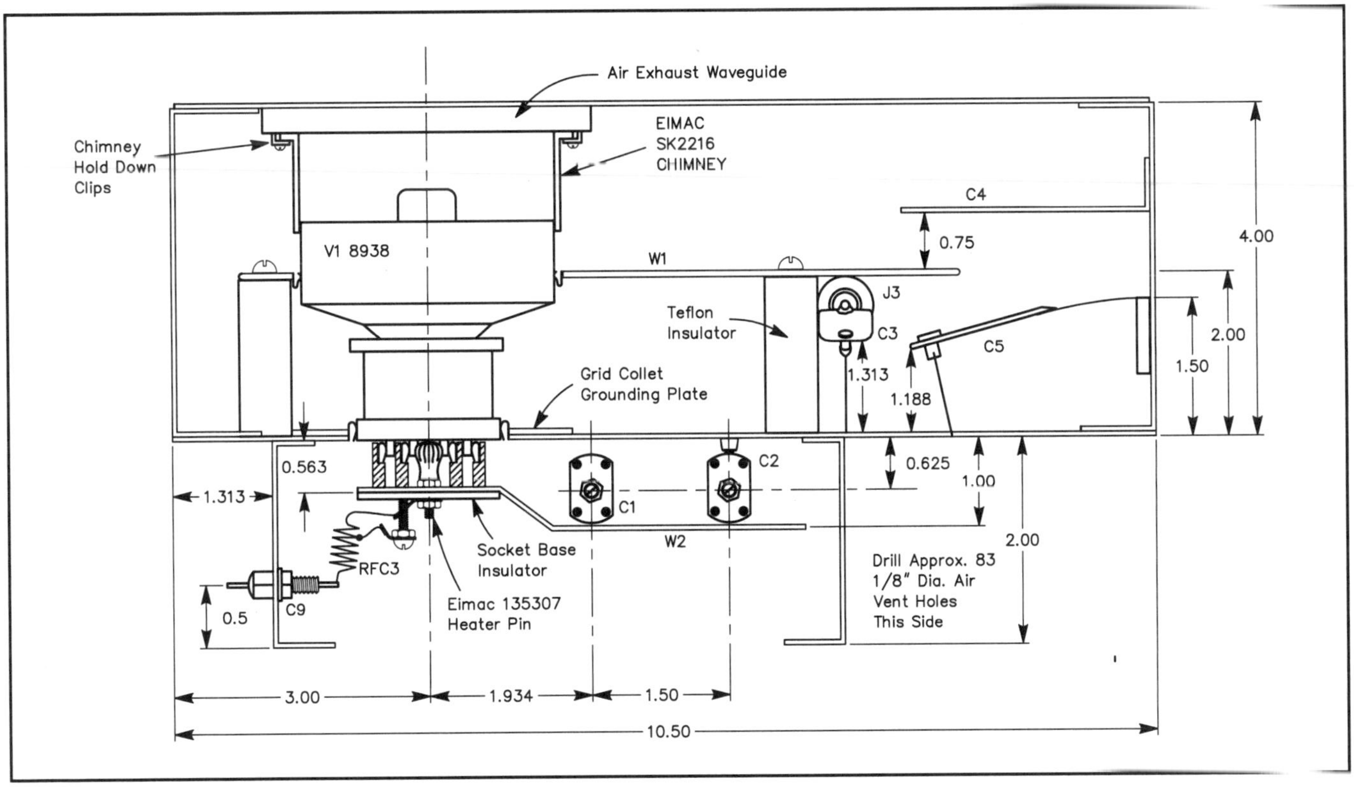
Air Exhaust Waveguide
Chimney
Hold Down
Clips
EIMAC
SK2216
CHIMNEY
C4
0.75
4.00
V1 8938
W1
J3
Teflon
Insulator
C3
C5
2.00
1.50
1.313
1.188
Grid Collet
Grounding Plate
0.563
0.625
1.00
C2
1.313
C1
W2
2.00
Socket Base
Insulator
RFC3
Drill Approx. 83
1/8" Dia. Air
Vent Holes
This Side
0.5
C9
Eimac 135307
Heater Pin
3.00
1.934
1.50
10.50

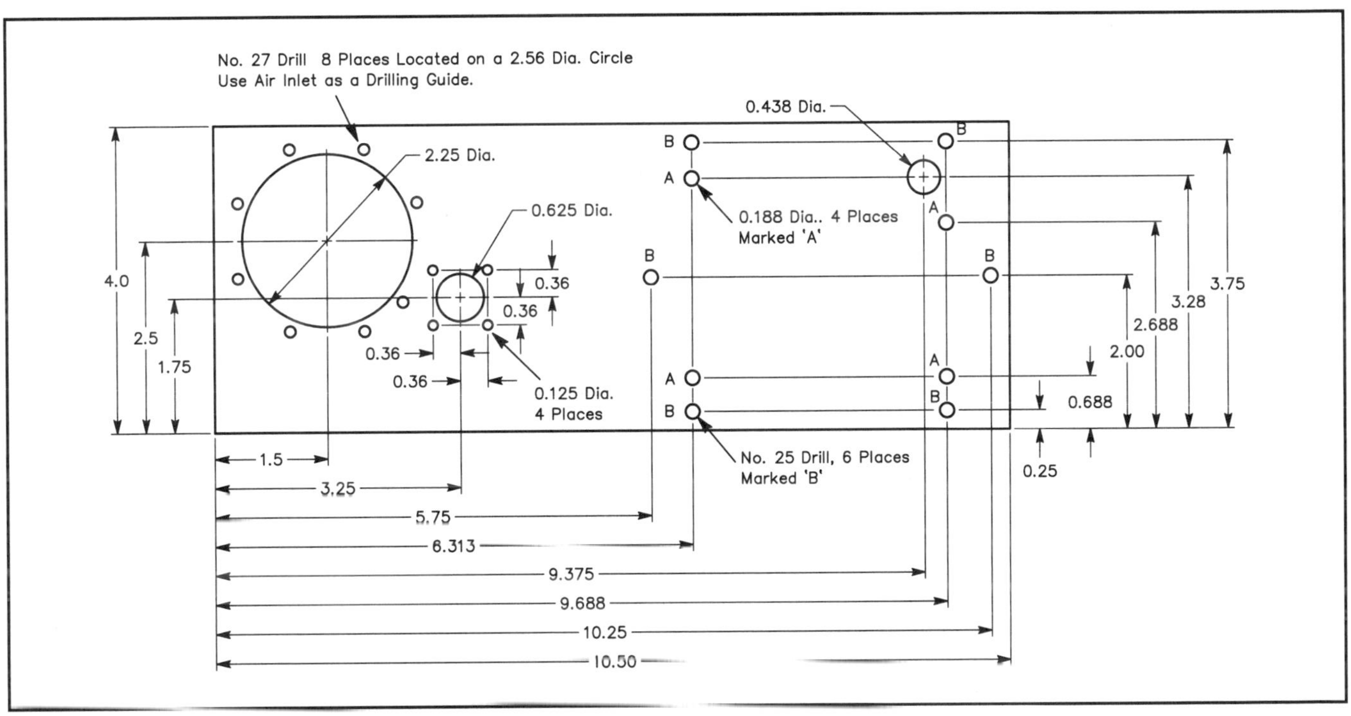
No. 27 Drill 8 Places Located on a 2.56 Dia. Circle
Use Air Inlet as a Drilling Guide.
0.438 Dia.
2.25 Dia.
B
A
0.188 Dia.. 4 Places
Marked 'A'
0.625 Dia.
4.0
0.36
0.36
3.75
3.28
2.688
2.00
2.5
1.75
0.36
0.36
0.125 Dia.
4 Places
0.688
0.25
No. 25 Drill, 6 Places
Marked 'B'
1.5
3.25
5.75
6.313
9.375
9.688
10.25
10.50

The 1500-W amplifier makes a compact, easy to use package.

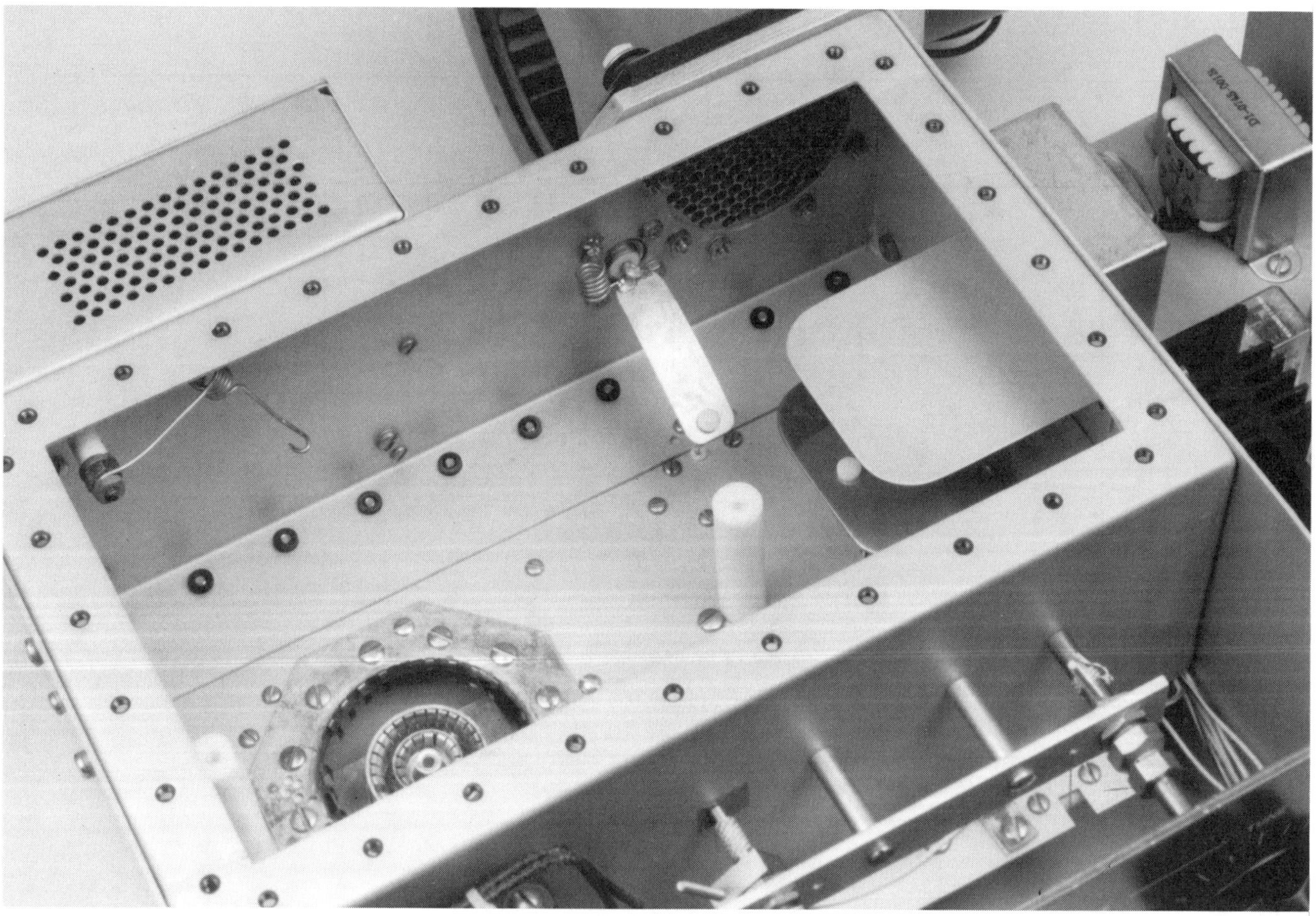

Plate compartment with plate line and tube removed. The large plates are the tuning capacitor; the small strip at top center is the ouput loading capacitor. The standoff supports the plate line.

Top view of the amplifier, showing the plate line and plate tuning flopper.

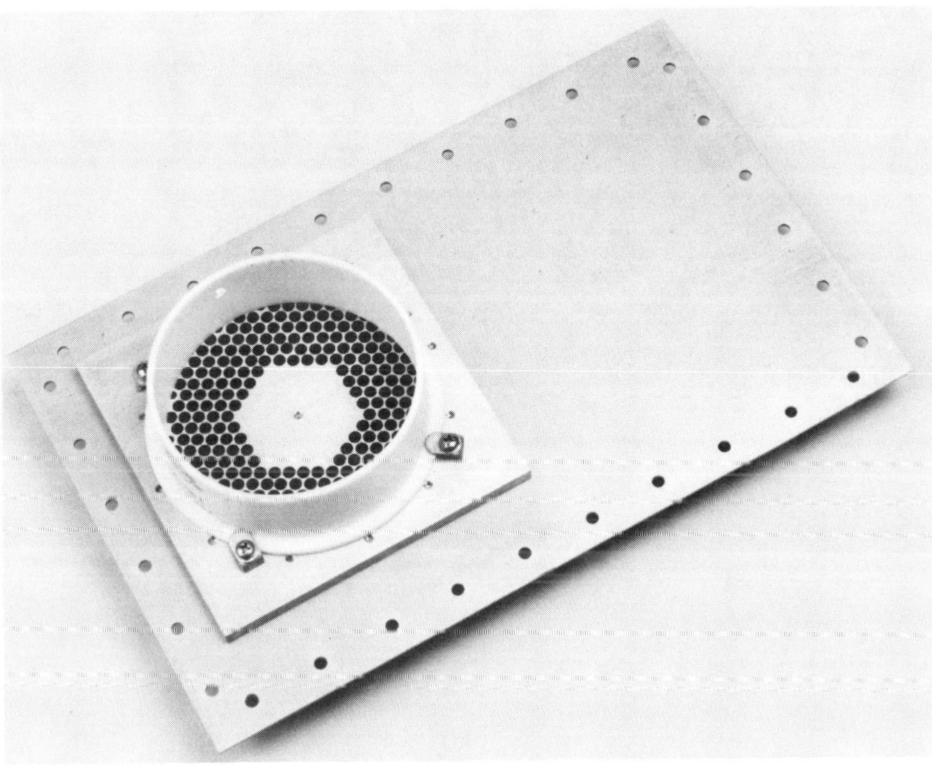

Top corner of the plate compartment showing the chimney and an outlet RF shield.

Top view of the amplifier with plate compartment cover in place.

Cathode compartment, showing the cathode line and tuning controls.

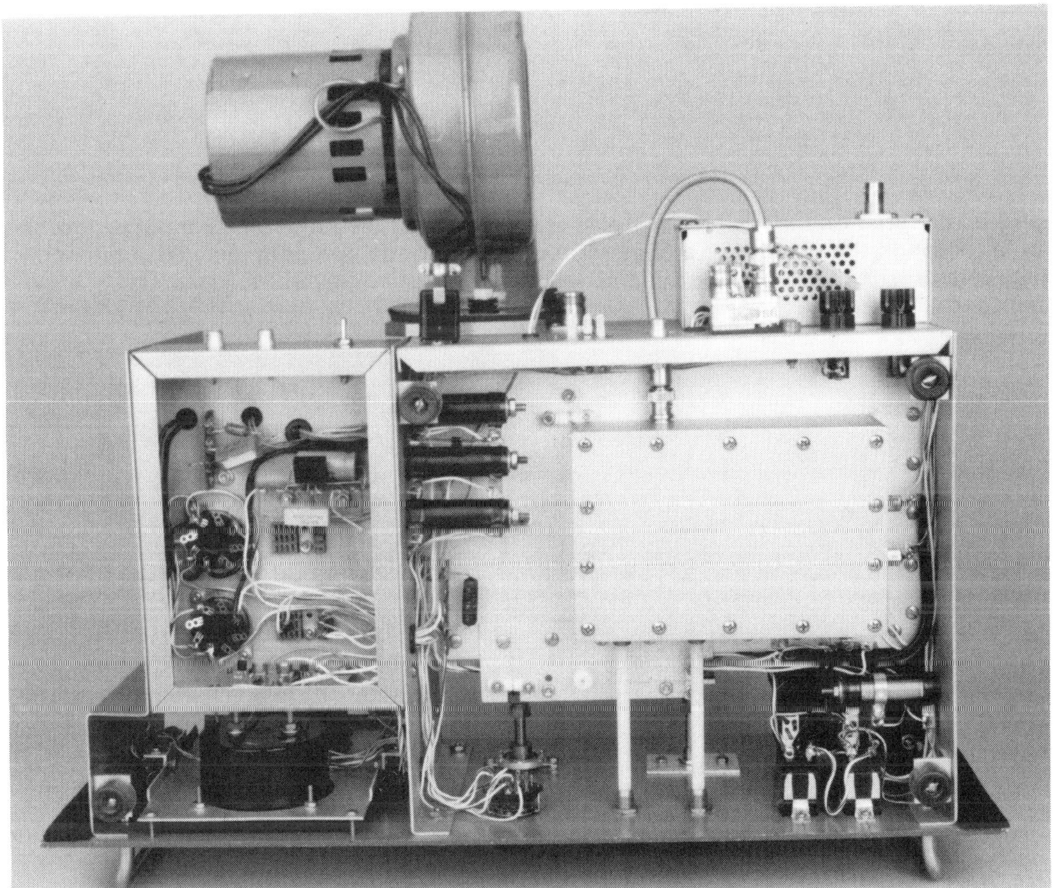

Underside view of the 8938 amplifier, with the cathode-compartment cover in place.

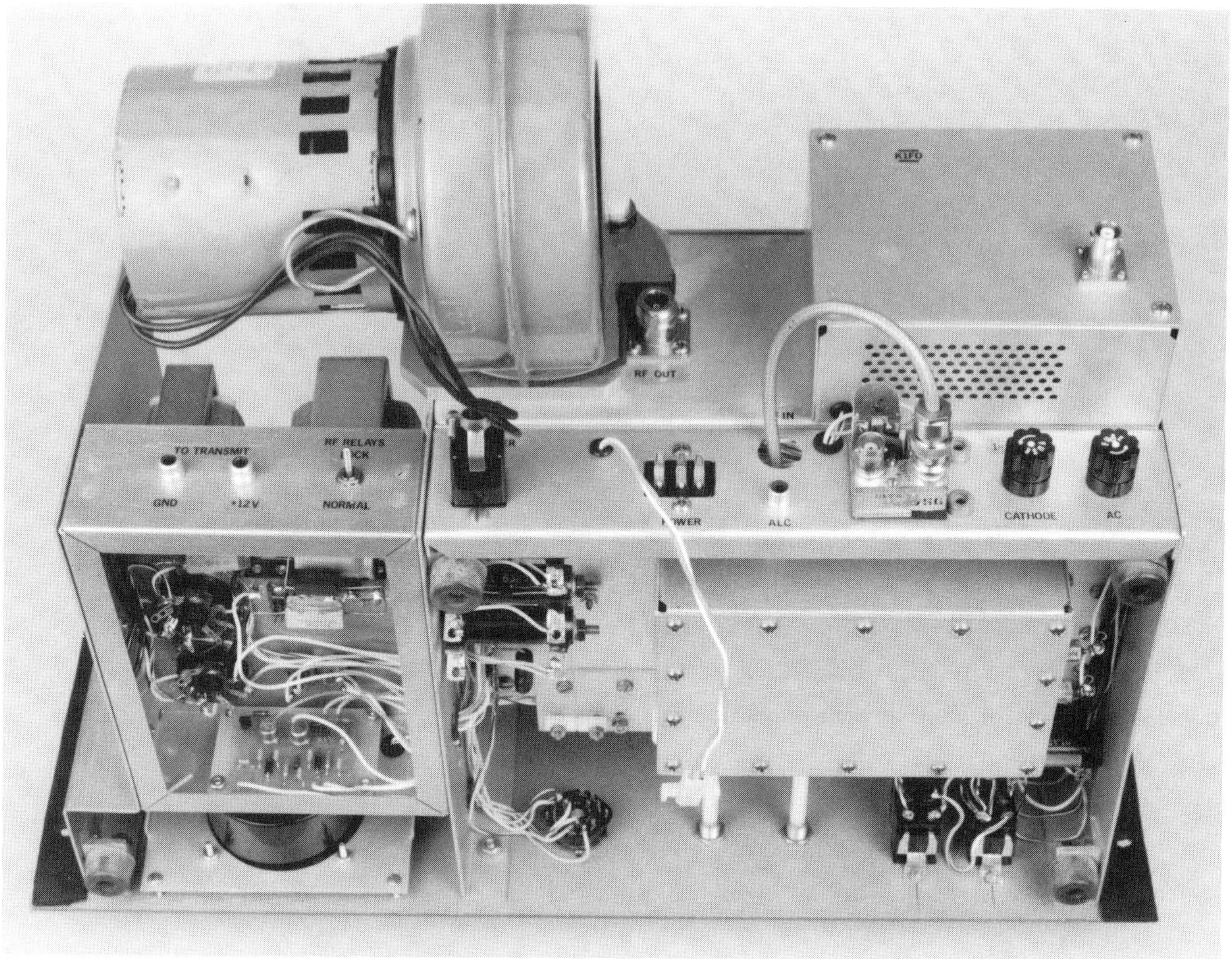

Rear view. The RF-input relay is mounted on the main chassis, to the left of the fuse holders. RF input to the amplifier is by the cable routed through the chassis to a connector on the cathode compartment. The connector on the subchassis above the relay carries high voltage.

A 1296-MHz Linear Power Amplifier

By Bill Olson, W3HQT

The linear amplifier shown in Fig 1 uses a Mitsubishi power amplifier module and a handful of other components to deliver up to 3 watts output from 1240 to 1300 MHz.

The M67715 module has 50-ohm input and output impedances. No impedance-matching networks are used, and no alignment is required. An input signal of 10 mW (10 dBm) will result in about 3 watts output. The M67715 is fairly rugged; Mitsubishi specifies it will tolerate an output SWR of 20:1.

Fig 2 is the schematic diagram of the amplifier. The module requires two positive supply voltages and a regulated bias supply. The higher supply voltage (V_{cc2}) can be up to 16 V dc, but the lower voltage (V_{cc1}) cannot exceed 9 V dc. A 7809 three-terminal regulator connected to the V_{cc2} supply provides the 9-volt V_{cc1} and bias voltages. Chip capacitors are used to bypass the V_{cc} and bias terminals directly at the module. Fig 3 shows an inside view of the amplifier.

U1 and U2 are fastened directly to a piece of 2-in. aluminum channel, which is in turn attached to a heat sink. Mitsubishi specifies a total efficiency of about 25% for this

Fig 1—A simple but rugged 1296-MHz linear power amplifier.

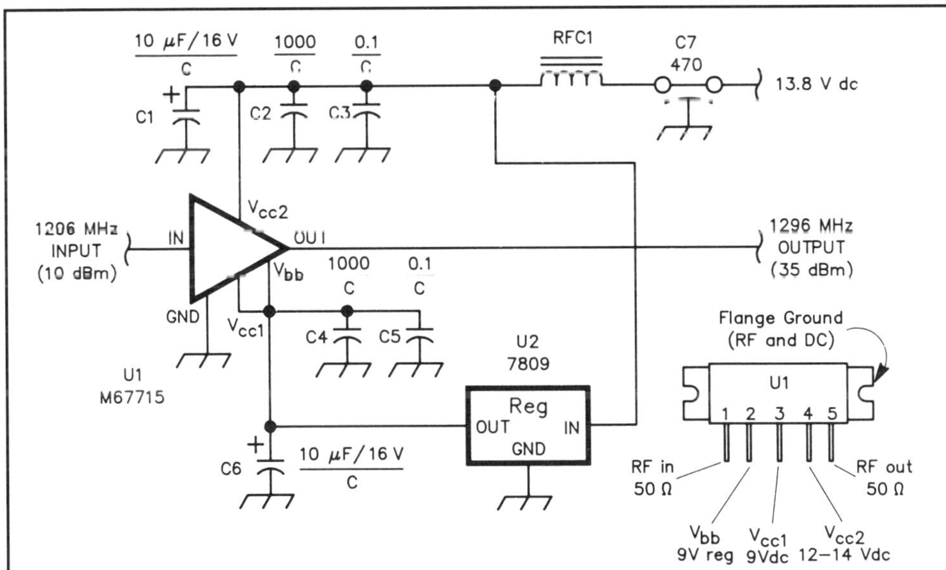

Fig 2—Schematic of the 1296-MHz linear power amplifier.

C1, C6—10-µF, 16-V dc chip electrolytic.

C2, C4—1000 pF, ceramic chip, NP0.

C3, C5—0.1 µF ceramic chip, NP0.

RFC1—5 + no. 28 bare or enam wire on FB-43-5111 ferrite bead.

module; for 3 watts output the module dissipates about 9 watts. The heat sink shown in Fig 1 may be conservative, but was of convenient size. A U-shaped aluminum shield should be used to cover the aluminum channel when the amplifier is in use.

Parts, kits and complete amplifiers are available from Down East Microwave.

Fig 3—Inside the amplifier. U1 is at the top and U2 at the bottom. Mount the circuit board to the heat sink before installing the Type-N connectors.

A Quarter Kilowatt 23-cm Amplifier

By Chip Angle, N6CA

(From *QST* for March and April 1985)

The amplifier project described here and shown in Figs 1 through 25 offers the following features.

1) Covers 1240 to 1300 MHz.
2) Linear operation.
3) Grounded grid 7289/2C39 cavity amplifier, single tube.
4) Power gain ranges from 12 dB to 20 dB depending on output power, input power, loading, anode voltage and grid voltage.
5) 50-Ω input and output—no stub tuner required.
6) Power output greater than 200 W with about 12-W drive.

The amplifier described here is a tried and proven design. It works well, is reliable and can be duplicated.

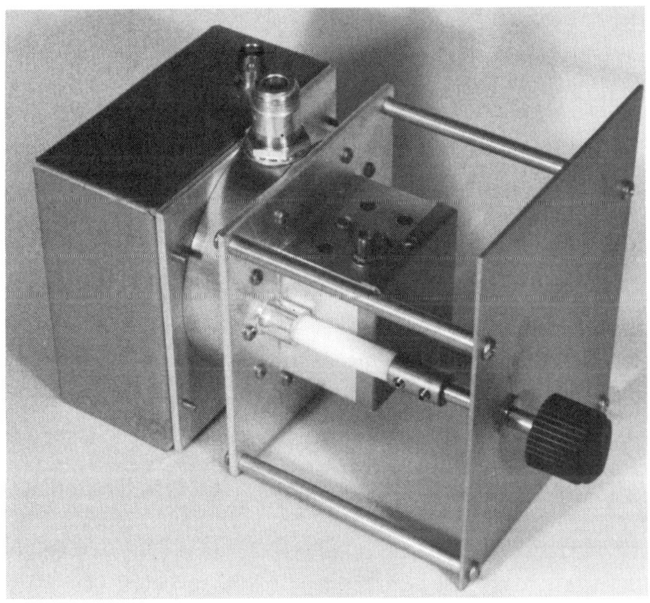

Fig 1—The completed 23-cm amplifier.

General Design Approach

A cavity amplifier is similar to a conventional amplifier designed for lower frequencies. The tube anode excites a resonant circuit, and power is in turn coupled into a load, usually 50 Ω. Instead of using coils and capacitors, as at lower frequencies, the cavity provides the resonant circuit necessary to tune the amplifier output.

The anode cavity of this amplifier is a squat cylinder. Cylinder height is set by mechanical tube requirements. The inside diameter of the cylinder sets the highest resonant frequency. Any capacitance added from the top to the bottom of the cavity will lower its resonant frequency, as will increasing the cavity diameter.

This amplifier uses ⅛-inch-thick copper plates for the cavity top and bottom and a thick-wall aluminum ring, cut from tubing, for the walls. This heavy construction virtually eliminates all resonant frequency variations caused by thermal and mechanical changes.

Fig 2 is the schematic diagram of the cavity amplifier. The circuit is simple. Filament voltage enters the RF deck through C4, C5, RFC1 and RFC2. High voltage is fed to the anode through RFC3. C8, the anode bypass capacitor, is homemade from Teflon dielectric sandwiched between a copper plate and the chassis.

The input pi network easily tunes the entire band at any power level. It is made from two Johanson piston trimmer capacitors and a "coil" made from copper wire. An input cavity is not necessary at 23 cm.

Output coupling is through a rotatable loop that serves as a variable loading control. This allows amplifier-tuning flexibility; it may be tuned for maximum gain or for maximum power. Light loading can produce stable power gains of up to 20 dB.

Amplifier tuning is accomplished with a homemade cylindrical coaxial capacitor with Teflon dielectric (C6). There are no moving metal parts to cause erratic performance. The Teflon rod/tube screws in and out of the coaxial capacitor, increasing or decreasing the capacitance by changing the amount of Teflon dielectric inside the cylinder. With the rod all the way in, the dielectric is all Teflon; with the rod all the way out, the dielectric is all air.

Teflon has a relative dielectric constant (relative to air = 1) of 2.05, which means that the value of the capacitor with the Teflon rod all the way in is twice the value of the capacitor with the rod all the way out. Full capacitance will pull the resonant frequency of the amplifier down to 1240 MHz. Use of only one

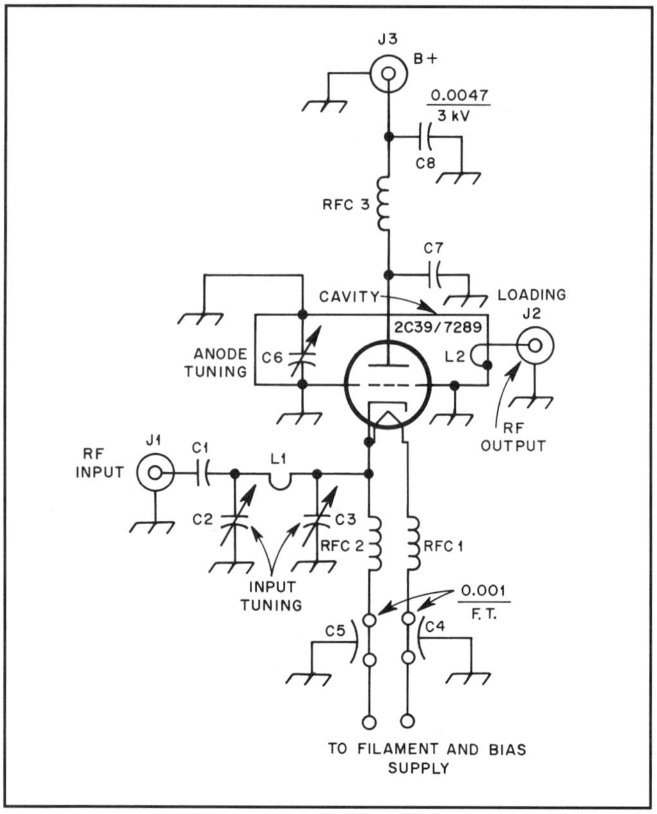

Fig 2—Schematic diagram of the 23-cm amplifier.
C1—3-pF dipped mica capacitor.
C2, C3—1- to 10-pF piston trimmer capacitor (Johanson no. 3957, 5201 or equiv.).
C6—Anode tuning capacitor. See text and Fig 12.
C7—Anode-bypass capacitor, 90 pF. Homemade from copper plate and Teflon sheet.
C8—Disc ceramic, 0.0047-µF, 3-kV capacitor.
J1—5-mm SMA connector, chassis mount, female.
J2—Modified Type-N connector. See text and Fig 8.
J3—Female chassis-mount BNC connector.
L1—Loop of no. 18 bus wire soldered between C2 and C3. See Fig 16.
L2—Output coupling loop. Part of output-connector assembly. See text and Fig 8.
RFC1, RFC2—5 t no. 20 tinned, 3/16-in. ID.
RFC3—3 t no. 20 tinned wound on a 20-Ω 1-W carbon-composition resistor.

anode tuning adjustment means the amplifier will have more gain because shunt capacitance has been minimized.

Thermal Considerations

The cavity walls are formed by a thick-wall aluminum ring, which is sandwiched between two thick copper plates. RF and thermal properties of these two metals are reasonably close, whereas brass is rather poor in both respects. The 7289/2C39 tube used in this amplifier is being run at 2 to 2½ times its normal dissipation rating; therefore it's important to have a cavity that remains thermally stable.

Most previously described amplifiers have used sheet brass in their construction. This has usually meant constant retuning of resonance to maintain output power at or near maximum.

The copper and aluminum construction in this amplifier has solved all thermal stability problems. The amplifier can easily be run key down for over an hour at 200-W output without retuning. This, of course, is obtained only with a good tube and water cooling.

Water cooling keeps the internal structure of the tube thermally stable. When air cooling is used for output levels of 100 to 150 W, output power fluctuations are a direct result of internal tube changes. These changes vary from tube to tube and must be tested for. In some cases, perfectly good RF tubes have had poor thermal stability. Such tubes can make good drivers at lower power levels.

Construction

Hand tools are great if you are skilled and patient. Most people want to hurry up and finish their new project. If that's you, then have a machine shop make all of the parts, leaving only the final assembly up to you. It should cost you about $200. The parts are not difficult to fabricate, but the process is time consuming. If you have the time and patience to do it yourself, this amplifier can be very inexpensive.

Gathering the Materials

All of the materials used in this amplifier are fairly common and should be available from suppliers in most metropolitan areas. Some suppliers have "short sale" racks where they sell odd pieces cut off standard lengths or sheets at reduced prices. The parts for this project are small enough that they may be fashioned from cutoff stock. Surplus-metal houses have some great buys, so start there if one is nearby.

The key to successfully completing this project is careful layout work before cutting or drilling any parts. Invest in a can of marking dye, a sharp scribe, an accurate rule, vernier calipers and several center punches. These tools are available at any machinists' supply shop. The marking dye will make cutting and filing lines much easier to see. Measure all dimensions as carefully as you can and then recheck them before cutting. Mark with a sharp scribe because the sharper the scribe, the finer the marked line, and the finer the marked line, the closer your cut will be to where it should be. Remember—the accuracy of your drilled holes is only as good as your center-punching ability, so use a fine punch for the first mark and then a bigger one to enlarge the mark enough for drilling.

Access to a drill press is a must. It's extremely difficult to drill holes accurately with a hand drill. Although not absolutely necessary, you should have access to a lathe or milling machine to do the best possible job.

Other tools that will aid you with this project are a nibbling tool, a set of files and some sharp drill bits. If you don't already have one, purchase a file card to clean aluminum and copper shavings out of your files as you work. Clean, sharp files are faster and more accurate to work with. You'll also need an assortment of sandpaper for the final finishing work.

The Template Approach

It's best to fabricate a single template for marking and drilling the anode plate, anode bypass capacitor, cavity ring, grid plate and front panel. The template shown in Fig 3 has all of the holes for these parts. If you use the template, you'll only have to make the careful measurements once — after that, it's simple to mark and drill the rest of the parts.

The template approach offers several other advantages. A template makes it much easier to maintain accuracy between the anode plate, cavity ring, grid plate and front panel; these parts will fit perfectly because they were all drilled from the same master. The template approach also makes it possible to set up a small production line if you decide to build more than one of these amplifiers and combine them for higher power, or if a friend wants to build an amplifier along with you.

See Fig 3 for complete template dimensions. Start with a piece of 1/16-inch-thick aluminum stock that is larger than you need and degrease it with soap and water. Dry it off and spray it with marking dye. Scribe a 4-inch square on the stock and cut the template to size. A shear will make this job much easier, but it can be cut with hand tools and filed to size.

Carefully measure and scribe all holes. Note that holes A and B are on the circumference of circles. Use a compass to scribe the circles, and then locate the holes. After you have marked and checked all holes, centerpunch and drill them. The holes should be drilled with a 1/16-inch or smaller bit. Recheck all measurements. If you goof, start again. The time you spend making the template as perfect as you can will save you much time and aggravation when you make and assemble the other parts.

When you finish the template, mark the front side for future reference. All plates that will be made from the template are marked and drilled from the front side (as viewed from the front panel).

Making the Copper Plates

Once you have completed the template, it will be easy to make the copper plates. The anode plate, grid plate and anode-bypass-capacitor plate are all made from 1/8-inch- thick copper. See Figs 4, 5 and 6 for the dimensions of these pieces.

Measure and cut the three plates to the proper dimensions. Carefully break (deburr) all sharp edges to avoid small cuts to your fingers and hands.

Clean the plates with alcohol and spray them with marking dye. Clamp the aluminum template to each plate and carefully scribe the correct holes. Remember that all plates do not have the same holes. The anode plate uses holes A, B, C, F and H; the grid plate uses holes A, C, D, G and H. The anode-bypass-capacitor plate uses holes B, E and H. Use a small center punch to punch all holes accurately and lightly. If they then look accurate, enlarge them enough for drilling.

Copper isn't the easiest metal to work with. It's very stringy, and drilling it can be frustrating. You'll need the proper drill bits for best results. Special drills can be purchased, but that's not really necessary. You can use a grinder to carefully

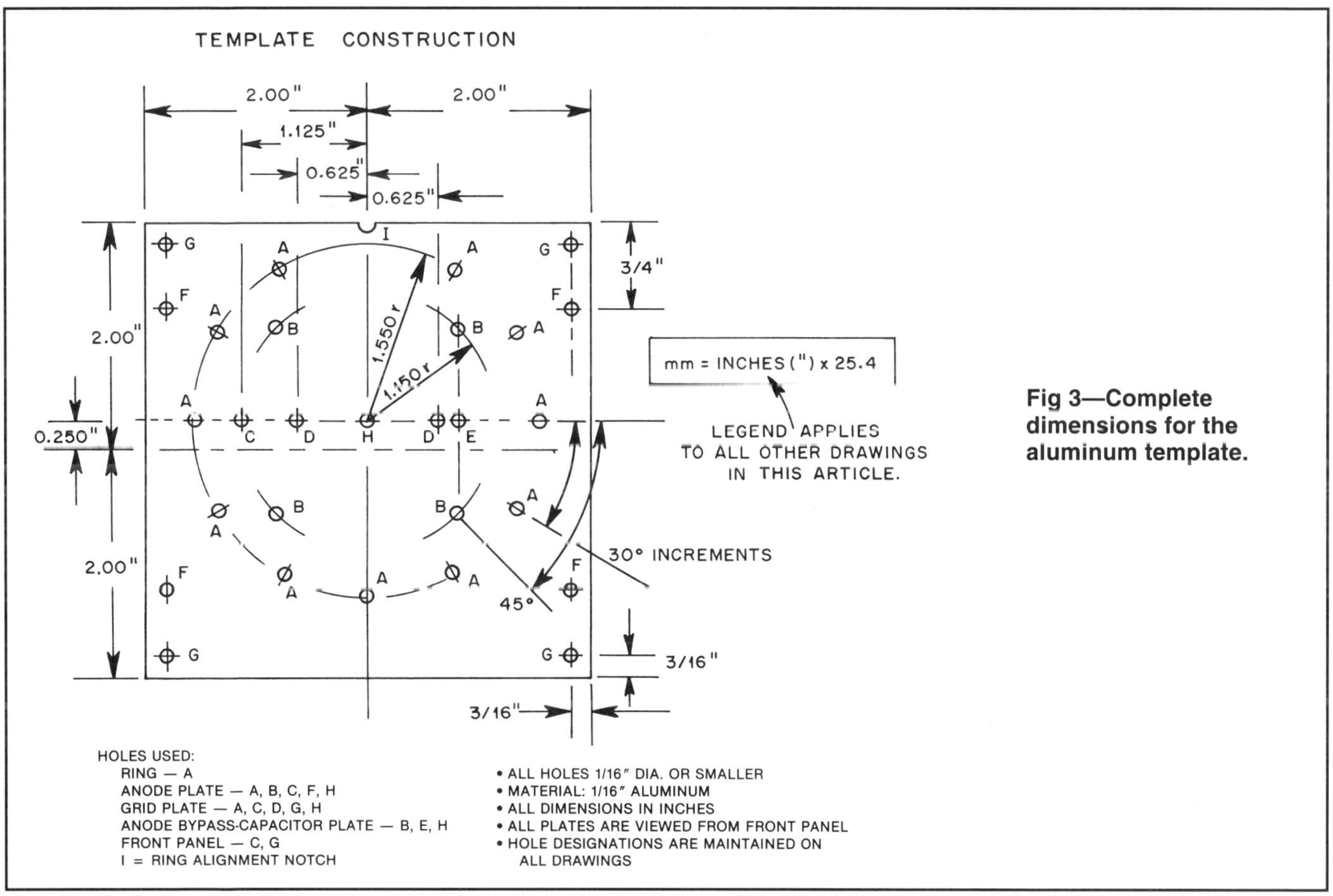

Fig 3—Complete dimensions for the aluminum template.

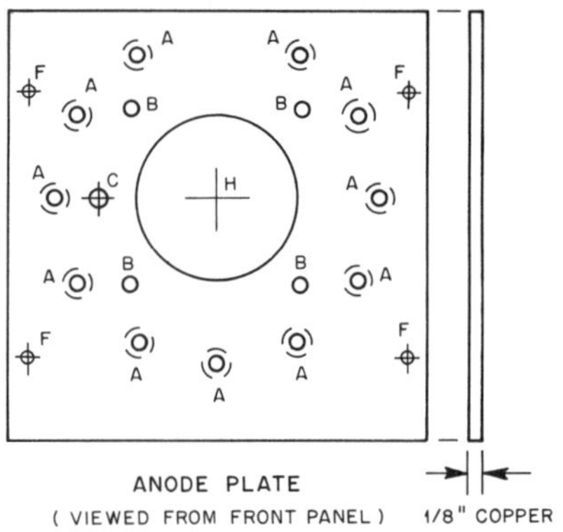

HOLE: A — 0.144-IN DIA. (NO. 27 DRILL). COUNTERSINK BACK SIDE TO 1/4-IN DIA. WITH A 100°C SINK.
B — 0.106-IN DIA. (NO. 36 DRILL), TAPPED NO. 6-32.
C — 5/32-IN DIA. (0.156). FOR ANODE TUNING POST, THIS SIDE.
F — 0.089-IN DIA. (NO. 43 DRILL). TAPPED NO. 4-40 FOR ANODE BOX MOUNTING.
H — 1.500-IN DIA.

Fig 4—Drilling details for the anode plate. See Fig 3 for additional information on hole location.

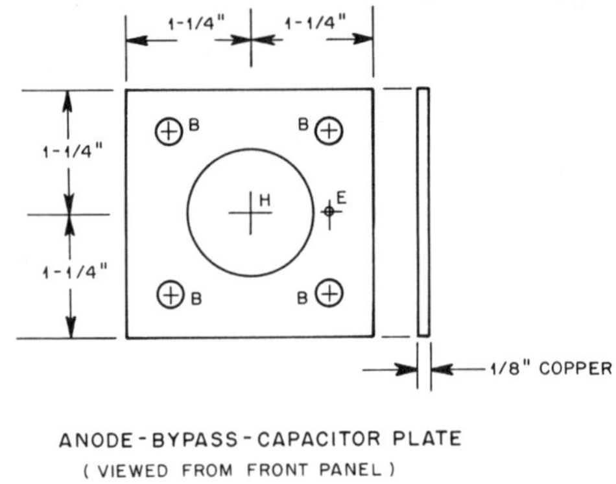

HOLES: B — 1/4-IN DIA.
E — 0.070-IN DIA. (NO. 50 DRILL). TAPPED FOR NO. 2-56 FOR HIGH-VOLTAGE LUG.
H — 1-1/4-IN DIA. FOR ANODE FINGER STOCK; FINGERS PROTRUDE OUT BACK SIDE.

Fig 6—Drilling details for the anode-bypass-capacitor plate. See Fig 3 for additional information on hole location.

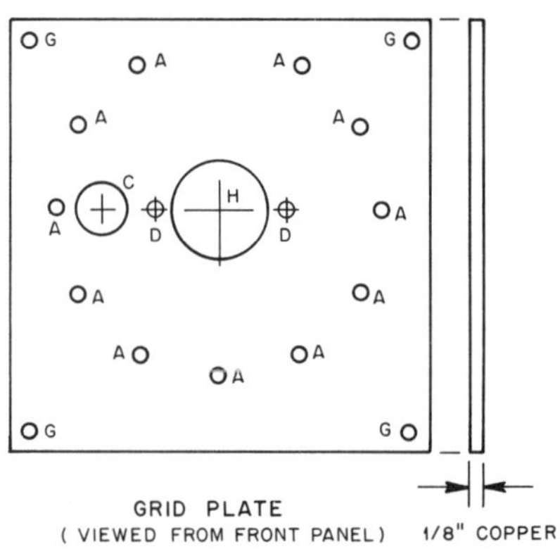

HOLE: A — 0.144-IN DIA. (NO. 27 DRILL).
C — 7/16-IN DIA. (0.438). ANODE TUNING COLLAR INSERTS FROM THIS SIDE.
D — 1/8-IN DIA. (0.125). FOR CATHODE-GRID MOUNTING.
G — 0.144-IN DIA. (NO. 27 DRILL). FOR FRONT PANEL MOUNTING.
H — 7/8-IN DIA. FOR GRID FINGER STOCK. FINGERS PROTRUDE OUT BACK SIDE.
NOTE: BRASS INPUT ENCLOSURE MOUNTS ON THIS SIDE.

Fig 5—Drill details for the grid plate. See Fig 3 for additional information on hole location.

remove the sharp points on the outer edge of the cutting surface of each side of a standard drill bit. This will eliminate any tendency for the copper to grab. Practice on an old bit and be sure to grind it symmetrically. Modified drill bits can still be used on aluminum and other metals.

Always start with a smaller drill and work up to the final hole size. It's safer and more accurate. The larger holes can be cut with a flycutter, or you can drill a series of smaller holes around the inside of a larger hole and file to finish. Either way is fine. Use lots of cutting fluid to lubricate the drill bit, and wear safety glasses and an old shirt. Remember, some cutting fluids are not to be used on aluminum.

Start with a no. 50 (0.070-in.) or smaller bit and drill pilot holes at each of your punched marks. The details for finishing each hole are listed in the drawings. Some holes are countersunk or tapped. Pay attention to the details and take your time.

When you are through drilling, you must deburr each hole. Copper is soft, so it tends to rise up around the hole during drilling and deburring. Use a flat file for the initial cut, and then remove any remaining material with a countersink. File the copper plates flat again because a flush fit on both sides of the aluminum ring is very important.

When all copper work is done, you should be able to stack the plates and see all pertinent holes align correctly. Enough tolerance is included in the dimensions to accommodate minor errors. After the holes are drilled, it can be difficult to tell which side of each plate is which, so mark the front side of each plate with a permanent marker.

Machining the Ring

The aluminum ring that forms the cavity wall is cut (sliced) from a length of 3½-inch OD tubing with a ⅜-inch wall thickness. See Fig 7. The tubing ID is approximately 2¾ inches. The dimensions of the ring are the most critical in this amplifier. Tolerance of the ring thickness is ± 0.005 inch to maintain full band coverage.

The ring can be hacksawed or bandsawed out of the tubing, but take extreme care to be accurate. Cutting tubing straight isn't easy. Clamp the tubing to prevent rotating on the bandsaw. The final finish cut is best done on a lathe or milling machine, but careful filing will work.

Once the ring is the correct thickness, deburr the sharp edges and spray it with marking dye. Notice that the outside and inside diameters are not concentric. This is normal for large tubing and is not a problem. Lay the ring flat and find the thickest wall section. Scribe a line across the wall at this point, across the center of the ring, and across the wall on the other side. The scribed lines on each side of the ring will be used to align the template. The output connector will be placed at the thick wall section.

Carefully align notch I on the template with the line scribed on the thickest wall section on the ring. Clamp the

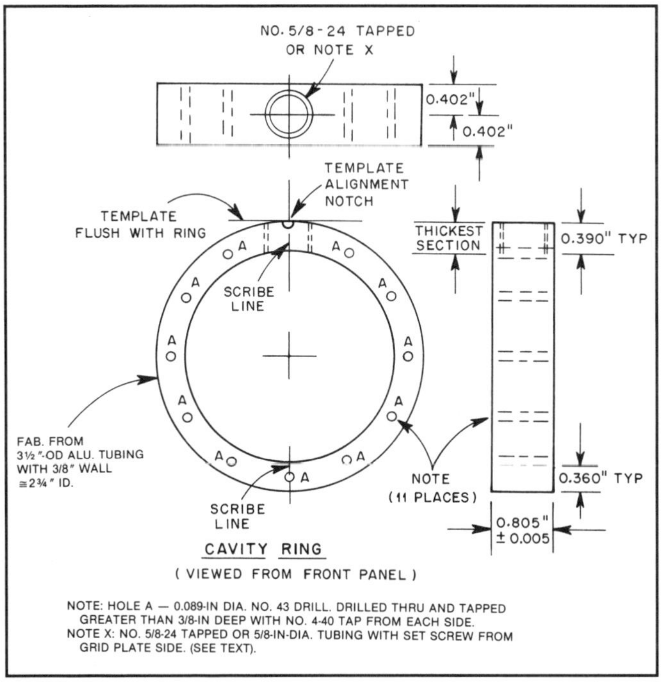

Fig 7—Details of the cavity ring. See Fig 3 for additional information on hole location.

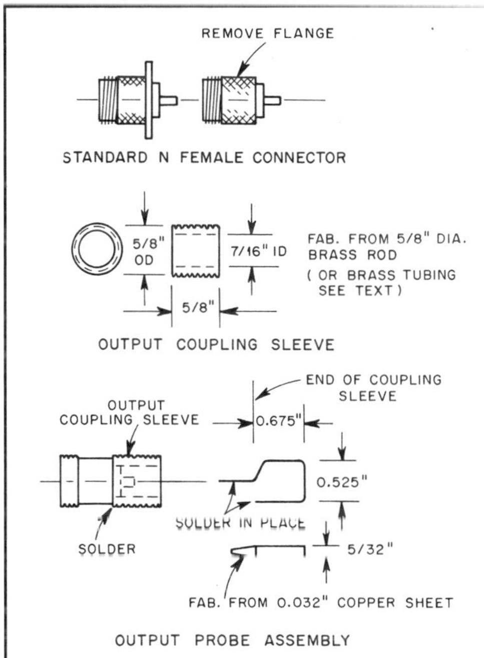

Fig 8—Output-probe/connector assembly details.

template onto the ring. Mark each of the 11 holes labeled A on the template. After you mark the holes and remove the template, check alignment with the copper plates just in case. If everything lines up, center punch all eleven holes on one side of the ring only, and drill each hole completely through the ring. Use lots of cutting fluid. File the ring flat before and after deburring, taking care not to change the wall thickness. Tap each hole to accept no. 4-40 machine screws. Each hole will have to be tapped to a depth of at least ⅜ inch from both sides because long taps don't exist. The inside of the ring doesn't need to be polished.

The hole for mounting the output connector can now be drilled. There are two ways to mount this connector, and either scheme works fine. Read ahead to the section on making the output connector for more information. The first method of mounting the output connector involves tapping the ring with a no. ⅝-24 tap and using a lathe to cut matching threads on the output connector coupling sleeve. Large taps are expensive, but both a tap and die for Type-N connectors come in handy if you do a lot of building.

If you don't have access to a lathe or a large tap, the second method is easier. Make the output connector coupling sleeve from ⅝-inch-OD brass or copper tubing and drill the ring to just clear it. Then drill and tap the grid-plate side of the ring above the output connector to accept a setscrew. Use the setscrew to secure the output connector.

Output Connector

A standard Type-N chassis-mount female connector (silver plated) is used for the output probe/connector. See Fig 8. First, remove the flange with a hacksaw and file flush with the connector body. Next, make the output coupling sleeve that is right for your application (threaded or unthreaded, depending on how you fabricated the ring). The sleeve will be the same length in either case. The output coupling loop is fashioned from a piece of 0.032-inch-thick copper sheet that is 5/32 inch wide. Bend it to the dimensions shown in Fig 8. We will solder the output connector together later.

Grid Compartment

The grid compartment measures 2 inches square by 1½ inches high. See Fig 9. It is made from brass and can be sawed out of square tubing or bent from sheet stock. The cover can be made from any sheet metal.

Two small PC boards (Fig 10) hold the finger stock that makes contact with the filament pin and cathode ring on the 2C39 tube. These boards are cut from 1/16-inch-thick, double-sided G-10 glass-epoxy stock. The copper pattern is identical for both sides of each piece. Mark and drill or file the holes first and then cut the boards to size. Small boards are difficult to hold while drilling them. Mark each side of each board and score the copper foil with a sharp knife.

The unwanted copper can be removed easily by heating the foil with a soldering iron and lifting it off. Use a flat file to deburr the boards. Do not use a countersink because the copper foil must be as close to the holes as possible to facilitate soldering the finger stock in place.

The input connector is a 5-mm SMA type. This is an excellent RF connector, especially for low-power UHF applications. Although an SMA is recommended, any small, screw-on connector will do. If you really feel you have to use a BNC then do so, but it's a lousy connector at frequencies above 200 MHz. Remember to move the connector hole to accommodate its larger size.

The input connector must be as close as possible to the first input capacitor. Lead length of the input dc blocking capacitor must be as short as possible. The 3-pF capacitor is

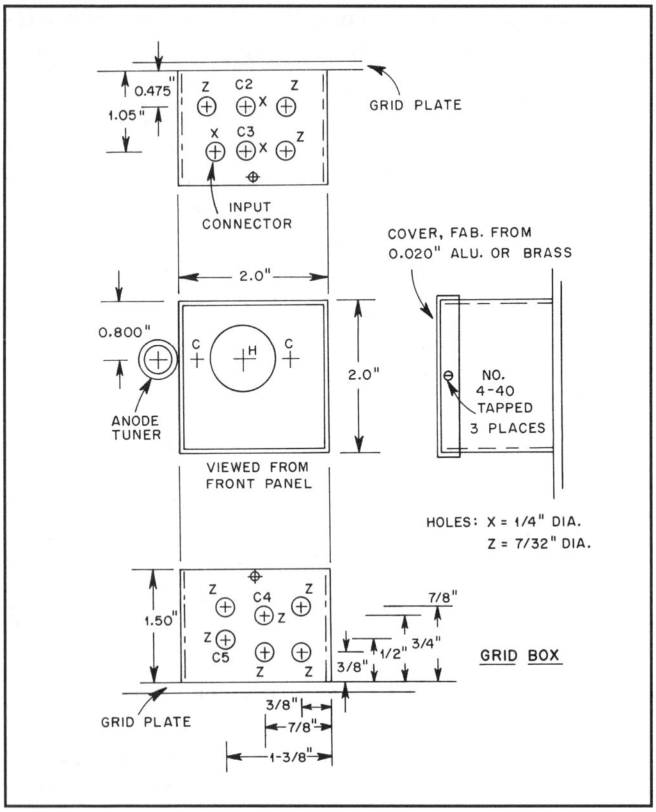

Fig 9—Input compartment details.

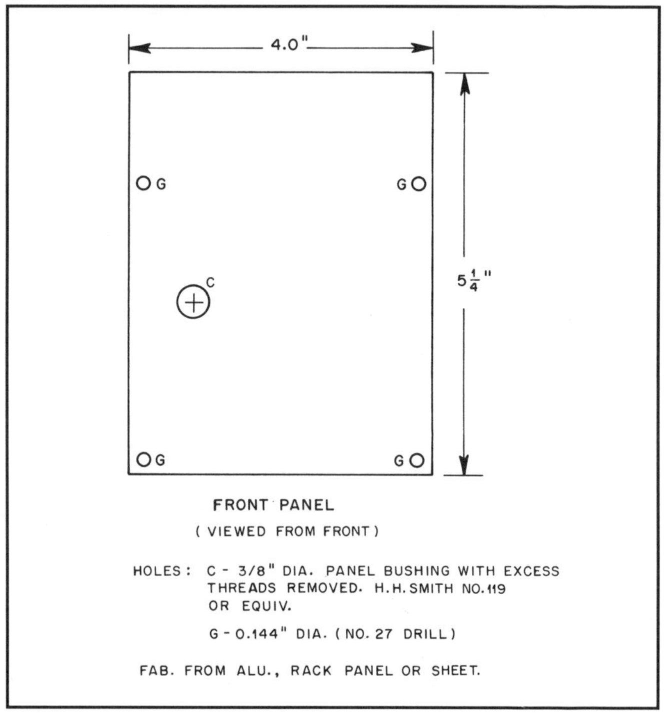

Fig 11—Front-panel details.

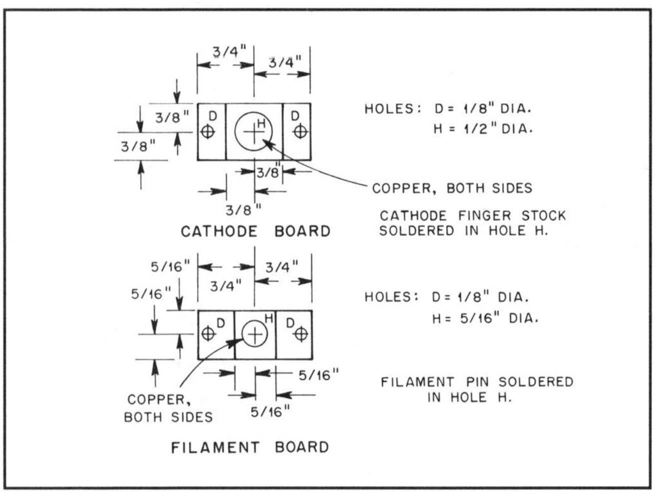

Fig 10—Cathode and filament PC-board details.

series resonant at 1200 MHz only with short (1/16-inch or less) leads.

Miscellaneous Bits and Pieces

There are still several small, but very important parts to fabricate. The front panel is shown in Fig 11. It is made from a piece of ⅛-inch-thick aluminum sheet. Some builders may wish to mount the amplifier on a rack panel. Wash and dry your front-panel material and spray it with marking dye. Clamp it to the template and mark the holes. Check the hole alignment with the copper grid and anode plates. If all lines up correctly, center punch and drill the holes. The only front-panel control is for the anode tuning capacitor, which is adjusted by a ¼-inch shaft protruding through a ⅜-inch panel bushing in hole C.

The anode tuning collar, shown in Fig 12A, is made from a piece of ½-inch-OD brass rod. This rod has a ⅜-inch hole drilled through its center, and it is turned down to 7/16-inch OD for half its length. The inside of the ½-inch-OD end is tapped to a depth of ¼ inch to accept ⅜-24 threads. This collar will be inserted into hole C on the grid plate. Fig 12B also shows the anode tuning post. It is simply a length of 5/32-inch-OD brass rod that inserts into hole C on the copper anode plate. This rod will form one plate of the anode tuning capacitor; the cavity wall is the other plate.

The anode tuner (Fig 12C) is machined from a piece of ⅜-inch-OD Teflon rod. One end of the rod is drilled out with a no. 21 drill. The outer wall of this end is threaded with a ⅜-24 tap. This is the end that will thread into the anode tuning collar and slip over the anode tuning post. The other end is turned down to fit inside a ¼-inch shaft coupler.

Fig 13 shows the remaining parts. The tuning shaft (A) is made from a piece of ¼-inch brass rod. A coupler (B) to connect the tuning shaft to the anode tuner may be purchased or made. This also applies to the front-panel spacers (C). The Teflon dielectric for the anode bypass capacitor (D) is made from 0.010-inch-thick Teflon sheet. Use the template to locate holes B and H. Teflon washers and inserts (E) are used to insulate the mounting hardware for the anode bypass capacitor from the chassis. The inserts are made from ¼-inch-OD Teflon rod. The washers are made from Teflon sheet. Sharpen a piece of ⅜-inch OD tubing and chuck it up in a drill press. This tool will cut neat, round washers from the sheet.

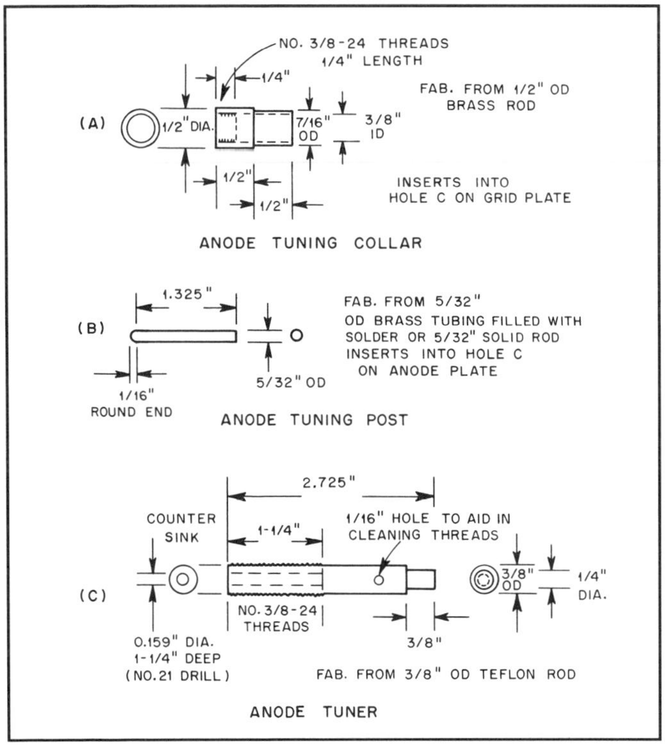

Fig 12—Anode-tuning capacitor details.

The box that encloses the anode compartment (Fig 14) is fabricated from a Bud AU-1083 utility cabinet. Clean the chassis and spray it with marking dye. Secure the template to the side of the enclosure that contacts the anode plate and scribe the holes labeled F. Make sure that these holes line up with the holes on the copper anode plate. If they do, center punch and drill them to size. If air cooling is used, the blower will mount to this box.

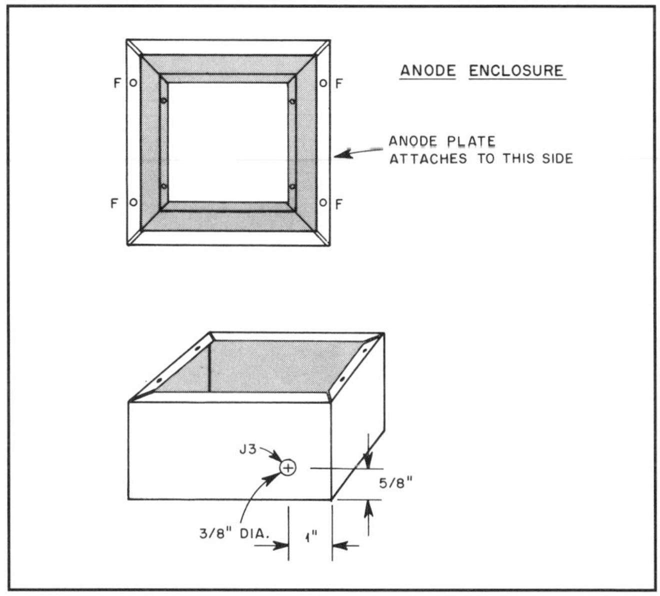

Fig 14—Anode enclosure details.

Soldering the Subassemblies

Once all copper and brass parts are drilled and deburred, they should be cleaned with alcohol and Scotch-Brite, a non-metallic pot cleaner, and washed in alcohol again. Set the pieces aside and avoid touching them. Fingerprints will inhibit soldering.

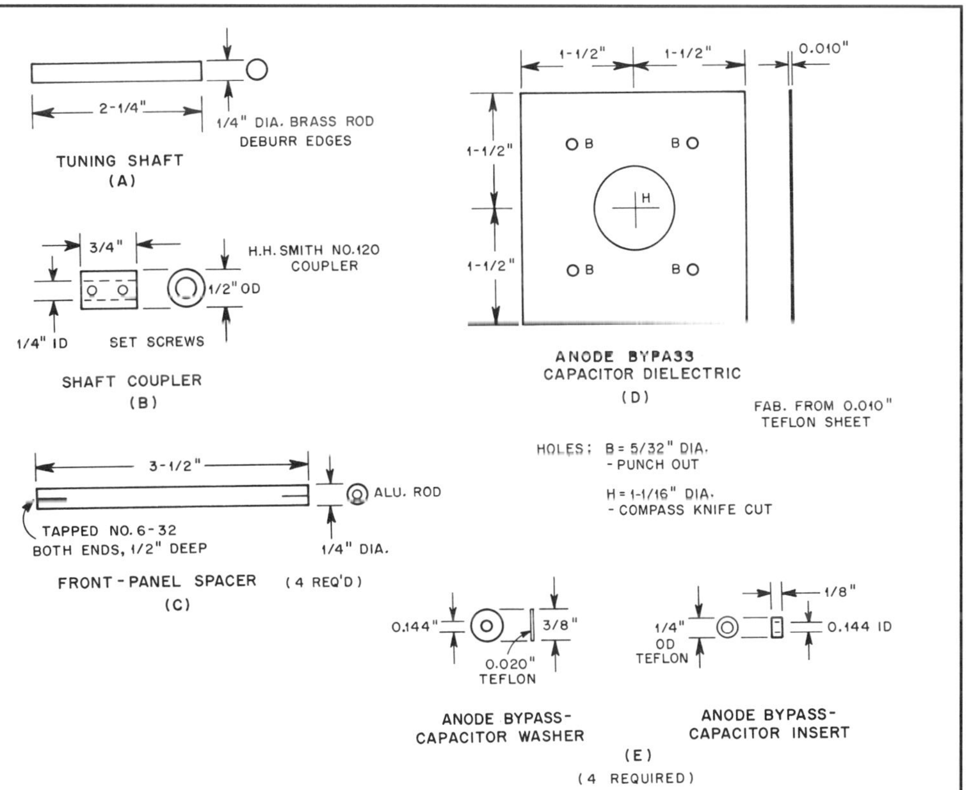

Fig 13—Miscellaneous parts necessary to complete the amplifier.

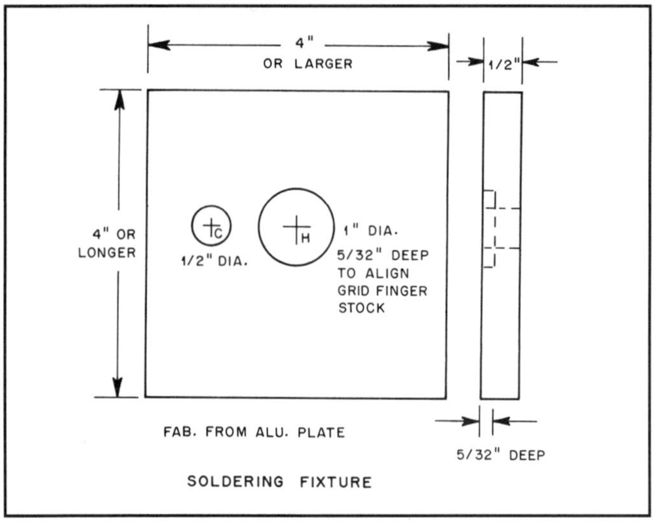

Fig 15—Dimensions of the soldering fixture. See Fig 3 for additional information on hole location.

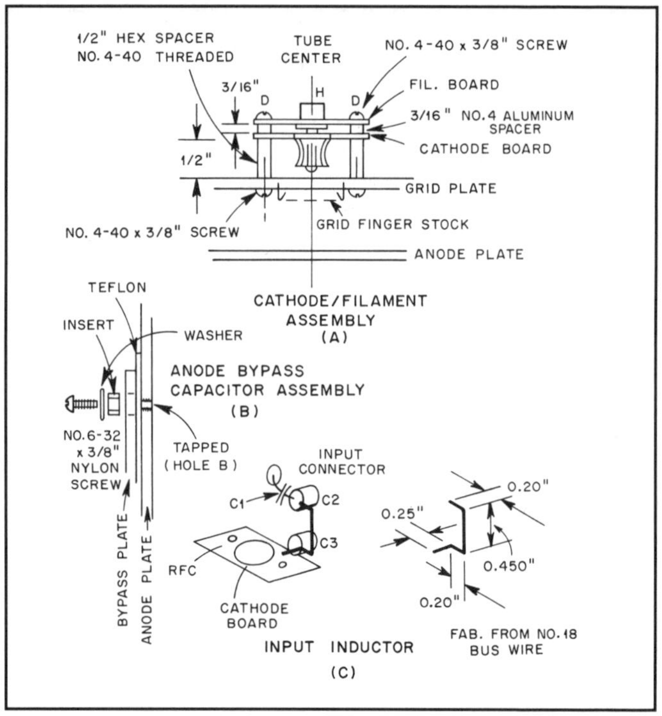

Fig 16—Assembly details for the filament and cathode boards (A), the anode-bypass capacitor (B) and the input pi network (C).

The best way to solder the heavy brass and copper parts is to first build the soldering fixture shown in Fig 15. This soldering fixture, made from ½-inch-thick aluminum plate, will evenly heat the entire assembly to be soldered. Even heating will allow you to do a much better soldering job than you could otherwise.

The aluminum soldering fixture should be preheated on a stove or hot plate until bits of solder placed on its surface just melt. At this point, reduce the heat slightly. Avoid excessive heat. If the copper parts placed on the fixture suddenly turn dark, the heat is too high.

Solder the grid plate assembly first. You will need the copper grid plate, grid finger stock, anode tuning collar and brass input compartment.[1] Look at the drawings again to be sure that you know which parts go where. Insert the grid finger stock into hole H on the grid plate. As viewed from the front-panel side, the curved fingers will protrude out the back side, away from you. Apply liquid or paste flux and set the grid plate in the soldering fixture. The finger stock will fit in hole H in the fixture, allowing the grid plate to rest flush with the surface of the fixture. Next, apply flux to the anode tuning collar and insert it in hole C of the grid plate. Part of the tuning collar will slip into hole C in the soldering fixture. Make sure the collar seats flush with the grid plate. The flux should start to bubble at this point. Carefully apply solder directly to the joints of the installed parts. The solder should melt almost immediately and flow bright and smooth. Next, place the square brass input compartment in place and apply flux. In a few seconds, it can be soldered by running solder around the joints, inside and outside. If you have trouble getting it to flow on both sides, merely tap the brass box aside (1⁄16 inch) and return it to its original position.

Now comes the hard part—getting the soldered assembly away from the heat without disturbing the alignment. A pair of forceps is recommended, but long pliers will do. Carefully lift the assembly off the soldering fixture and set on a cooling rack. Do this without moving any part. The cooling rack can be any two pieces of metal that will allow clearance for the protruding parts. You can expedite cooling by using an ordinary hair dryer in the "cool" position to gently blow air across the assembly. While the grid assembly is cooling, assemble the output connector. See Fig 18. Place the modified Type-N female connector, threaded end down, on the soldering fixture. Apply flux to the top and install the output coupling sleeve. Allow both parts to heat before applying solder. Carefully remove the soldered output connector from the fixture. When it has cooled, solder one end of the loop to the center pin of the N connector and the other to the output coupling sleeve. Now place the anode plate on the soldering fixture and allow to heat. Apply flux to hole C. Insert the anode tuning post (5⁄32-inch-OD brass tube) and allow to heat; apply solder. Remove the parts and cool as before. This completes the work with the soldering fixture. Be sure to let it cool off before handling! Save the fixture for future construction; you never know when you might want it again. The anode plate and the anode-bypass-capacitor plate must be filed and then sanded flat on their butt surfaces to assure that there are no solder bumps or sharp points to puncture the Teflon dielectric. This must be done after soldering. The Teflon sheet is adequate insulation for many times the anode potential of this amplifier, but only if the surfaces it separates are smooth! Next, clean the cathode and filament PC boards. Install the finger stock in hole H of the cathode board. Apply flux to both sides of the board. Heat with a hot iron and apply solder around the circumference of hole H, soldering the finger stock on both sides of the board. Use the same technique to install the filament pin. After all parts have cooled, use a spray can of flux remover to clean them. Slight scrubbing with

"Scotch-Brite" pot cleaner will finish them nicely. Congratulations: You have finished the pieces and are now ready to bolt the amplifier together.

Silver Plating

Over the years, many people have pushed silver plating as the only way to go. You may wish to silver plate the amplifier components before soldering them together, but it is not necessary. The RF skin conductivity of aluminum and copper is pretty good at 23 cm; these materials are much better than brass. In actual testing with four amplifiers, there was no difference in performance among tin-plated, silver-plated and unplated versions. A nickel-plated amplifier exhibited 3-dB less gain than the others.

Assembly

After fabrication of all parts, assembly is simple. Figs 16 through 18 show assembly details. Loosely fasten the grid and anode plates to the ring. Mount the input connector and capacitors on the input compartment. Loosely install the cathode and filament boards and their respective spacers. See Fig 16A.

Now insert a 7289/2C39 tube. This will center up all finger stock. Place the Teflon anode tuner in its collar on the grid plate and screw it most of the way in. Now tighten all of

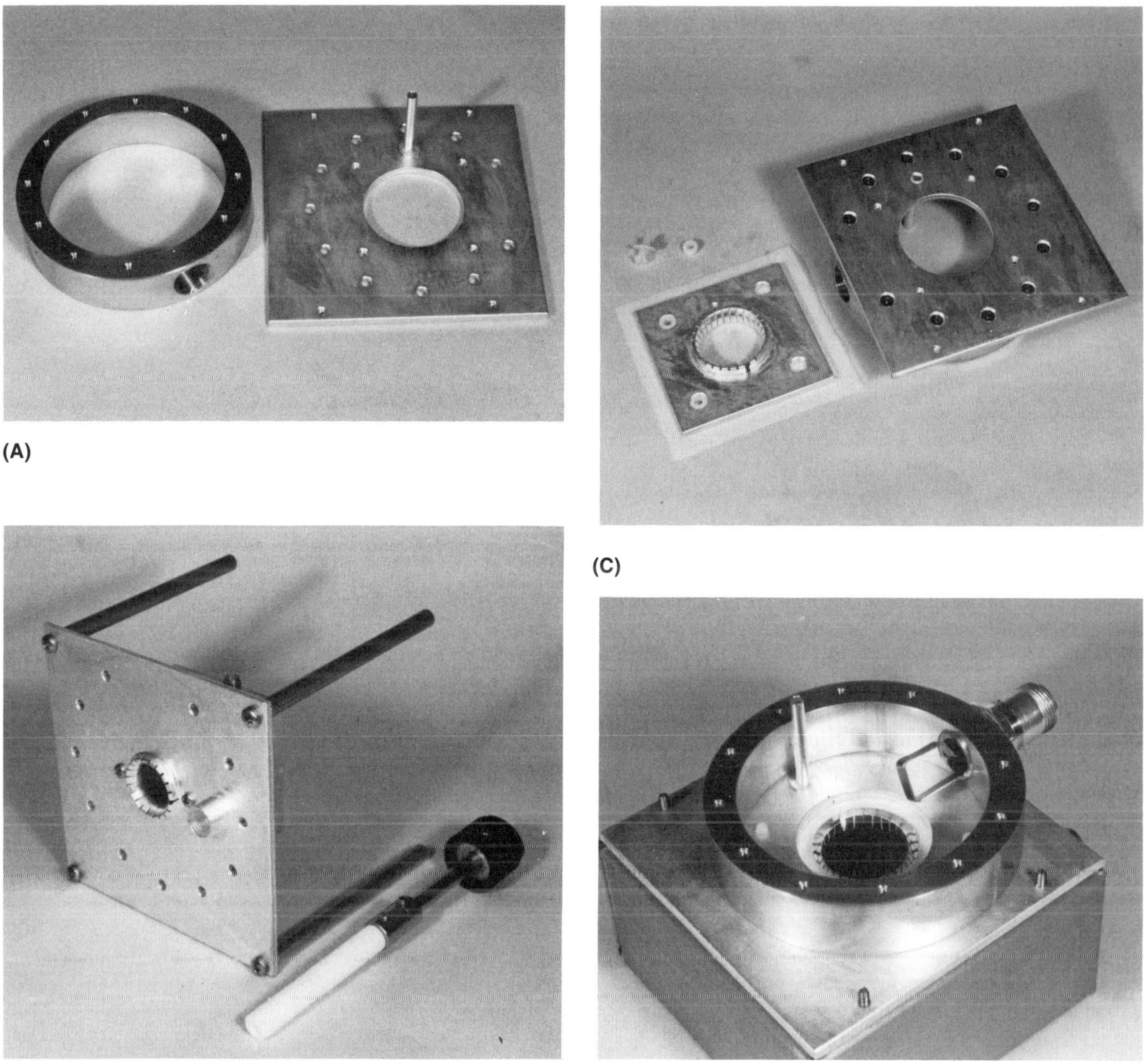

(A) (B) (C) (D)

Fig 17—The completed cavity ring and anode plate with anode tuning post soldered in place are shown at A. The photo at B shows the grid plate with finger stock, input compartment and anode tuning collar soldered in place. The completed anode tuner is at the right. C shows the cavity ring attached to the anode plate. The anode-bypass capacitor is ready for installation. At D, the interior of the cavity as seen from the grid plate side is visible. The output probe/connector assembly is installed. The anode-bypass capacitor and anode enclosure had been installed on the anode plate.

(A) (B)

Fig 18—At A, the interior of the completed input compartment is visible. The photo at B shows the interior of the anode compartment with the anode bypass, RFC3, C8 and J3 installed.

the screws. The 7289/2C39 tube should slide in and out snugly, and the anode tuner should screw in and out smoothly. The Teflon sheet and anode bypass capacitor plate can be installed now (Fig 16B). Assemble the remaining input components, the filament feed through capacitors and RFCs (Fig 16C). Screw the output probe into the cavity ring (or push in the probe and tighten the setscrew, depending on which method you chose). Install the high-voltage connector and other parts in the anode box. Mount the amplifier on the front panel and install the anode tuner shaft. This completes the assembly.

Power Supplies

The filament and bias supplies for the cavity amplifier are shown schematically in Fig 19. The manufacturer's specification for the 7289/2C39 filament is 6.0-V ac at 1 A. The use of a standard 6.3-V ac, 1-A transformer only slightly increases the tube emission without much loss of tube life. The filament should be allowed to warm up before operating the amplifier, so the filament, bias and high-voltage supplies incorporate separate primary switches.

Biasing

Many biasing schemes have been published for grounded-grid amplifiers. Fig 19 shows a bias network that satisfies all of the following operating requirements:

1) External bias supply referenced to ground.
2) Low-power components.
3) Variable bias to accommodate tube-to-tube variations.
4) TR switchable with relay contact or transistor to ground.
5) Bias-supply protection in case of a defective or shorted tube.

U2 provides a variable bias-voltage source, adjustable by R1. The output of U2 drives the base of Q1, which is used to increase the current-handling capability of the bias supply. Q1 must be mounted on a heat sink. J1 is connected to the station TR switching system so that R1 is grounded on transmit and disconnected on receive. The approximate range of the bias supply is 6 to 20 V. Z1 and Z2 provide protection for R1 in case of a shorted tube. The amplifier can be run without Z1 and Z2 if you keep the anode voltage below 1100 V.

High-Voltage Power Supply

A safe, reliable high-voltage power supply is described here. Of course, you can use any readily available HV supply; keep in mind, however, that the 7289/2C39 anode potential should never exceed 1400-V dc at full load and that the amplifier will withstand 1900-V dc at low cathode current and cut-off-bias conditions. For maximum power output, assuming adequate drive power is available, anode voltage under full load should be about 1200- to 1400-V dc. Fig 20 is a schematic diagram of the high-voltage supply. A power transformer (T2) that delivers 900- to 1050-V ac is ideal. The type of rectifier circuit used will depend on the type of transformer chosen. Each leg of the rectifier is made from two 1000-PIV, 3-A silicon diodes connected in series. Each diode is shunted with a 0.01-muF capacitor to suppress transient voltage spikes, and a 470-kΩ equalizing resistor. Filtering is accomplished with a string of four 360-µF, 450-V electrolytic capacitors connected in series. R3-R6 equalize the voltage across each capacitor in the string and serve as bleeder resistors. Of course, a single oil-filled capacitor may be used here if available. Whatever type of filter you use, the total capacitance should be about 80 µF at a voltage rating of at least 1500-V dc. This value allows

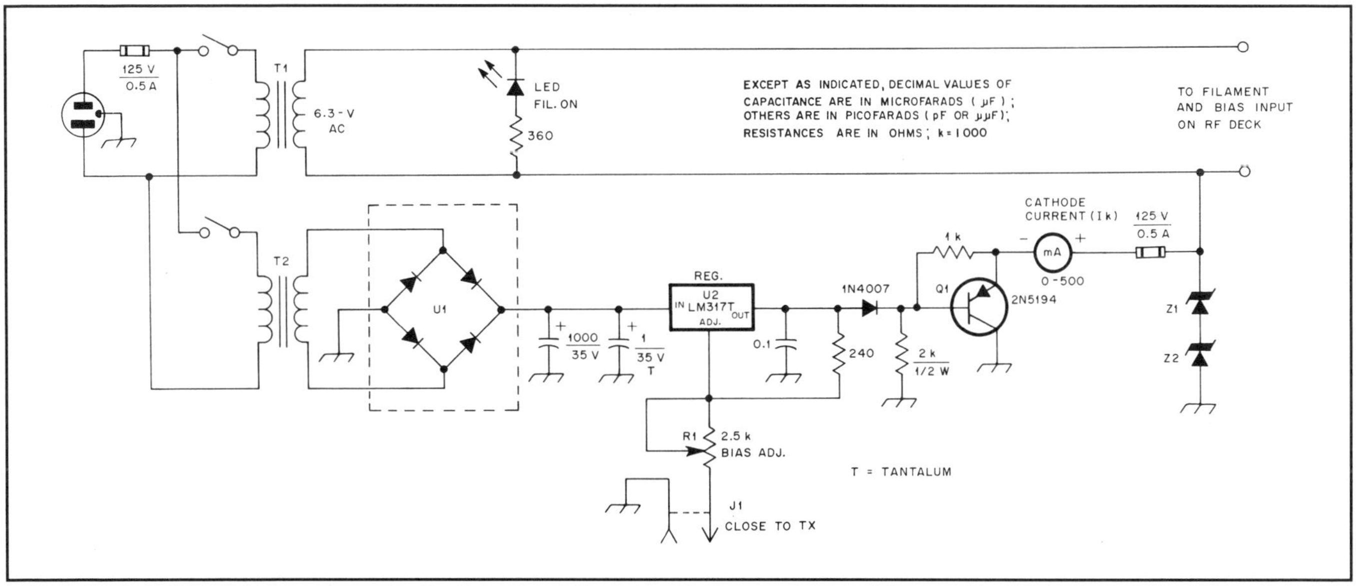

Fig 19—Schematic diagram of the cavity amplifier filament and bias supplies. All resistors are ¼-W carbon types unless otherwise noted.

J1—Female chassis mount phono connector.
T1—Filament transformer. Primary 117-V, secondary, 6.3 V at 1 A.
T2—Power transformer, Primary 117 V; secondary, 24 to 28 V at 50 mA or greater.
U1—Bridge rectifier, 50 PIV, 1 A.
U2—Adjustable 3-terminal regulator (LM317 or equiv).
Z1, Z2—20-V unipolar metal-oxide varistor (General Semiconductor SA20 or equiv.) or two 20-V, 1-W Zener diodes.

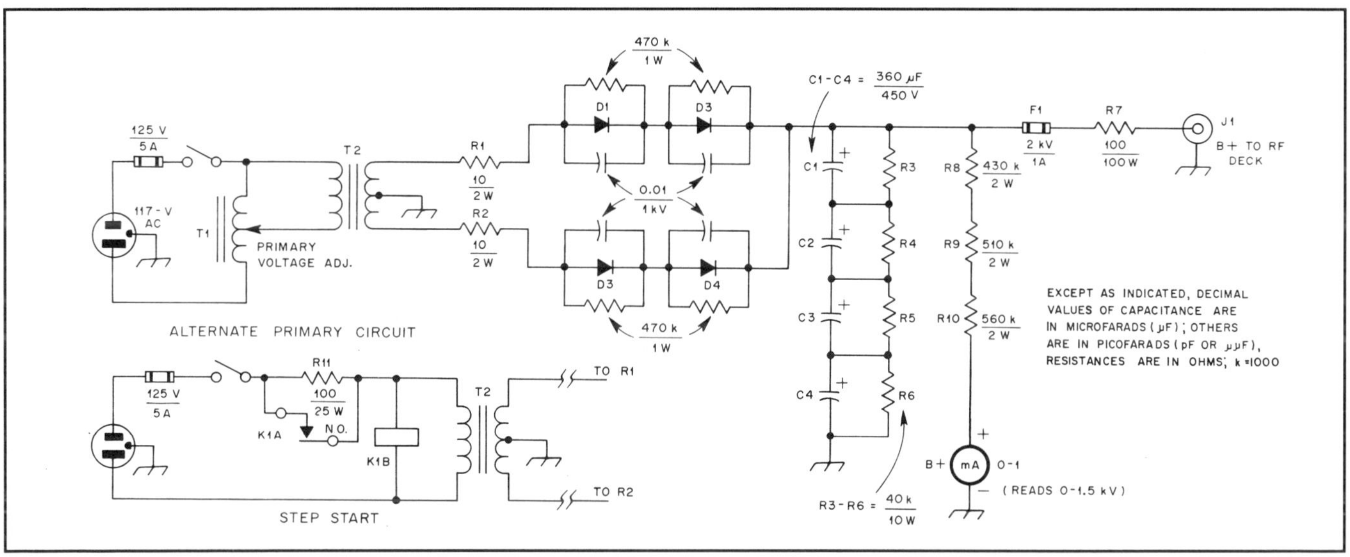

Fig 20—Schematic diagram of the amplifier high-voltage supply.

C1-C4—Electrolytic capacitor, 360 μF, 450 V.
D1-D4—Silicon rectifier, 1000 PIV, 3 A.
F1 High voltage fuse, 2 kV, 1 A.
J1—Chassis mount female BNC or MHV connector.
R3-R5—Wirewound resistor, 40 kΩ, 11 W.
T1—Variable autotransformer, 500 VA.
T2—High-voltage transformer, Primary 117 V; secondary, 900 to 1050 V at 500 mA.

adequate "droop" of the anode voltage under high-current loads to protect the amplifier in case of RF overdrive or a defective tube.

Protective Circuitry

Some type of start-up protection should be incorporated in the primary. Fully discharged filter capacitors look like a dead short at supply turn-on. Initial surge current (until the capacitors charge) may be high enough to destroy the rectifiers. R1 and R2 provide some surge-current limiting, but either of the two primary configurations shown in Fig 20 should be used. T1, a variable autotransformer (Variac and Powerstat are two common trade names), is ideal. In addition to allowing you to bring the primary up slowly, (and charging the capaci-

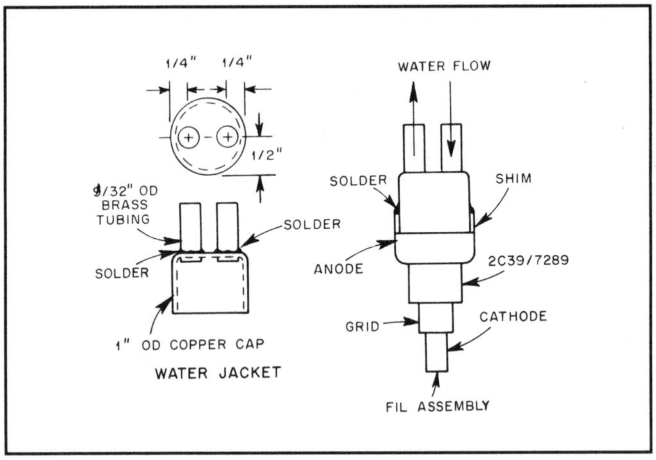

Fig 21—Details of the solder-on water jacket.

tors gradually), it also allows full control of amplifier output power by varying anode voltage. The second method, a "step-start" system, uses a resistor in the T2 primary to limit the turn-on surge current. When the capacitors have charged, K1 is energized, shorting out R11 and applying full voltage to the T2 primary. F1 and R7 protect against high-voltage arc-overs or short circuits. If sustained overcurrent is drawn, F1 will open and remove B+ from the RF deck. Use a high-voltage fuse here; standard fuses may arc when blown and not interrupt the B+. R7 provides current limiting to protect the amplifier and power supply in case of a high-voltage arc.

Safety

An HV meter should always be used to monitor the status of the power supply. The values for R8-R10 shown in Fig 20 will give a 1500-V dc full-scale reading on a 0-1 mA meter. RG-58 or -59 coaxial cable should be used for the high-voltage interconnection between the power supply and the RF deck. Ground the shield at both ends for safety and a good dc return. Safety must be observed when working with all power supplies. These voltages are lethal! Always disconnect ac power and then discharge the filter capacitors before working on the power supply. Never guess or make assumptions about the status of a power supply. Assume it is hot.

Metering

Cathode-current monitoring is all that's really necessary for observing amplifier dc performance. Cathode current (IK) is the sum of the plate (IP) and grid (IG) currents. Normally, when this amplifier is driven to 300- or 400-mA IK, the grid current will be around 40 to 50 mA. The inclusion of a grid-current meter is not really necessary and only makes biasing and TR switching complicated. Cooling desired output power and the level of drive power available will dictate what type of cooling to use. For intermittent duty (SSB, CW) at output levels less than 50 W, air cooling is satisfactory. Any small blower may be easily mounted to the aluminum box surrounding the tube anode. For high-duty-cycle modes and/or output levels greater than 50 W, water cooling is highly recommended. Greater than twice the normal air-cooled output

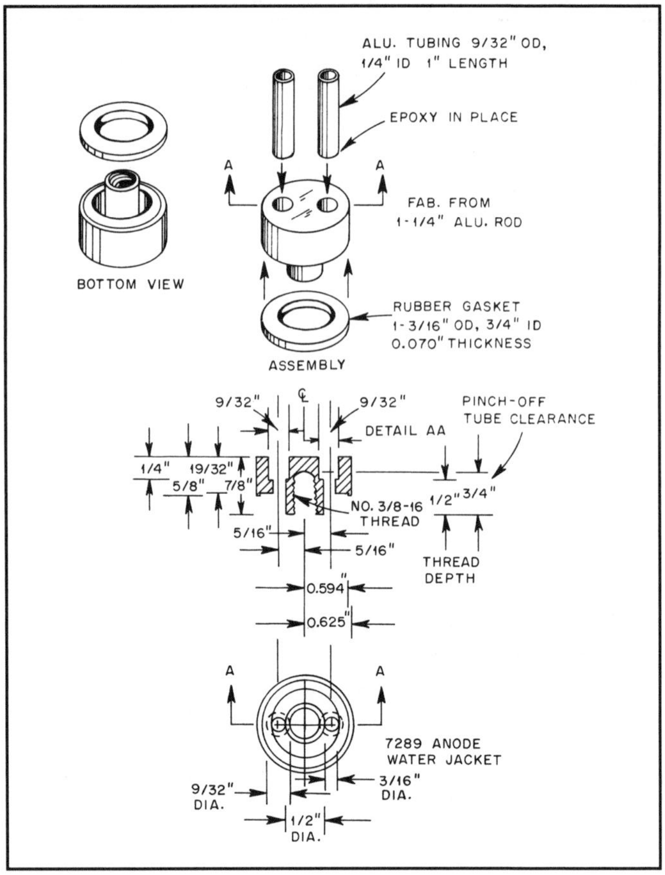

Fig 22—Details of the screw-on water jacket.

power can be obtained from a water-cooled tube, and water cooling is quiet.

Tube Modification and Water Jacket

The first step is to remove the air radiator from the tube. The air radiator screws on, so it may simply be unscrewed without damage to the tube. First, place a hose clamp around the tube anode. Secure the radiator fins in a vise and grip the hose clamp with a pair of large pliers. Gently unscrew the tube from the radiator. If the hose clamp slips slightly, tighten it. Some 7289/2C39 tubes use an air radiator that is attached with setscrews. To remove the radiator, simply remove the setscrews and pull the radiator off. The air radiator will be replaced with a water jacket that allows water to be circulated past the tube anode and through a heat exchanger, where it is cooled and circulated past the tube anode again. Two different types of water jackets are described here. The water jacket shown in Fig 21 will work with any type of 7289/2C39. It is fabricated from a 1-inch-OD copper tubing cap and two short pieces of 9/32-inch OD brass tubing. The copper tubing cap should be available from a local hardware store or plumbing supply house. Brass tubing is available from many hobby stores and metal supply houses. Mark and drill the copper cap so that the brass tubing is a snug fit. Thoroughly clean the parts until they shine. Push the tubing into the holes in the end cap and degrease the assembly with alcohol. Use plenty of flux and solder the seam around each section of tubing. Allow the jacket assembly to cool. Meanwhile, thoroughly clean the 7289/2C39 anode to a bright finish. Check the water jacket for fit. In some cases, you'll have to use a 0.005- to 0.010-in.-thick copper shim to fill the gap between the copper cap and the tube anode. This shim helps eliminate pin holes in the solder. Using plenty of flux, solder the water jacket to the tube anode. Solder it quickly with a hot, high-wattage iron. Allow the tube to cool in the air after soldering to avoid thermal shock and possible breakage. After the tube has cooled, use plenty of alcohol to remove all traces of flux from the tube and water jacket. The second type of water jacket is shown in Fig 22. This jacket will work only with 7289/2C39 tubes that have a screw-on air radiator. It is designed to thread onto the tube anode just like the air radiator did. This jacket is machined from a piece of 1¼ inch aluminum rod. The water inlet and outlet tubes are made from ⁹⁄₃₂-inch-OD, ¼-inch-ID aluminum tubing that is epoxied in place. A rubber gasket seals the jacket against leaks. If you have access to a lathe, you should have no trouble duplicating the jacket. You could have one made up at a local machine shop. Complete screw-on water jackets are also available from Angle Linear.[2] After you unscrew the air radiator from the 7289/2C39, check for and remove any burrs from the tube anode. The anode surface must be flat if the rubber gasket is to be effective. Screw the water jacket onto the tube. Tighten by hand only. Do not use any tools, or you could damage the tube or jacket! Do not use the water inlet and outlet tubes for leverage—they have thin walls and break easily.

WATER-COOLING SYSTEM

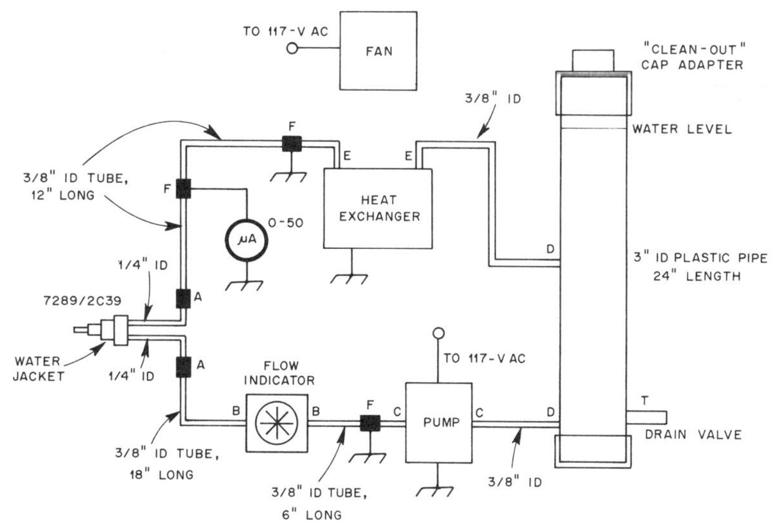

PART	DESCRIPTION
A	1/4" TO 3/8" REDUCER.
B	1/2" NPT TO 3/8" TUBE ADAPTER.
C	3/8" TUBE TO PUMP ADAPTER.
D	3/8" OR 1/2" NPT TO 3/8" TUBE ADAPTER.
E	3/8" TUBE TO EXCHANGER ADAPTER.
F	3/8" OD COPPER TUBE W/WIRE SOLDERED TO IT. (SEE TEXT).

Fig 23—Details of the water-cooling system. Recommended pumps are (1) Little Giant Pump Co. Model 1-42A or larger, available from most hardware stores; or (2) Calvert Engineering, Cal Pump Model 875S (160 gal/h), available from Calvert, 7051 Hayvenhurst Ave., Van Nuys, CA 91406, tel. 818-781-6029. The flow indicator (Model 15C; requires two ½-in. NPT connectors) is available from Proteus Industries, 240 Polaris Ave., Mountain View, CA 94043, tel 415-964-4163.

Water System

Fig 23 depicts the complete water-cooling system. Recommended pumps and accessories that have proven reliable and effective are listed in the caption. Any small pump, such as a fountain pump, that can deliver 160 to 200 gallons per hour can be used here. Most inexpensive pumps are not self-priming, which means that they won't pump water if they have air in the rotor. Although water can be forced through the pump for the initial prime, my system uses gravity priming. The water reservoir is a 2-foot length of 3-inch-OD plastic pipe that is available from hardware or plumbing stores. It is usually sold for use in residential sewer systems. The outlet is at the bottom and the inlet about halfway up the column. The inlet is located here to eliminate aeration that ionizes the water and reduces its effectiveness. The outlet feeds the pump directly. The pump and the reservoir outlet port should be mounted in the same plane. The pump should be oriented so that air bubbles will rise into the impeller output port and can be blown out once the pump starts running.

Flow Indicator and Heat Exchanger

Water cooling is best described as "super quiet." There is no noisy fan to hear to reassure you that the tube is receiving adequate cooling. If water flow is reduced or cut off during amplifier operation, tube damage is virtually assured. Flow interlocks and switches to shut down the amplifier if water flow is reduced are hard to find and expensive. Flow indicators, however, are inexpensive and reliable. A flow indicator has a spoked rotor that turns as water passes through the unit. If the wheel is turning, there is water flow; if not, you have a problem. Changes in flow rate can be observed by watching for speed changes in the rotor. A small lamp illuminates the flow indicator, making it easy to see rotation. The flow indicator should be mounted where it can be seen from the operating position and monitored during operation. Heat exchangers, or radiators, remove the heat from water as it passes through. For this application, a small automobile transmission-oil cooler works great. Most auto-parts stores and speed shops have a good selection. Some come with mounting brackets. Look for a cooler with the input and output ports on the top so air bubbles will rise to the top and move on without becoming trapped. Trapped air degrades cooler performance. If you use the amplifier for high-duty-cycle modes such as ATV or FM, or for long, slow-speed CW transmissions (EME, for example), you should use a small axial Whisper fan to increase the effectiveness of the heat exchanger. A fan isn't necessary during normal operation, or even for sustained operation at moderate power levels, but is highly recommended if you plan prolonged operation at maximum power. Locate the fan so the warm exhaust air won't heat up other equipment.

Hoses and Fittings

Most hardware stores carry a complete line of brass fittings and adapters that can be used for this project. Brass, however, will eventually corrode and pollute the water supply. Plastic fittings are cheaper and don't corrode, but they are harder to find. Recreational vehicle suppliers are my main source for these parts. They are used extensively in drinking water systems for mobile homes and travel trailers. Procure the fittings when you have the rest of the parts in hand, as there are many variables to consider. You can use any relatively soft, thin-wall vinyl tubing for all water lines. The main runs are made from ⅜-inch-ID hose, while ¼-inch-ID stock is used to connect to the 7289/2C39 water jacket. The ¼-inch-ID tubing fits snugly over the 9/32-inch-OD inlet and outlet tubes on the water jacket, so no clamps are required. All other hose connections should be secured with stainless-steel clamps to prevent leaks. Any leaks mean air in the system and deterioration of cooling performance.

Safety

The anode of the tube, and hence the water jacket and water, are in direct contact with the high-voltage supply, so some safety precautions must be observed. Approximately 12 to 18 in. of tubing should run between the 7289/2C39 jacket and any other component in the cooling system. This will allow enough resistance in the water to provide adequate current limiting, should the water contact any components that are grounded. It is best to ground the water supply at the pump. This can be accomplished by replacing a short section of the tubing that runs to the flow indicator with a piece of brass or copper tubing. Solder a wire to this metal tubing and connect the other end of the wire to your station ground. Use at least 24 in. of vinyl tubing between the anode cooling jacket and the ground point. On the warm-water side of the 7289/2C39, run 12 in. of vinyl tubing to a small metal fitting or short section of metal tubing, and then another 12 in. of vinyl tubing to a grounded point (this can be at the heat exchanger). You can measure the water leakage current to ground by placing a microammeter between the metal fitting that connects the two vinyl hoses and ground. Leakage current should be less than 10 mA with clean water and an anode potential of 1 kV. As the water ages, the leakage current will rise; when this happens, replace the water. Grocery stores carry distilled water for use in steam irons. It may be deionized and not truly distilled, but it works fine for about four to six months in this application. Filters can be purchased from scientific supply houses, but it's not really worth it because deionized water is so cheap. Do not use tap water under any circumstances! When you turn on the water system for the first time, run a gallon of water through it for half an hour to wash out fabrication impurities. Replace with clean water before using the system to cool the amplifier. Water was chosen because it's inexpensive, nontoxic, nonflammable, and easy to clean up if you have a leak. Better liquid coolants are available, but they are toxic. Don't use them!

Cooling Performance

Fig 24 is a graph of several transmit/receive cycles on a water-cooled, 500-W output, 23-cm power amplifier. For this test, two of the amplifiers described here were coupled with a hybrid combiner. This particular cooling system used one gallon of water. Experiments indicate that, during extended operation, the water temperature rises only 30 to 35 F° above ambient room temperature. Typically, the tube anode and water average 10 to 15F° above ambient during casual operating. Flow rates in this system are typically ⅓ gallon per minute per tube, which is more than adequate. At this rate, more than 300 W of dissipation from a single inefficient 7289/2C39 were required to boil the water in the water jacket. The water should not be allowed to boil because this will heat the rubber gasket.

Tubes

It is not really necessary to buy a new 7289/2C39. Used

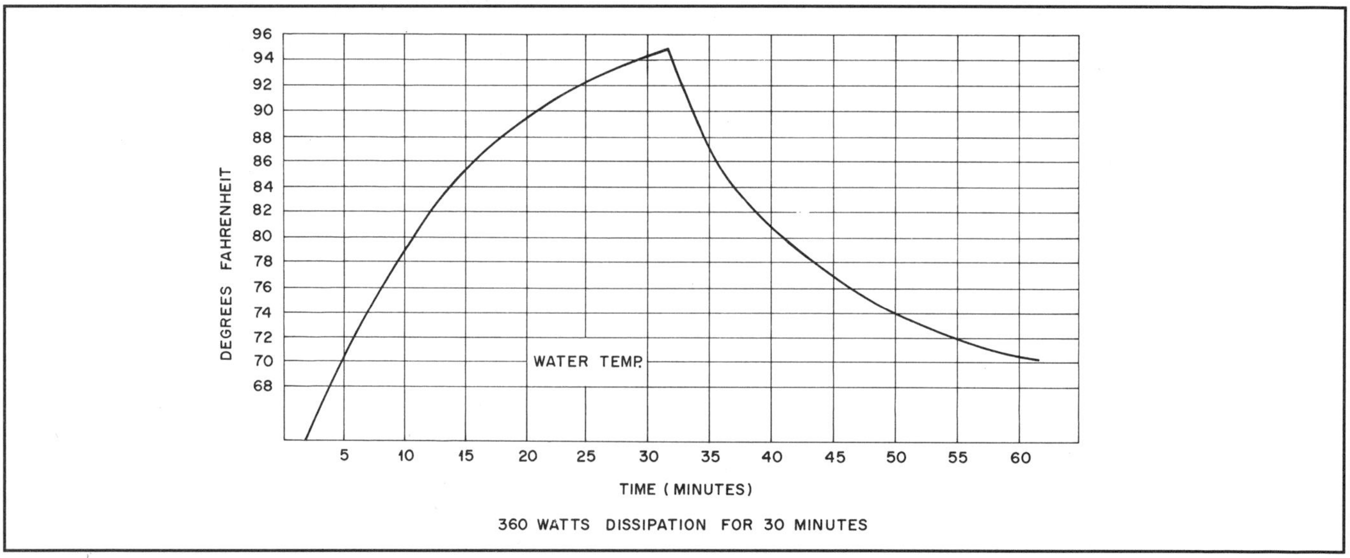

Fig 24—Performance graph of the water-cooling system.

tubes can be found surplus for a few dollars and, in many cases, will perform as well as a new tube. Most used tubes have been sitting around for several years, so it's a good idea to run them through the dishwasher to clean them up and then run the filaments for about 24 hours. This will restore operation in many cases. If you buy a new tube, you should be aware that the 7289/2C39 is being run far in excess of its ratings in this amplifier. The manufacturer's warranty will not cover tubes run in this application. Contrary to popular opinion, glass tubes will work. Physically, they are not as rugged as the ceramic version, but the glass-to-metal seal seems to provide better shelf life than the ceramic seal. The glass tubes make great driver tubes and will work fine for power levels up to 100-W output. Pulse tubes (7815, 7211) are not recommended because of their poor thermal stability at high power levels. Also, they generally are 30 to 40 MHz lower in resonant frequency in this amplifier compared to the 7289/2C39. Some 7289 tubes can be as much as 30 MHz lower in frequency. Minor length adjustment of the anode-tuning post may be required to accommodate amplifier and tube differences.

Tube Insertion

Extreme care must be exercised when inserting the 7289/2C39 tube. Never force the tube in place as damage (bending) of the cathode finger stock may result. Observe the layout of the finger stock to get an idea of how the tube inserts. Carefully position the tube so it is straight as you gently push. It should slide in snugly without any solid resistance.

Testing

After you have completed all of the parts for the amplifier, it's time to test everything before hooking it all together. Test the water-cooling system by turning it on and watching for steady water flow as indicated on the flow meter. The tube and water jacket can be removed from the cavity amplifier for this test. Check all of the power-supply voltages first without connecting them to the RF deck. Then, without the tube in place, hook the bias and filament supplies to the cavity and check the voltages again at the tube finger-stock connections. Connect the high-voltage supply to the RF deck and bring the voltage up slowly with a variable autotransformer. Monitor the high-voltage on the anode bypass capacitor plate, and look and listen for any possible arcing between the anode bypass capacitor plate and ground. Use extreme care when measuring and testing the high-voltage supply. If everything looks okay with the power supplies, shut them off and disconnect them. You can make a safe, low-power test of the cavity resonance without applying any voltage. With the tube in place, insert a 2-inch-long coupling loop on the end of a piece of coaxial cable between the spring fingers of the anode down into the cavity. Connect the amplifier output probe/connector to a device capable of detecting low-level RF at 23 cm (for example, a spectrum analyzer or microwattmeter). Feed a signal from an L-band signal generator into cable attached to the wire coupling loop that you inserted into the cavity. Set the signal generator for various frequencies in the 23-cm band and tune the amplifier anode tuner. There will be a sharp peak in output at cavity resonance. This testing method can be used to determine cavity tuning range, anode-bypass-capacitor effectiveness and resonance of various tube types for use in this amplifier. Any cavity amplifier can be tested completely without ever applying high voltage. The better your test equipment, the easier the amplifier is to test. If all dimensions were followed strictly, the amplifier will tune as designed.

Amplifier Hookup

Installation and operation of this amplifier is relatively straightforward, but as with any amplifier, several precautions must be followed. If these are adhered to, it will provide years of reliable service. The amplifier is designed to be operated in a 50-Ω system and should never be turned on without a good 50-Ω load connected to the output connector. Never operate it into an antenna that has not been tuned to 50 Ω! Drive power to the amplifier should never exceed 15 W. Never apply drive power in excess of 1 W unless all operating voltages are present and the tube is biased on. Otherwise, the tube grid-dissipation rating will be exceeded and you will probably ruin it. As in all TR-switched systems, some type of interlock or sequencing of transmit and receive functions should be incorporated. In most systems, the sequence for going into transmit is something like this: First, switch the antenna changeover relay from the receiver to the power amplifier. Next, bias the power amplifier on. Last, key the exciter and apply drive to the amplifier. To go to receive, unkey the exciter, remove operating bias from the amplifier and switch the antenna relay back to the receiver. If the antenna relays are switched while the power amplifier is operating and putting out power, damage to the relay contacts and/or the amplifier is likely. If there is a momentary removal of the antenna while the power amplifier is biased on, it may oscillate. This can damage the TR relay, the tube, or even the receive preamplifier.

Tune Up and Operation

This is it—the big moment when you will see your project come to life! Connect an accurate UHF power meter and a 50-Ω antenna or load to the amplifier output connector. A Bird Model 43 wattmeter with a 100- or 250-W, 400- to 1000-MHz slug will give reasonable accuracy, depending on the purity of the drive signal. Apply filament power and tube cooling, and allow 3 to 5 minutes for the filaments to warm up. Turn on bias supply (the amplifier will draw maximum current if the anode voltage is applied without bias). Apply 300 to 400 V to the anode. There should be no current flowing in the tube as indicated on the cathode-current meter. Ground J1 on the bias supply to apply transmit bias and observe cathode current. As R1, the bias control, is turned clockwise, quiescent idling current should increase. Set for about 25 mA. Apply 1 W of RF drive power. Turn the anode tuner while observing the RF output power meter and tune for maximum output. The output should go through a pronounced peak at cavity resonance. Adjust C2 and C3 on the input tuning network for maximum amplifier output. If possible, use a directional wattmeter between the driver and the amplifier input to check that best input SWR and maximum amplifier output occur at roughly the same setting. Depending on the amount of drive power available, you may want to tune the amplifier for maximum power output or maximum gain. Fig 25 shows what you can expect from different drive levels. Once the amplifier is tuned

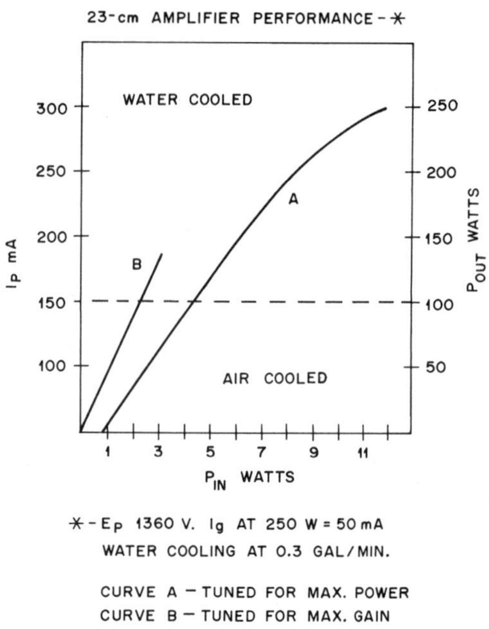

Fig 25—Performance of the cavity amplifier under different drive and plate-current conditions.

for best input SWR and maximum output with 1 W of drive, anode voltage and drive power can be increased. Increase both in steps; be sure to keep the anode tuner peaked for maximum output power. When you get to the 100-W output level, very carefully readjust the input circuit for maximum output. The input capacitor closest to the cathode is critical and should need to be rotated less than 90 degrees maximum. Maximum output power will be roughly coincident with best input SWR. Increase the drive power and keep the anode tuner peaked for maximum output. Increase the drive until you reach the desired output level, but do not exceed 400-mA IK! At 1300-V dc and 350-mA IK, output power with a good tube should be about 230 to 250 watts. At lower anode voltages, IK will be higher for the same output power. Higher anode voltages result in higher gain, lower drive levels, lower grid current and lower plate current for a given output power. The anode tuner's tuning rate is approximately 5 MHz per turn. Clockwise rotation of the tuner lowers the resonant frequency of the cavity. This control will require readjustment as you make large frequency excursions within the 23-cm band (for example, if you go from 1296 weak-signal work to the 1269-MHz satellite segment). You should also check the input SWR if you move more than 15 MHz. Generally, amplifier tuning does not change much after initial setup. You should be able to turn it on and use it without retuning as it heats up. Slight adjustments may be necessary, however, depending on cooling, inherent thermal differences from tube to tube and duty cycle of the operating mode. Always keep the anode tuner peaked for maximum output, and check it from time to time, especially while you are first learning how the amplifier operates.

The output loading control is the output connector and probe assembly. Loading is changed by minor rotational adjustment of the Type-N connector. First loosen the jam-nut (or setscrew) slightly. While observing output power and keeping the anode tuner peaked, rotate the loading control ±30 degrees maximum for greatest output power. This should be done only once and should not need repeating unless another tube is installed. Even then it may not be required.

Notes

[1]The finger stock for this project is available from Instrument Specialties, P.O. Box A, Delaware Water Gap, PA 18237. Contact them for the name of the closest distributor. The part numbers for this amplifier are: anode bypass capacitor plate, no. 97-70A; grid plate, no. 97-74A; cathode board, no. 97-420A; filament board, no. 97-280A

[2]25309 Andreo Ave., Lomita, CA 90717.

A 2304-MHz 80-Watt Solid-State Amplifier

By David T. Hackford, N3CX

(From *Eastern VHF/UHF Conference Proceedings*, 1992)

It's always been difficult to generate more than 10 watts on the 13-cm band. The surplus TRC-29 cavity will produce 10-15 dB gain, but is hard to find. To generate this much power, you were limited to forced-air or water cooling. Water cooling provides better stability, but is difficult mechanically, and is not conducive to portable operation.

Fortunately, microwave power transistors are now available at reasonable cost. This amplifier uses two 40-W Class-C amplifier assemblies combined with 3-dB hybrid couplers. This is not a project for beginning builders. The enclosures are made out of milled brass stock, and a channel is milled in the box for the transistor flanges. The milled channel for the flanges is made deep enough that the circuit board can be soldered to the brass box with the transistor leads resting on the traces. Rivets through the circuit board at grounding points are required for good RF grounding.[1] The Sage Wireline 3-dB hybrid couplers (octave bandwidth) are easily made. If you're not familiar with this line, write for the *Designers Guide to Wireline & Wirepac*.[2] Also see the article by W3HQT [3].

Specifications for the single 40-W amplifiers I built are given in Table 1.

These amplifiers saturated at about 45 watts output. At 28 V, maximum current consumption is about 6 amps. Gains up to 8 dB are possible with some devices.

After building an amplifier you always want to do the big operational test. I ran mine in a January VHF Sweepstakes, and took it to the '91 June VHF outing. Both the single and combined amps worked normally. Running the dual amp, you'll need a good 28-V, 15-amp supply. With the losses in the phasing lines and the combiners adding up to approximately 1.5 dB, 15-20 W drive is needed to get 80 W out for that weak new grid or to get those birds off your new EME array. Fig 1 shows the art work I used for a single SD1870/TTC2223-18. Fig 2 is a schematic diagram of the single-device amplifier. Fig 3 is a photograph of two amplifiers combined with Sage Wireline com-

Table 1

Typical Specifications for Amplifier Using Two SD 1870/TCC2223-18 Transistors

Power In (W)	Power Out (W)	Gain (dB)
V_{cc} = 25 V		
8	30	5.75
10	36	5.55
V_{cc} = 28 V		
8	36	6.55
10	40	6.00

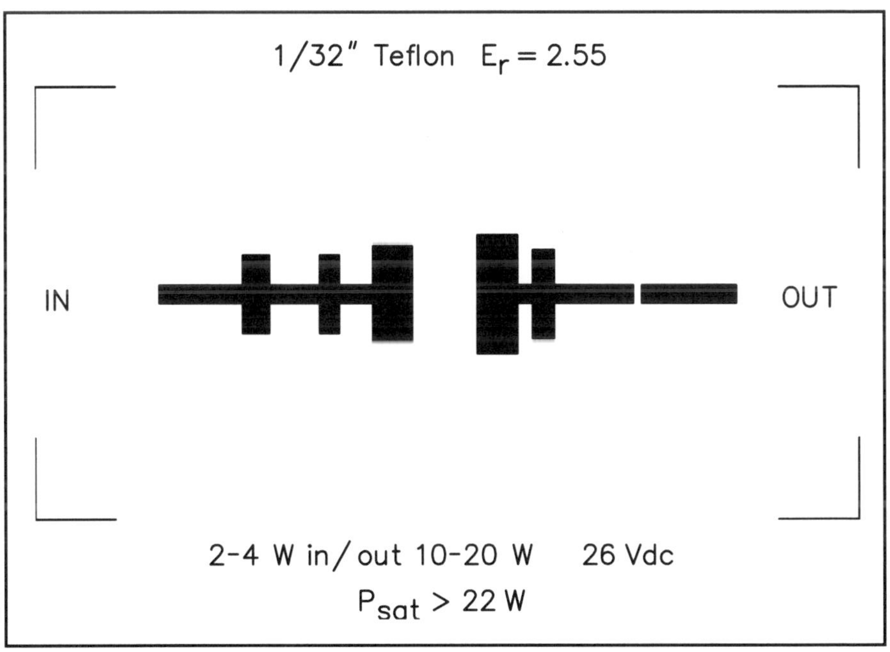

Fig 1—Full-size artwork for the amplifier circuit board.

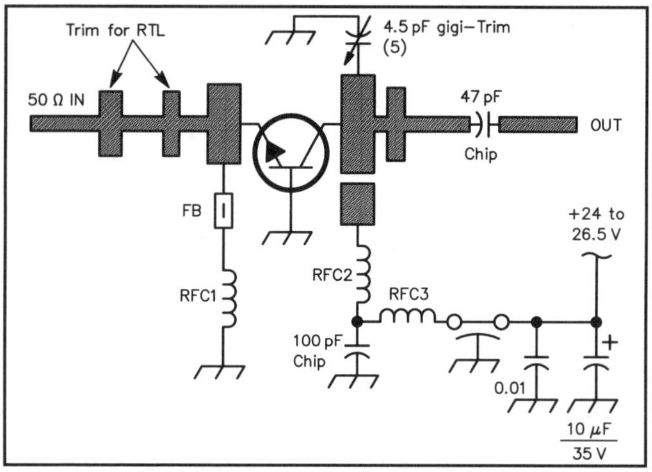

Fig 2—Schematic diagram of an amplifier stage. Two 40-watt stages are combined at their inputs and outputs to derive 80-watts total output.

biners. Fig 4 shows the power divider.

I'd like to thank David Mascaro for his help in testing the amps and transistors in the prototypes. Also Dick Comley, N3AOG, for his vertical mill workmanship and Paul Drexel, WB3JYO for his encouragement.

Notes

[1]Rivets are available from Frontier Microwave, RD 1 Box-467, Ottsville, PA 18942.

[2]SAGE Laboratories Inc., 11 Huron Drive, Natick, MA 01760-1314 (ask for sample of the line and a copy of *Designers Guide to Wireline and Wirepac*).

[3]B. Olson, "Solid-State Construction Practices," *QEX*, Jun 1987, p 11.

Fig 3—Photograph showing two amplifiers with inputs and outputs combined with Sage Wireline combiners.

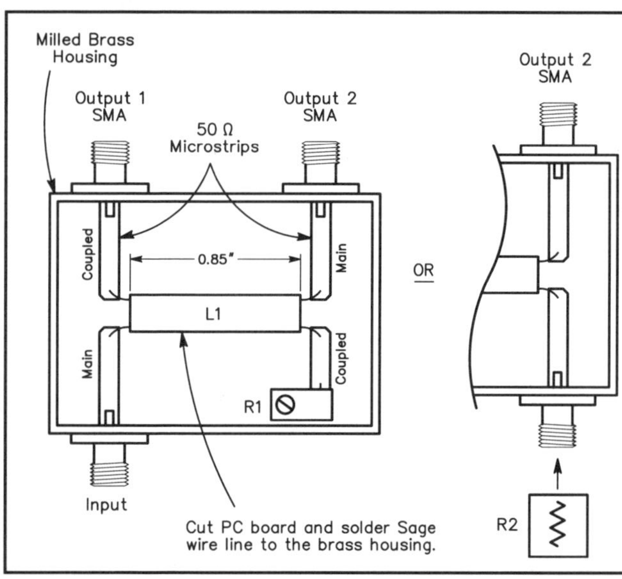

Fig 4—WA3JUF's two-way 2304/2160-MHz power divider.
L1—Sage 3-dB wireline
R1—50-Ω microstrip type termination
R2—50-Ω termination

A 7289 Amplifier For 3456 MHz

By B. W. Malowanchuk, VE4MA

(From *Microwave Update 1991*)

The recent rise in 3456-MHz activity due to the availability of surplus material and the no-tune transverters, has created a need for a reproducible high-power amplifier. This article describes a power-amplifier design for 3456 MHz, using a water-cooled 7289 or 7211 triode. The prototype amplifier has produced output powers of 35 W, with a gain of 10 dB and plate efficiencies of 10%. The detailed amplifier design has been previously published.[1] This paper focuses on the physical construction and operating characteristics of the amplifier.

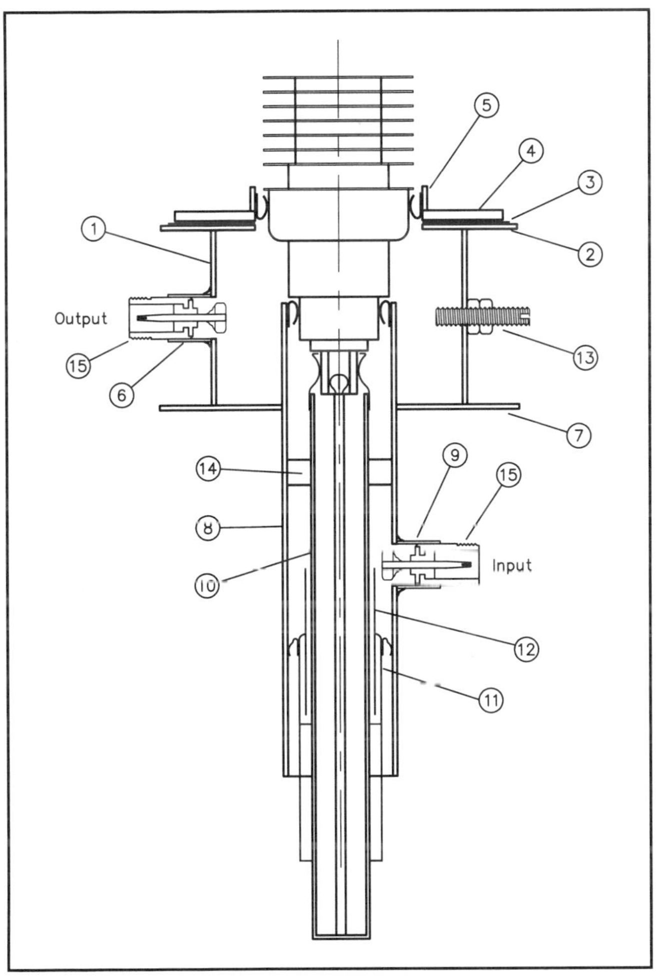

Fig 1—Cutaway diagram of the amplifier. Numbered components refer to the steps given in the text.

Amplifier Design

The amplifier design followed previous cavity amplifier work.[2,3] The anode cavity is a three-quarter wavelength capacitively tuned line. The output coupling uses a capacitive probe. A five quarter wavelength cathode cavity is used in order to have it physically extend outside the anode cavity. The cathode timing uses a sliding RF short. Cathode coupling is again by capacitive probe.

Amplifier Construction

The full-size cross section drawing of the amplifier is shown in Fig 1. Most of the material required for construction should be fairly easy to locate, although some parts may require machining, or improvising with surplus material you have available. Referring to Fig 1, the construction of the numbered parts of the amplifier discussed in sequence:

1. Part No. 1 is the outside wall of the anode cavity. The material is copper or silver-plated brass stock with a 2-in. outside diameter and ¹⁄₁₆-in. wall. The anode cylinder is 1.5 in. high and is drilled 0.67 in. from one end, to accept the output coupling probe (Part 6) and anode tuning capacitor (Part 13). These pieces are on opposite sides of the cavity (Fig 2).

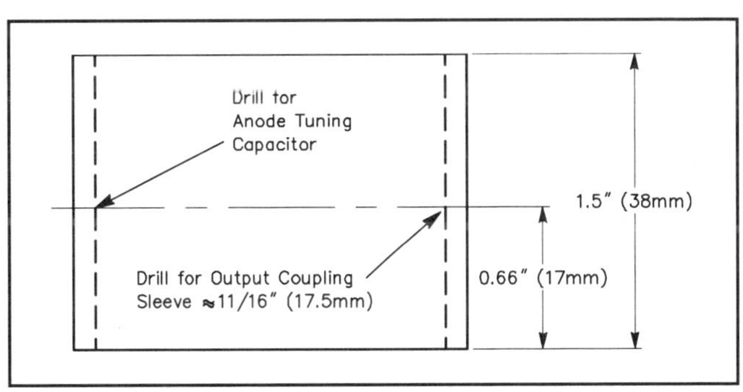

Fig 2—Anode cavity outer conductor.

2. Part No. 2 is the top plate of the anode cavity (See Fig 3). It is a 3-in. square plate of copper or silver-plated brass, 1/16-in. thick. It is drilled and tapped to accept four machine screws, which secure parts 2, 3 and 4 of the anode-bypass capacitor together. Cut a 1⅜ in. hole in the center to provide electrical clearance from the anode.

3. Part No. 3 is the plastic insulator for the anode-bypass capacitor (Fig 4). It is 3-in. diameter and 0.0035-in. thick. The material used was "drafting" Mylar, which gives a much higher capacitance than Teflon. Higher dielectric-constant materials such as Kapton are desirable. The material should be as thin as possible to minimize radiation hazards. The center hole is 1.125-in. diameter.

4. Part No. 4 is the upper plate of the anode-bypass capacitor. As shown in Fig 5, it is a 2.5-in. diameter disk of 1/16-in. copper or silver-plated brass. The center-hole diameter will depend on your choice of finger stock for the anode contact. I used 1.25-in; see Part No. 5.

5. Part No. 5 is a support collar for the anode finger stock. Very small finger stock, such as Instrument Specialties 97-380, can be soldered directly to the inside of the center hole in Part No. 4. This finger stock is fairly fragile, so take care during tube insertion or removal. A support collar adds considerable strength.

Another option is to use larger finger stock and support it above Part No. 4 using Part No. 5. If this is done in a normal manner, the anode cavity length will have to be compensated for or you may be able to mount the finger stock upside down without compensation, to achieve good results.

Fig 6 reflects construction with this latter method and the dimensions for Parts 4 and 5 also reflect this choice. The finger stock used was similar to Instrument Specialties 97-135.

6. Part No. 6 is the sleeve for the output coupling probe and is shown in Fig 7. It is cut from a coupling for ½-in. copper water pipe, which conveniently is a good fit over a flangeless UG-58 A/U type-N female connector. It is slotted with a hacksaw to allow a small hose clamp to tighten the sliding joint (see discussion on Part 15). The sleeve should be made longer if extended input/output couplers are used.

7. Part No. 7 is the anode cavity bottom plate. It is a plate similar to Part No. 2, but no machine screw holes are required (See Fig 8). The center hole diameter of approximately ⅞-in. is a tight fit over Part No. 8.

The interfaces between Part Nos. 1, 7 and 8 are in a high-current area, so that connections should be soldered using a silver solder if possible.

8. Part No. 8 (Fig 9) is the inside conductor of the anode cavity and outside conductor of the cathode cavity. It uses a 3-⅞-in. length of ¾-in. copper water pipe. A length of 97-380 finger stock is soldered to one end for the grid connection. At 1⅝-in. from the other end, the pipe is drilled for the input probe.

9. Part No. 9 is the sleeve for the input coupling probe (Fig 10). This part is different from the output sleeve because of the small diameter of Part No. 8. The sleeve should be made longer if extended input/output couplers are used.

10. Part No. 10 is the inner line of the cathode cavity. It is not detailed here because it is fairly complex to build. It contains the filament and cathode connections. The important dimensions are the diameter of 7/16-in. and overall length of approximately 5 in.

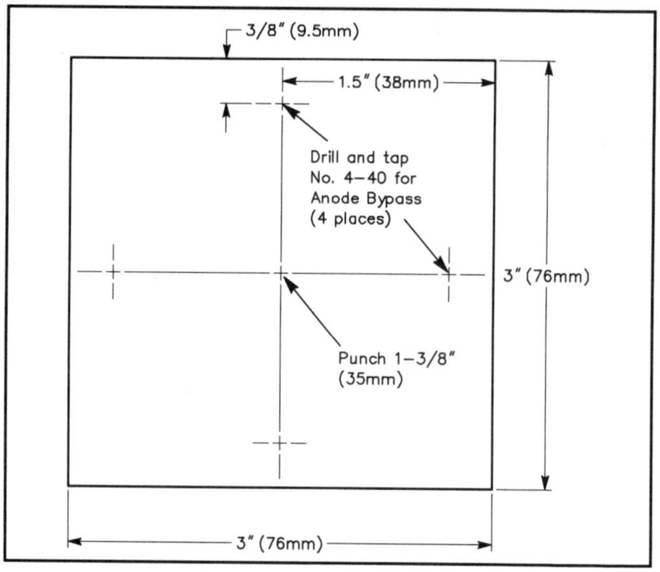

Fig 3—Anode cavity top plate.

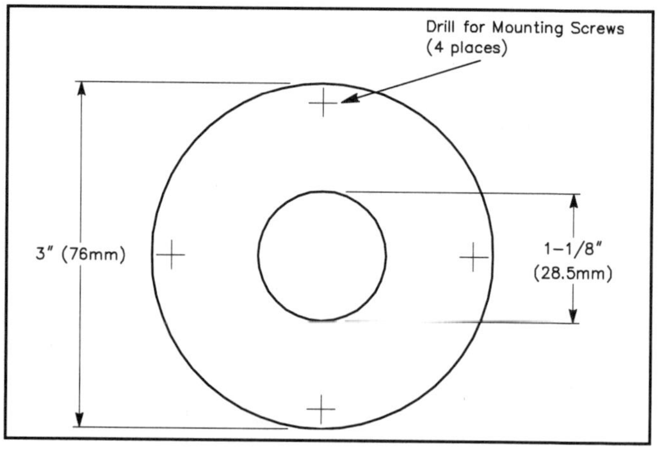

Fig 4—Anode bypass insulator.

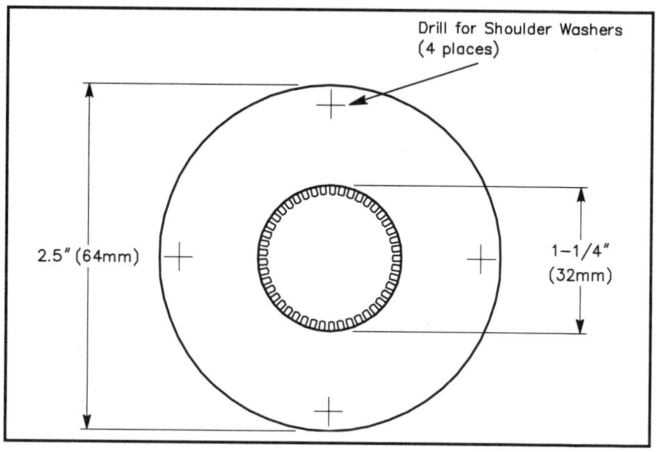

Fig 5—Anode bypass capacitor plate.

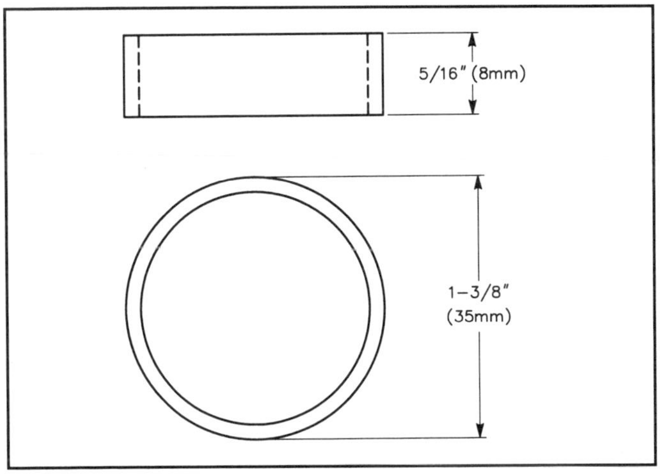

Fig 6—Anode finger-stock support ring.

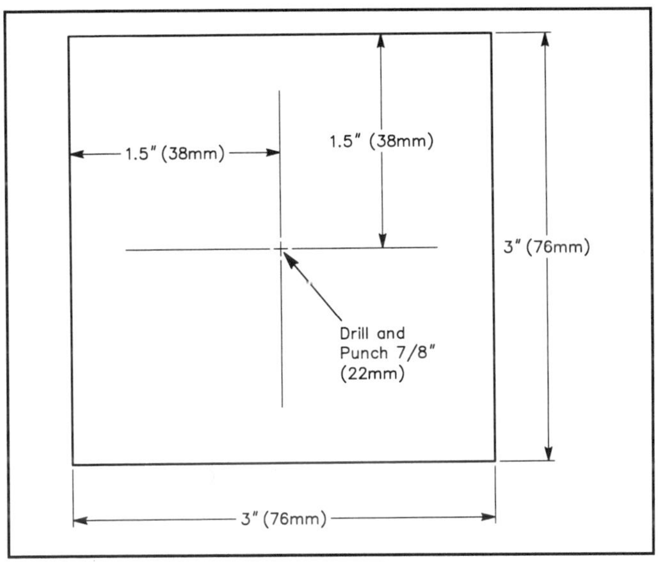

Fig 8—Anode cavity bottom plate.

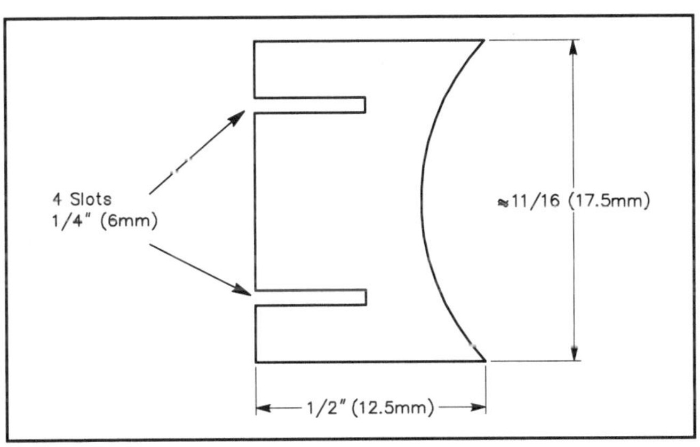

Fig 10—Input probe mounting sleeve.

This assembly is probably best constructed from surplus parts. 11 and 12. Parts Nos. 11 and 12 form the tunable cathode-RF short. They really are a moving RF-bypass capacitor. Part No. 12 is a half wavelength piece of ½-in. brass tubing, which slides along Part No. 10, which has a layer of Teflon or Mylar

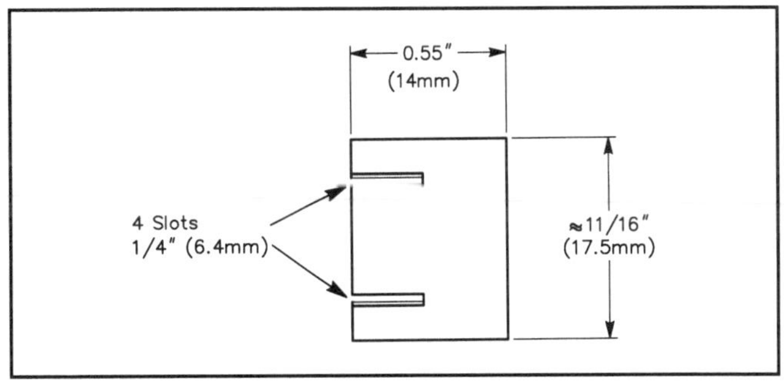

Fig 7—Sleeve for output coupling probe.

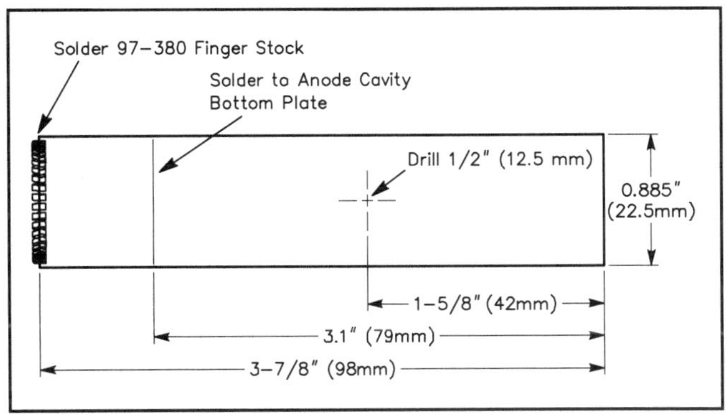

Fig 9—Anode cavity inner conductor and cathode cavity outer conductor.

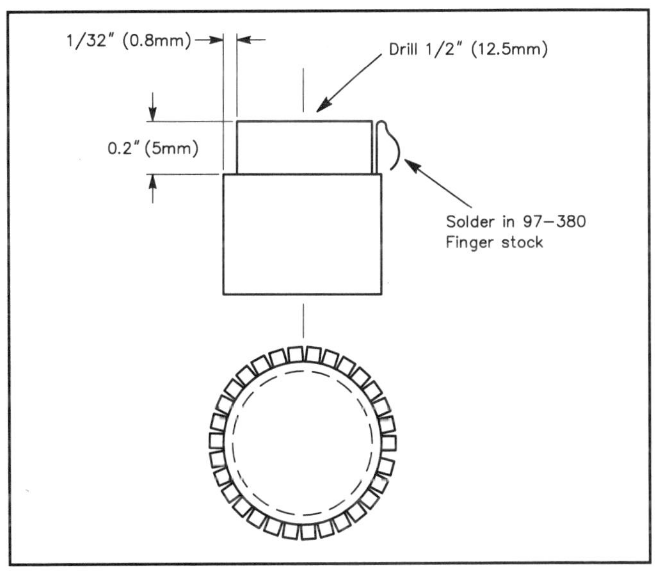

Fig 11—Tunable cathode short.

tape wrapped on part of its length (1-4 in.) from the outside end. This forms a coaxial capacitor.

The center of the sliding capacitor is connected to the inside surface of Part No. 8 by Part No. 11 (Figs 11 and 12). Part No. 11 is created from a ½-in. copper water pipe "end cap." The closed end of the cap is drilled ½-in. to accept Part No. 12. The closed edge of Part No. 11 is turned down on a lathe (or in a drill press with a file) for a depth of approximately

1/32 in. and a length of approximately 0.2 in. The depth of the groove is chosen to let the finger stock make connection to Part No. 8 without excess mechanical tension.

A piece of 97-380 finger stock must be soldered to this groove. The soldering operation is delicate since you do not want to overheat the finger stock and Part No. 12 must be held concentrically while it is soldered.

The tunable RF short is extended outside the cathode cavity by two push rods soldered to the outside open end of Part 11. These can be made of 1/16-in. brass, and should be connected to a crossbar to ease adjustment.

13. Part No. 13 is the anode-tuning capacitor. This part deserves special attention, since the tuning will be rather sharp. I used a surplus 3/8-in. diameter brass bolt, which was threaded for 1-in. with 24 threads per inch (tpi). This worked reasonably well, but 32 tpi would be a better choice.

A threaded bushing 1/4-in. long is soldered into or onto the wall of Part No. 1. A similarly threaded nut with a split ring washer to provide tension on the threads will be required. The anode capacitor parts should be silver plated if possible, to minimize tuning noise as the metals oxidize.

14. Part No. 14 (Fig 13) is a Teflon insulator which holds Part No. 10 concentric inside Part No. 8. It can be held captive inside Part No. 8 by lightly center punching the walls.

15. Parts No. 15 are the input/output capacitive coupling probes. These can be made from UG-58A type-N connectors, with the mounting flange removed. This arrangement does not give much adjustment range, due to the short body length (Fig 14). A larger probe could be made by extending the UG-58 connector with a short length of 1/2-in water pipe and 1/4-in. rod.

The length of the extension should be made in multiples of a half wavelength so that the impedance variations will not be critical. The capacitive probe is completed by attaching a 10-32 brass nut to the end of the connector. The nut should protrude out from the coaxial section by 0.2 in.

Operating Results

The amplifier was tested using several GE 7289s of known quality with the following results:

Power Output	35 W
Power Input	3.5 W
Power Gain	10 dB
Anode Voltage	1300 V
Anode Current	269 mA
Grid Current	11 mA
Anode Efficiency	10%

A power gain of 10 dB was also measured at 20-W output, so saturation was not evident. The idling current was set to 100 mA, as thermal drift is minimal at that level.

Keep grid current low. High grid current causes heat that warps the grid, which changes amplifier tuning and makes the tube prone to arcing.

The amplifier is capable of more output power with additional driver power and anode current. High anode voltage would improve the output power and gain, but at a great risk of arcing.

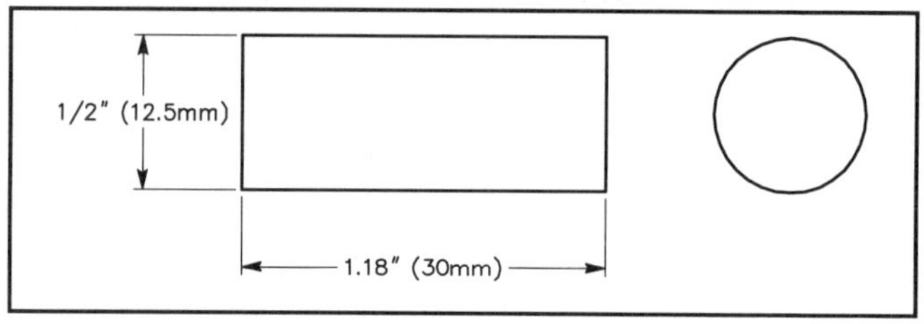

Fig 12—Half-wave stub.

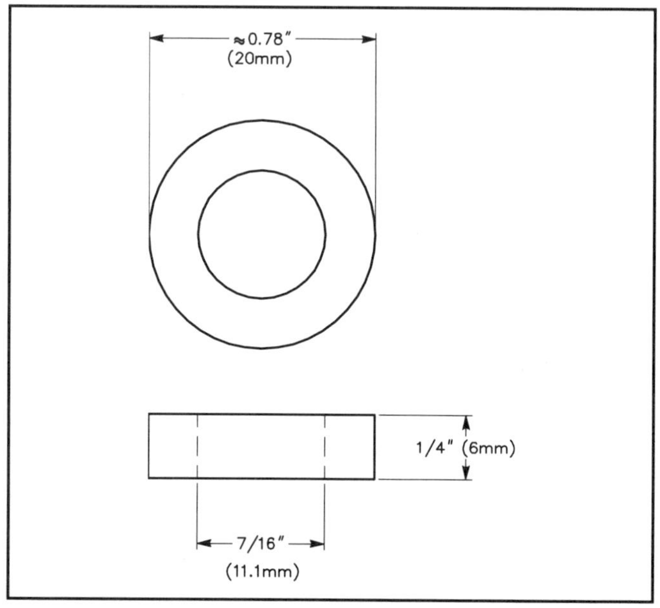

Fig 13—Cathode insulator.

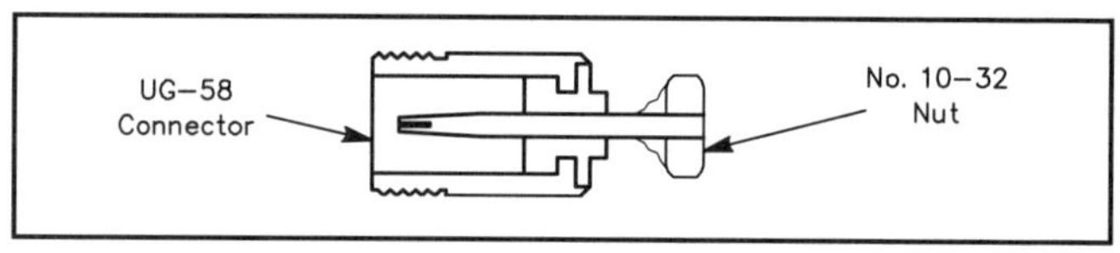

Fig 14—Input/output coupling probes.

Further Work Required

Although the amplifier met the original design objectives, an article by DJ6EP[4] suggests a major improvement in anode efficiency and output power by using an improved anode-bypass capacitor. This technique needs to be evaluated.

The ultimate output power of the amplifier needs to be explored fully, as EME contacts appear practical with this type of amplifier and a reasonably sized dish.

Notes

[1]B. Malowanchuk, "High Power Tube Amplifiers for 2304 MHz," *Proceedings of Microwave Update '89*, pp 88-98.

[2]Buzz Miklos, "Coaxial Cavity Amplifiers," *Proceedings of the Central States VHF Conference*, 1985.

[3]B. Malowanchuk, "2304 MHz Power Amplifier Using 7289 or Similar Tube," *Proceedings of Microwave Update '87*, pp 108-112.

[4]R. Wesolowski, "9-cm Band Tube PA Stage," *VHF Communications*, Issue 4, 1989.

A 125-Watt Amplifier for 902 MHz

By Ken Schofield, W1RIL

(From *QEX* for April, 1988)

This amplifier is a scaled version of an old friend, originally designed for 23 cm. A later version using water cooling, appeared in the Crawford Hills VHF Club *Technical Report*. In addition to scaling the dimensions to the 33-cm band, other changes to the original design include:

1. Additional reactance loading in the anode cavity.
2. Changes to output circuit with additional sliding drawer to facilitate easy loading adjustment.
3. Waveguide-beyond-cutoff air ducts to reduce RF leakage.
4. Use of aluminum wall stock instead of brass.
5. Use of micrometers as tuning devices.

Power output on 33 cm is 125 W with 10 W drive, a gain of just over 10 dB. Efficiency is about 40%. The amplifier is thermally stable with air cooling. During a 4-hour operating stint, power remained at the 125-W level, with no tuning adjustments necessary throughout the period.

You don't need a machine shop to construct this amplifier. Much of the drilling can be done with a small electric drill, but a small drill press makes the work easier. A drill press is especially handy for "fly cutting" the holes for finger stock. These holes can be opened up by drilling a series of small holes and finishing with a file, if necessary.

A drill press and a compound vise can serve as a small milling machine. The addition of a "dead center," bolted to the press table, turns it into a small vertical lathe. The Teflon shoulder washers for the amplifier can be easily made in this manner. It's just a matter of learning to work vertically, instead of horizontally!

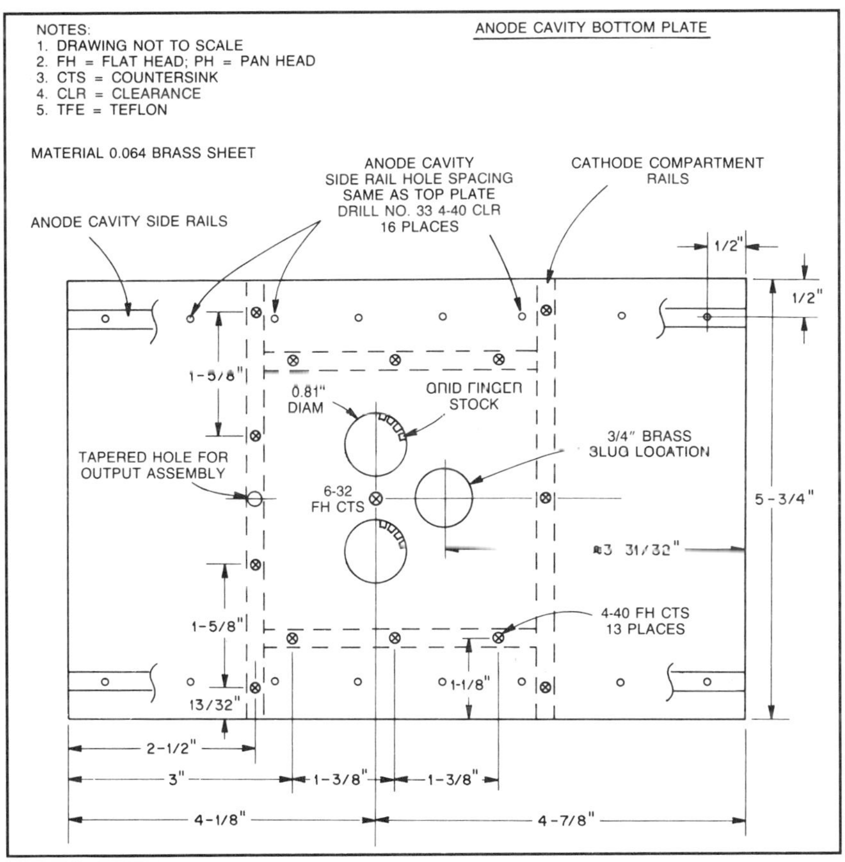

Fig 1—Construction information for the anode cavity bottom plate.

Input Circuit

An electrical λ/2 stripline is used in the cathode circuit. A portion of the cathode circuit includes the internal construction of the 7289 tubes. The line is adjusted capacitively on one end, and adjustable capacitive coupling provides input matching.

Output Circuit

The anode cavity is a capacitively loaded λ/2 circuit, tuned with a sliding drawer on one end. Loading adjustment is accomplished by a variable loop, formed by the output stud and the second sliding drawer on the other end of the cavity. Micrometers make tuning and loading adjustments smooth and postive.

Cooling and Mounting

The amplifier is adequately cooled by air pressure forced through ⅜-in. OD, beyond-cutoff, aluminum tubes in the sides of the cathode and anode compartments. Air is also blown across the plate fins of the tubes via a plenum. Both cooling and mounting can be successfully combined on a pressurized chassis, by mounting the amplifier on its side with the 6 air ducts protruding through the chassis. An additional cutout in the chassis allows air to be directed by the plenum to cool the finned anode plates.

You can make a suitable plenum from clear plastic mounted on ⅜-in. studs, which replace the 4 corner screws on the top plate. This cover also prevents accidental contact with the HV bypass capacitor. I use IPS *Weld-On 16* to cement the plastic.

Construction Considerations

Refer to Figs 1-14. It is important to keep the tube-socket locations in the several layers in proper alignment. Perhaps one of the best ways to accomplish this is to lay out the top bypass plate, drilling small pilot holes for the tube centers. Then, you can use this plate as a template to make like holes in the cavity top and bottom plates. Location of the Teflon shoulder washer and other critical holes are done in like manner. The cathode line can be held in place with its support hole and the tube centers located easily.

The air ducts are pressed into the aluminum walls in a vise. First, drill out the walls with a smaller drill, then ream the holes to accept the beveled end of the ⅜-in. tube approximately halfway through the wall. When pressed in flush, the tubes lock in firmly and make good electrical contact with the wall stock. The slippery characteristic of Teflon is used as a bearing surface to convert the rotary motion of the tuning micrometers to the push-pull motion required to move the tuning and loading drawer slides. Very little play is experienced

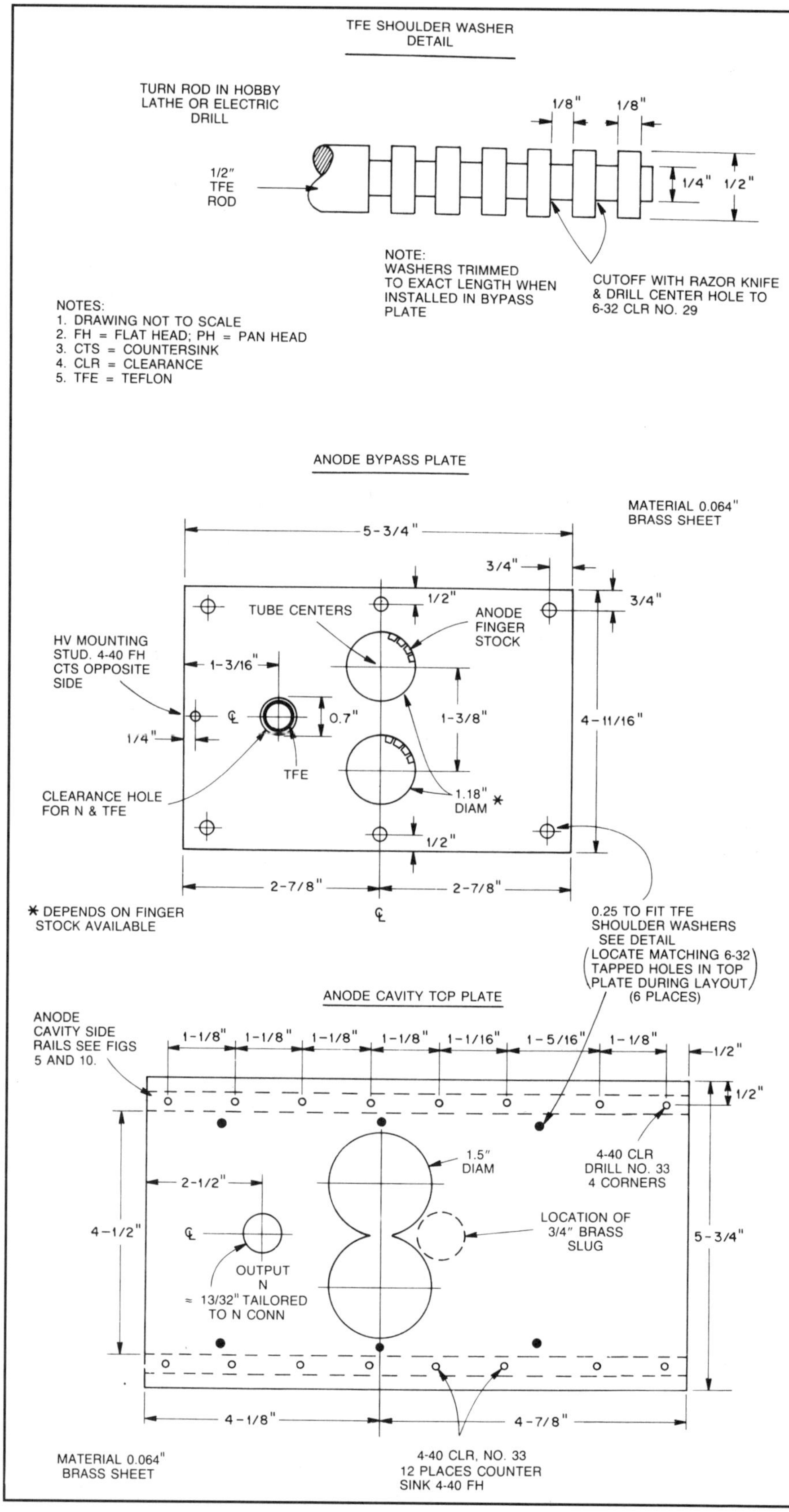

Fig 2—Details of the anode bypass plate and anode cavity top plate.

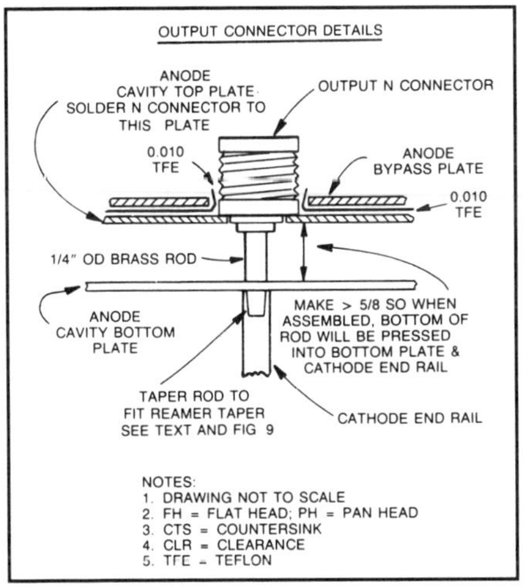

Fig 3—Output connector details.

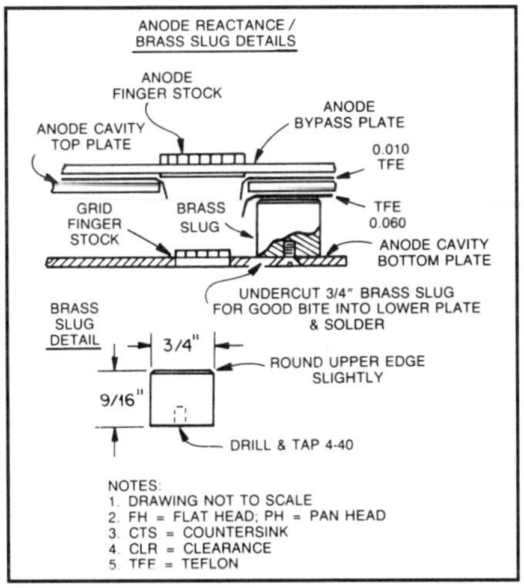

Fig 4—Construction and assembly of the brass slug that provides additional reactance in the anode cavity.

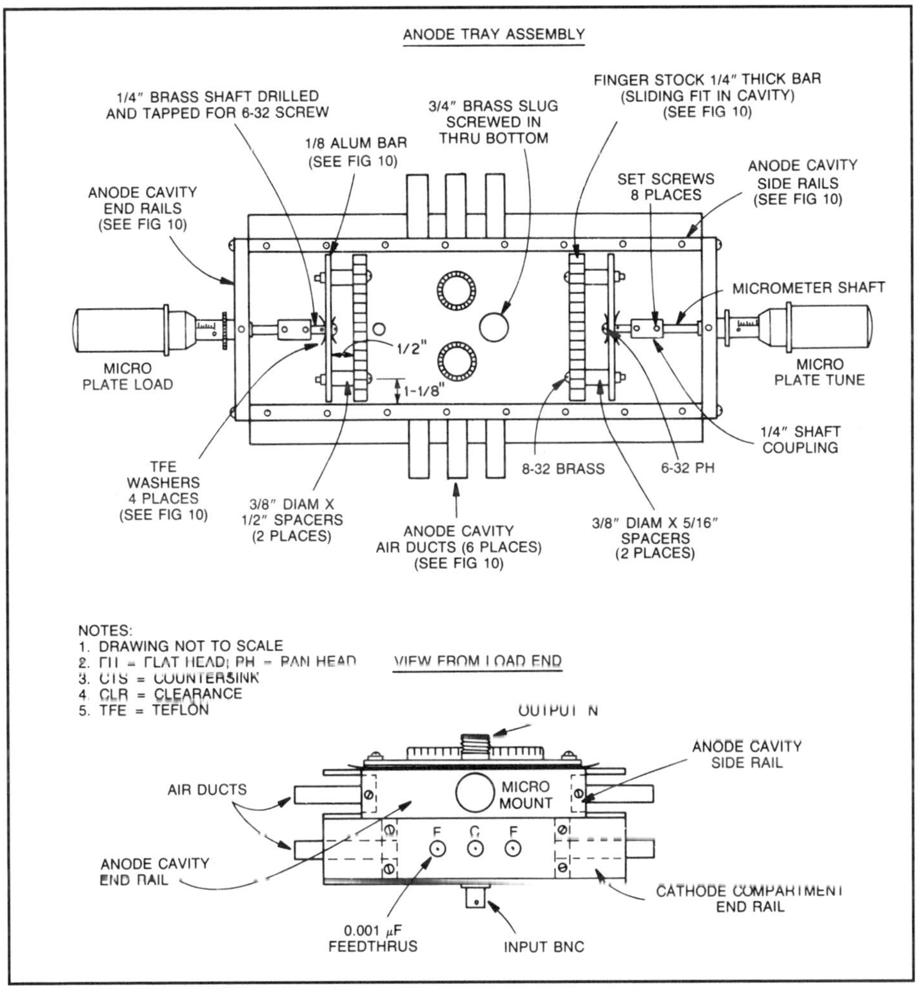

Fig 5—Assembly of the anode cavity.

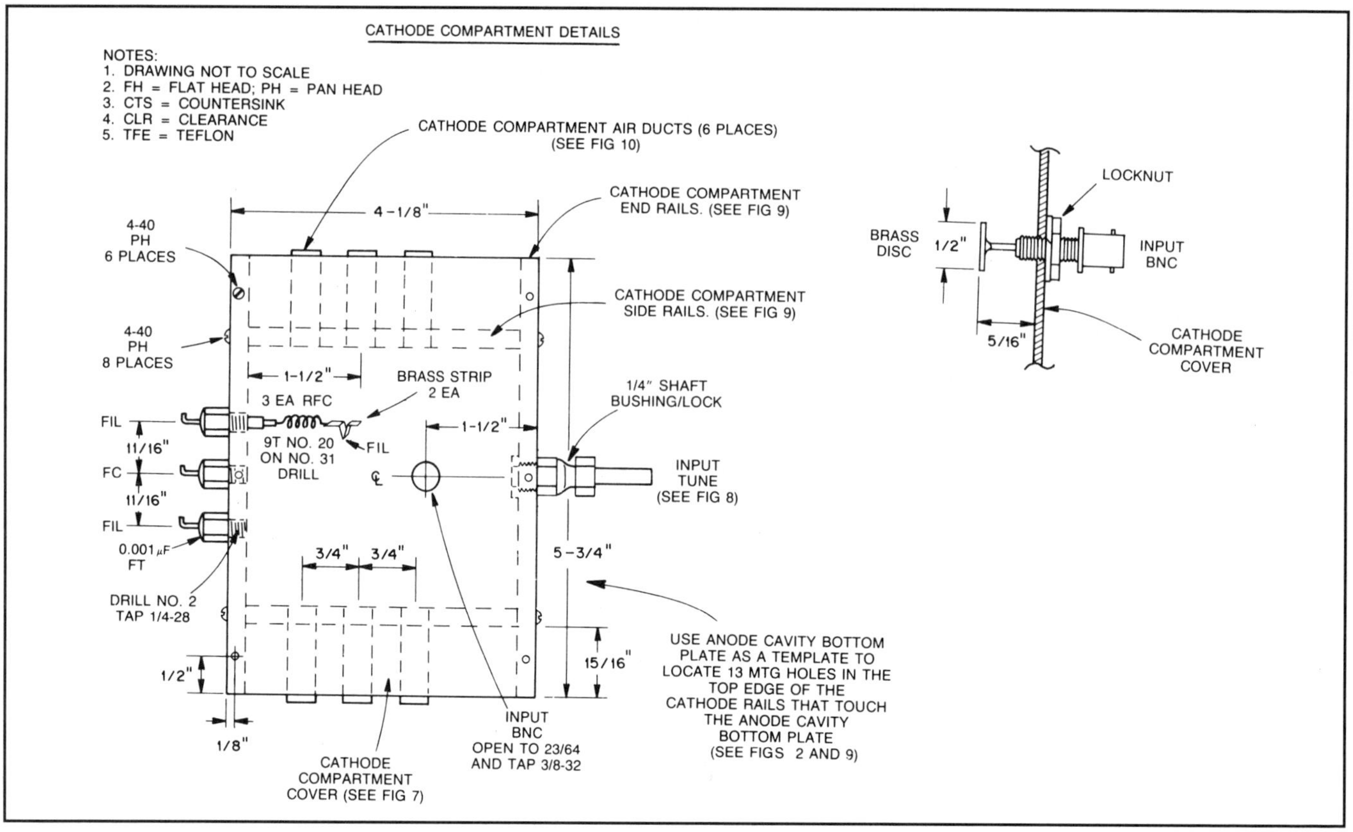

Fig 6—Cathode compartment details.

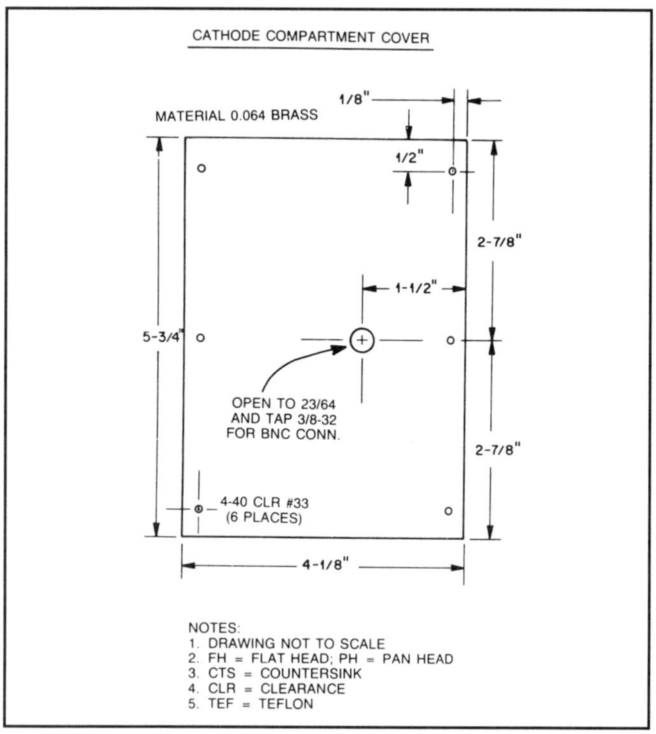

Fig 7—Cathode compartment cover drilling information.

using this method. Finger stock used in the grid and cathode areas is Instrument Specialties no. 97-251. You can use the same stock on the tuning slides. The anode finger stock was obtained at a flea market, and was formed into a ring. This ring is 7/16-in. high and can be formed from Instrument Specialties no. 97-139 stock.

Do not use nylon shoulder washers on the plate bypass capacitor, as they will break down! Teflon washers work well, even with voltages greater than 1400.

Tuneup and Results

Preset the load drawer to approximately 1/4-in. from the load stud, and the tune drawer to approximately 3/4 of the way out. Apply 500 V to the anode and adjust the bias circuit for a no-drive cathode current of 50 mA. Apply drive and adjust the cathode circuit for maximum cathode current. At the same time, adjust tuning and loading for maximum power output. Slowly increase anode voltage while maintaining no-drive cathode current at 50 mA by adjusting the bias, while repeaking all controls.

Adjust input SWR if necessary by variying the input BNC connector disc spacing. At the 10-W drive level, with 1100 plate V, anode current should be about 290 mA and output power should be about 125 W.

A consumer-grade microwave leakage detector indicated no RF leakage near the air ducts. When coupled tightly to the tube anodes, the detector's meter barely moved.

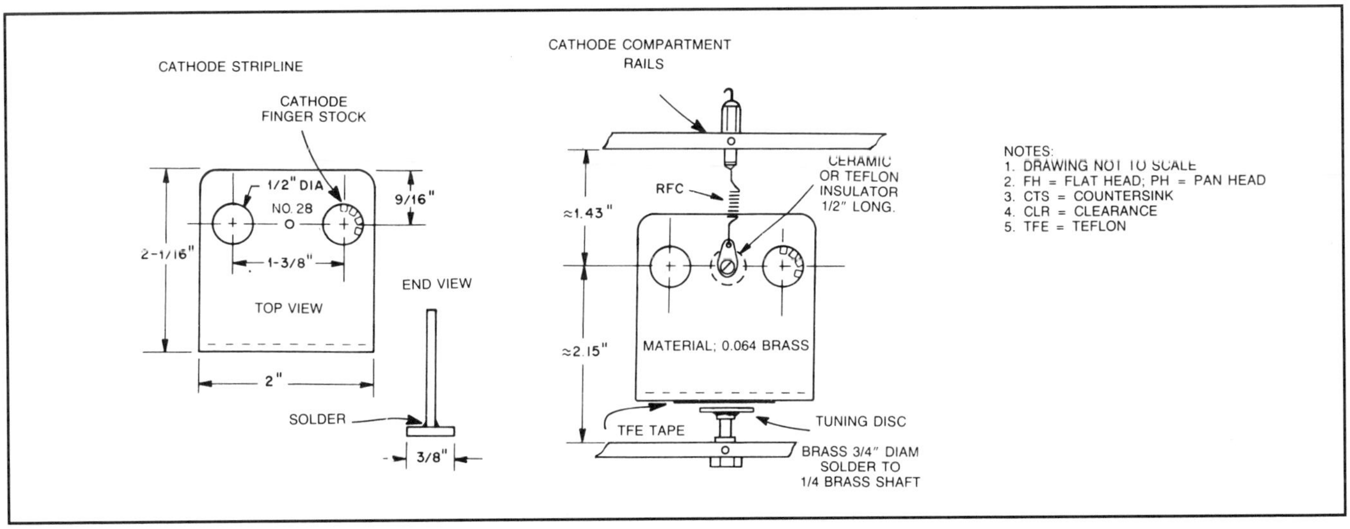

Fig 8—Construction and assembly of the cathode stripline.

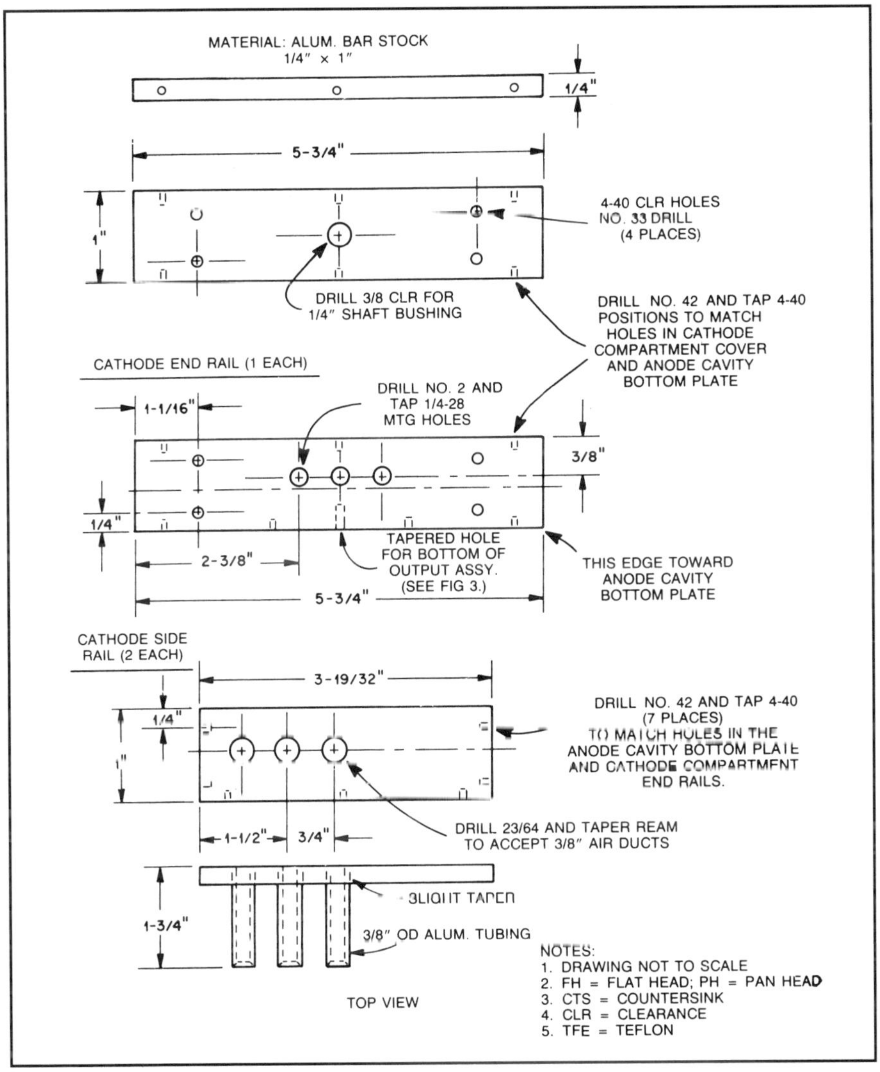

Fig 9—Construction information for the cathode compartment end rails, side rails and air ducts.

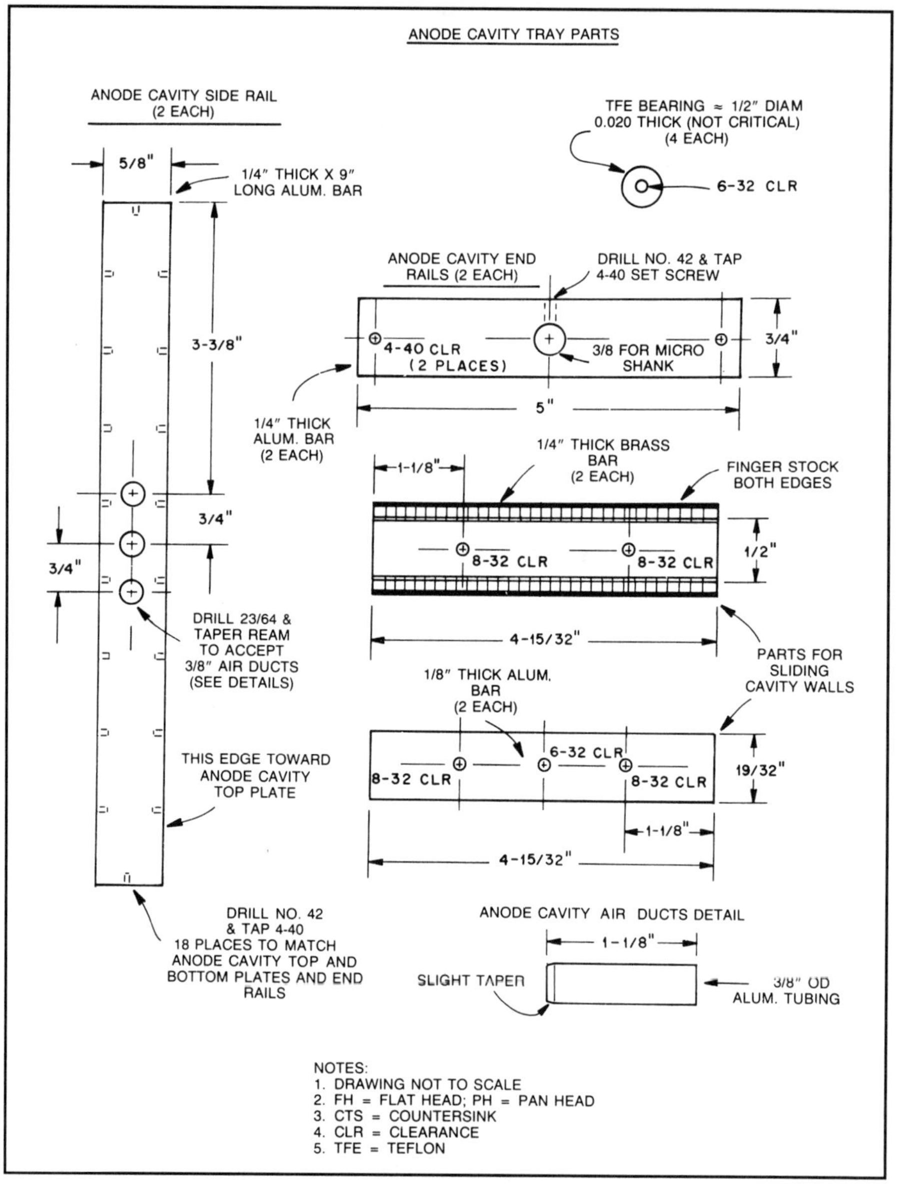

Fig 10—Construction information for the anode cavity parts.

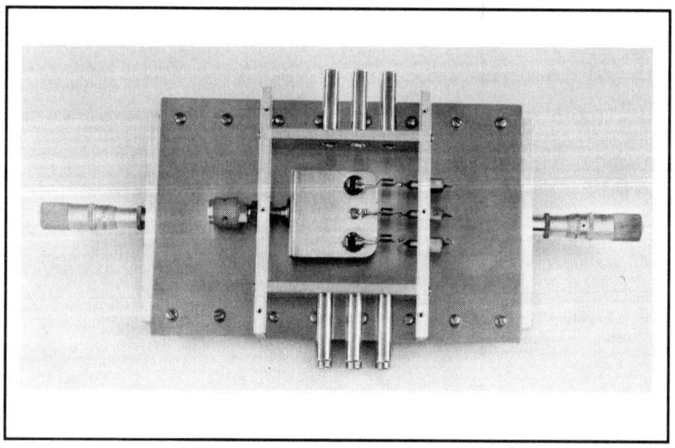

Fig 11—View of the anode cavity with the anode bypass plate and anode cavity top plate removed.

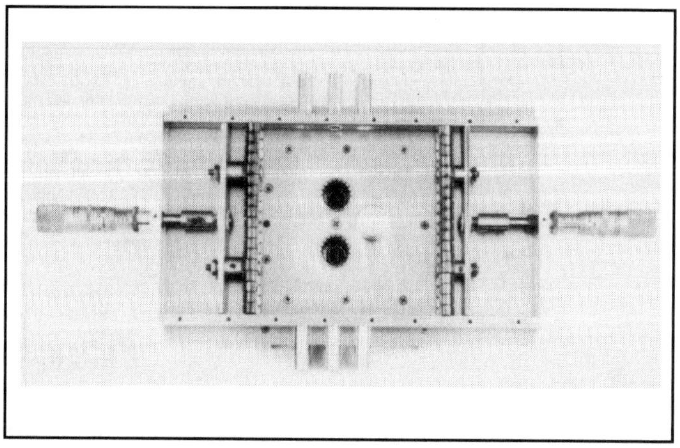

Fig 12—View of the cathode compartment with the cover removed.

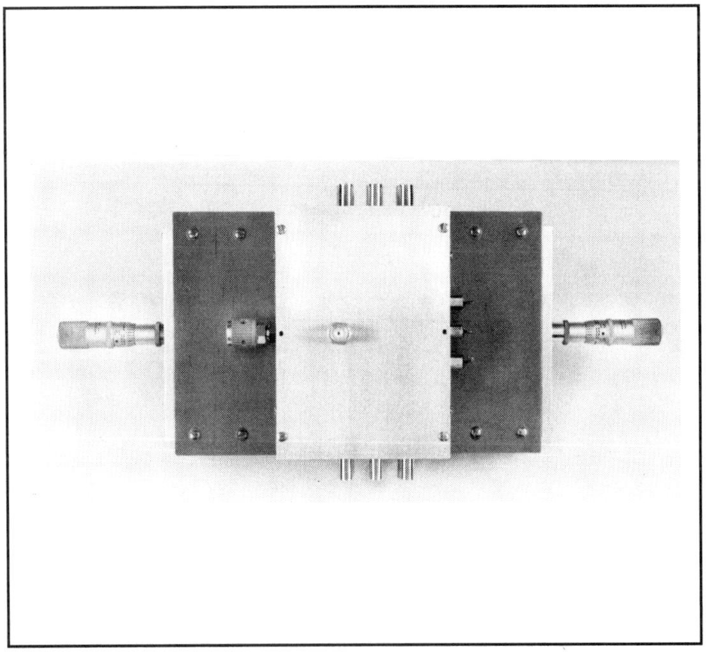

Fig 13—Underside of amplifier with the cathode compartment cover in place.

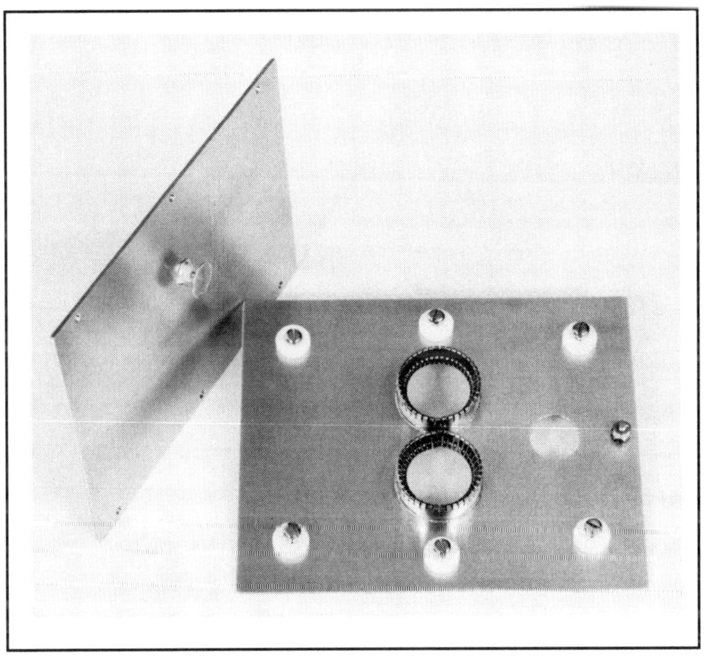

Fig 14—Cathode compartment cover and input coupling disk (left) assembled anode bypass plate (right).

9

Transmission Lines

Contents

Easier BNC and Type-N Connector Installation

By Zack Lau, KH6CP

(From *QST*, June 1989)

Getting all the cable-shield wires through the clamp can be difficult during installation of BNC and Type-N connectors; usually, a few braid wires end up getting squashed under the clamp. Solution: Hold the wires down by wrapping them with electrical tape. Don't tape beyond the braid wires onto the coax outer jacket; just tape the braid itself.

Connectors for 1-Inch Hardline

By Ray Benny, N6VR

(*From* QST, *May 1992*)

This item first appeared in the *National Contest Journal.*

The following is a description of the system I use to install female coax connectors on 1-inch Hardline. The only parts needed are the Hardline (available from many cable-TV companies), a ½-in. male copper pipe adapter, and a standard PL-258 barrel connector that can be disassembled (Amphenol 83-1J is the type I use). Figs 1 and 2 are provided for reference.

1. Make a square cut at one end of the Hardline with a fine-tooth hacksaw.

2. Use either a hacksaw or tubing cutter and cut off 9/16 in. of the plastic jacket and aluminum tubing. Do not cut or nick the center conductor. Completely remove the plastic jacket and aluminum outer conductor.

3. Using a sharp knife, remove the foam from the copper center conductor. Use a fine file to deburr the end of the center conductor.

4. Tin the center conductor with a thin film of solder.

5. Cut off evenly about ½ in. of the outside plastic jacket.

6. Disassemble the coax barrel connector by removing the internal snap ring. Note the end of the barrel from which the snap ring came.

7. Apply solder flux to the snap ring end of the barrel and the inside of the male adapter. Place the fluxed end into the ½-inch male adapter. Tap it gently to seat it. Next, place the two pieces upright in a vise. Use a propane torch to solder the pieces together.

8. Slip the barrel center coupler about halfway onto the tinned center conductor of the Hardline. It may take some experimenting to find the best position for proper spacing of this coupler. Solder the coupler onto the center conductor.

9. Apply a small amount of silicone grease (I use the non-hardening type) on the foam, around the center conductor. Apply a small amount of aluminum/copper conductive grease (Noalox, Penetrox or equivalent) to the outside of the Hardline's aluminum jacket.

10. Hold the prepared end of the Hardline pointed upward and slip half of the clear plastic barrel spacer over and onto the barrel center coupler. The end with the outer recess goes upward (the same way it came apart). The second half of the plastic spacer is not used. Place the male adapter/barrel assem-

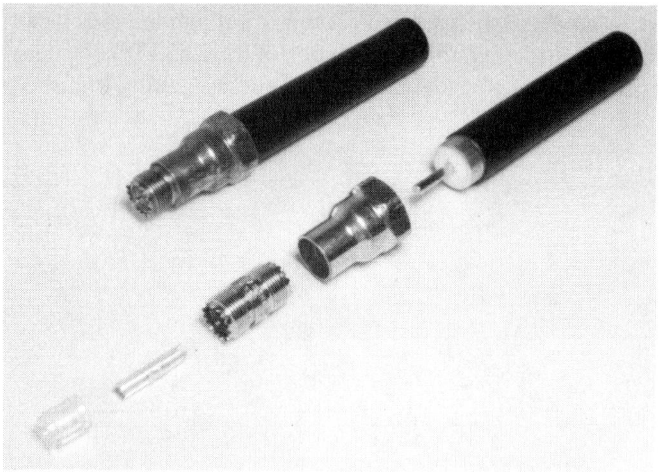

Fig 1—At the top, an assembled Hardline connector. At the bottom, the pieces of the barrel connector and the plumbing fitting are lined up, ready for installation on the cable.

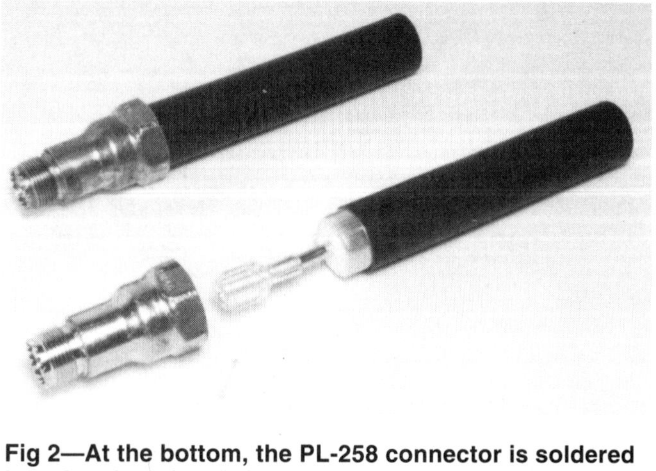

Fig 2—At the bottom, the PL-258 connector is soldered into the plumbing fitting. Half of the barrel's insulator is installed and the barrel center coupler is soldered to the Hardline center conductor. All that's left to do is screw the connector assembly onto the Hardline and weatherproof the connections.

bly over the center coupler soldered to the Hardline. Use a large adjustable wrench to screw this assembly onto the aluminum jacket. Screw it on until the plastic barrel center spacer just touches the inner ridge of the barrel.

11. There, done! When this connector is mated with its other half, I wrap it with several layers of black tape to provide a water-tight seal. Another thing that I do is to trim/tune the coax for a multiple of a half wavelength at the operating frequency.

Once tuned, the input impedance present at the coax input will be the same at the output of the coax. (See ON4UN's *80-Meter DX Handbook*, or *Low-Band DXing*,[1] which describe how to use an SWR meter and a dummy antenna to trim the coax line.) I have used this 1-in. CATV Hardline as described on several long runs (200 to 300 feet) in place of standard RG-213 cable. Although I have not measured the losses with a load and wattmeter, I have definitely noticed a difference in both the receive and transmit signals on the higher HF bands. I have also replaced a 60-foot run of RG-213 on 450 MHz with about 50 feet of electrically cut 1-in. Hardline (with RG-213 jumpers on both ends for flexibility) and noticed a much-improved signal. This 1-in. Hardline is very expensive if purchased outright, but my Hardline was free! The trick is to call your local TV cable company and ask them if they would like to unload their "reel ends" (large reels with only 100 to 300 feet remaining). My local company was in the process of cleaning their yard, so I was able to take as much as I wanted. If they are not too happy about giving it away, it may be worthwhile to offer them payment to at least offset the scrap value they could receive from a surplus dealer.

[1] J. Devoldere, *Antennas and Techniques for Low-Band DXing* (Newington: ARRL, 1994).

Using Type-N Connectors on Half-Inch Hardline

By Dave Mascaro, WA3JUF

I have used standard Type-N connectors on half-inch aluminum Hardline for over 10 years. Performance and weather resistance is as good as or better than commercial connectors. Most commercial connectors have a push-fit center conductor connection that turns green after six months outside.

This procedure allows the use of ordinary Type-N connectors on half-inch aluminum Hardline with a solid copper center conductor. Refer to Fig 1.

1. Strip off the plastic covering. Be careful not to nick the aluminum.
2. Bore the connector nut to $^1/_2$-in.
3. Slide the nut onto the Hardline.
4. Slide the ferrule onto the Hardline (see Fig 1). The ferrule is made from a 0.2-in. piece of $^1/_2$-in. soft copper pipe, split to fit snugly around the aluminum outer conductor.
5. Flare $^1/_8$ in. of the end of the aluminum jacket to form a clamping ring inside the connector housing.
6. Cut the center conductor flush with the end of the outer conductor.

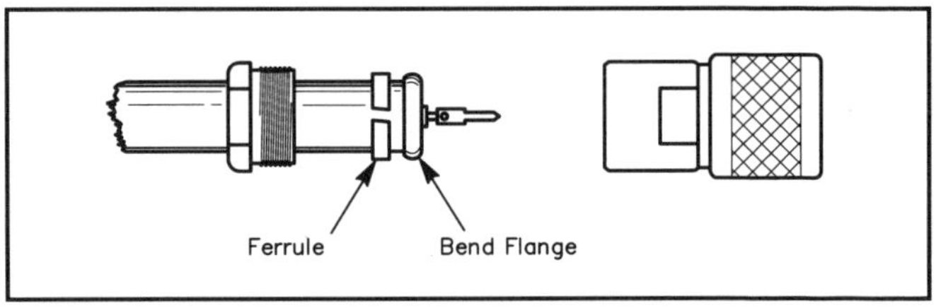

Fig 1—Assembling a Type-N connector to ½-in. Hardline.

7. Drill a hole in the center conductor to accept a piece of 18-gauge wire.
8. Solder the connector center pin and wire to the center conductor.
9. Coat the ferrule-aluminum connection with Oxiban.
10. Assemble the connector. After tightening the nut, the connector housing should not turn. If the connector is loose or the nut bottoms out, use a wider ferrule.
11. Coat the connector with silicone sealant.

Semi-rigid 0.141-Inch Coax to BNC Adapter

By Wes Atchison, WA5TKU

This approach to fitting a BNC connector to the popular 0.141-inch semi-rigid coax takes advantage of the 0.156-in.-ID of 3/16-in.-OD brass tubing. This technique was developed on UG-88 BNC connectors. The connector dimensions were taken from the Amphenol connector catalog. Other manufacturers' connectors should work as well. In fact, the technique should be adaptable to any connector by simply using the appropriate-sized brass tubing.

The ID of the UG-88 locking nut is a nominal 0.214 in. This results in a difference between the coax OD and the locking nut ID of 0.073 in. This difference is taken up by two pieces of brass tubing (1/32-in. wall thickness). A 1/2-in. and 7/16-in.-long piece of 3/16-in. and 7/32-in.-OD brass tubing (respectively) are slid onto the coax as shown in Fig 1. A trial fit of the 7/32-in.-OD tubing is necessary before assembling the connector, since the nominal dimensions of the locking nut hole and the tubing OD are a 0.004-in. force fit. Depending on the tolerances of your materials, the nut may have to be drilled out with a no. 2 or 7/32-in. drill bit. An alternative method would be to sand down the OD of the brass tubing.

The following steps are taken to assemble the connector:

1. Strip the shield and center insulation of the coax back far enough to install the center pin of the connector. Do *not* install the pin.
2. Slide the drilled-out nut onto the coax.
3. Slide the 3/16-in. brass tubing onto the coax.
4. Slide the 7/32-in. brass tubing onto the 3/16-in. tubing.
5. Slide a no. 6 washer onto the coax.
6. Adjust the washer, 3/16-in. tubing, and the 7/32-in. tubing so that the assembly is flush with the end of the coax shield.
7. Solder the washer, 3/16-in. tubing, 7/32-in. tubing and the coax shield as shown in Fig 1.
8. Solder the center pin to the coax center conductor.
9. Slide the connector housing onto the coax.
10. Slide the nut into the housing and tighten.
11. Use an ohmmeter to check for shorts.

Brass tubing can be found in 12-in. lengths at most hobby shops (for about a dollar a piece). The no. 6 washer can be found at your local hardware store for about 15 cents.

Another approach I have developed is shown in Fig 2. This technique is not as "pretty" as the one described above, but it does work. The backing washer used to install the connector on RG-58 coax is cut with a side cutters. The inside diameter is reduced to fit the coax with a pair of pliers. It is necessary to cut the overlapping ends of the washer to fit the assembly inside the connector housing. The washer is soldered to the coax shield after the insulation has been stripped to allow the installation of the center pin. The housing is now installed and the nut tightened.

This technique is not recommended for cables that will be used outside.

I prefer the first method because it supports the coax better and appears to be more professional. I use both methods in my 1296-MHz system.

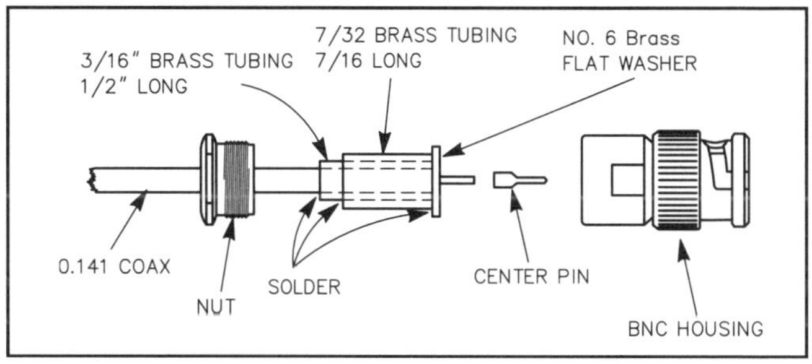

Fig 1—WA5TKU's method of attaching a BNC connector to 0.141-in. semi-rigid coax. See text for details.

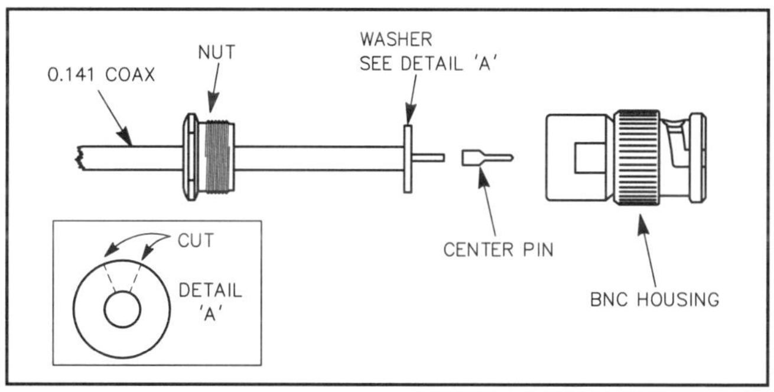

Fig 2—An alternate method of fitting a BNC connector to 0.141-in. semi-rigid coax. See text and Fig 1.

Adapting Commercial Circulators and Isolators for Amateur Use

By Kent Britain, WA5VJB

(From *Proceedings of the Central States VHF Society Conference '91*)

Ah, those little things you see in all the commercial microwave equipment, but almost never in ham gear. I now have over a dozen of these in my stations . . . they solve so many problems!

Think of these ferrite devices as a fluid check valve, or a doggie door that only lets the pooch go out (Fig 1). An isolator will absorb 5 to 10% of the signal going from 1 to 2, but if you send a signal backward through it, 99 to 99.9% of the signal is absorbed. Boy, if you could put one of these in your 20-meter station, the antenna could fall down and the transmitter would never know the difference.

These work great, Stage 1 never knows Stage 2 exists. You can tweak away without interaction between stages. Unstable stages see a 50-ohm load from dc to the GHz; preamps and LOs are happy. So why don't hams use isolators more often?

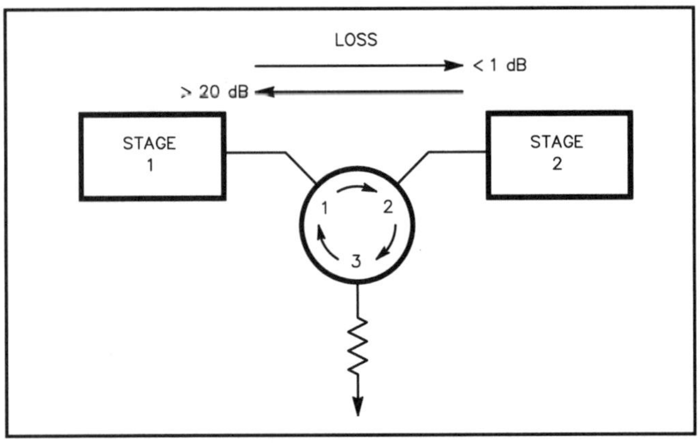

Fig 1—A circulator isolates Stage 2 from Stage 1 by more than 20 dB, while allowing almost unattenuated signal transfer from Stage 1 to Stage 2.

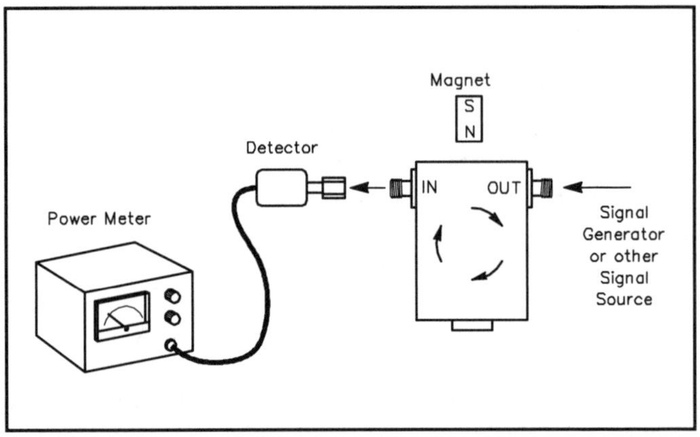

Fig 2—Using a magnet to tune a circulator. Once the desired isolation is obtained, glue the magnet or magnets in place.

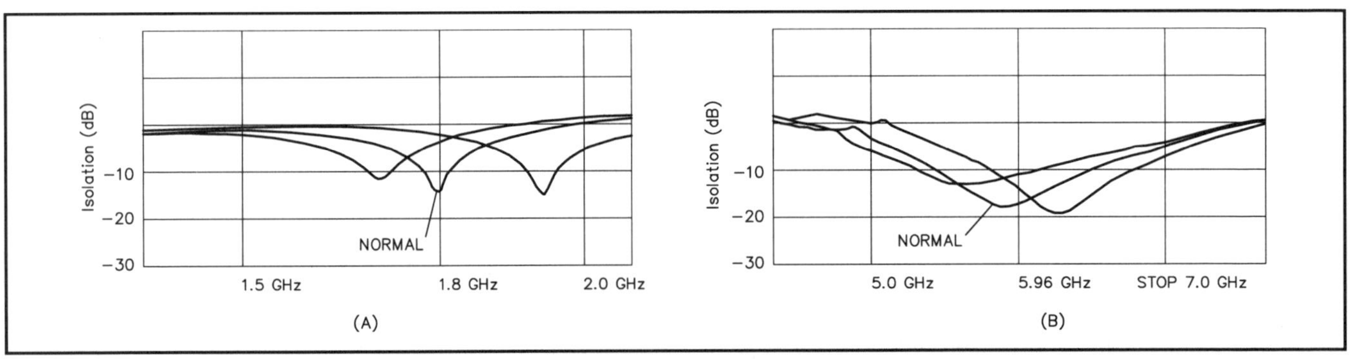

Fig 3—Plots of two isolators. At A, a 1.8-GHz isolator; at B, a 5.9-GHz isolator. The center plots show the normal plots for the isolators. The left and right plots were obtained when the isolators were approached by the poles of a magnet.

It is very rare to find a surplus isolator or circulator that is made for one of our ham bands. Fortunately, WA3RMX has shown us that the ferrite devices can be retuned with just a magnet (Fig 2). Just feed a signal backward through the isolator (on the frequency you want!), then move a magnet around on the outside of the isolator. If loss goes down, flip over the magnet. One magnet works well? Try two! Keep experimenting until you get as much loss/isolation as you can, then glue the magnets in place.

The plots in Fig 3 (made with the help of WB5LUA) show a 1.8-GHz and a 5.9-GHz isolator. As you can see, isolators move up in frequency better than they move down. You also can move isolators about 10% of their frequency. So, a 2.0-GHz isolator can probably be moved up to 2304 MHz, a 6.0-GHz isolator can be moved down to 5760 MHz, a cellular-telephone isolator can be moved to 902 MHz and a 9-GHz or 11-GHz isolator can be moved to 10368 MHz.

Make Your Own Waveguide Transitions

By Kent Britain, WA5VJB

(From *Proceedings of Microwave Update '88*)

In putting together several 10-GHz stations out of surplus parts, the first thing you notice is the different sizes of moding, WR-112, WR-90, WR-75, and WR-62. These can be identified by measuring the width of the opening. WR-112 is 1.12 in., WR-90 is 0.9 in., WR-75 is 0.75 in. and, of course, WR-62 is 0.62 in. across. (All are 200-ohm impedance.)

Well, very quickly you'd like to hook a WR-90 thingamagig, to a WR-75 whatchamacallit. I had two commercial adapters and they had a simple milled step, changing the opening from one waveguide size to another. I contacted our local waveguide expert (K5SXK) and asked Harold what was going on. He explained that the impedance of a waveguide is the ratio of width to height, so if you go from one size to another it's like connecting RG-8 to RG-58.

So, I redrilled some flanges and started making some loss tests. At first I had trouble getting consistent results because of reflections. The isolator/circulators proved to be necessary to get consistent numbers when measuring tenths of a dB. The results are shown in Table 1. Note that these measurements are actually for *two* waveguide size transitions, one on either end.

In short, while I would not claim these numbers as the absolute loss values, they show that simply bolting the different sizes together introduces very little loss. Alignment was also noncritical. Misalignment had to reach a point where one of the openings was being partially blocked before loss rose above a few tenths. Get out your drills and start sticking this stuff together!

Table 1

Measured Loss for Various Waveguide Combinations

	Losses in dB			
	Sample WR-62	WR-75	WR-90	WR-112
Flange				
WR-62	<0.1	0.3	0.3	0.4
WR-75	0.3	<0.1	0.1	0.1
WR-90	0.3	0.1	<0.1	<0.1
WR-112	0.4	0.2	0.1	<0.1

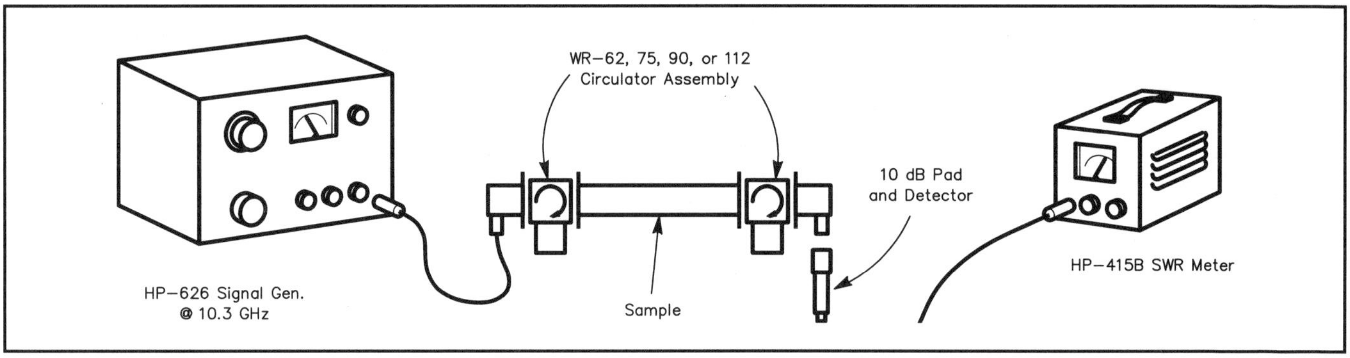

Fig 1—Test set-up for measuring insertion loss of waveguide transitions.

Using WR-62 Waveguide on 10 GHz

By Kent Britain, WA5VJB

(From *Proceedings of Microwave Update '91*)

WR-62 is perhaps the most readily available surplus waveguide. Designed for 12 to 18 GHz, it is almost never seen on 10 GHz. With care, it can be used in the upper portion of our 3-cm band. Figs 1 and 2 compare WR-62 and WR-90 waveguide.

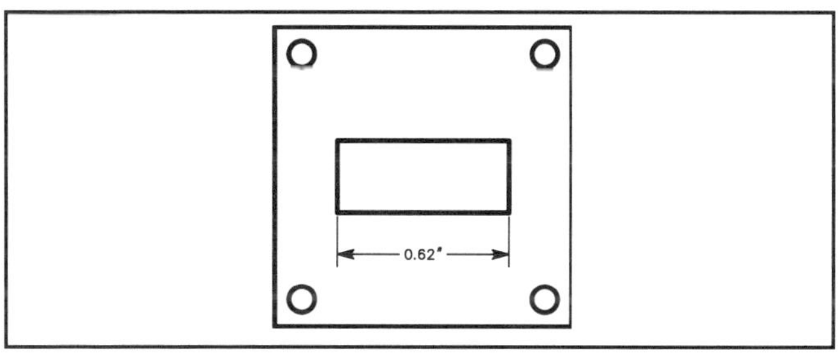

Fig 1—WR-62 waveguide dimensions.

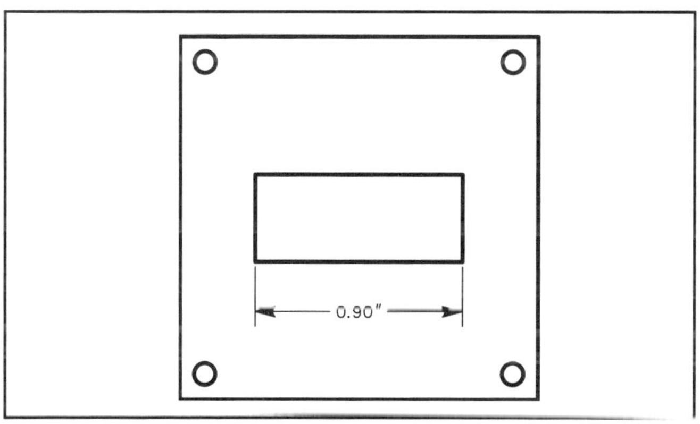

Fig 2—WR-90 waveguide dimensions.

The textbook says WR-62 has an ultimate cut-off at 9.5 GHz. I took a large pile of WR-62 and started bolting it together. I ended up with 8 ft. of waveguide containing 18 flanges and 21 bends. The graph in Fig 3 shows how much loss I had through 39 bends and flanges.

As the textbooks say, WR-62 really doesn't work at 10 GHz, but it works just fine at 10.368 GHz! On average, this means a foot of WR-62 with a few bends and flanges has only 0.2-dB loss. The only problem seems to be the WR-62 to-coax transitions. These are usually centered in the middle of the 12-18 GHz band. Several had about 0.5-dB loss, due to mismatch at 10.368 GHz. A three-screw tuner easily matched a transition.

WR-62 really works: I heard my first 10-GHz signals off the moon through a WR-62 Waveguide Switch.

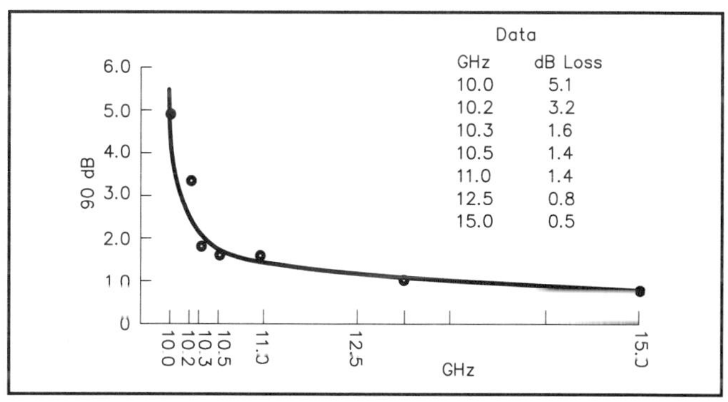

Fig 3—Plot of insertion loss for an 8-foot length of WR-62 waveguide sections having 10 flanges and 21 bends!

A Simple Rectangular/Circular Waveguide Transition for 10 GHz

By Sam Popkin, K2DNR, and Kent Britain, WA5VJB

(From *Proceedings of Microwave Update '89*)

Introduction

Sam noticed that if you squeeze ¾-in. copper water pipe into a rectangular shape it's the same size as a WR-90 waveguide. Sam built several samples and sent them to Kent for loss measurements and other tests. The results were very encouraging!

Theory

What Sam has come up with is a TE-10 to TE-11 or a rectangular or circular waveguide transition. Even constructed with simple tools, the test section only had about 2% loss due to reflections and mismatches. The inch or so of bent pipe between the rectangular flange and the circular pipe forms a tapered transition section with excellent matching.

Construction

Just start out by squeezing the end of the ¾-in. copper pipe in a bench vise down to a 0.4-in. opening. Then rotate the pipe 90° and squeeze the other side to a 0.9-in. opening. Back and forth, back and forth, until you form a 0.4- × 0.9-in. rectangular opening. A pair of large pliers or Visegrips can be very useful in forming the corners. Sam prefers to make his own WR-90 flanges out of sheet copper, while Kent likes to reuse old flanges from various pieces of junk. These flanges are then soldered onto the formed end of the copper water pipe. A few quick strokes with a file on the insides and you have WR-90 flanges on a circular waveguide. When you mount the second flange on the other end, be very careful to align the flanges with each other. There should be zero rotation misalignment.

Use

Circular waveguide has less loss than rectangular waveguide. Copper is an excellent conductor, so loss should be about 2 or 3 dB per 100 feet for your home-built circular waveguide.

There is one big limitation of using circular waveguide: Any protrusion, bend, or discontinuity will cause the wave to rotate. There just isn't anything to keep the E and H field aligned with the walls. Running the signals around a 45° plumbing bend rotated the polarization 20-30°. A 90° bend rotated the polarization about 90°, and with both samples the amount of polarization twist varied with frequency.

Simply use your homebrew waveguide for straight runs then transition back to regular WR-90 for any twists or bends.

One 40-in. section of this homebrew waveguide is even in use in WB5LUA's 10-GHz EME station.

A Quick Reference Guide for Circular Waveguide

Ron Neyens, NØCIH

(From *Proceedings of the Central States VHF Society Conference '91*)

Amateur Band	*EIA Type Designation*	*Inside Diameter in Inches (Centimeters)*	*Useful Upper and Lower Frequency Range in GHz*
902 MHz	WC 992	9.915 (25.184)	0.800 1.100
1296 MHz	WC 724	7.235 (18.377)	1.100 1.510
	WC 618	6.181 (15.700)	1.290 1.760
2304 MHz	WC 451	4.511 (11.458)	1.760 2.420
	WC 385	3.853 (9.7870)	2.070 2.830
3456 MHz	WC 281	2.812 (7.1420)	2.830 3.880
	WC 240	2.403 (6.1040)	3.310 4.540
5760 MHz	WC 175	1.750 (4.4450)	4.540 6.230
	WC 150	1.500 (3.8100)	5.300 7.270
10 GHz	WC 94	0.938 (2.3830)	8.490 11.600
	WC 80	0.797 (2.0240)	9.970 13.700
24 GHz	WC 44	0.438 (1.1130)	18.200 24.900
	WC 38	0.375 (0.9530)	21.200 29.100
	WC 33	0.328 (0.8330)	24.300 33.200
48 GHz	WC 22	0.219 (0.5560)	36.400 49.800
	WC 19	0.188 (0.4780)	42.400 58.100
	WC 17	0.172 (0.4370)	46.300 63.500

10

Antennas

Contents

An Optimum Design for 432-MHz Yagis

By Steve Powlishen, K1FO
(From Dec 1987 and Jan 1988 *QST*)

I spent nearly two years perfecting two different 10.5-wavelength 432-MHz Yagi designs (24-foot boom length) and presented the results of those efforts in 1986 at the major VHF/UHF conferences. Frank Potts, NC1I, successfully used one of the 10.5-wavelength designs in a 16-Yagi EME array. Frank's success encouraged me to plan a 26.0-dBd-gain array using eight of those Yagis. The antennas would be stacked two wide by four high and located 80 feet high for tropo and EME use. After thoroughly researching this planned array, however, I came up with a different solution.

This 22-element, 6.1-λ, 432-MHz Yagi was originally designed for use in a 12-Yagi EME array. Many designs were constructed and evaluated on a homemade antenna range. The resulting design is based on W1EJ's computer-optimized spacings.

The attention paid to the design process has been worth the effort. The 22-element Yagi not only has exceptional forward gain (17.9 dBi), but has an unusually "clean" radiation pattern. The measured E-plane pattern is shown in Fig 1. Note that a 1-dB-per-division axis is used to show pattern detail.

Like other log-taper Yagi designs, this one can easily be adapted to other boom lengths. Versions of this Yagi have been built by several amateurs. Boom lengths ranged between 5.3 λ (20 elements) and 12.2 λ (37 elements).

The size of the original Yagi (169 in. long, 6.1 λ) was chosen so the antenna could be built from small-diameter boom material (⅞ in. and 1 in. round 6061-T6 aluminum) and still survive high winds and ice loading. The 22-element Yagi weighs about 3.5 lb and has a wind load of approximately 0.8 sq ft. This allows a high-gain EME array to be built with manageable wind load and weight. This same low wind load and weight lets the tropo operator add a high-performance 432-MHz array to an existing tower without sacrificing antennas on other bands.

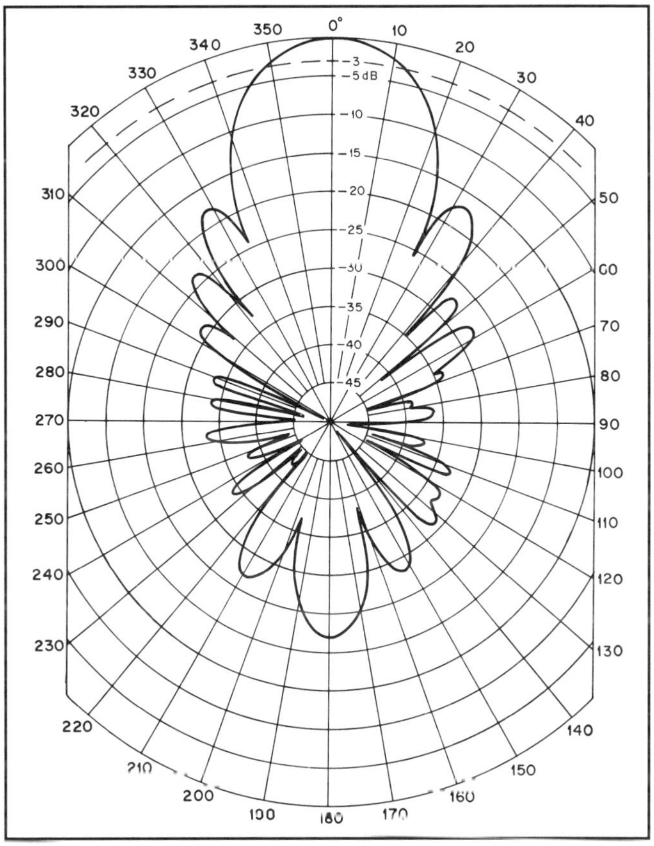

Fig 1—Measured E-plane pattern for the 22-element Yagi. Note: This antenna pattern is drawn on a linear dB grid, rather than on the standard ARRL log-periodic grid.

Table 1 lists the gain and stacking specifications for the various length Yagis. The basic Yagi dimensions are shown in Table 2. These are free-space element lengths for 3⁄16-in.-diameter elements. Boom corrections for the element-mounting method must be added in. The element-length correction column gives the length that must be added to keep the Yagi's center frequency optimized for use at 432 MHz. This correction is required to use the same spacing pattern over a wide range of boom lengths. Although any length Yagi will work well, this design is at its best when made with 18 elements or more (4.6 λ). Element material of less than 3⁄16-in. diameter is not recommended because resistive losses will reduce the gain by about 0.1 dB, and wet-weather performance will be worse.

Quarter-inch-diameter elements could be used if all elements are shortened by 3 mm. The element lengths are intended for use with a slight chamfer (0.5 mm) cut into the element ends. The gain peak of the array is centered at 437 MHz. This allows acceptable wet-weather performance, while reducing the gain at 432 MHz by only 0.05 dB.

The gain bandwidth of the 22-element Yagi is 31 MHz (at the –1 dB points). The SWR of the Yagi is less than 1.4:1 between 420 and 440 MHz. Fig 2 is a network analyzer plot of the driven-element SWR versus frequency. These numbers indicate just how wide the frequency response of a log-taper Yagi can be, even with a simple dipole driven element. In fact, at one antenna gain contest, some ATV operators conducted gain versus frequency measurements from 420 to 440 MHz. The 22-element Yagi beat all entrants including those with so-called broadband feeds.

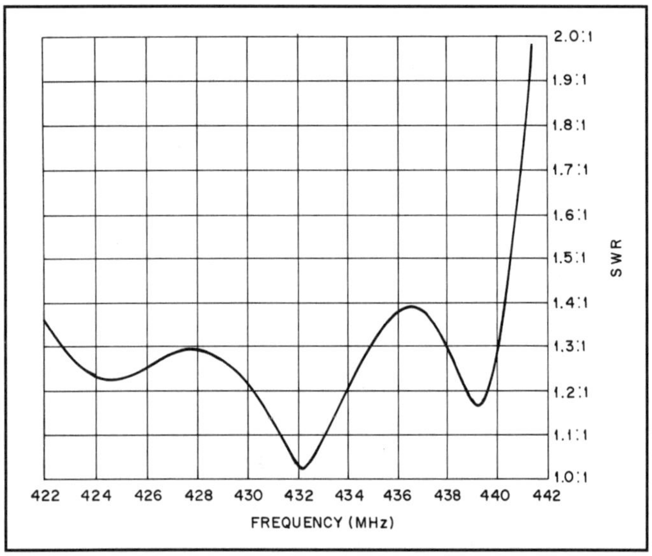

Fig 2—SWR performance of the 22-element Yagi in dry weather.

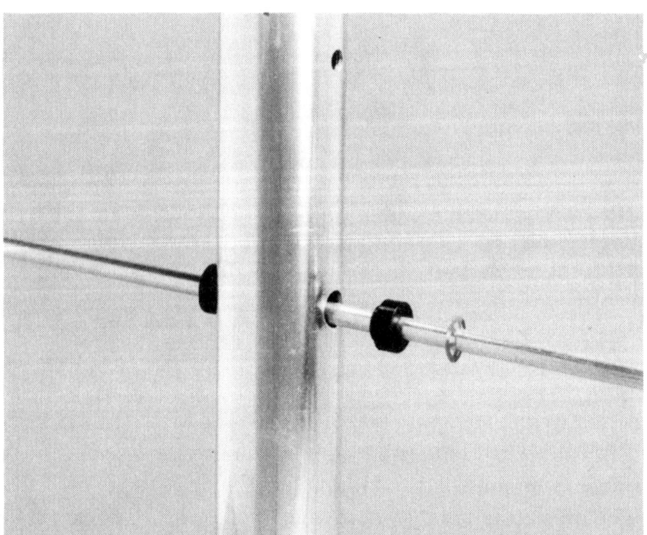

Fig 3—Element-mounting detail. Elements are mounted through the boom using plastic insulators. Push-nut retaining rings hold the element in place.

Table 1

Specifications for 432-MHz Yagi Family

No. of El	*Boom length (λ)*	*Gain (dBi)**	*FB ratio (dB)*	*DE impd (ohms)*	*Beamwidth E / H (°)*	*Stacking E / H (inches)*
15	3.4	15.67	21	23	30 / 32	53 / 49
16	3.8	16.05	19	23	29 / 31	55 / 51
17	4.2	16.45	20	27	28 / 30	56 / 53
18	4.6	16.8	25	32	27 / 29	58 / 55
19	4.9	17.1	25	30	26 / 28	61 / 57
20	5.3	17.4	21	24	25.5 / 27	62 / 59
21	5.7	17.65	20	22	25 / 26.5	63 / 60
22	6.1	17.9	22	25	24 / 26	65 / 62
23	6.5	18.15	27	30	23.5 / 25	67 / 64
24	6.9	18.35	29	29	23 / 24	69 / 66
25	7.3	18.55	23	25	22.5 / 23.5	71 / 68
26	7.7	18.8	22	22	22 / 23	73 / 70
27	8.1	19.0	22	21	21.5 / 22.5	75 / 72
28	8.5	19.20	25	25	21 / 22	77 / 75
29	8.9	19.4	25	25	20.5 / 21.5	79 / 77
30	9.3	19.55	26	27	20 / 21	80 / 78
31	9.7	19.7	24	24	19.6 / 20.5	81 / 79
32	10.2	19.8	23	22	19.3 / 20	82 / 80
33	10.6	19.9	23	23	19 / 19.5	83 / 81
34	11.0	20.05	25	22	18.8 / 19.2	84 / 82
35	11.4	20.2	27	25	18.5 / 19.0	85 / 83
36	11.8	20.3	27	26	18.3 / 18.8	86 / 84
37	12.2	20.4	26	26	18.1 / 18.6	87 / 85
38	12.7	20.5	25	25	18.9 / 18.4	88 / 86
39	13.1	20.6	25	23	18.7 / 18.2	89 / 87
40	13.5	20.8	26	21	17.5 / 18	90 / 88

* Gain is approximate real gain based upon gain measurements made on six different-length Yagis.

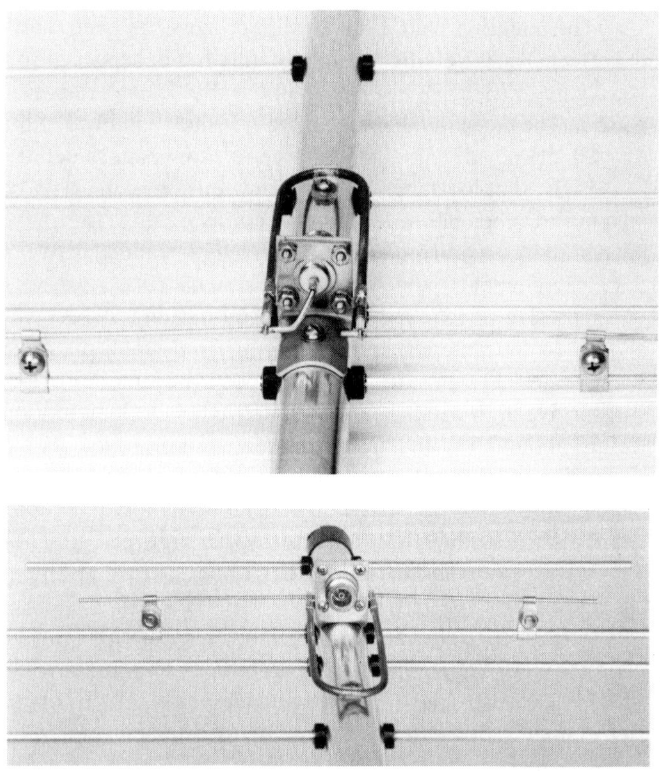

Fig 4—Two views of the driven element and T match. (A) (B)

To peak the Yagi for use on 435 MHz (for satellite use), you may want to shorten all the elements by 2 mm. To peak it for use on 438 MHz (for ATV applications), shorten all elements by 4 mm. If you want to use the Yagi on FM between 440 MHz and 450 MHz, shorten all the elements by 10 mm. This will provide 17.6 dBi gain at 440 MHz, and 18.0 dBi gain at 450 MHz. The driven element may have to be adjusted if the element lengths are shortened.

Although this Yagi design is relatively broadband, I suggest that you pay close attention to copying the design exactly. Metric dimensions are used because they are convenient for a Yagi sized for 432 MHz. Element holes should be drilled within ±2 mm. Element lengths should be kept within ±0.5 mm. Elements can be accurately constructed if they are first rough cut with a hacksaw and then held in a vise and filed to the exact length.

The larger the array, the more attention you should pay to making all Yagis identical. Elements are mounted on shoulder insulators and run through the boom (see Fig 3). The element retainers are push nuts, preferably made from stainless steel. These are made by several companies, including Industrial Retaining Ring Co in Irvington, New Jersey, and AuVeco in Ft Mitchell, Kentucky. Local industrial hardware distributors can usually order them for you. The element insula-

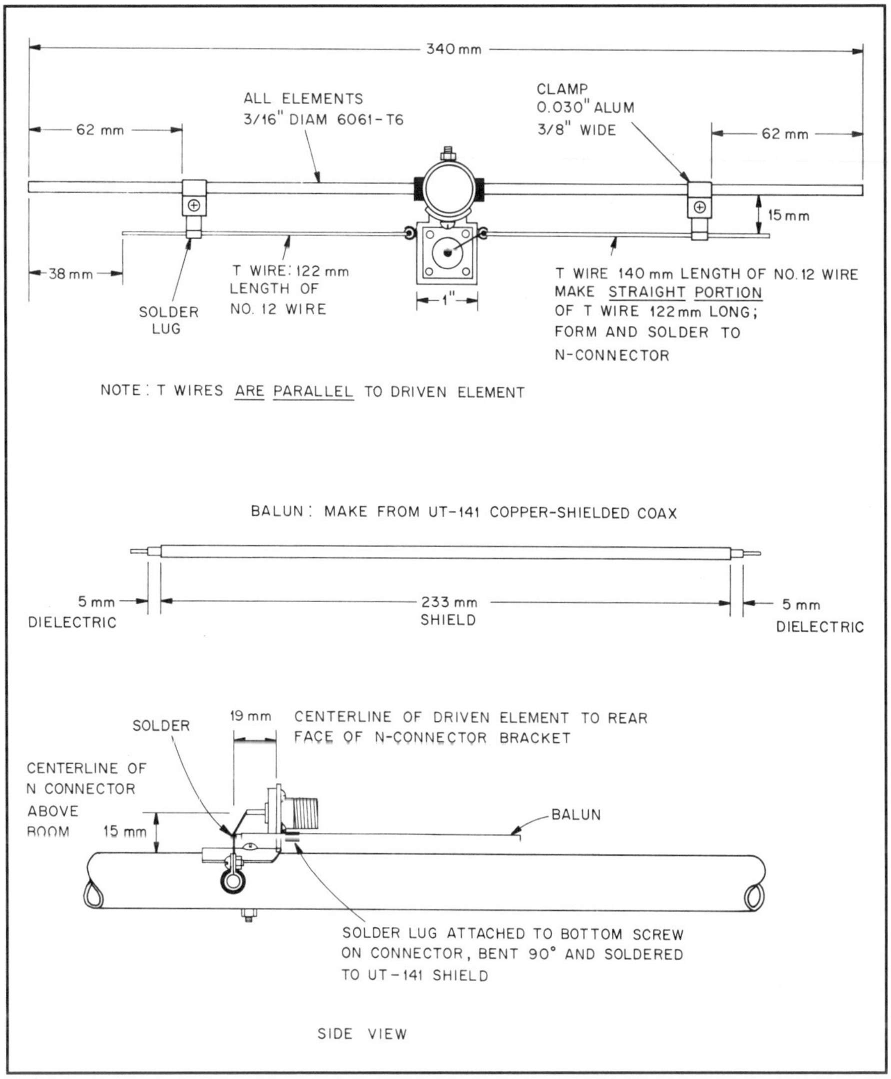

Fig 5—Details of the driven element and T match for the 22-element Yagi. Lengths are given in millimeters to allow precise duplication of the antenna. See text.

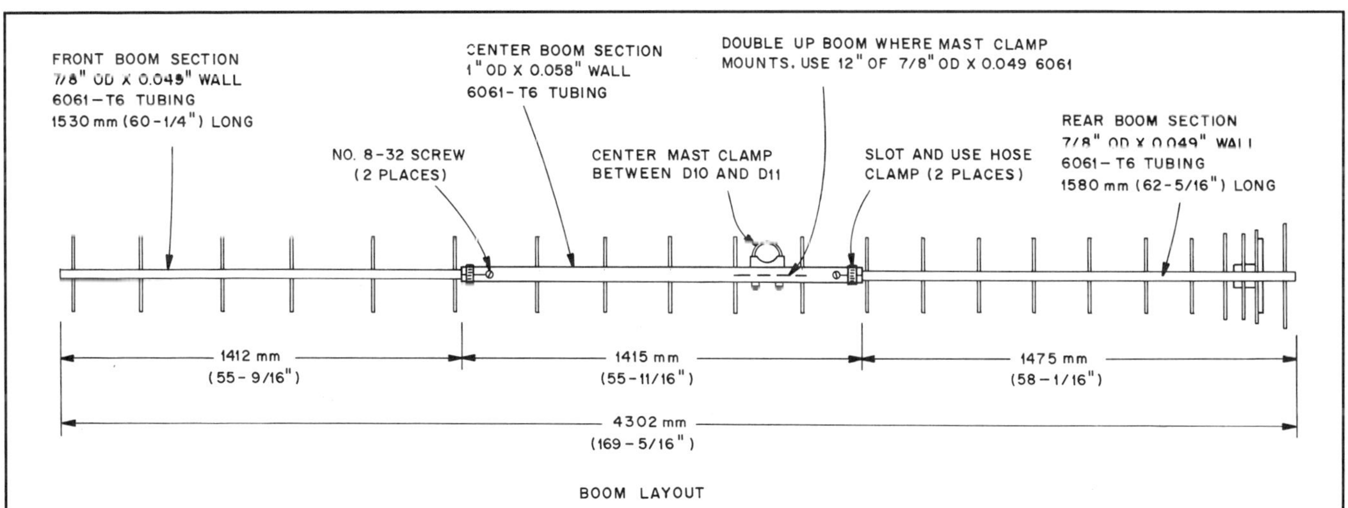

Fig 6—Boom-construction information for the 22-element Yagi. Lengths are given in millimeters to allow precise duplication of the antenna. See text.

Table 2

Free-Space Dimensions for 432-MHz Yagi Family

Element lengths are for 3/16-inch-diameter material.

El No.	*Element Position (mm from rear of boom)*	*Element Length (mm)*	*Element Correction**
REF	0	340	
DE	104	334	
D1	146	315	
D2	224	306	
D3	332	299	
D4	466	295	
D5	622	291	
D6	798	289	
D7	990	287	
D8	1196	285	
D9	1414	283	
D10	1642	281	−2
D11	1879	279	−2
D12	2122	278	−2
D13	2373	277	−2
D14	2629	276	−2
D15	2890	275	−1
D16	3154	274	−1
D17	3422	273	−1
D18	3693	272	0
D19	3967	271	0
D20	4242	270	0
D21	4520	269	0
D22	4798	269	0
D23	5079	268	0
D24	5360	268	+1
D25	5642	267	+1
D26	5925	267	+1
D27	6209	266	+1
D28	6494	266	+1
D29	6779	265	+2
D30	7064	265	+2
D31	7350	264	+2
D32	7636	264	+2
D33	7922	263	+2
D34	8209	263	+2
D35	8496	262	+2
D36	8783	262	+2
D37	9070	261	+3
D38	9359	261	+3

*Element correction is the amount to shorten or lengthen all elements when building a Yagi of that length.

Table 3

Dimensions for the 22-Element 432-MHz Yagi

Element Number	*Element Position (mm from rear of boom)*	*Element Length (mm)*	*Boom Diam (in)*
REF	30	346	
DE	134	340	
D1	176	321	
D2	254	311	
D3	362	305	7/8
D4	496	301	
D5	652	297	
D6	828	295	
D7	1020	293	
D8	1226	291	
D9	1444	289	
D10	1672	288	
D11	1909	286	
D12	2152	285	1
D13	2403	284	
D14	2659	283	
D15	2920	281	
D16	3184	280	
D17	3452	279	
D18	3723	278	7/8
D19	3997	277	
D20	3272	276	

tors are not critical. Teflon or black polyethylene are probably the best materials. The Yagi in the photos is made with black Delryn insulators, available from Rutland Arrays in New Cumberland, Pennsylvania.

The driven element uses a UG-58A/U connector mounted on a small bracket. The UG-58A/U should be the type with the press-in center pin. UG-58s with center pins held in by "C" clips will usually leak water. Some connectors use steel retaining clips, which will rust and leave a conductive stripe across the insulator. The T-match wires are supported by the UT-141 balun. RG-303 or RG-142 Teflon-insulated cable could be used if UT-141 cannot be obtained. Fig 4 shows details of the driven-element construction. Driven-element dimensions are given in Fig 5.

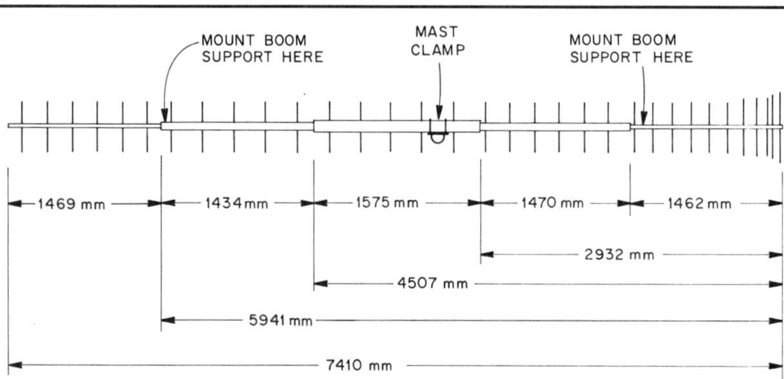

Fig 7—Boom-construction information for the 33-element Yagi. Lengths are given in millimeters to allow precise duplication of the antenna.

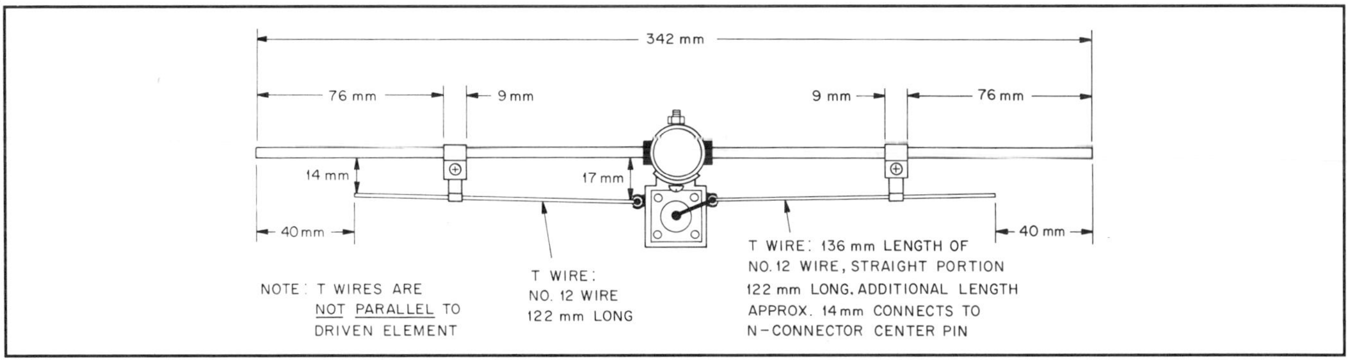

Fig 8—Details of the driven element and T match for the 33-element Yagi. Lengths are given in millimeters to allow precise duplication of the antenna.

Dimensions for the 22-element Yagi are listed in Table 4. Fig 6 details the Yagi's boom layout. Element material can be either 3/16-in. 6061-T6 aluminum rod or hard aluminum welding rod.

A 24-foot long, 10.6-λ, 33-element Yagi was also built. The construction methods used were the same as the 22-element Yagi. Telescoping round boom sections of 1, 1⅛ and 1¼ in. diameter were used. A boom support is required to keep boom sag within acceptable limits. At 432 MHz, if boom sag is much more than two or three inches, H-plane pattern distortion occurs. Greater amounts of boom sag reduce the gain of a Yagi. Table 4 lists the proper dimensions for the antenna when built with the previously given boom diameters. The boom layout is shown in Fig 7, and the driven element is described in Fig 8. The 33-element Yagi exhibits the same clean pattern traits as the 22-element Yagi (Fig 9). Measured gain of the 33-element Yagi is 19.9 dBi at 432 MHz. A measured gain sweep of the 33-element Yagi gave a –dB gain bandwidth of 14 MHz with the –1 dB points at 424.5 MHz and 438.5 MHz.

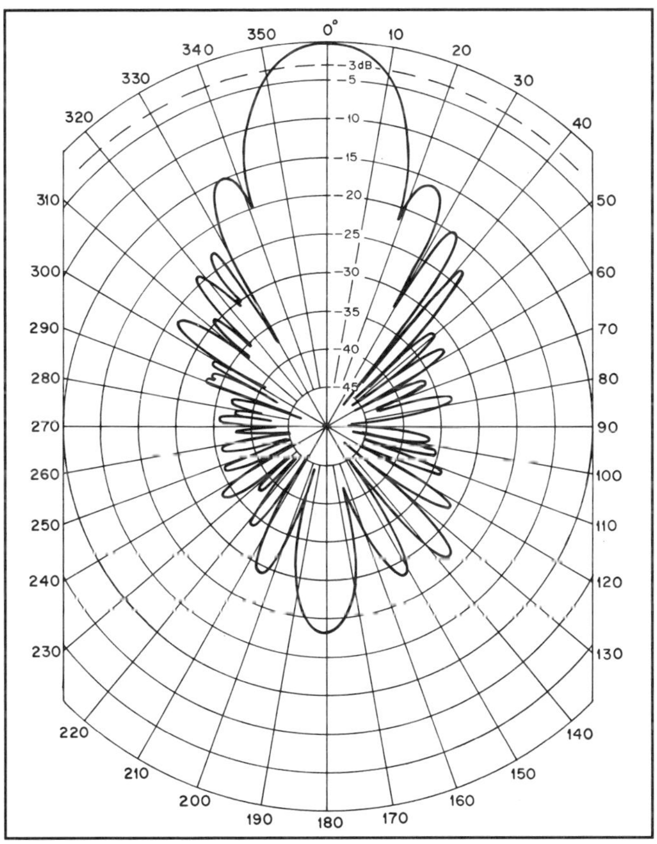

Fig 9—E-plane pattern for the 33-element Yagi. This pattern is drawn on a linear dB grid, rather than on the standard ARRL log-periodic grid.

Table 4
Dimensions for the 33-Element 432-MHz Yagi

Element Number	*Element Position (mm from rear of boom)*	*Element Length (mm)*	*Boom Diam (in)*
REF	30	348	
DE	134	342	
D1	176	323	
D2	254	313	
D3	362	307	
D4	496	303	1
D5	652	299	
D6	828	297	
D7	1020	295	
D8	1226	293	
D9	1444	291	
D10	1672	290	
D11	2909	288	
D12	2152	287	
D13	2403	286	1⅛
D14	2659	285	
D15	2920	284	
D16	3184	284	
D17	3452	283	
D18	3723	282	1¼
D19	3997	281	
D20	4272	280	
D21	4550	278	
D22	4828	278	
D23	5109	277	1⅛
D24	5390	277	
D25	5672	276	
D26	5956	275	
D27	6239	274	
D28	6524	274	1
D29	6809	273	
D30	7094	273	
D31	7380	272	

Loop Yagis for 23 Cm

By Chip Angle, N6CA

(From *The ARRL Handbook for Radio Amateurs*)

Described here and shown in Figs 1 through 5 are loop Yagis for the 23-cm band. Several versions are described, so you can choose the boom length and frequency coverage desired for the task at hand. Mike Walters, G3JVL, brought the original loop Yagi design to the amateur community in the 1970s. Since then, many versions have been developed with different loop and boom dimensions.

Three sets of dimensions are given. Good performance can be expected if you follow the dimensions carefully. Recheck all dimensions *before* you cut or drill anything. The 1296-MHz version is intended for weak-signal operation at 1296 MHz, while the 1270-MHz version is optimized for FM and Mode L satellite work. The 1283-MHz antenna provides acceptable performance from 1280 to 1300 MHz.

These antennas have been built on 6- and 12-foot booms. Results of gain tests at VHF conferences and by individuals around the country peg the gain of the 6-footer at about 18 dBi, while the 12-foot version provides about 20.5 dBi. Swept measurements indicate that gain is about 2 dB down from maximum gain at ±30 MHz from the design frequency. SWR, however, deteriorates within a few megahertz on the low side of the design center frequency.

The Boom

The dimensions given here apply only to a ¾-in.-OD boom. If you change the boom size, the dimensions must be

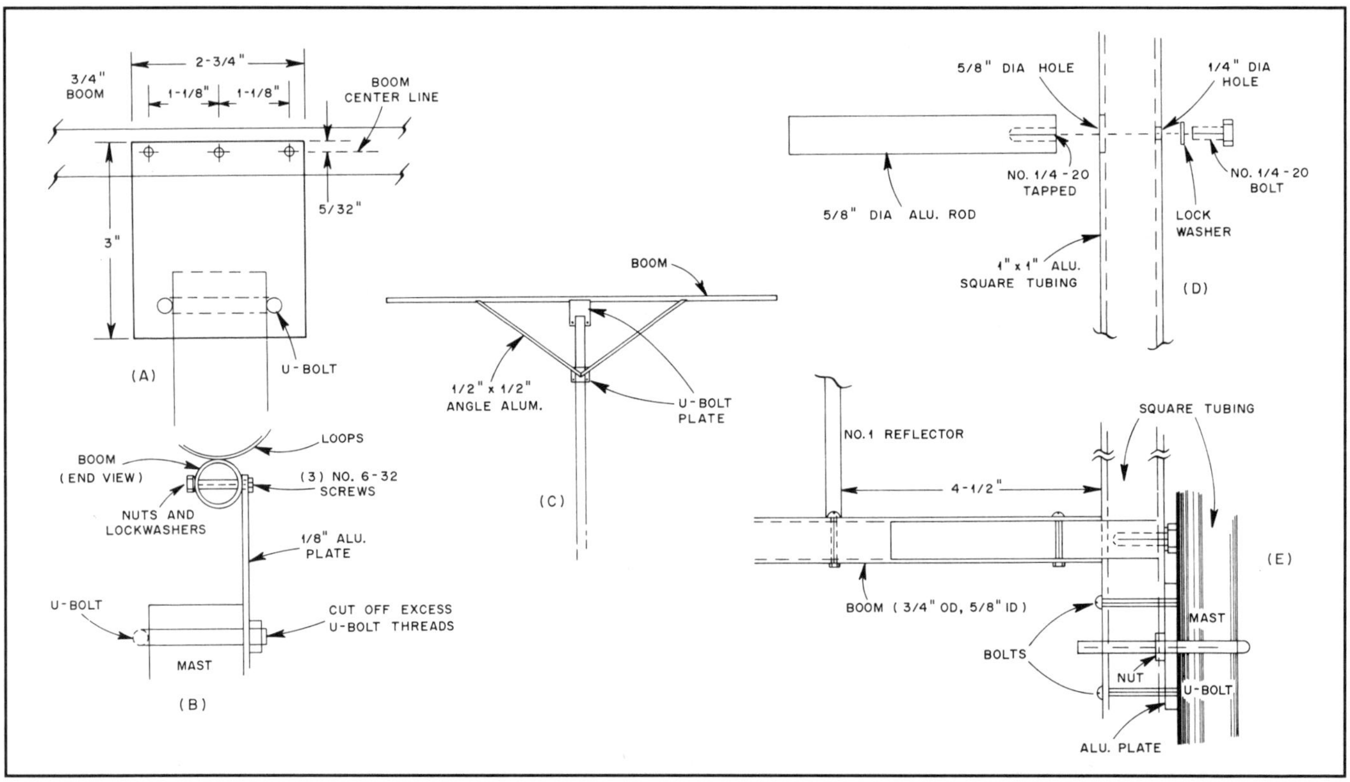

Fig 1—Boom-to-mast plate details are given at A. B shows how the Yagi is mounted to the mast. A boom support for long antennas is shown in C. The arrangement shown in D and E may be used to rear mount antennas up to 6 or 7 feet long.

scaled accordingly. Many hardware stores carry aluminum tubing in 6- and 8-foot lengths, and that tubing is suitable for a short Yagi. If you plan a 12-foot antenna, you should find a piece of more rugged material, such as 6061-T6 grade aluminum. Do not use anodized tubing. The 12-foot antenna must be supported to minimize boom sag. The 6-foot version can be rear mounted. For rear mounting, allow 4.5 in. of boom behind the last reflector to eliminate SWR effects from the support.

The antenna is mounted to the mast with a gusset plate. The plate mounts at the boom center. See Fig 1. Drill the plate-mounting holes perpendicular to the element-mounting holes, assuming the antenna polarization is to be horizontal.

Elements will be mounted to the boom with no. 4-40 machine screws, so a series of no. 33 (0.113-in.) holes must be drilled along the center of the boom to accommodate this hardware. Fig 2 shows the element spacings for different parts of the band. Tolerances should be followed as closely as possible.

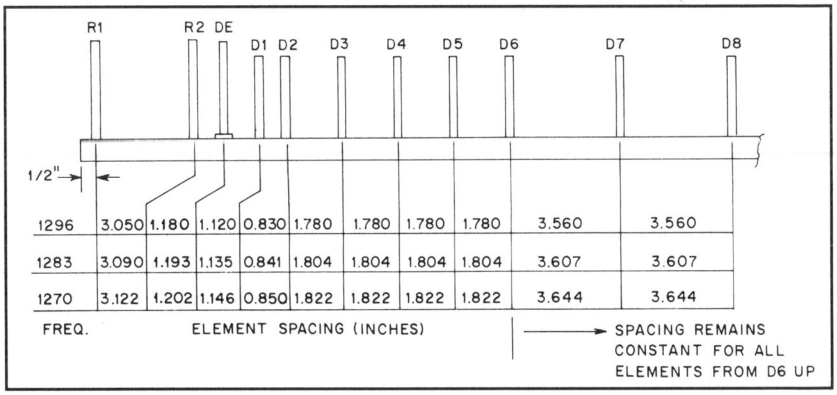

Fig 2—Boom drilling dimensions. Pick the version you want and follow these dimensions carefully. Spacing is the same for all directors after D6. Use as many as needed to fill up your boom.

Parasitic Elements

The reflectors and directors are cut from 0.0325-in. thick aluminum sheet and are ¼-in. wide. Fig 3 indicates the lengths for the various elements. These lengths apply only to elements cut from the specified material. For best results, the element strips should be cut with a shear. If you leave the edges sharp, birds won't sit on the elements.

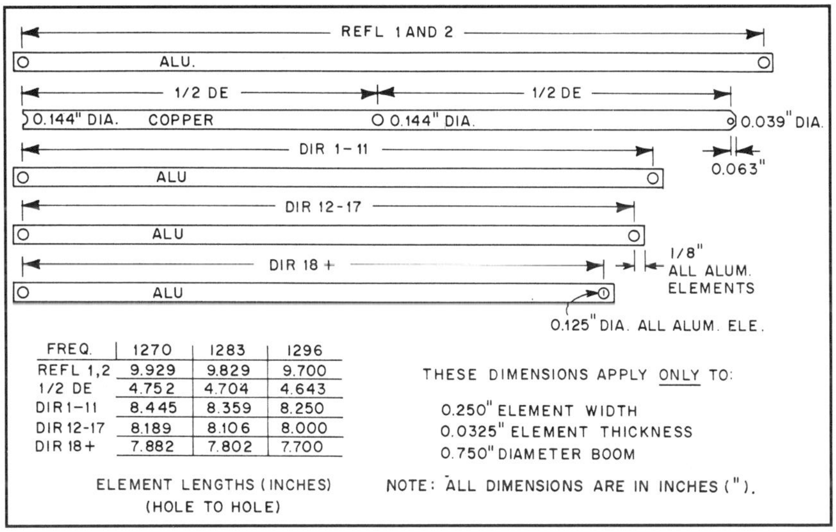

Fig 3—Parasitic elements are made from aluminum sheet. The driven element is made from copper sheet. The dimensions given are for ¼-in. wide by 0.0325-in. thick elements only. Lengths specified are hole-to-hole distances; the holes are located 0.125 in. from each element end.

Drill the mounting holes as shown in Fig 3. Measure carefully! After the holes are drilled, you must form each strap into a circle. This is easily done by wrapping the element around a round form (a small juice can works well).

Mount the loops to the boom with no. 4-40 × 1-in. machine screws, lock washers and nuts. See Fig 4. It's best to use stainless-steel or plated-brass hardware for everything. Although the initial cost is higher than for ordinary plated-steel hardware, stainless or brass hardware won't rust and need replacement after a few years. Unless the antenna is painted, the hardware will definitely deteriorate.

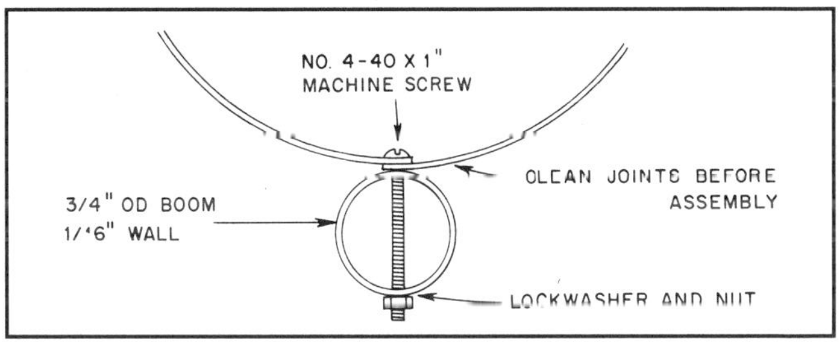

Fig 4—Element-to-boom mounting details.

Driven Element

The driven element is cut from 0.0325-in. copper sheet and is ¼-in. wide. Drill three holes in the strap, as detailed in Fig 3. Trim the ends as shown and form the strap into a loop similar to the other elements. This antenna is like a quad; if the loop is fed at the top or bottom, it will be horizontally polarized.

Driven-element mounting details are shown in Fig 5. A mounting fixture is made from a 1.4-20 × 1¼-in. brass bolt. File the bolt head to a thickness of 0.125 in. Bore a 0.144-in. (no. 27 drill) hole lengthwise through the center of the bolt. A piece of 0.141-in. semi-rigid Hardline (UT-141 or equiv) will mount through this hole and connect to the driven loop. The point at which the UT-141 passes through the copper loop and brass mounting fixture should be left unsoldered at this time to allow for matching adjustments when the antenna is completed, although the range of adjustment is not that great.

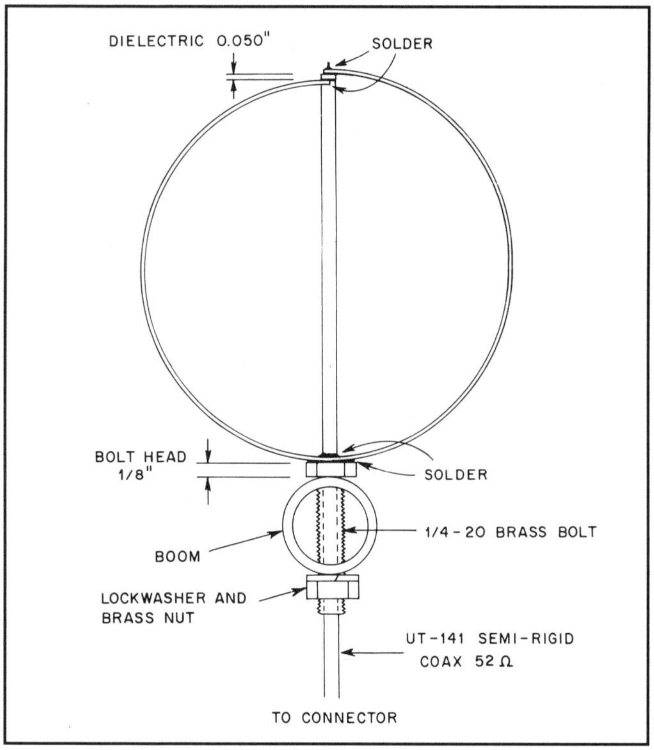

Fig 5—Driven element details. See Fig 3 and the text for more information.

The UT-141 can be any convenient length. Attach the connector of your choice (preferably Type N). Use a short piece of low-loss RG-8 cable (or better yet, ½-in. Hardline) for the run down the boom and mast to the main feed line. For best results, your main feed line should be the lowest-loss 50-ohm cable obtainable. Good ⅞-in. Hardline measures at 1.5 dB per 100 feet and virtually eliminates the need for remote mounting of the transmit amplifier.

Tuning the Driven Element

If you built the antenna carefully to the dimensions given, the SWR should be fine. Just to be sure, check the SWR if you have access to test equipment. You must be sure that your signal source is clean, however; wattmeters are often fooled by dirty signals and can give erroneous readings. If you have a problem, recheck all dimensions. If they look okay, a minor improvement may be realized by changing the shape of the driven element. Slight bending of reflector 2 may also help optimize SWR. When you have obtained the desired match, solder the point where the UT-141 jacket passes through the loop and brass bolt.

Quite a few people believe that practices common on lower frequencies can be used on 1296 MHz. That's the biggest reason they don't work all the DX that there is to be worked.

First, when you have a design that's proven, copy it exactly—don't change it. This is especially true for antennas. Use the best feed line you can get. Why build an antenna if you are going to use lossy cable?

Here are some realistic measurements of common coaxial cables at 1296 MHz (loss per 100 feet): RG-8, 213, 214—11 dB; ½ in. foam/copper Hardline—4 dB; ⅞ in. foam/copper Hardline—1.5 dB.

Mount the antenna(s) to keep feed line loss to an absolute minimum. Antenna height is less important than keeping the line losses low. Do not allow the mast to pass through the elements, like on antennas for lower frequencies. Cut all U-bolts to the minimum length needed—a quarter wavelength at 1296 MHz is a little over 2 in. Avoid any unnecessary metal around the antenna.

Microwave EME Using A Ten-Foot TVRO Antenna

By Dave Hallidy, KD5RO

(From *Microwave Update '89*)

INTRODUCTION

Since I moved to Upstate New York in the fall of 1987, I wanted to experiment with EME on the bands above 900 MHz. In this endeavor, I was faced with two major problems. First, I was renting my home, so I didn't want to upset what was a good landlord/tenant relationship with a request for permission to erect a large dish. Second, and maybe most important, I didn't have much money to spend on the project! What follows is a detailed account of the work I did to achieve success in this area with a minimum of effort and money.

Background

I was aware that a small dish had the potential of being a usable EME antenna, because in 1987 WA5TNY and I became the first amateurs to successfully work another station off the moon on the 3.4 and 5.7 GHz bands, using a 3m (10 ft) TVRO dish and modest power.[1] Calculations showed that, properly fed, a dish of this size would produce detectable echoes on 903 and 1296 MHz using a couple of hundred watts of power and a good preamp at the feed and respectable echoes on the higher bands with at least a hundred watts at the feed. In the spring of 1988, I was able to purchase a new 10-ft dish through a dealer who was also an amateur. In April of that year, I managed to assemble the dish and install it on its polar mount (temporarily) and worked F1HDI on 1296 during their expedition to the large radiotelescope near Nancy, France. This convinced me that the antenna could work, since at that time my feed was linearly polarized and my preamp was in the shack, at the end of 75 feet of Heliax. At that time, I set a goal of being operational on at least one band in time for the ARRL EME Competition in 1988.

Fig 1—View of complete antenna installation.

DISH MECHANICS

In October of 1988, I began assembling the hardware for the "real" installation, after spending most of the summer collecting the necessary pieces (see Fig 1). As mentioned before, the dish is a 10 footer purchased new for about $250, including the polar mount. This dish is designed for use through Ku band, and so should be good for all bands through 10 GHz, should the opportunity arise to try some EME there. I recommend checking with local dealers, and even neighbors, as many people have lost interest in their TVRO systems (now that most satellite programming is scrambled)—you might get lucky enough to find a very cheap dish. I mounted my dish on

Fig 2—View of elevation system, showing modified TVRO mount.

Fig 3—View of azimuth rotation system, showing prop-pitch motor and position sensor.

a single section of Rohn 25G tower. The base is set in one cubic yard of concrete and is very solid, though in certain parts of the country this might not be sturdy enough—don't underestimate the forces such an antenna will exert on the mount in a high wind, especially with some ice built up on it. I used just one section of tower to provide me with easy access to the feed using an eight foot stepladder, rather than a custom work platform.

My azimuth rotator is a small prop-pitch motor I obtained at a hamfest for $35. I power it with 24 V ac and it rotates one full revolution in about two minutes. This provides a slow enough rate of turn that the dish can be postioned accurately, but it doesn't take forever to move it across the sky. I couple this motor to the dish mount with a piece of 3-inch pipe which is aligned with the motor by a heavy duty flange bearing on the top plate of the tower. The original dish mount is then bolted to the top of the pipe as in TVRO installations.

As I mentioned earlier, the dish came with a polar mount, and at first I thought I would try to position the dish using the mount and suitable modifications to allow for easy declination adjustment. This idea was not feasible in my case (but it's worth checking out—different dishes have different style mounts). So, I modified the mount to turn the dish pivot points horizontal (see Fig 2), causing the dish to lift vertically when the arm on the TVRO tracking motor was extended. This has provided an easy, reliable method of elevation. Running the tracker at its rated voltage of 36 V dc allows full 0-90 degree elevation change in less than five minutes. Again, this is slow enough for accurate positioning without taking forever.

From the outset, I decided that digital readout of both azimuth and elevation would be desirable. I located a Digital Multimeter Module (Datel model DM-LX3) at a local surplus house and proceeded to wire it up per instructions given by Larry Molitor W7IUV.[2] This unit is a 3½-digit multimeter and provides readout to within one degree in azimuth and 0.1 degree in elevation. The azimuth sensor (seen in photo 3) consists of a multi-turn linear pot (value not critical—mine is 10k) coupled to the main shaft by a hard rubber wheel I found at the same surplus store that had the DMM and the pot. Although I have had good success with this mechanism, a better approach would be to enclose the pot in a weatherproof container. The elevation sensor cannot be seen, as it is a part of the motorized TVRO tracker. I was fortunate to find a tracker with a 10k multi-turn pot built in (used by the satellite receiver to detect the antenna's position), so I merely coupled this pot to the elevation circuit. (See Fig 4, which is a schematic of the entire drive/readout system.)

FEEDING THE DISH

I decided that the first band I would make ready would be 903 MHz. There were several reasons for this decision. First, if the system worked at the lowest frequency at which I planned to use it, it should do better at the higher frequencies where dish gain would be higher. Second, there was a DX record to claim! WB5LUA and I had discussed this for some time and decided that we would try to claim the EME DX record on the band to spur others to get on and increase activity.

With that goal in mind, I built a Dual-Dipole EIA-type feed for 903 MHz, which was a scaled version of a 2.3 GHz feed designed by Don Hilliard, WØPW.[3] Since the only other stations active on 903 EME were also using linear polarization, I felt I should follow suit. Though circular polarization is

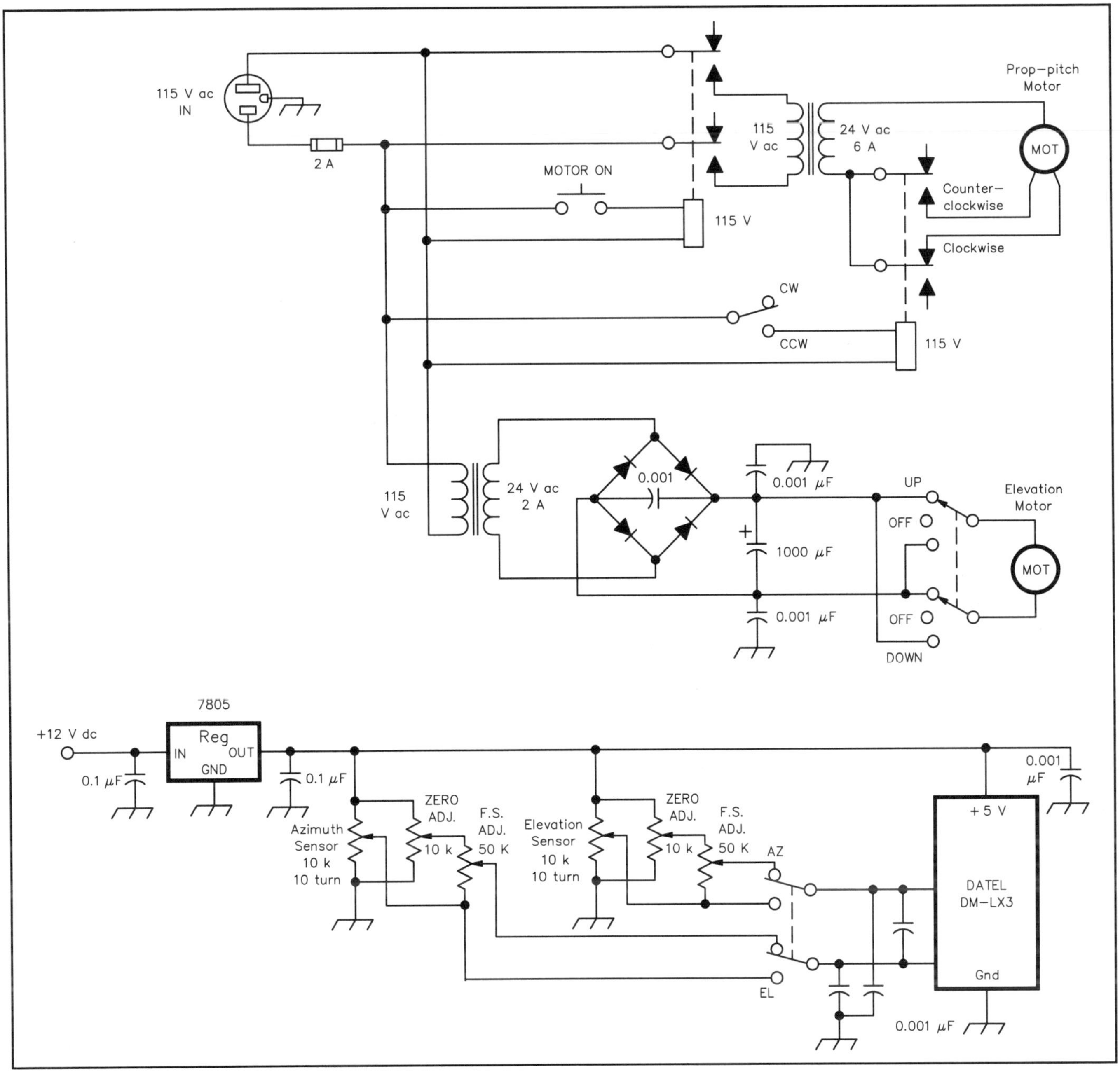

Fig 4—Schematic diagram of the dish control system. The Datel DM-LX3 is a digital meter, used to indicate azimuth and elevation angles.

surely the way to go at these frequencies, the large size of the feed at 903 MHz would have caused significant blockage of my dish and, I suspect, would have nullified any advantage due to circularity. Also, the 3dB linear-to-circular loss might have reduced chances for a successful QSO. As more stations become operational on the band, circular polarization will likely become the standard.

For 1296, I felt that since my station was pretty marginal anyway, I would put forth the extra effort to build a circularly polarized feed, in order to maximize my chances of success. I wanted to build a VE4MA-type feed[4] for my dish (0.4 f/D), but I had no access to the necessary materials. So, at first I built a version of the "old standard" coffee-can feed, using two three-pound coffee cans stacked end to end and with two 1/4-wave-length feed monopoles spaced 90 degrees apart and a quarter of a guide wavelength from the bottom of the rear can (see photo 4). I found a commercial 90-degree hybrid coupler for 900-2000 MHz at a fleamarket and used it to generate the necessary phase shift for the feed. This feed was not optimum for the f/D of the dish (resulting in under-illumination of the reflector), but it did allow me to get on and work a lot of stations. I finally did construct a copy of the VE4MA design and have observed a very significant improvement in performance. This feed optimally illuminates the surface of the dish by permitting adjustment of the scalar ring to alter the feed angle required by dishes of different depths. This improvement has provided about 1-2dB more antenna gain (affecting both transmit and receive). Note: the f/D (focal length to Di-

Fig 5—View of feed, showing coffee-can feedhorn and hybrid coupler.

ameter) ratio of a dish is important. The higher the f/D, the flatter the curve of the dish, and the easier it is to feed properly. Typical TVRO dishes range from about 0.3 to 0.4 f/D. They tend to have a deeper curve to them to reduce pickup of undesired signals or noise from terrestrial sources at the possible sacrifice of a little gain. While noise reduction is certainly desirable from the EME'ers standpoint as well, maximizing gain and simplifying feed design is probably more important. If you have a choice, opt for a higher f/D ratio. VE4MA's feeds are optimum for dishes in the 0.3 to 0.5 range. The W2IMU horn, another popular feed for 1296, is optimum for dishes in the 0.5 to 0.6 range. It is fairly difficult to properly feed a dish with an f/D below 0.3.

For 2304, I scaled the VE4MA 1296 design and found that the body of the feed horn could be made from two one pound coffee cans very nicely. I constructed the scalar ring for this feed from brass sheet stock, and after optimizing its position, soldered it into place on the body of the feed. I also use a commercial 90-degree hybrid coupler (2-4 GHz) for phase shift on this band.

Sun noise measurements are always a rather subjective test of receive system performance, due to variations in the Sun's energy output (especially true during Solar Cycle 22!), but when measured during December of 1988 using preamps of about the same performance on both 903 and 1296 (0.43 dB NF/ 15.5-dB gain and 0.48 dB NF/ 15.0-dB gain respectively), I measured about the same amount of sun noise on each band (10.5 dB). The dish should have exhibited about 2 dB more gain at 1296, resulting in 2 dB more sun noise. This similarity of reading indicated that the 1296 feed was in fact, not optimum, due to the under-illumination I mentioned earlier. Another somewhat subjective test of dish performance is listening for one's own lunar echoes. On 903 MHz, I could hear my echoes consistently 1-2 dB above the noise, every time I tested. On 1296, using the coffee-can feed, I could detect my own echoes just above the noise, and only at lunar perigee. Changing to the VE4MA feed has resulted in a 1-2dB improvement in *both transmit and receive gain.* This means a 2-4 dB improvment in echoes, meaning they can now be heard any time I listen for them. On 2304 MHz, with lower RF power, but 5dB more antenna gain, lunar echoes are easily heard. I should note that hearing one's own echoes is not a prerequisite for EME success. There are plenty of stations with large enough antennas to more than make up for any of a small stations' deficiencies.

THE REST OF THE STATION

Obviously, to be successful at EME requires a lot of attention to detail. This is particularly true when using a small dish because there just isn't enough margin to allow for great errors. Using the lowest loss feedlines possible is one important detail. When I began this project, I used ⅞-in. Andrew foam Heliax on transmit on all bands, accounting for 1 dB of loss on 1296, but *almost 2 dB* at 2304. By the time I had added smaller flexible cables at the feed and transmitter ends, I had almost 2 dB loss at 1296 and *3 dB* at 2304. Half my power was wasted heating up the cables on 2304! I acquired a run of 1-⅝-in. Andrew foam Heliax, which is now what I am using. This has improved the power situation significantly, reducing transmit losses by a dB or more on all bands (Note: locating the transmitter as close as possible to the dish solves this problem, but that wasn't feasible in my case). On receive, I use ½-in. Andrew foam Heliax and with the gain of the preamps on each band, this causes no degradation in receive performance. The preamps are another area of great importance, more so than the feedlines. I use homebrew preamps on 903, 1296, and 2304. All units use Avantek ATF10135 GaAsFETs and were designed by Al Ward WB5LUA. Their fine performance has already been noted.[5] My transmit power amplifiers are areas that can stand some improvement. On 903, I used a borrowed HI-SPEC amplifier which used two 7289s and was air-cooled. It produced about 250 W with about 15-20 W drive. On 1296, I use a single 7289, water-cooled, in an N6CA designed, homebrew cavity.[6] This amplifier also produces about 250 W with about 10 W drive and does a very nice job. On 2304, I use a surplus TWTA (Travelling Wave Tube Amplifier) which is rated at 100 W, but produces nearly 200 W output with 100 mWof drive. On 903 I use a homebrew transverter with a GaAsFET front end (about 0.5 dB NF). On 1296 I use an ICOM IC-1271A transceiver, with a homebrew GaAsFET front end. The transceiver's total noise figure is about 0.8 dB. On 2304 I use a homebrew transverter with a noise figure of

1 dB. On both 903 and 2304, I use an ICOM IC-271A as the tunable IF. I have modified both the 1271 and 271 to allow me to defeat the AGC. I have found this to be helpful when receiving weak signals.

Conclusions

It doesn't take a lot of equipment or money to become EME-capable on the microwave bands. My total investment in the antenna system, including the dish and mount, feed, preamp, and feedlines was less than $700. By shopping wisely, this price can probably be reduced by several hundred dollars. Though small in size, the compromises have been few, resulting in many enjoyable hours working stations off the moon on 903, 1296, and 2304 MHz. With future improvements and the addition of several more bands, this enjoyment will increase many times. This additional mode of operation at my station has had the benefit of allowing me to work many more Grid Squares, and to add significantly to my Worked All States total. Hopefully, in the not too far distant future, I will be able to add Worked All Continents to my list of accomplishments on 1296. The obvious technical challenge in this type of activity shows no sign of abating.

REFERENCES

[1] D.Hallidy, KD5RO, "Microwave EME Experiments, Spring 1987," *Proceedings of the 21st Conference of the Central States VHF Society*, 1987.

[2] L. Molitor, W7IUV, "Antenna Position Indicator" *Proceedings of the 20th Conference of the Central States VHF Society*, 1986.

[3] D.L. Hilliard, WØPW, "A 2.3GHz Feed For .43 f/D Reflectors" *Proceedings of Microwave Update*, 1986.

[4] B. Malowanchuk, VE4MA, "Use of Small TVRO Dishes for EME" *Proceedings of the 21st Conference of the Central States VHF Society*, 1987.

[5] A. Ward, WB5LUA, "Simple Low-Noise Microwave Preamplifiers" *QST*, May 1989 (reprinted in this book).

[6] E.R. (Chip) Angle, N6CA, "A Quarter Kilowatt 23-cm Amplifier" *ARRL Handbook*, 1986-1993.

[7] B. Malowanchuk, VE4MA, "Selection of an Optimum Dish Feed," *Proceedings of the 23rd Conference of the Central States VHF Society*, 1989.

A Dual-Probe Coffee-Can Feed

By Sam Popkin, K2DNR

(From *Microwave Update '91*)

This is the method I used to put a separate CW transmitter and receive converter on 2304 MHz without using a good coaxial relay.

I was working with a Multipoint Distribution System "cigar" type antenna, which consists of a typical "coffee-can" feed followed by a disc Yagi. It occurred to me that a second feed probe in the can would let me connect the transmitter directly into one feed and the converter to the other. Fig 1 shows the general arrangement, after adding the second probe.

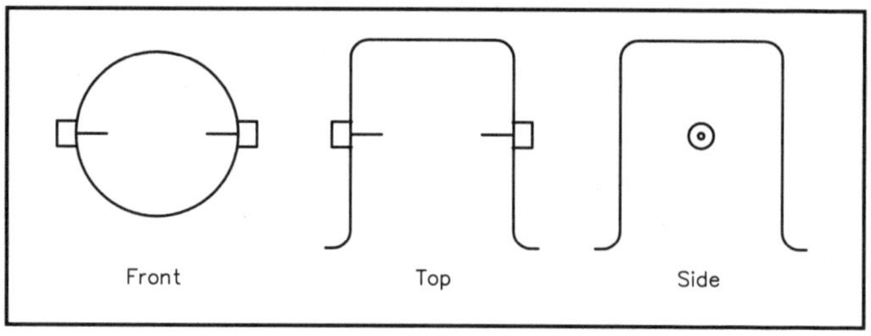

Fig 1—Three views of the dual-probe feed.

Before actually subjecting the converter to all that RF floating around inside the can, I tried one measurement. I applied 15 mW to one probe terminal and measured the power at the output of the second probe. It was a bit more than 10 dB down. The particular receive converter that I used has a bipolar device, a 2N6603, and that device didn't seem to mind that power level for sustained periods of time. I would not recommend this method for significantly higher power levels or where GaAsFETs may be employed in a preamp or converter front end. In general, the system performed as well as could be expected with the discs in place.

A Triband Microwave Dish Feed

By Tom Hill, WA3RMX

(From Aug 1990 *QST*)

Many hams today think of microwaves as the province of wizards and experimenters. Fortunately, that stigma is disappearing now that commercial transverters, amplifiers, antennas and other needed accessories have become widely available, and now that no-tune-transverters[1,2] and simplified construction ideas are in wide use. This construction article is intended for everyone; no machine tools or exotic materials are needed. In fact, this project doesn't require anything expensive or hard to obtain. After the parts are gathered, you can accomplish the construction work in one concentrated evening, or over a leisurely weekend. It's a great project for groups interested in starting up some microwave work for events such as the ARRL VHF/UHF contests.

A triband microwave antenna? Why not? If we can do it for HF, we can do it for the microwaves! This feed covers the 2304, 3456 and 5760-MHz bands in one unit that's designed for use in a single dish—with no adjustments—while you switch among the three bands. If you want to build your own dish feed, the simplicity of this design makes it a real winner. And, it's available—ready to mount to your feed line or connector—from Down East Microwave.[3] Dish antennas are available at hamfests and from other sources.[4]

The Past

I used to use separate feeds for each band. These were the well-known, dipole/splash-plate feeds, which are excellent performers. These feeds require machine work to build, and often quite a lot of fussing to change one for another when switching bands. The alternative is to have a separate dish for each band, which is not very practical for a fixed station, and is simply unworkable for portable operation. Another disadvantage is the cost of the dipole/splash-plate feeds: After the large initial investment ($70 or so each if purchased new), the possibility of damage still looms. When the wind picks up, as it is wont to do, the dish may fall over on its face, squashing the feed.

Initial Requirements

After building a new triband SSB rig for the microwaves, I decided that a new feed arrangement was also needed to simplify setup and operation. I developed a list of requirements based on my past experience with dishes:

1. The new system must cover all three bands at once. (No more feed changing during DX pileups!) Also, no adjustments should be needed when changing bands, such as moving the mounting hardware to accommodate differing phase centers.
2. Complete broadbandedness is not desirable; the antenna should reject signals outside of the amateur bands, such as radar, satellite-TV uplinks, etc.
3. It should be easy to build (a PC board would be best). No machine work should be required; hand tools only.
4. It should be cheap. (Not so much crying when it's broken during a windy mountaintop expedition, as spares can be on hand.)
5. No exotic materials should be needed. It must be made from commonly available parts.
6. Must be rugged and easy to carry (flat is good).
7. Must not be affected too adversely by rain. My dipole/splash-plate feeds have a section of transmission line open to the air. Here in Oregon, the frequent rain fills up those feeds, rendering them useless until emptied (and often leaving them corroded as well).
8. Performance must not be more than 2 dB worse than the single-band feeds, and hopefully not even that bad.

The resulting triband feed meets all of these goals. The feed is etched on both sides of a standard FR4 (G-10) circuit board, then sprayed with clear urethane coating to protect the copper against weathering. The board requires no holes, nor any connections between the front and back sides of the circuit board. See Fig 1. The finished assembly is shown in Fig 2, mounted in a dish with semirigid coax.

How it Works

The triband feed is a variation of the original dish feeds, which consisted of a dipole driven element and a single reflector. The triband feed has three driven elements attached to a common feed point, and three separate reflectors. The lengths of the three driven elements have been adjusted to compensate for proximity to, and the detuning effects of, the other elements. The elements are positioned for a common phase center for all three bands. This obviates the need for adjustments

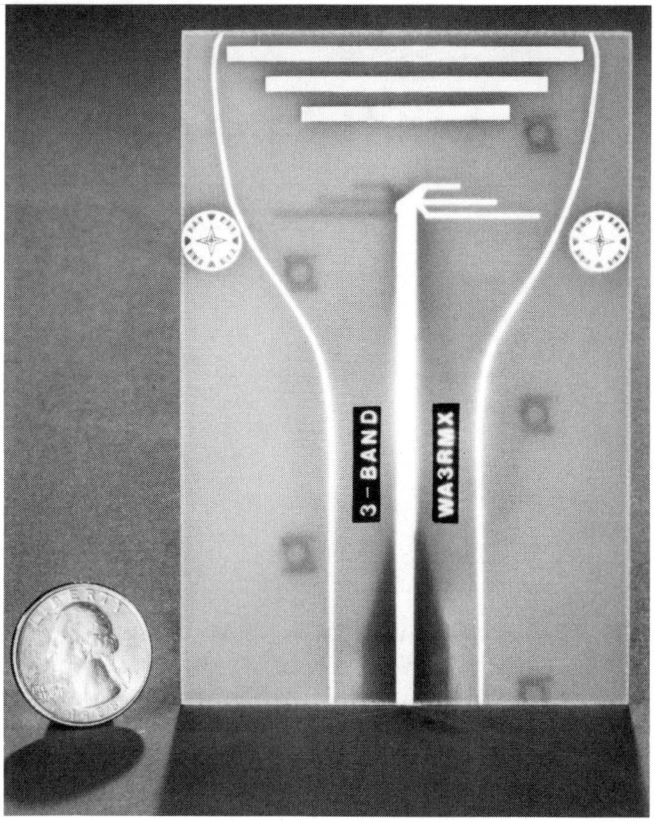

Fig 1—The triband dish feed is etched on common FR4 PC-board material.

Fig 2—The first prototype triband feed, mounted in a 19-inch dish, has performed well from my rooftop for almost two years.

when changing bands. The 5760-MHz dipole is farthest from the dish surface. The lower-frequency dipoles distort the 5760-MHz dipole's radiation pattern more than if the order was reversed, but this placement facilitates the common phase center.

The balun is a simple, slowly tapered transition from microstripline at the end of the board where the connector mounts, to balanced line as it moves toward the dipoles. Although it's a compromise to allow printing the whole assembly, this technique is surprisingly effective.

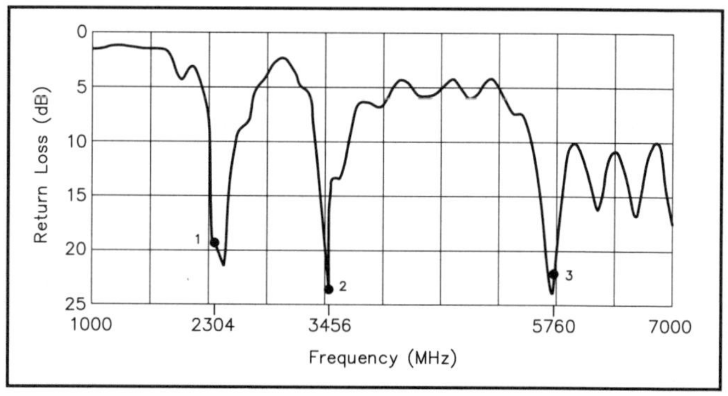

Fig 3—A network-analyzer plot showing the return loss (another way of expressing SWR) of a prototype feed. The greater the return loss, the better the SWR. For more information on return loss, see R. Schetgen, Ed., *The 1994 ARRL Handbook* (Newington: ARRL, 1993), p 35-38, and W. Hayward and D. DeMaw, *Solid-State Design for the Radio Amateur*, second printing (Newington: ARRL, 1986), pp 154-155.

Material and Results

The SWR is typically better than 1.5:1 on the two lower bands and better than 2:1 on 5760 MHz. See Fig 3. SWR varies somewhat from one feed to another, and is affected slightly by the mounting method. The feed helps to reject some of the extraneous signals sometimes found outside the amateur bands on the same mountaintops that you'll occupy for your DXpeditions!

The board material dramatically affects these results. I used a 3- × $4\frac{1}{2}$-inch piece of $\frac{1}{16}$-in.-thick FR4 material with 1-ounce copper on both sides, and presensitized with positive-acting resist. (Although I used Injectorall type 40P board from the local parts store, PC-board material without the positive resist shouldn't affect performance.) Other manufacturers' FR4 boards will probably work fine, but if you use another material (phenolic, Teflon, epoxy-paper, etc), or one of a different thickness, you will have to make major changes to the dimensions of both the transmission line and the elements.

I have tested only the configuration shown here. It works so well, both electrically and mechanically, that I didn't try others. Teflon (either woven mat or the amorphous types, such as Rogers Duroid) would exhibit slightly lower loss, but at the price of much less mechanical strength. I used a protective spray coating from the local parts store (Fine-L-Kote UR spray urethane conformal coating, made by Tech Spray, stock no. 1711-16S). I suspect that many other plastic spray coatings will work fine, but I have not tried any, so, as in all microwave work, if you change the recipe, you are on your own. Extensive

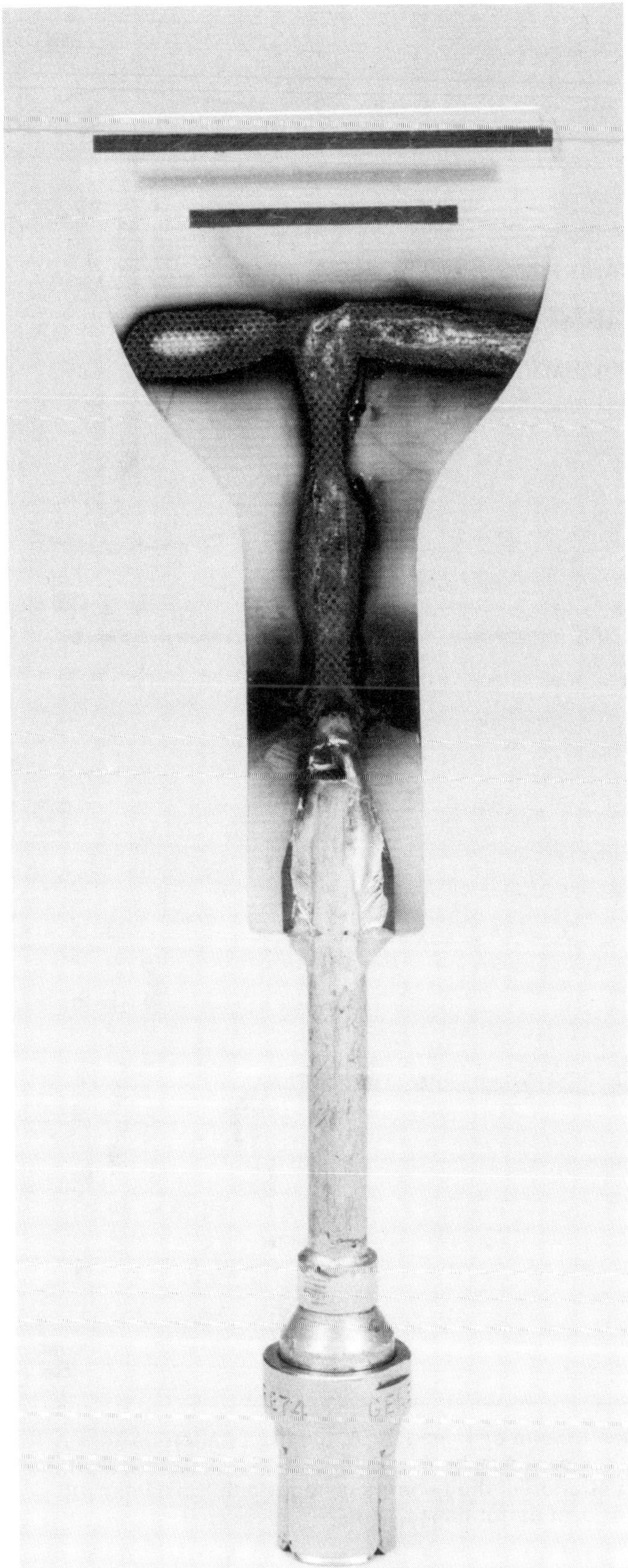

Fig 4—This triband feed was tested with 200 W of 2304-MHz RF; it failed. Don't apply more than 20 W to this feed!

Table 1

Gain Differences Between PC-Board Triband Feed and Optimized Dipole/Splash-Plate Feed (0.31 f/D dish)

Frequency	*Triband Feed Loss*
2304 MHz	0.75
3456 MHz	1.25
5760 MHz	1.5 dB

tests have shown that the resonant frequencies of the feed move less than 25 MHz after the spray is applied, causing only negligible changes in SWR. The copper patterns have been designed to account for the presence of the plastic coating, but you can omit it without performance problems. Still, I recommend the plastic coating—especially around the edges of the board, which can absorb water if left unprotected—for outdoor use.

When tested in a 1-meter dish (f/D = 0.31), and compared to dipole/splash-plate feeds, the triband feed's gain, although somewhat less than that of optimized single-band units, is still very good. See Table 1. Some of this loss is due to suboptimal dish illumination and some is a result of dielectric loss in the circuit-board material (more on this later). The three reflectors are not optimal (they replace three separate round discs), and the three driven elements "argue" with each other. But that's the price of a PC-board, do-all feed! For critical long-haul, weak-signal work, consider keeping a set of optimized, single-band feeds handy.

This feed handles enough power for a very effective mountaintop station, and I have regularly put 20 W through it with no trouble. But there is a limit: Al Ward, WB5LUA, tested a copy of this feed with 200 W of 2304-MHz RF for an extended time during the January 1990 ARRL VHF Sweepstakes contest. Even a small amount of dielectric loss was enough to precipitate spectacular heat damage, as shown in Fig 4. The board was burned so badly that all the copper fell off, predictably causing a rather severe SWR increase. I recommend not applying more than 20 W to this feed!

Construction

To build the triband feed, first etch the patterns on the front and back of the board. The full-size patterns are shown in Fig 5. The front pattern is dimensioned so that you can easily check the board size before etching. Fig 6 is an X-ray view of the board showing the feed's apparent phase center. Accurate registration of the front and back patterns is fairly critical; testing has shown that the two sides should be aligned to within at least 0.030 inch for good performance and low SWR. Pattern alignment can be achieved either by drilling two pilot holes in the unetched board at the locations of the two bullseyes on the artwork, or by carefully taping the edges of the artwork together and clamping the unetched board between them before exposure. (I prefer the drilling method.) After etching and washing the board, cut away the excess board material just inside the cut lines on the front artwork. Be sure to remove all metal from the cut lines.

Mount a connector to the board in whatever arrangement suits your dish. I suggest you use semirigid 50-Ω coax, either 0.250 or 0.375 inch in diameter, with a connector to fit the feed line to the rig (see the title photo). Prepare the board and the coax to fit each other as shown in Fig 7. You can also mount a connector right to the board, as shown in Fig 8, trimming off the excess copper on the top. In any case, use the shortest leads possible: Even 1⁄16 inch or less excess wire kills performance at 5760 MHz!

I built one triband feed using the popular 0.141-inch-diameter semirigid coax (UT-141), as this is readily available at hamfests and from suppliers that cater to microwave hobbyists. Although this works electrically, UT-141 is too weak to hold the feed in place except with very gentle handling. If UT-141 is all that you have available, use it; it works fine for light-duty service. If you do use such small Hardline, flatten the part of the shield to be soldered to the back side of the board. The board is thicker than the spacing from the center conductor to the shield, and the center conductor shouldn't be as long as would be necessary for it to wrap around to the front side of the board.

I use a GR874 connector on my portable version of the feed, which allows quick assembly and disassembly, but it also allows water to enter the connection if it's used in heavy rain. My permanently mounted versions all use N connectors for water-resistant, low-SWR connections. If you expect to use the feed in the rain, it is a good idea to mount it in the dish at a slight angle from the horizontal to allow runoff. (If water pools on the flat upper surface of the feed, dielectric loading will detune the antenna.)

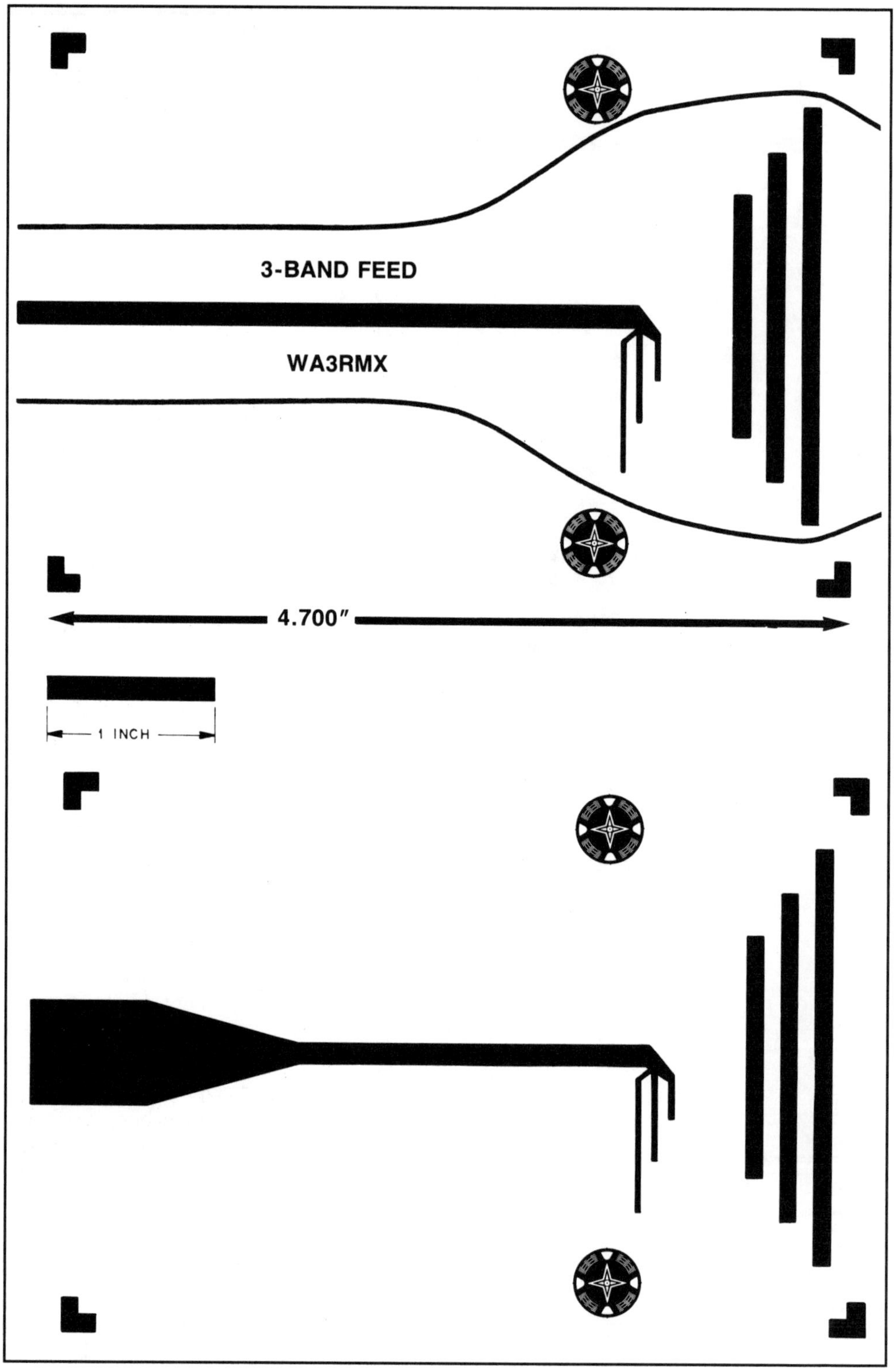

Fig 5—PC-board pattern for the triband dish feed. At A, the front pattern is shown. It includes a rule for checking the board size before etching. At B, the back side of the board is shown. Accurate registration of the patterns is important; the alignment should be within 0.030 inch for best performance.

After the mechanical and electrical assembly is finished, spray the plastic coating onto the board. After it dries, apply some electronics-grade (non-acidic) silicone sealant to the area of the solder connection to seal out moisture, especially if the feed is to be used extensively outdoors. For this purpose, I use Dow no. 3145 silicone sealant.

Now, mount the feed in the dish, with the apparent phase

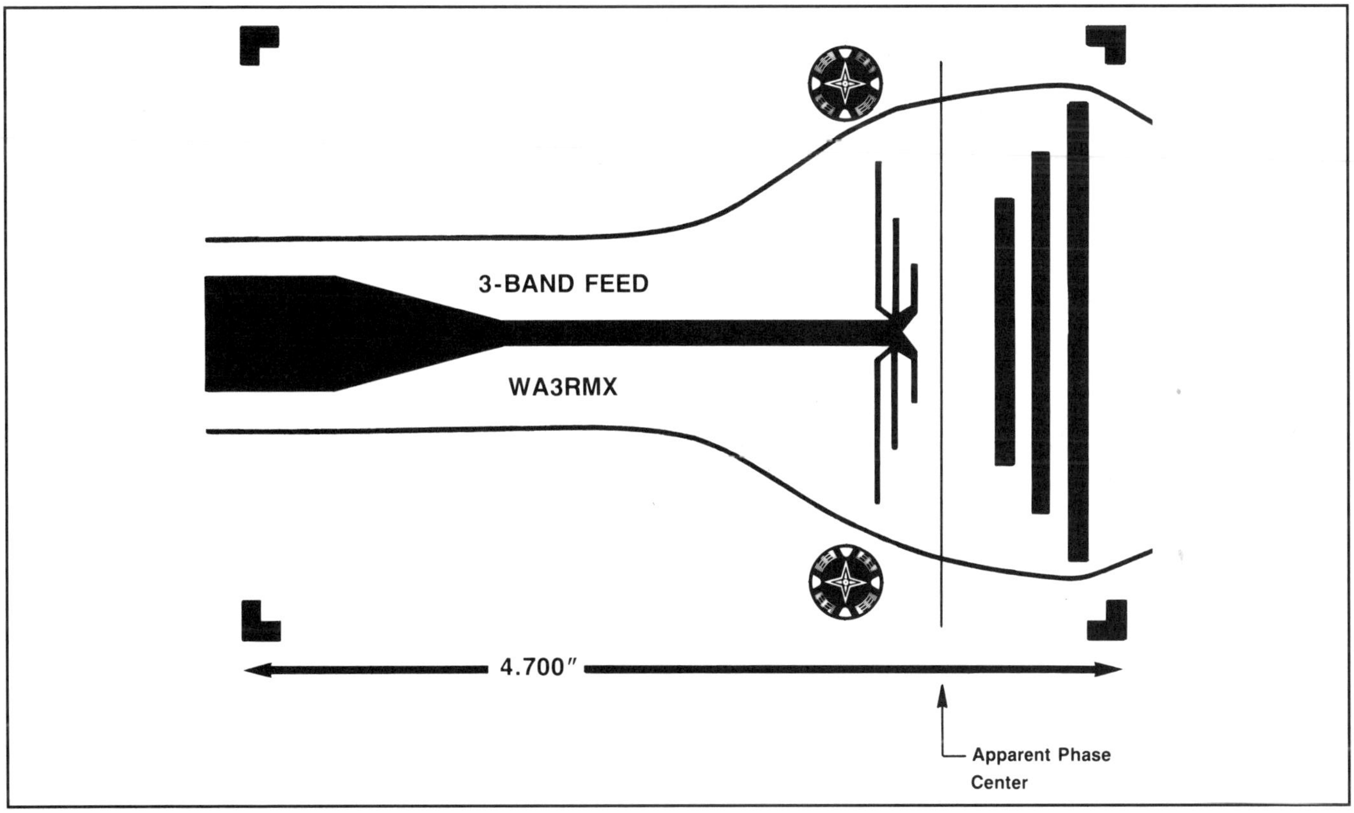

Fig 6—An X-ray view of the PC board, showing the apparent phase center of the feed. Place this part of the feed at the focal point of your dish, as described in the text.

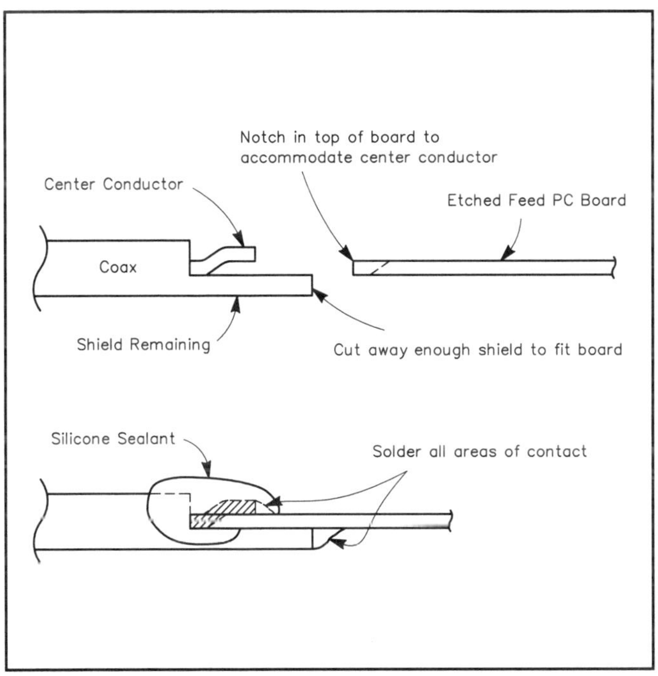

Fig 7—Prepare the Hardline for mounting to the dish-feed PC board by cutting away half of the shield and dielectric from the cable over a distance long enough to ensure a solid mechanical connection to the board. Then, bend the center conductor slightly and fit it into a notch in the PC board, as described in the text.

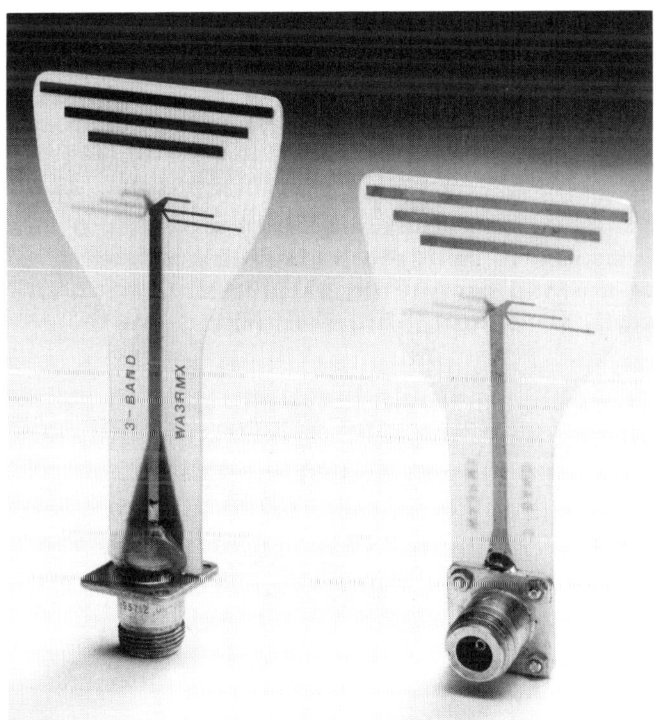

Fig 8—Two prototype triband feeds with N connectors mounted directly to the PC boards. When mounting connectors this way, be sure to minimize stress on the feed resulting from feed-line strain.

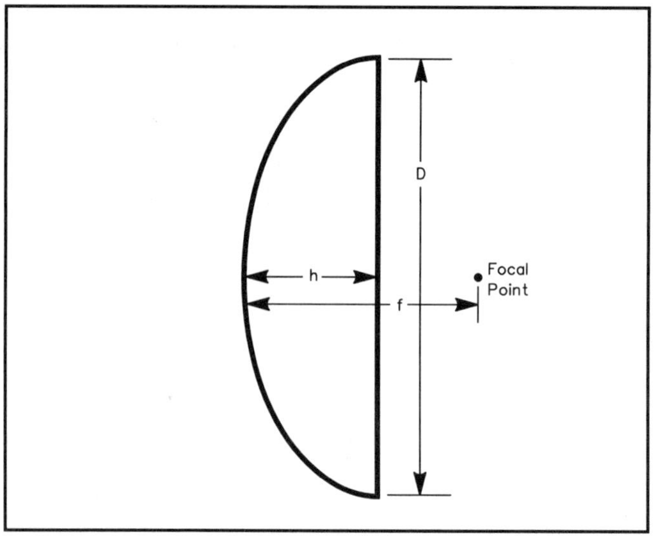

Fig 9—Measure the proportions of your dish and calculate the location of the focus as described in the text.

center (Fig 6) at the focal point of the dish. My tests showed that, although it's a compromise, this is the best feed placement for all three bands. To find the focal point of a dish, use the equation $f = D^2 / 16h$, where D is dish diameter and h is dish depth (see Fig 9). The triband feed works well with dishes having f/D ratios from 0.25 to 0.4, which is the range over which I've tested the feed.

Operation

With one of these feeds in a 30-inch-diameter dish, I have made 80-mile 2304-MHz FM contacts with 100 mW—with plenty of signal to spare. On SSB, greater range can be obtained, or much lower power used. Lynn Hurd, WB7UNU, and I have made 115-mile contacts with 50 mW of SSB on 2304 MHz using these feeds at each end—stuck straight up into the air, without dishes! In a test to see what is possible using minimal power over a line-of-sight path, we set up 50-mW SSB rigs with 29-inch dishes at each end, each equipped with a triband feed. Communicating over a 66-mile path, we added attenuators in the feed line to one rig until the SSB signal was just barely readable. With 50 dB of added attenuation, we still had readable signals. (This corresponds to ½ microwatt of transmitter power to the antenna!) We were pleased.

Roger McCoy, W7ADV, and I made a 130-mile 2304-MHz contact over an obstructed path using 29-inch dishes equipped with triband feeds at each end, by using the tip of Mt Hood (12,000 feet above sea level) as a knife-edge diffractor. We first tried this with 20 W and had fine signals, so we also tried CW and managed a weak contact with only 100 mW. At the extreme, I have even used a triband feed to make 10-mile contacts with 25 mW on 10 GHz. Although far from optimal, the feed still radiates a signal at 3 cm! Three VUCC awards each on 2304, 3456 and 5760 MHz have been earned with copies of this feed. WB7UNU (grid CN85NL) and I (CN85NM) each use these feeds in 19-inch dishes on our roofs, and regularly chat on 2304 and 3456 MHz (5760-MHz signals don't make it too well through the grove of trees on my neighbor's property).

I hope this article helps generate more activity on the microwave bands. With the triband feed and the simple transverters now on the market, it's easier than ever before to put together lots of roving stations for the VHF and microwave contests. VUCC, here we come!

Acknowledgments

Thanks to Deane Kidd, W7TYR, for his help with the artwork and his untiring encouragement. Also, thanks to WB5LUA for his high-power tests, and to W7ADV and WB7UNU for their help field-testing the prototypes.

Notes

[1]R. Campbell, "A No-Tune Transverter for 3456 MHz," *QST*, Jun 1989, pp 21-26.

[2]No-tune transverter kits for 903, 1296, 2304 and 3456 MHz, as well as preamps and accessories, are available from Down East Microwave, 954 Rte 519, Frenchtown, NJ 08825, tel 908-996-3584, fax 908-946-3072.

[3]See Note 2.

[4]Spun-aluminum dishes are available in several sizes (24, 36, 48, 60 and 72 inches) from Antenna Center, 505 Oak St, Calumet MI 49913, tel 906-337-5062.

Mounting the WA3RMX Tri-Band Feed

By Kent Britain, WA5VJB

(From *Microwave Update '91*)

I have had excellent results with Tom Hill's dish feed, described in the previous article, using one for all five grids in my 5760-MHz VUCC. For outside mounting I suggest changing Tom's trim line. The point allows dew and rainwater to collect away from the driven element (Fig 1). For the usual bird problems, I mounted several nylon bolts through the feed, then sharpened the ends. This gives a really rude surprise to the local mockingbirds (Fig 2).

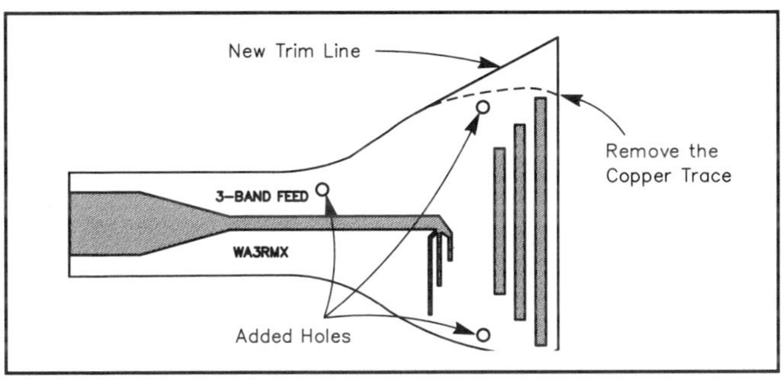

Fig 1—Using a sharp corner, instead of a rounded one, causes rainwater to run away from the dish feed driven element.

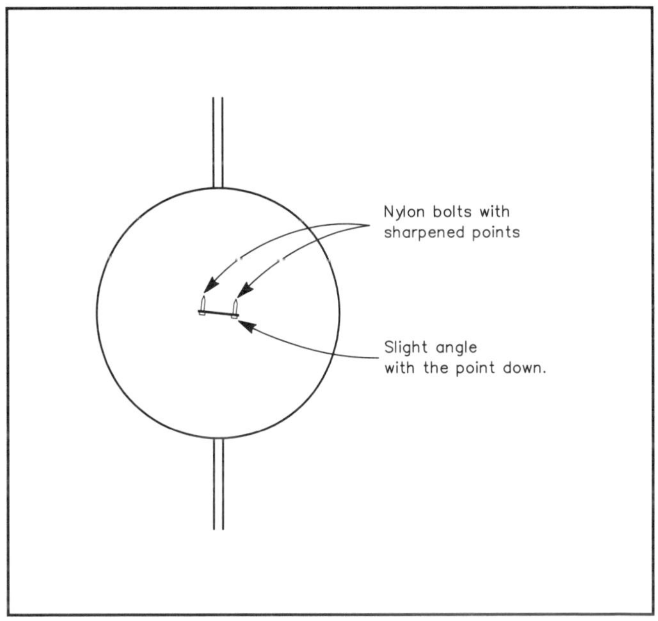

Fig 2—The feed is installed at a slight angle, with the point down. Sharpened nylon bolts installed through the feed discourage birds from sitting on the feed.

A Simple Dual-Mode (IMU) Feed Antenna for 10.368 GHz

By Richard H. Turrin, W2IMU

(From *Microwave Update '91*)

The small-aperture dual-mode reflector-feed antenna can be simply constructed from readily available copper water pipe plumbing fittings, as shown by the cross-sectional drawing. The circular waveguide section is standard ¾-in. Type-M copper water pipe which will support only the dominant TE11 mode. The tapered section is a ¾- to 1½-in. fitting reducer. This part may be constructed differently by different manufacturers. Try to find a fitting reducer with abrupt corners at both ends of the taper. Some parts are made with a well-rounded corner at the large end. While this part is satisfactory, it requires a slightly different positional length of the 1½-in. copper pipe coupling, for proper mode phasing.

While the correct flare angle of the conically tapered section should be 30°, the copper fitting reducer will have a flare angle of about 40°. The ⅛-in. soft copper tubing ring fitted inside the fitting reducer compensates for the difference in taper angle.

A simple, inexpensive, efficient and very low SWR coupling may be made from a standard ¾-in. coupling, which is slotted axially (six slots about ½-in. long), so that a hose clamp may be used as a securing device.

Form the soft copper ring to fit snugly inside the fitting reducer, and sweat it in place first. Clear all burrs from the inside of the pipe ends and sweat the parts together as per Fig 1, with a minimum of solder and paste. Wash the assembly thoroughly with a detergent, to remove all traces of soldering paste, and if desired, spray all surfaces with clear Krylon to help preserve the bright copper finish.

This feed antenna will have a –10 dB beamwidth of –44 degrees, with very low (less than –30 dB) side and rear radiation. The E- and H-plane main radiation patterns are virtually identical, as in the original dual-mode design. The resulting main radiation is circularly symmetric and will support any polarization.

This feed antenna is ideally suited to a rear-fed cassegrainian antenna reflector system, where the hyperbolic subreflector is designed to match the main reflector f/D ratio.

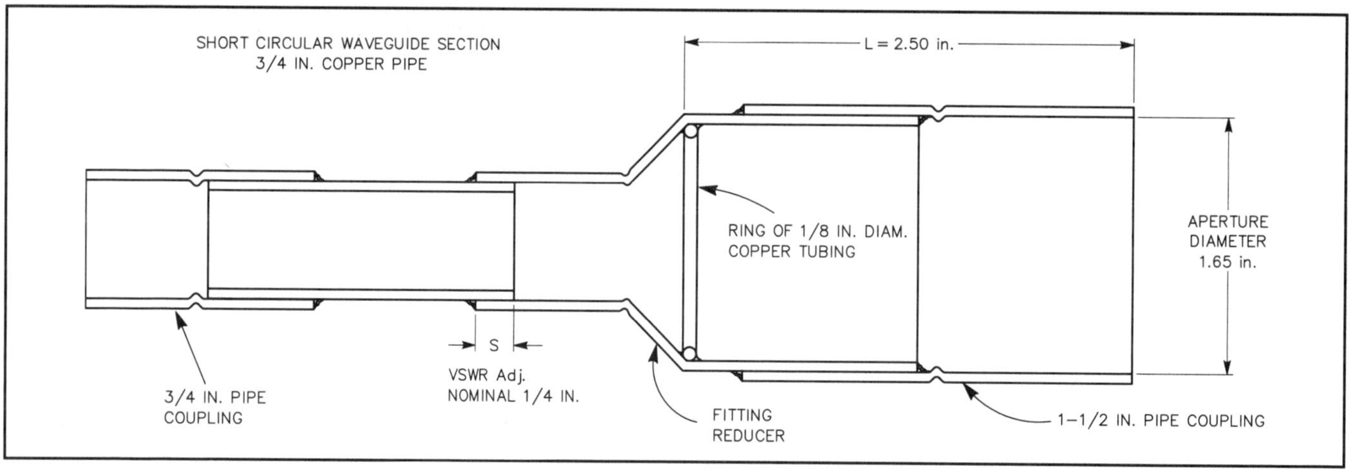

Fig 1

10-GHz IMU Feedhorn Update

By Kent Britain, WA5VJB

I have had excellent results with Dick Turrin's dual-mode feedhorn. I now use the simple dual-mode feed on several antennas, including the source antenna for the Central States VHF Society's 10-GHz antenna range.

In Texas, the local source for 3/4-in. to 1½-in. adapters is Nibco. They use a different technique for molding the adapters with much less taper in the transition. It was very nice of them to transition over λ/2 at 10 GHz, so the 1/8-in. ring that Dick added in his horn was unnecessary.

Watch that taper! If the reducer transitions over about ½ to 5/8 in., you can use the coupler as is.

The horn can get a bit heavy. One trick is to cut back the 1.5-in. coupler. Not only do you shave off a little weight, but some plumbing companies make the coupler long enough to let you build two antennas from that $3 coupler.

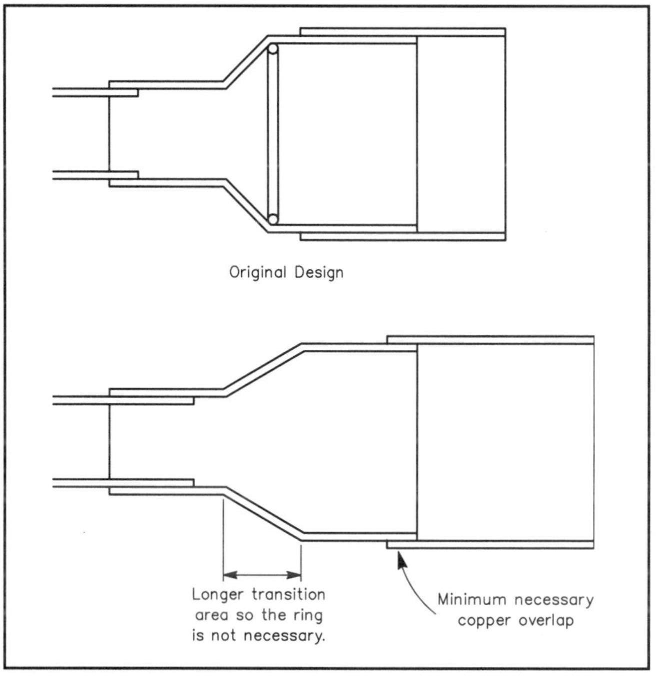

Fig 1—Details of the 10-GHz dual-mode feed-horn.

Big Dish Feed for 10.3 GHz

By Chuck Steer, WA3IAC

(From *Microwave Update '91*)

This paper describes an easy-to-build dish feed for 10.368 GHz. A feed of this type can be made using a piece of copper water pipe with a piece of brass as the back wall. The feed should be within a proper size for the frequency being used.

Using 0.75-in. ID copper pipe, my initial results were less than desired. This was with a probe about 0.2-in. from the back wall and 0.2-in. long. This type of feed was hard to reproduce with good results. I was able, however, to "tune" for best match by adding a 2-56 × 1-in. screw, 180° from the feed. A typical return loss was –13 to –15 dB.

Recently, I read about a 10-GHz feedhorn built by G4DDK. The odd thing about this feed is that the ID was 1.125-in., the probe was 0.030-in. long, and it was placed about 0.375-in. from the back wall.

Given $TE\ 01 = \frac{9034.85}{d\ (inches)}$ and $TE\ 11 = \frac{6917.26}{d\ (inches)}$

where d is the ID of the feed, we get:

TE 01 = 7227.88 MHz and

TE 11 = 5533.81 MHz

Given the above calculations, the feed should not work at 10.3 GHz, but rather much lower in frequency. I then wondered if there was not a mistake in the drawings from G4DDK. My first try at duplicating this feed was using pipe with 1.25-in. ID and a probe length of 0.2-in., 0.375-in. from the back wall. My probe was just the center conductor of the 0.141-in. semirigid line, about 3 in. long, with an SMA connector on the other end. This produced a return loss of about –5 dB at 10.3 GHz. The best return loss was at about 12.5 GHz, *higher*, not lower in frequency as calculated. For my next try, I made the probe 0.3 in. long and measured a return loss of –18 dB (1.22:1 SWR). Not bad!

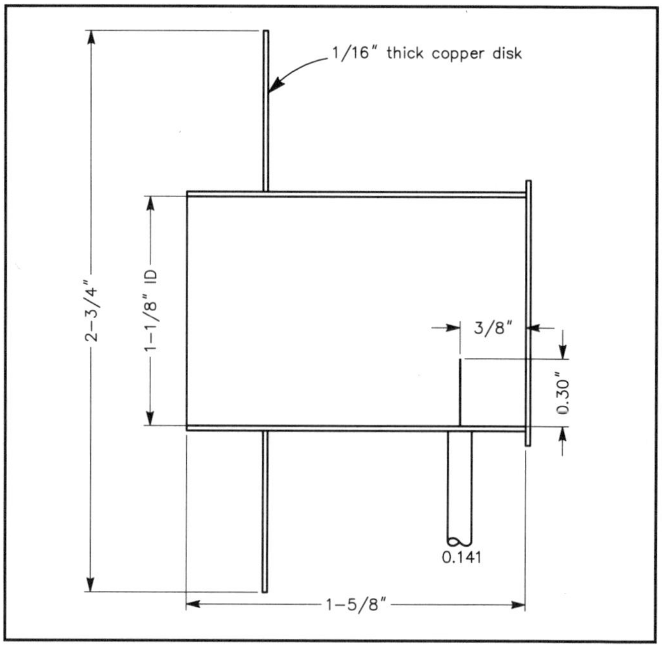

Fig 1—X-band feed for large dish antennas, made from 1.25-in. ID copper pipe.

I have since made three feeds this way and they all worked well. The worst return loss was –15 dB and the best was about –30 dB.

Next came the choke ring. I stayed with the G4DDK feed and made the outside diameter of the choke plate 2.75-in. and the inside diameter to just fit the OD of the copper pipe. The next step is to adjust the choke ring position for best gain. Start with the choke plate about 0.5-in. from the open end of the feed and move it away from the open end.

Horns for 10 GHz and Up

By Kent Britain, WA5VJB

(From *Microwave Update '89*)

When that waveguide gets really small, it can be a real challenge to assemble a horn antenna. This assembly method uses a short piece of waveguide and a sheet of hobby brass or sheet tin. File down the inside lip of the waveguide, align the top and bottom pieces and add a drop of "Super Glue" letting it wick under the edges. After the glue sets for a few minutes, bend the top and bottom pieces up a bit, then align and glue on the side pieces.

Next, line up the edges of the opening and spot solder the corners. When you have things pretty well square, solder the outside seams together. Soldering the outside edges helps you keep the lossy lead and tin off the inside surfaces.

Horns have been constructed this way on WR-62 and WR-42 for 10 to 24 GHz, and the method should work up to 50 GHz.

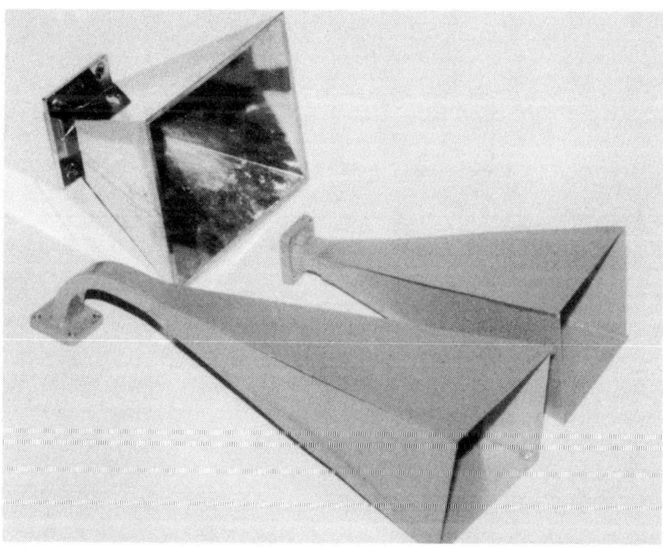

Fig 1—A commercial 17-dBi, 10-GHz horn, and two homebrew 22 dBi, 24-GHz horns.

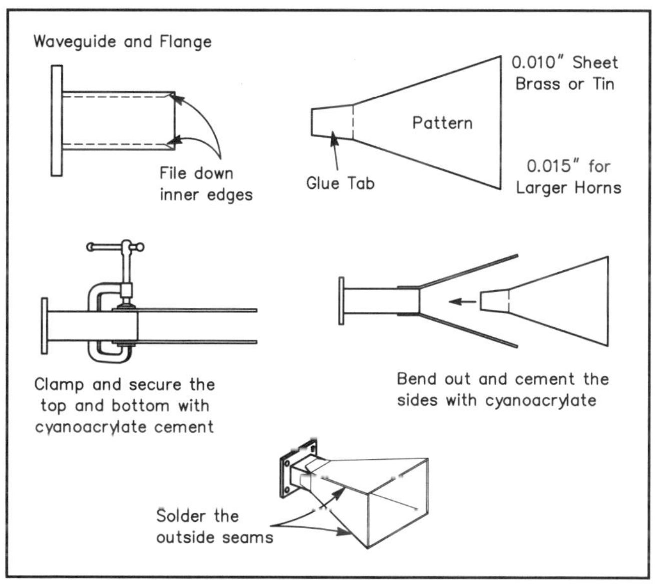

Fig 2—Details of homebrew horn construction.

Dual-Mode Feedhorn for 24 GHz

By Kent Britain, WA5VJB

(From *Microwave Update '91*)

I really like Dick Turrin's (W2IMU) latest Dual-Mode Feedhorn for 10 GHz, built from plumbing fittings. A little math and a trip to a plumbing supply house (Boy, has that sales clerk learned to stop asking me questions!) and I came up with a 24-GHz version of Dick's very popular 1296-MHz EME dish feed.

The end of the 3/8-in. coupling can be formed to the same size as a WR-42 flange. Soldering is a bit tricky, since all three are butt joints. I held the entire assembly in a long clamp, soldering everything at once with a torch on a low flame. The end was cut off with a pipe cutter after it cooled off.

Testing went well. Gain was a little over 12 dBi and the 10 dB-down points (suggested dish illumination at the edges) were ±45°. The E and H beamwidths were within 5°, and no sidelobes were noted. At the 1991 Central States VHF Society antenna contest, I fed a 12-in. 0.42 f/D dish with this feed. Performance was very close to theoretical gain.

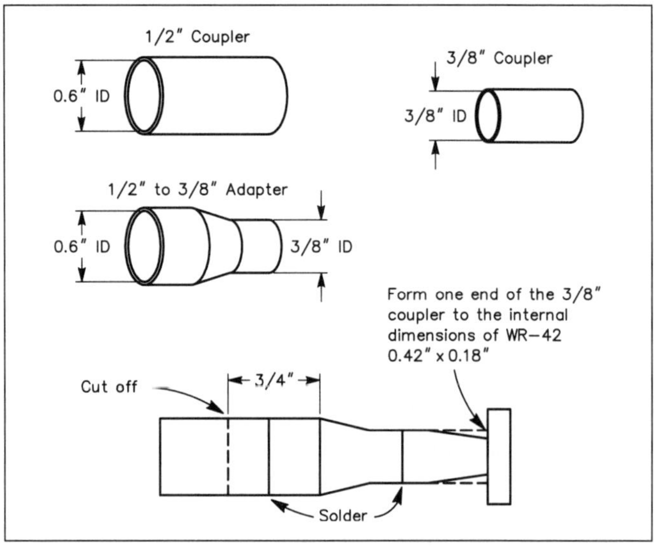

11

Test Equipment

Contents

Testing GaAsFETs with a VOM

By Kent Britain, WA5VJB

When Al Ward, WB5LUA, first suggested the possibility of checking GaAsFETs with an ohmmeter, I gasped! The test works fine, however, if you take a few safety precautions. Use a regular VOM powered by a 1½-V battery (don't use a 120-V ac-powered DVM!) with a 10-kΩ resistor connected in series with the VOM's probe. Don't forget the usual static electricity precautions. A diagram of the test setup and typical measured values is shown in Fig 1.

The 0.01-mA current doesn't stress the transistor's gates, and 1½ V is well below the transistor's breakdown voltage. With different manufacturers using different gate designations, this test is useful in determining transistor pinouts.

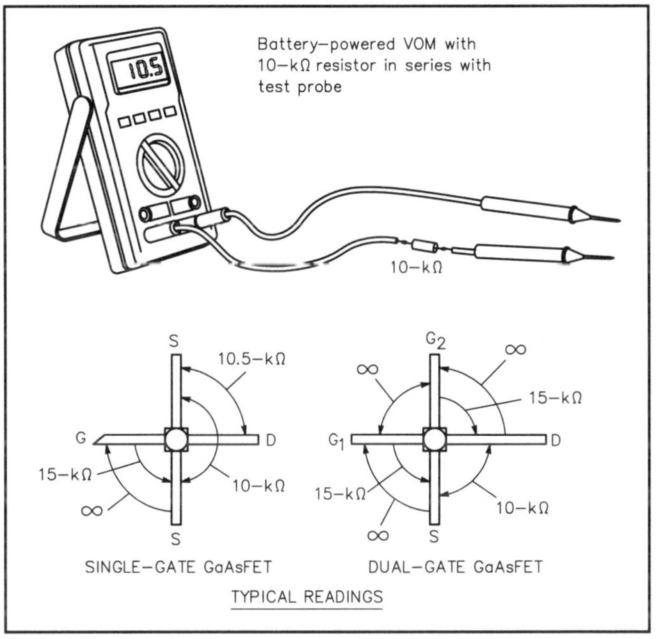

Fig 1—Test setup for testing GaAsFETs with a VOM. Typical resistance values are shown.

Microwave Absorber Notes

By Henry Burger, K7PSS

When the uses of microwave absorber materials are discussed, it's often stated that these materials are too expensive for hams to use. That's true—if the materials are purchased from a manufacturer. They are expensive indeed for a professional, building a quiet anechoic chamber for exacting tests, but they need not be so for the experimenter.

First, samples may be available from manufacturers, some of whom are hams themselves. Second, some industries may have scraps of old, unserviceable pieces that will be fine for amateur applications. Any ham that is interested in microwaves in the first place can probably find such a source. Third, the first absorbers were made in garage shops using hardware-store materials, and while not very good by today's standards, they can still be used for loads, and to dampen reflections in equipment boxes and in local areas.

The first absorbers were animal hair and wood shavings, coated with carbon-filled rubber. Flat black latex paint is a similar material. It is easy to get and handle, and its conductivity can be increased by mixing lampblack, graphite, or some other conductive material into it. Black powder toner from the waste container of a copy machine is another possibility, as is iron filings or powder. Several coats of such a mixture poured over a sheet of open-fiber packing or upholstery material should make a reasonably effective absorber for large surfaces.

You can make another type of absorber at home by loading any kind of plastic polymer with carbon or conductive material, and casting it to shape in a mold. Candidate materials are automobile body plastic, fiberglass resin from a fiberglass repair kit, and potting resins that hobbyists use to encapsulate objects for display. Again, lampblack, graphite, copy machine toner and iron filings could supply the conductivity required. Mix in the loading material before adding the catalyst. After mixing the catalyst, pour it (or cram it) into the desired mold. After curing, most of these materials can be shaped further by cutting or filing. This procedure can be used to make loads for waveguides and small cavities where it is desirable to absorb microwave energy.

These home-made absorbers can be tested in two ways. First, if a waveguide-type SWR setup is available, simply place a piece of absorber material inside the waveguide and observe the result as compared to a short circuit. Sheet materials can be tested with an open-end waveguide or small horn (the presence of the absorber in front of the open end should not increase the SWR, and should actually reduce the SWR compared to a metallic reflector).

The second way to test absorbers is to use a microwave oven. (The primary user of the oven may have something to say about this, but be firm.) Put a piece of absorber in the oven with a cup of water, and heat them for about 30 seconds. The water protects the oven if the candidate absorber material is not very lossy. If, after 30 seconds, the absorber does not get warm, forget it—if it smokes, however, you have a winner. (Better monitor the test closely for signs of overheating. You don't want to start a fire.) The faster the material heats up, the better it is as an absorber. It ought to heat up faster than the water.

One use for solid absorbers is as a waveguide load. The object is to shape the load so that it absorbs all the energy incident upon it, with none reflected. Starting with a reason-

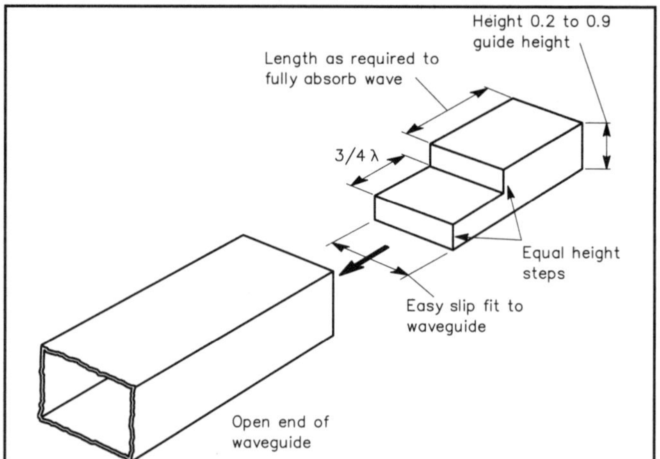

Fig 1—Construction details of a stepped-load for waveguides using microwave absorber materials. A "plug" of microwave absorber material is inserted into the open end of the waveguide. If the absorber material is formulated and shaped correctly, no microwave energy will escape. See text.

ably good absorber, shape the absorber according to Fig 1. The step is a one-quarter wavelength transformer to match the impedance of the waveguide to the impedance of the absorber. I have built waveguide loads this way using bar absorber stock.

Absorber can be applied on almost any surface where a bothersome reflection may occur, such as an unavoidable metal object near an antenna under test. For best results, the incident energy should be perpendicular to the absorber, but incidence at an angle will do. The absorber material should be several wavelengths thick, and more if the absorber is inefficient.

A dummy load for an antenna can be made this way by lining a box or a can with absorber and placing it over the antenna. (If everything is okay, the SWR on the antenna should not change when you cover it, and no signal should radiate outside the cover). This is a common practice in the industry for small antennas.

The solid absorbers can also be placed along the inside walls of circuit enclosures to absorb unwanted microwave energy. For receivers built in a can, this may be necessary for proper functioning, because all waveguide joints leak energy (Murphy's Law says this energy will always get in the way.) The enclosure for the circuit may resonate at a critical frequency, and the use of an absorber will spoil the Q of the resonance. This is also standard practice in the microwave industry.

Attenuators

By Bob Atkins, KA1GT

(From Jul 1988 *QST*)

Microwave experimenters have a wide range of applications for attenuators. Attenuators can be used in determining amplifier gain and noise performance, and they can be used to "force" an impedance match to interacting components such as diode multipliers, noise sources and load-sensitive oscillators. Attenuators can also be used in conjunction with small amplifiers to simulate isolators.

Although it is quite easy to make accurate, low-SWR attenuators for use at HF, it is considerably more difficult to do this for VHF and microwave frequencies. There are two reasons for this difficulty. First, the physical layout of the attenuator components is unlikely to provide a constant impedance through the attenuator. This causes a high SWR to exist on the transmission line to the attenuator. The magnitude of the SWR is a function of the impedance and electrical length of the mismatched section. The SWR is minimum at zero electrical length, rises to a maximum at λ/4, and falls again to a minimum at λ/2. Thus, if the length of the mismatched attenuator is a very small fraction of the wavelength of operation (less than 0.1%), as will be the case at low frequencies, then the resultant SWR increase will be negligible.

The second reason for the difficulty in building attenuators for use at microwave frequencies is the presence of reactive components in addition to the desired resistance in resistors. Component leads have inductance whose reactance rises with frequency. Also, inter-resistor stray capacitance causes reactance that decreases with frequency. This can result in unwanted coupling. Despite these problems, it is possible to construct attenuators with reasonable performance up to the lower microwave bands. In addition, many of the measurement applications use attenuators in the IF section of a receiving system where the frequency is much lower, and attenuator performance is more predictable.

Two resistor configurations are commonly used to form attenuators, as shown in Fig 1: the T arrangement and the pi arrangement. A list of resistor values for given attenuation values can be found in Chapter 25 of recent editions of the ARRL *Handbook*. The resistors should be of carbon composition. (Don't use wire-wound resistors—they make great inductors at RF!) To provide the best impedance match to a microstrip circuit, use the technique shown in Fig 2 to connect an attenuator to a source. This arrangement uses the pi configuration of Fig 1, with the single resistors to ground replaced by pairs of paralleled resistors to minimize inductance, increase power-handling capability and create a symmetrical structure. A binary sequence of attenuation values (1 dB, 2 dB, 4 dB, 8 dB and 16 dB) will yield the largest range of possible attenuation levels with the minimum number of attenuators. To minimize stray coupling problems, no single attenuator should provide more than about 20 dB of attenuation.

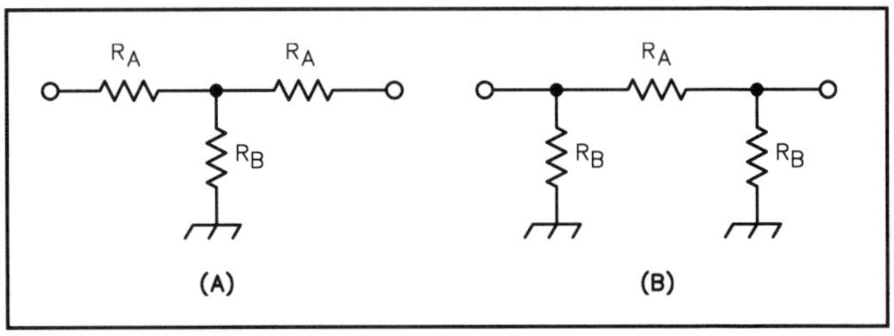

Fig 1—T (A) and pi (B) attenuator configurations.

Resistor values for various attenuation levels are given in Table 1. The values in Table 1 do not correspond exactly to commonly obtainable resistor values. One way to deal with this problem is to connect resistors in series and/or parallel to approximate the desired resistance. This approach is not desirable if the attenuator is to be used above VHF, because increased physical attenuator size and reactances can cause problems. A second alternative is to select resistors that have

Table 1
Pi-network Attenuator Resistor Values

Attenuation (dB)	R_A *(ohms)*	R_B *(ohms)*
1	5.8	869.5
2	11.6	436.2
4	23.8	221.0
8	52.8	116.1
16	153.8	68.8

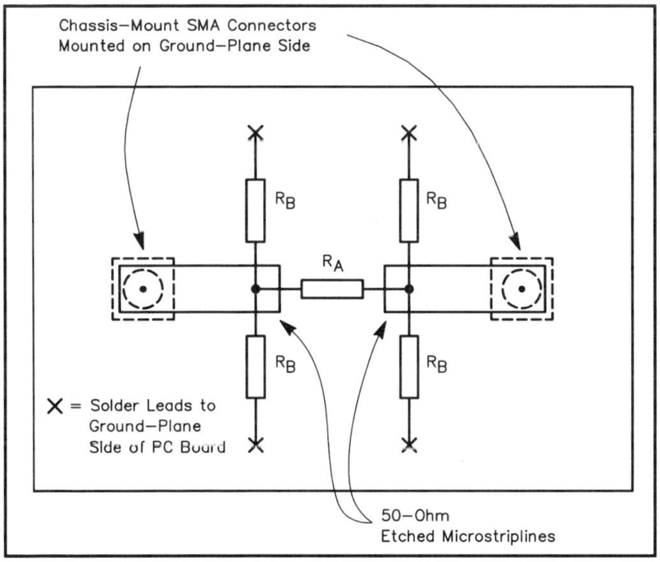

Fig 2—Circuit-board layout for a 50-Ω, pi-configuration microstrip attenuator for VHF use. Microstripline width should be 0.1 in. if double-sided, 1/16-in.-thick glass-epoxy PC-board material is used. Keep all resistor lead lengths as short as possible to minimize reactances at high frequencies.

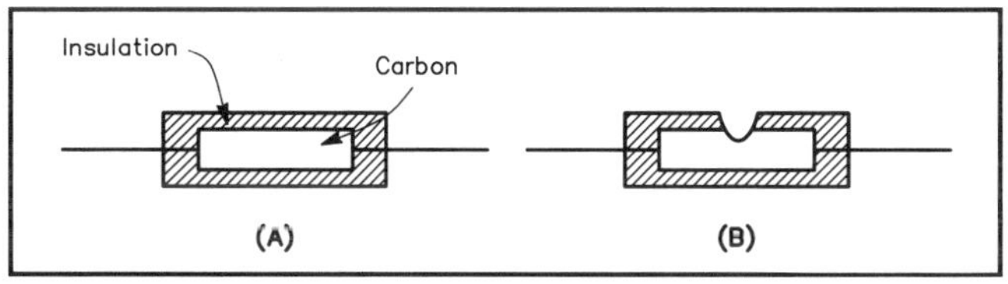

Fig 3—Cross-section of a carbon-composition resistor before (A) and after (B) filing to increase resistance. Be sure to seal the body of the resistor after filing (use a few drops of paint or nail polish) to prevent contamination and subsequent resistance-value drifting.

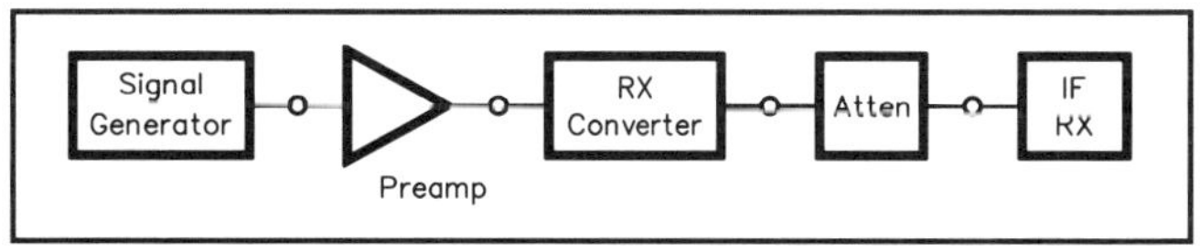

Fig 4—Preamplifier-gain measurement setup. See text.

slightly lower resistance than that required. A small round file can then be used to remove a small amount of the resistor, as shown in Fig 3. This will increase its resistance. After the resistor has been adjusted to the required value, a small dab of paint should be applied to cover the exposed carbon. If high power-handling capacity is not required, ⅛-W resistors may be used. The smaller the physical size of the resistors used, the larger the frequency range over which the attenuator will function as designed. Chip resistors are ideal for use in UHF attenuators, if the correct values are available.

Resistive attenuators built as described should work from dc up to the frequency at which impedance mismatch and/or reactances become significant. It follows that attenuators can be characterized at dc, and their RF performance can be inferred from this. To check the impedance match, the input and output dc resistance to ground should be 50Ω. Attenuation can be calculated as follows: Terminate one port of the attenuator with a 50-Ω load. Apply a small dc voltage to the unterminated attenuator port, and measure the dc voltage at the terminated port. Attenuation is given by:

$$A\ (dB) = 20 \log (V_{in} / V_{out}) \qquad (Eq\ 1)$$

Remember that this is only a dc check—how these values will change at RF depends on the factors described earlier. For accurate RF attenuation measurements, more complex testing is required. Alternatively, attenuators can be calibrated by comparison with attenuators of known accuracy.

Attenuator Uses

A typical setup for measuring preamplifier gain using attenuators is shown in Fig 4. A signal source is fed into a converter and the output of the converter fed into an IF receiver, typically at 144 MHz or below. The signal level at the IF receiver (S-meter reading or audio output voltage) is then noted. The preamp is then put in line between the signal source and the converter and an attenuator is placed between the converter and IF receiver. The attenuator is then adjusted until the IF signal level duplicates that measured previously. At this point, the attenuator is canceling the preamplifier gain (assuming the converter is operating in its linear region). The preamp gain (in decibels) is then given by the total attenuation in decibels. Since the signal level at the IF receiver is the same in both cases, nonlinearities (such as AGC action) do not affect the result. An attenuator can also be placed ahead of a converter, if the attenuator is known to be accurate at a higher frequency.

Noise-figure (NF) comparisons can be made with the arrangement shown in Fig 5. A noise source is connected to a

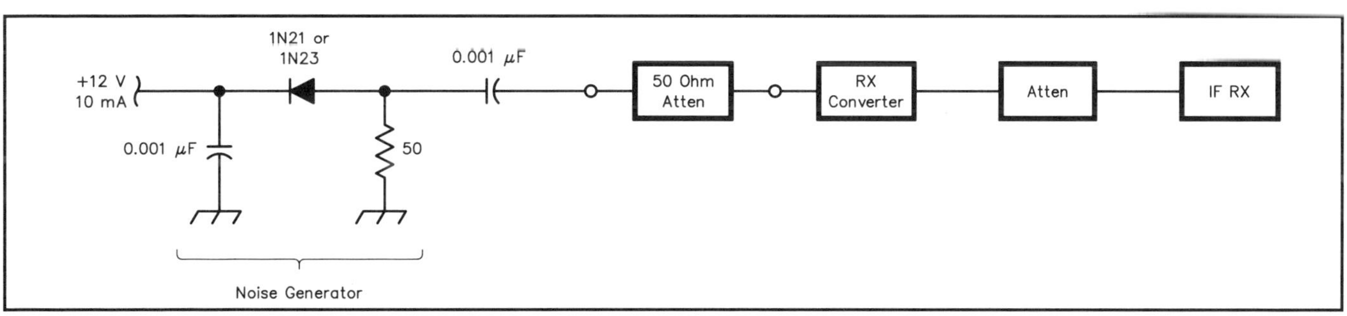

Fig 5—Receiver-noise measurement system. The attenuator immediately after the noise generator is for impedance-matching purposes. See text.

converter via an attenuator. The attenuator provides a 50-Ω match between the noise source and converter. The output of the converter is then fed to the IF receiver via a variable attenuator. The receiver output (S-meter reading or audio voltage) is noted with the generator off and the IF attenuator set to 0 dB. The noise generator is then turned on and the IF atten-uator adjusted for the same reading as previously noted. The amount of attenuation required is related to the converter noise figure—the more attenuation required, the better the NF. The front end of the converter may then be tuned to yield the best noise figure, or two different converters may be compared.

Simple diode noise generators are useful for NF comparison tests, but are not reproducible enough to make accurate measurements. If a stable, calibrated noise source is available, this method may be used to determine actual NF.[1]

Note

[1] Schetgen, ed., *The ARRL Handbook for Radio Amateurs*, 71st ed. (Newington: ARRL, 1993), p 12-3.

Thermistor Power Metering

By Bob Atkins, KA1GT

(From *QST*, Aug 1988)

When building microwave equipment, it is often desirable to measure the relatively low output power of local oscillators or mixers. One cheap, simple and nearly foolproof way of doing this is by using a thermistor power meter. A thermistor is a device with resistance that depends on temperature. The power meter shown in Fig 1 uses a thermistor to sense the change in temperature of a 50-Ω, ⅛-W resistor used as a load for the power being measured. The 50-Ω load resistor dissipates applied RF in the form of heat, causing the resistor temperature to rise. The thermistor is in contact with the resistor, secured by a spot of epoxy glue. As the temperature of the thermistor rises, its resistance changes (usually, it decreases). When the resistance reaches a steady value, the reading is noted and the RF power removed. A dc voltage is then applied to the resistor. The voltage is adjusted so that the thermistor resistance stabilizes at the same value as that obtained when the RF power was applied to the resistor. Under these conditions the dc power and RF power applied to the resistor are equal and given by:

$$P = V^2 / R \qquad \text{(Eq 1)}$$

A few points to note: The physically smaller the load resistor, the higher and more rapidly its temperature will rise for a given input power, hence, the greater the sensitivity of the measurement system. With a ⅛-W resistor, power levels of 10 mW can be easily measured. The smaller the thermistor bead and leads, the less heat it will absorb and the better it will track the resistor temperature.

There are, of course, some problems with this technique. It assumes that all of the power supplied by the generator is dissipated in the resistor. Any reflected power will not be measured, so a good match is required. Perhaps the best physical layout would be a 50-Ω microstrip terminated by a small 50-Ω chip resistor.

The temperature reached by the resistor/thermistor combination depends not only on input power, but also on ambient temperature. If the circuit is in an enclosed box, some drifting of readings may occur if there is too much heat buildup in the box. Because of this, it is better to measure higher power levels (more than 50 mW) using calibrated attenuators ahead of the

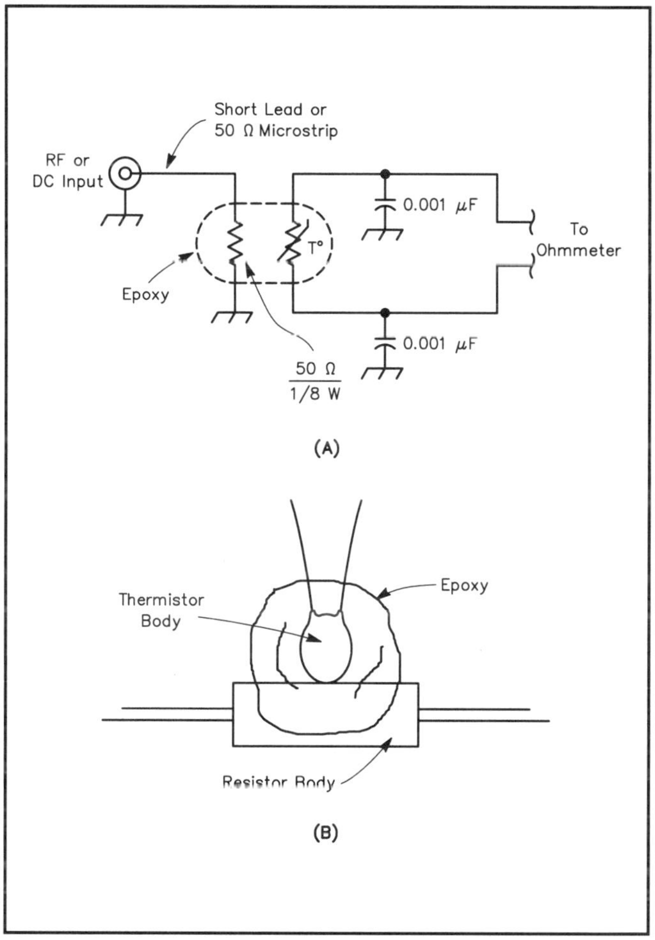

Fig 1—The schematic diagram of a thermistor power meter is shown at A. An ohmmeter is connected to the thermistor leads and resistance drop measured while power is applied to the load resistor, after the load temperature stabilizes. The resistance change of the thermistor from room temperature is then calculated. A dc signal that causes the same resistance change is then applied, and the power level calculated. The smaller the load resistor, the more precise the measurement can be. At B, the method of physical attachment of the thermistor to the load resistor is shown.

power meter. Finally, the system response time is slow (several tens of seconds, depending on the input power and thermal capacity of the system), so it cannot be used as an indicator to "peak" circuits. Nevertheless, I have found this type of instrument quite useful, and it can be constructed for only a few dollars.

Thermistor prices depend on the thermistor used (its stability, materials, construction and so on), but are typically in the $5 range. They may be obtained through local electronic parts distributors. One thermistor I have used is the Fenwell Electronics type 112-503JAJ-B01. Because it has a 0.040-inch bead with very thin lead wires, its thermal mass is small. Its resistance at room temperature is about 50 kΩ, and drops by about 7 kΩ when the device is in contact with a ⅛-W, 50-Ω resistor dissipating 10 mW. There are undoubtedly many other thermistor types and brands that would work equally well. A final point to remember is that this measurement technique measures the total power delivered to the resistor, including power at the desired frequency and that at all other frequencies (such as harmonics). This should not usually cause problems unless a particularly "dirty" source is being measured!

Zap Insurance—A Simple RF Detector for Operator Safety

By W. O. Troetschel, K6UQH

As higher levels of microwave power become available, amateurs must seriously consider a method of detecting dangerous levels of stray RF. The "Zap Insurance" detector (Fig 1), designed and constructed by Bill Troetschel, K6UQH, can help you locate sources of stray energy. When placed at head level, it provides a positive indication that your operating area is safe.

Circuit Details

The schematic diagram is shown in Fig 2. RF energy from a probe antenna is fed to a 50-ohm input circuit, consisting of R1 and R2. Germanium diode D1 is a simple RF detector. An RC filter follows, consisting of C1, C2, R3 and R4. The resulting dc is fed to M1 via potentiometer R5, which provides calibration adjustment. The frequency response of this simple circuit is extremely flat from 28 to 1296 MHz, and more than adequate up to 3456 MHz.

Fig 1—View of the K6UQH "Zap Insurance" detector, with a 1296-MHz probe antenna in place.

Construction

Fig 3 shows the parts layout. The entire circuit is built on a $\frac{1}{16} \times 1\frac{3}{4} \times 3$-inch piece of single- or double-sided glass-epoxy circuit board. A hole is drilled for the Type-N connector, which mounts on the underside of the circuit board (Fig 3C). A shim is placed between the connector flange and circuit board so that the ground "sleeve" on the connector is flush with the parts side of the board. Four screws are used to hold the connector in place, but a pair of thin brass shims must also be positioned through the circuit board. These should be soldered to both ground planes (or to one ground plane and the

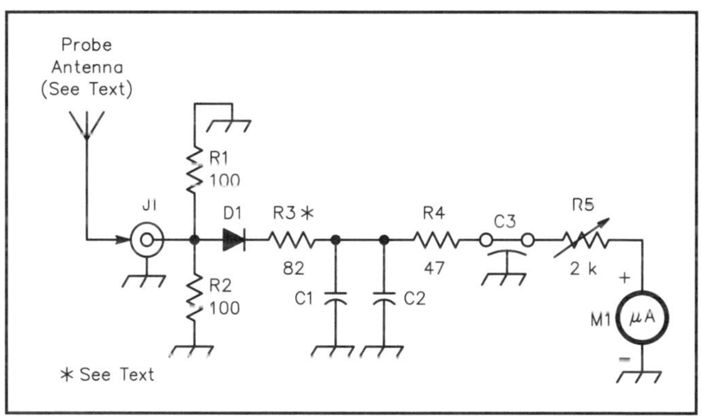

Fig 2—Schematic diagram of the zap detector. All resistors are ¼-W, carbon film.

C1—15- to 27-pF chip capacitor.
C2—560- to 910-pF chip capacitor.
C3—1000-pF feedthrough capacitor.
D1—1N82AG-C (International Rectifier Co).
J1—Chassis-mount female Type-N connector.
M1—0-200-μA panel meter.
R5—Miniature trimmer potentiometer.

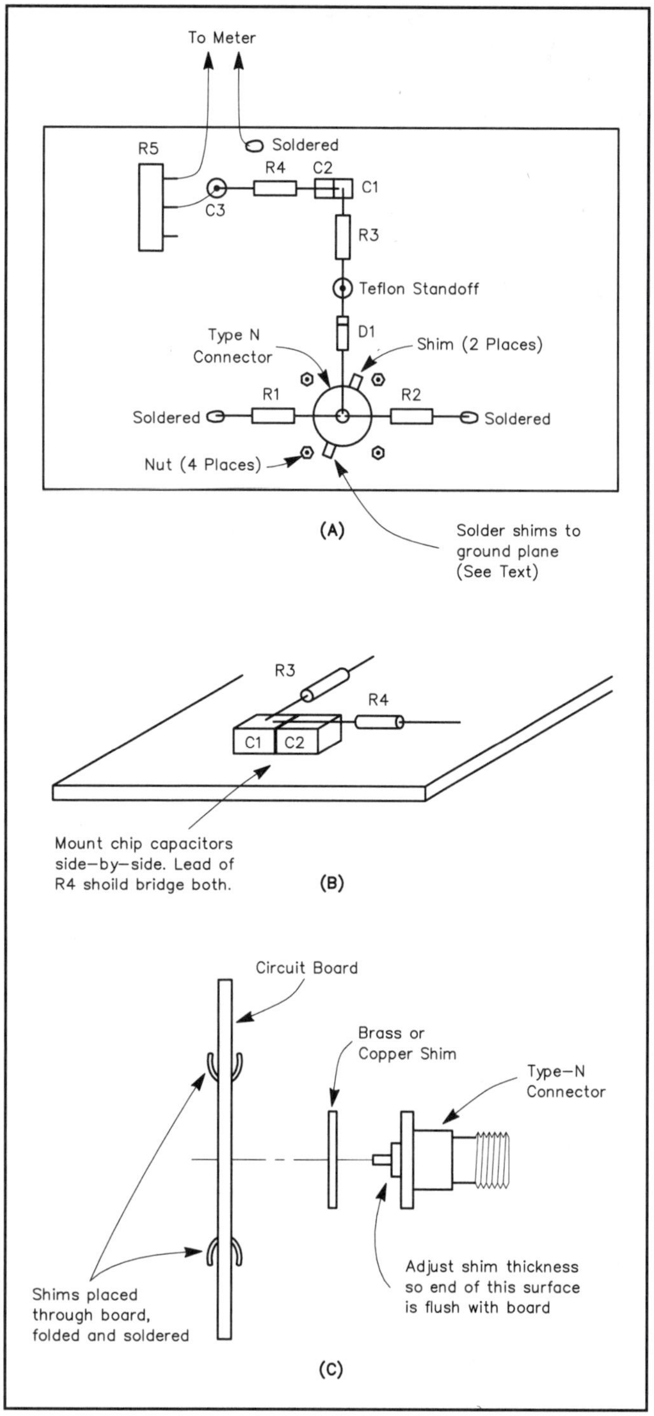

Fig 3—Top view of the circuit board (A). Part placement is not critical, with the exception of C1 and C2 (B). All leads should be kept as short as possible. Connector mounting detail is shown in C.

connector housing if single-sided board is used) to provide a long-lasting RF connection. Holes are also drilled for mounting the feedthrough capacitor and Teflon standoff insulator. The latter hole should be small enough so that the standoff is a press fit.

Keep all component leads as short as possible to prevent unwanted resonances. In addition, pay particular attention to the mounting of C1 and C2 (Fig 3B). The circuit-board assembly and meter are mounted together in a 4 × 2¾ × 2-inch LMB chassis box. A hole should be made in the box to provide access to the calibration trimmer potentiometer.

Calibration

The detector is calibrated by applying a 2-mW, 1296-MHz signal to the Type-N connector. Adjust R5 for a full-scale meter deflection. Reduce the drive level to 1 mW and note the reading. Using this point as a reference, apply a 1-mW, 28-MHz signal and check this reading against the reference. Should they coincide, no further adjustments are necessary for the range of 28 to 1296 MHz. If the 28-MHz reading is less than the 1296-MHz reading, increase the value of R3. If the 28-MHz reading is greater than the 1296-MHz reading, decrease the value of R3. The proper value of R3 is both diode and load dependent, and may vary from approximately 68 to 120 ohms. Whenever R3 is adjusted, always return to the beginning of the calibration procedure.

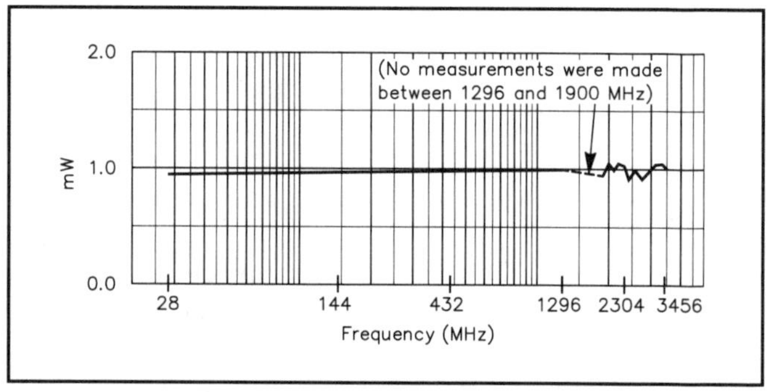

Fig 4—Relative response of the zap detector with 1 mW of RF applied.

Fig 4 shows the measured response of the unit from 28 to 3456 MHz. Notice the minor resonances occurring above 2000 MHz. These are minimized by a secondary calibration procedure. The object is to "tune" out stray inductance and capacitance by adjusting the height of D1 and R3 above the ground plane. With the proper equipment and a little patience, it's possible to extend the useful range of this instrument to well above 1296 MHz, as shown in the graph. The only possible exception is if a substitute diode is used for the 1N82AG-C. Many of the modern-day substitutes are silicon, rather than germanium. If a silicon diode is used, varying the value of R3 and R5 should allow operation up to 1296 MHz, but no guarantee of higher-frequency operation is given. Differences in diode junction capacitance and "catwhisker" inductance may produce unwanted resonances that can't be tuned out.

When calibration is completed, you can change the meter scale to indicate 2 mW full-scale. Alternatively, it can be calibrated in microwatts, as shown in Fig 1.

Operation

Because the zap detector is a simple device, you should not expect it to be as accurate as a true power-density meter. However, when checked against a commercial meter of this type, the accuracy was found to be surprisingly good. The zap detector's meter is a bit generous, which provides an additional, "built-in" safety factor.

Operation is simple. A piece of heavy bus wire is cut to a quarter wavelength on the band you wish to monitor, and inserted in the Type-N connector. A quick check can be made by inserting a 2304-MHz antenna and placing the unit near the door of an operating microwave oven. In most cases, the meter should read about 500 microwatts. (If it reads higher, have your oven checked for radiation leaks!) The detector should be placed near head level in the shack as an RF "sniffer" during transmission.

A Milliwattmeter for HF to 1296 MHz

By W. O. Troetschel, K6UQH

One of the most exasperating problems many hams face is accurately measuring power levels from HF to 1296 MHz, and knowing how the power readings taken at different frequencies relate to one another.

This device gets around those problems, and thus provides a method of measuring filter insertion loss, amplifier stage gain, oscillator/multiplier output power, and of course, serves as a field strength meter, etc.

The version shown in Fig 1 measures power in two valuable ranges: 0 to 10 mW and 0 to 100 mW. It also can be constructed with a single power scale, and either version can be easily extended to higher power ranges by appropriate external attenuators, directional couplers, or a combination of the two.

Different types of power meters have their advantages and disadvantages. The Milliwattmeter offers portability, quick power measurements, and ease of construction. It also measures everything from HF to 1296 MHz and up! Thus, accurate measurements require clean signals.

This problem shows up when measuring the output from an oscillator/multiplier chain. You might read 100 mW initially, and 40 mW after you install a filter on the output. Ignoring filter insertion loss, the difference represents power at other "unwanted" frequencies! However, once you know this, it becomes a manageable problem, and is in itself quite revealing.

In Fig 2, the detector's frequency response is compared to that of a commercial HP-423A detector head (for use up to 12.4 GHz). If you can get your hands on one of these—or its equivalent—to calibrate your meter scale, you'll really have a piece of equipment!

To build an accurate Milliwattmeter you must be willing to make your own meter scales. By doing this, you ease the problem of finding a suitable meter, and increase the accuracy of the unit (equal to your calibration standard). Buy or borrow some drafting pens and India ink, locate some white paper or cardboard that doesn't soak up the ink like a blotter, and "have at it." It's really not that difficult after the first dozen tries! Rub-on transfers complete the job.

In the dc and digital world, diodes are commonly viewed as simple-minded on/off switches. In the RF world, diodes are

Fig 1—The K6UQH Milliwattmeter shown here is accurate and easy to build. Note the dual-range homemade meter face.

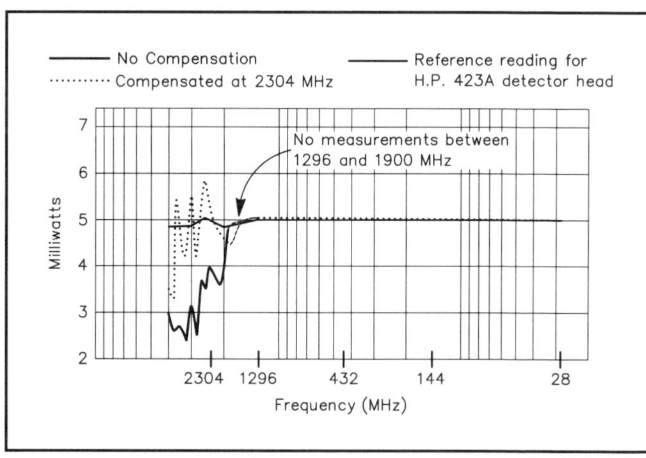

Fig 2—A comparison of the K6UQH Milliwattmeter versus the Hewlett-Packard HP423A. The response of both units is similar up to 1296 MHz.

Fig 3—The K6UQH Milliwattmeter can be used with two different detector heads, labeled 1 and 2 in the photo. See text.

anything but simpleminded. In fact, they're quite subtle in terms of behavioral characteristics. The "physics" of the junction type and the diode's physical construction must be considered for RF applications. The junction capacitance and the lead inductance can create unwanted resonances at microwave frequencies. This is one reason why the wattmeter's upper usable limit is set at 1296 MHz (using the hot-carrier diode). Although the input circuit represents a good impedance match at 1296 MHz—it's not perfect, and the input SWR starts to act up in the 2- to 4-GHz region.

Fig 3 shows two different "detector heads." Detector head no. 1 is made with the most readily available parts. Note the circuit-board shim used to keep the Type-N fitting flush with the parts side of the single-sided, 1⁄16-inch circuit board. Also note the "carry-through" shim strips from the ground plane to the Type-N connector.

The basic Milliwattmeter can also be used to measure levels of stray RF in your shack (up to 1296 MHz). If you see levels greater than 5 mW, you'd better shut down and add improved shielding to your gear, or relocate your antenna to ensure your safety.

Before we get into the nuts and bolts, there are a few timesavers worth mentioning. The internal resistance and current rating of the meter will affect the value of the calibration resistor. Make sure the calibration pot has enough range to accommodate the meter.

Flat-plate capacitors have to be flat—otherwise, regardless of the dielectric material, the measured value of the capacitor will be less than its calculated value. The flat-plate capacitor was around long before the ceramic chip type became available, and still represents a good, low-inductance capacitor.

The 100-Ω input resistors should be mounted flat against the ground plane, with a minimum lead length. This provides a wide-band 50-Ω termination.

Be sure the input resistors are the carbon-film type—not the standard composition type.

Table 1
Detector Head Construction

Detector Head No. 1	*Detector Head No. 2*
Base plate—1⁄16-in. single-sided glass-epoxy circuit board. Size = 3 × 3-in.	Base plate—Brass plate, 0.05-in.thick. Size = 3 × 2¼-in.
Capacitor plate—Brass plate, 0.02- to 0.032-in. Size = 1 × 2-in.	Capacitor plate—Brass plate, 0.02- to 0.032-in.-thick. Size = 1.2 × 2.3-in.
Input fitting—Type-N.	Input fitting—Type N.

Construction

To build the Milliwattmeter you'll need standard UHF construction tools. Table 1 lists the materials used for the detector heads. Fig 4 shows a typical parts layout for the detector head, and Fig 5 shows the schematic diagram.

The detector heads are designed to provide a flat plate capacitance value of about 800 to 1100 pF, depending upon the type and thickness of your mica stock. In any case, a value of 600 pF or greater seems to work well.

The resistor in series with the diode is quite important. For the hot-carrier diode, its value will range from 39 to 82 Ω. Its effect is most noticeable at the low frequency calibration point.

This resistor will also interact with the calibration resistors, so if you change its value, start the calibration process over. When you find the right resistance value for your meter's diode, the frequency response will be acceptably flat from 28 to 1296 MHz.

Calibration

First, you'll need a power reference standard—usually

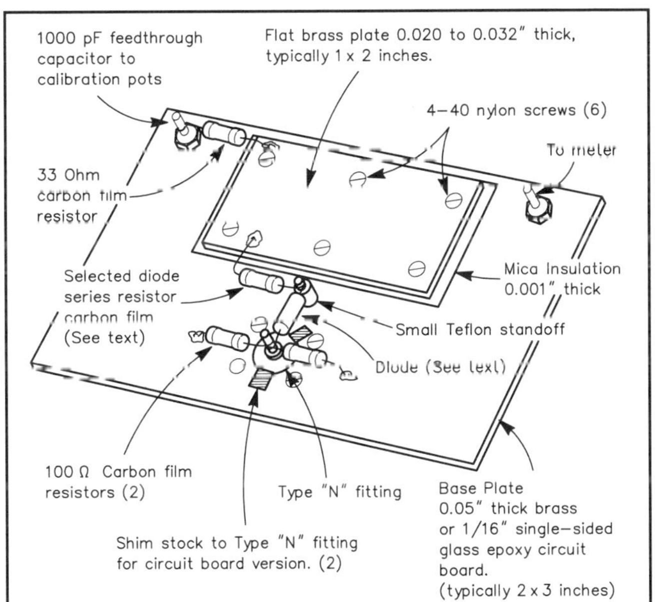

Fig 4—Construction details of a typical detector head. See text and Fig 3.

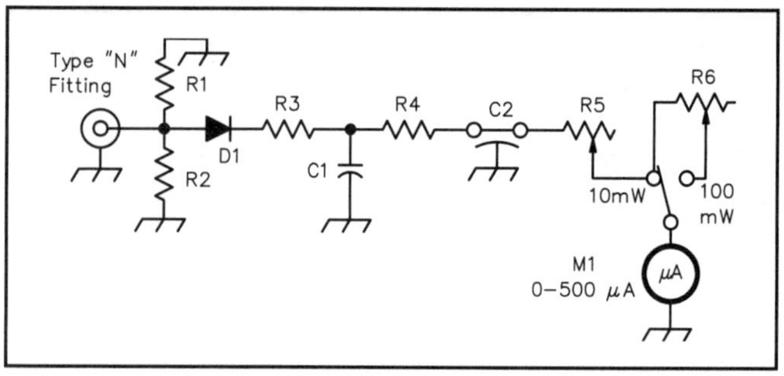

Fig 5—Schematic diagram of the Milliwattmeter. Fixed resistors are carbon film, not carbon composition.

R1, R2—100-Ω.

R3—Selected carbon-film resistor (see text).

R5—5- to 10-kΩ calibration potentiometer (for 10-mW range).

R6—25-kΩ calibration potentiometer for 100-mW range.

D1—HP 5082-2835.

C1—Flat-plate capacitor, approximately 800 pF. See text.

C2—100-pF feedthrough capacitor.

available at surplus stores, flea markets or from a friend. This meter was calibrated at 1296 MHz using an HP-430C power meter. For the dual-range meter, calibrate the 10-mW range first, since its calibration trimmer is in series with the trimmer used for the 100-mW range. All measurements of the detector response are made using a 5-mW variable frequency source, referenced to the meter scale reading.

The frequency sources can be your station's exciters, properly "padded," to provide a 5-mW signal to your power standard (and then switched to your meter). Be sure to use a clean signal for this purpose—use high-Q filters if possible.

A PRD Type S712A Signal Source was used to measure the frequency response from 1900 to 4000 MHz.

Comments

Schottky-barrier, hot-carrier diodes were selected for the Milliwattmeter because of their consistent characteristics from HF to 1296 MHz. Similar results can be obtained using germanium point-contact diodes such as the 1N82AG-C.

Between 1900 and 4000 MHz, the diode junction capacitance, the inductance of the "catwhisker," the load presented to the diode, and the input circuit all have considerable effect on stray resonances that can occur at these high frequencies. If you have the equipment and the patience, you can take advantage of these characteristics to obtain a useful detector above 1296 MHz. A sweep frequency generator will make this job much easier.

The response curve of the detectors above 1296 MHz will vary considerably, depending upon which frequency the meter is optimized for. With no adjustment, the detector response typically falls off rapidly above 1296 MHz, and then rises between 3.5 to 4 GHz (the limit of my measurements). If you choose to make the adjustments, the hot-carrier diode needs more stray capacitance, and the point-contact diode needs less stray capacitance to compensate the detector for the frequency you select.

This is accomplished by moving a capacitive "tab" near the diode series resistor.

By using these adjustment tricks, it's possible to obtain a properly-calibrated response at 2304 or 3456 MHz.

Surplus stores and flea markets are a source of many hard-to-get parts. I use a Micromatch 40-dB throughline directional coupler (rated at 400 watts at 1296 MHz) in conjunction with a low-power, 10-dB attenuator (both flea-market purchases) with the Milliwattmeter to monitor the output power of my 1296-MHz amplifier (typically 400 watts).

1296 Power and SWR Indicator

By Bob Atkins, KA1GT

(From *QST*, Nov 1980)

Many power meters used on the lower bands are unsuitable for use on 1296 MHz. A constant (50-Ω) impedance must be maintained in the transmission line to make meaningful SWR and power measurements. The stripline forward and reverse power indicator shown in Figs 1 and 2 will enable you to make such measurements at low cost. Since all the lines on the PC board are 50-Ω microstrip lines with a width of 2.6 mm, they can be laid out using standard ¹⁄₁₀-in. wide (2.54 mm) PC drafting tape as the etch-resistant material, with very little resultant error. This is somewhat easier for most constructors than the use of photoresist etching techniques. You can calibrate this instrument by comparison with a power meter of known accuracy at 1296 MHz, if one is available. The unit may be used as an indicator for tuning up a low-power transmitter (maximum forward power) or for tuning an antenna (minimum reflected power).

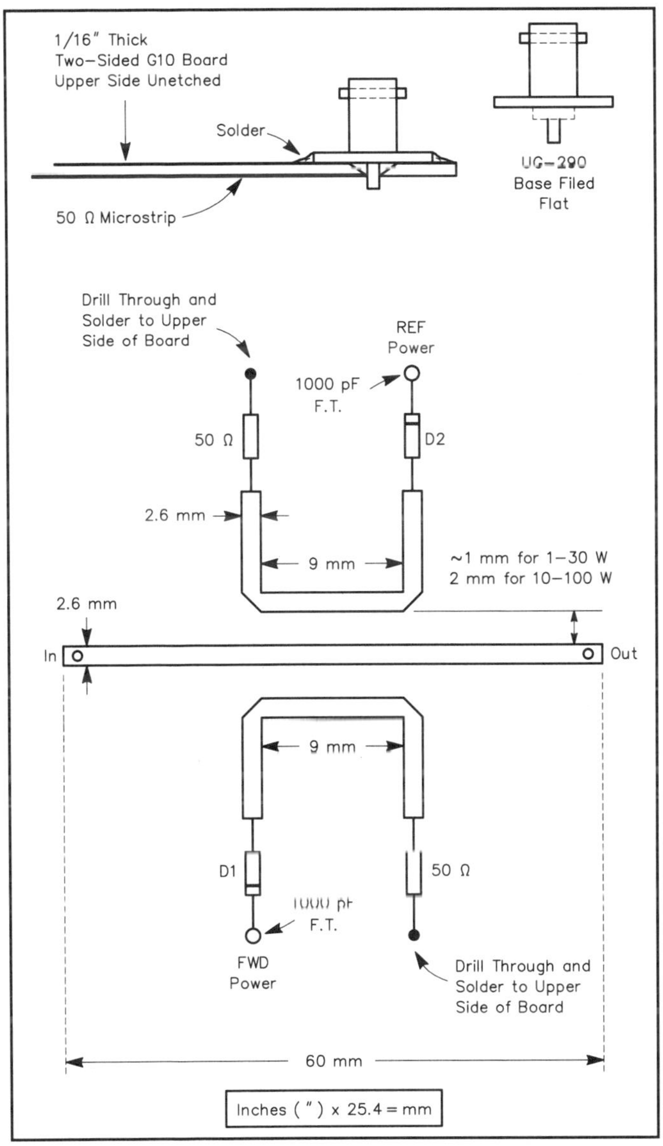

Fig 1— Configuration and design data for a 1296-MHz power and SWR indicator.

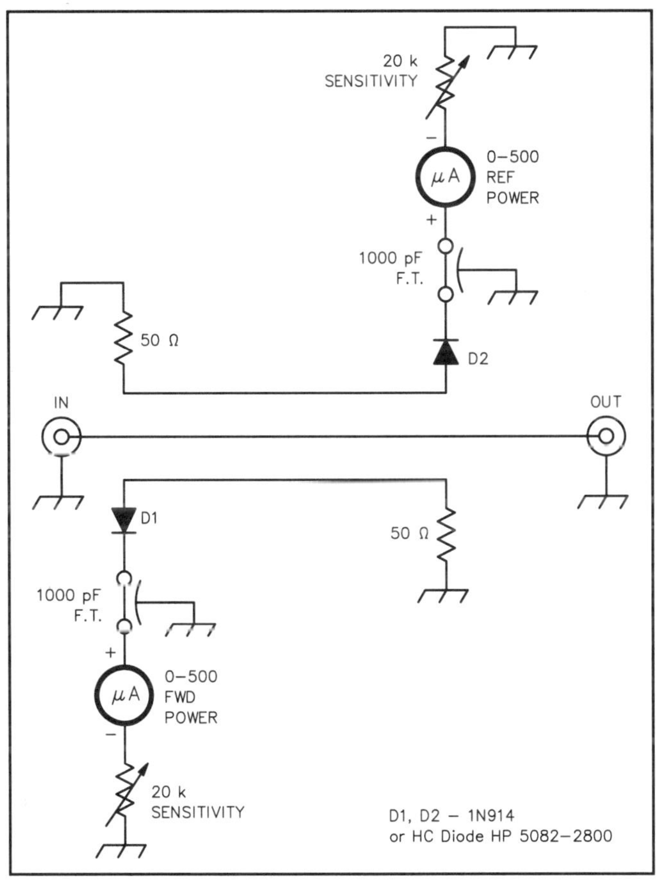

Fig 2— Schematic diagram of a power and SWR meter suitable for use on 1296 MHz.

5.7-GHz Waveguide Probes

By James D. Green, K5JG

A wealth of mathematical and practical data is available on coax-to-waveguide transitions. These usually take the form of a shorted section of waveguide with a coaxial connector and probe. Very little information is available on the insertion of a probe in the main transmission guide, even though they are quite often needed for sampling without disturbing the transmission system. A simple means of sampling the output of your transmitter or microwave test gear to measure frequency or relative power without interfering with the signal being transmitted, would certainly be valuable. The probe arrangements described here will accomplish this handily.

With the trend toward miniaturization, the SMA-type connector was chosen for coaxial connection to the waveguide probe. This offers several advantages; particularly, it presents only a very small obstacle in the main guide. The specific connector used here is an E. F. Johnson type JCM-142-0294-001. The JCM connector is a semi-precision miniature (3-mm) connector fully compatible with all SMA types. When used with semi-rigid coax its performance is satisfactory well beyond 6 GHz. Even though it is gold plated, the price is reasonable. As with most connectors of this type it has a characteristic impedance of 50 Ω.

The object of this project was to accumulate sufficient data that the average amateur microwave experimenter could sample energy from any WR-137 waveguide without the usual expensive "cut and try" approach. The curves shown in Fig 1 were derived by first assembling three short-flanged sections of WR-137 waveguide. Holes to accommodate the coax connector were drilled in each as per spacing indicated. The connectors were measured and averaged before mounting. These measurements are shown in Fig 2. The probe length referred to here is measured from the flat nut portion of the connector to the tip that coax would normally be sol-

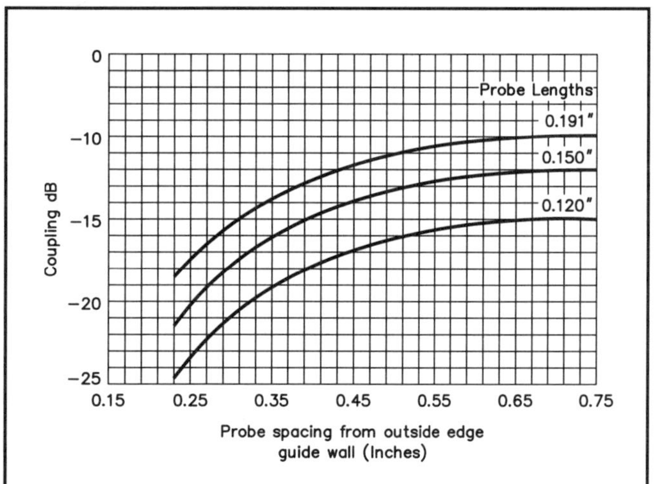

Fig 1—Graph of probe coupling for various spacings and probe lengths.

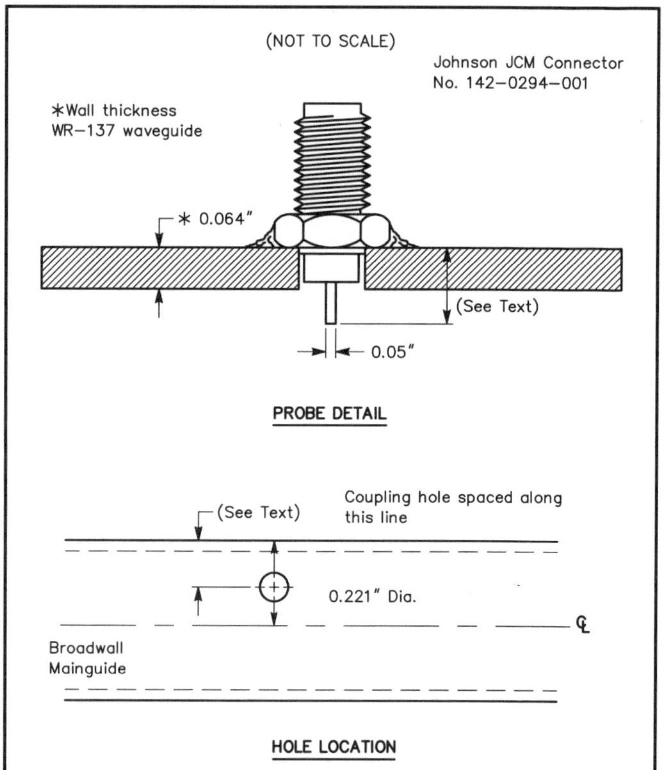

Fig 2—Probe construction and mounting.

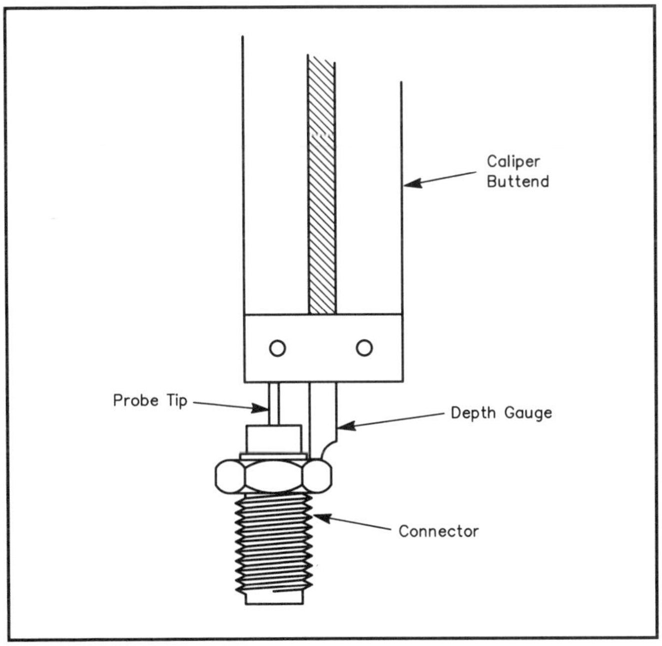

Fig 3—Probe length measurement.

dered (see Fig 3). This was done because of ease in measuring this distance with the depth measuring facility of a standard caliper. Each connector was lightly soldered into position with the probe (tip) length as they came from the factory.

Measurements were made on each section for coupling, frequency response of each probe and return loss and frequency response through the main guide. A standard sweep-reflectometer setup was used to conduct these measurements. Connectors were removed and probe lengths shortened successively and measurements conducted for each length and spacing. The results of these measurements are summarized in the curves shown in Fig 1. From these you can decide the spacing and probe length for the coupling desired. For example, if –20 dB coupling is desired, the curves indicate that a spacing of 0.32 in. from the main guide side wall and a probe length of 0.120 in. would be satisfactory. It can also be seen that a probe spacing of 0.25 in. with a probe length of 0.150 in. would also supply –20 dB of coupling.

The probe protrusion into the main guide has much less effect on transmission in the main guide than might be expected. This is mainly due to its small size. A Type-N connector, for example, would have a much greater effect for the same degree of coupling. The worst case is where the longest probe length is used and mounted in the center of the guide (0.75 in. spacing). This effects only a 24-dB return loss across the band. This is equivalent to approximately 1.14:1 SWR. As the probe is shortened and moved toward the side wall, the return loss gets even better. Frequency response across the band is rather good, exhibiting no more than 1.0-dB variation regardless of the combination used.

Construction

Location of the probe in line with the long dimension of the main guide is unimportant and may be placed at the most convenient point. After this location is determined measure and scribe the location of the probe along the perpendicular axis by measuring with a caliper from the outside wall toward the guide center. This hole should be marked with a center drill and then drilled with a no. 2 (0.221-in. diam.) drill. Clear the hole of any burrs.

If some dimension other than the normal length of the connector is desired, file the tip carefully. Hold the connector with the threaded portion down and probe tip up. Seat the butt end of the caliper on the probe tip. Run the sliding arm of the caliper depth gauge down until it barely touches the flat side of the nut-shaped portion of the connector (Fig 3). This will give you the probe length. File only a small portion of the probe at one time and remeasure. The tip is small, delicate and easy to over file. When the probe length is correct the connector may be soldered into the main guide. Use only soft solder and do not apply heat directly to the connector body. Heat applied to the waveguide will be conducted sufficiently to the connector for good solder flow. Clean up any flux residue and the probe is ready for use.

Conclusions

The probe provides one of the simplest methods of allowing one to check the frequency or power of a transmitter while it is in operation without disturbing the transmitted signal. Determine what level RF is required. By using the data supplied here a simple probe can be installed that will do the job without the usual "cut and try" approach. If it is desired to use more than one probe in the main guide this should not present any problem.

Waveguide Loads

By Kent Britain, WA5VJB

(From *Proceedings of Microwave Update '89*)

Simple, quick, and common stuff: that was the idea. I went through about two dozen designs and these are the two that worked best. The absorber is that black foam stuff you store ICs in. The foam conductivity varies quite a bit. Put your ohmmeter probes on the foam, about an inch apart. *H*ighly conductive (approx 10 kΩ), *M*edium (approx 50 kΩ and *L*ow/Static Dissipative (>200 kΩ). This carbon-loaded foam usually comes in thin sheets, so you may have to build the load up with several layers.

My commercial loads showed from 32- to 45-dB return loss so these foam versions worked pretty well. For the purist, Eccosorb worked best, but isn't commonly available. Finally, the lowest SWR occurs if you leave the back end of the waveguide open, but there will be some leakage.

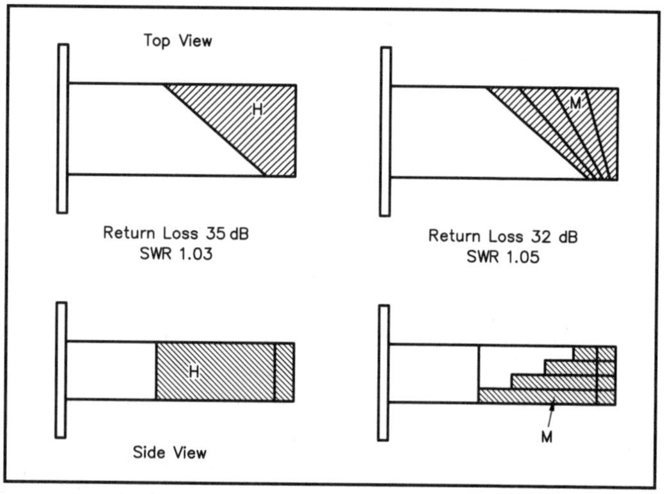

Fig 1—Various methods of making waveguide loads from conductive foam. H = Highly conductive, M = Medium conductive, L = Low conductive (see text).

A Single-Knob, Single-Crystal VHF/Microwave Calibrator

By Bill Troetschel, K6UQH

This unit, shown in Fig 1, provides a highly accurate set of RF signals from VLF to the microwave spectrum—directly referenced to WWV. It can be used to calibrate and align HF or VHF receivers and provide accurate reference signals in the VHF-microwave spectrum—simultaneously.

Error sources that normally affect the accuracy of frequency measurements are typically:

- Basic receiver calibration and tracking
- Crystal oscillators in converters
- General confusion in terms of the myriad ways receivers and transmitters internally use "offset" oscillators for various modulation modes.
- The inability of the human ear to hear a true subaudible zero-beat signal.

The first three items generally represent "fixed" conditions, while the last illustrates the flaw in the general technique used to identify the others.

Nearly all amateur receivers have very little audio response below 100 Hz, and few people have good audio perception below 20 Hz. Therein lies the problem with obtaining an accurate zero beat.

Most hams use a transfer oscillator to zero-beat WWV, and then use its harmonic to zero beat at the desired receive frequency. A few hams have "error-free" frequency counters calibrated to WWV. The problem with either approach is that an audible WWV zero beat is generally a "dead space" on the dial, typically 30 Hz wide. Assuming you use a 10-MHz WWV signal to calibrate a 2304-MHz signal, a 20-Hz error at the fundamental means an error greater than 4.6 kHz at the harmonic. The visual method described in this article is much more accurate.

The Single-Knob, Single-Crystal Calibrator provides an easy-to-adjust signal with visual subaudible zero-beat indicator. (You watch the receiver S meter as the WWV signal and the calibrator signal combine in the receiver.)

Circuit Description

A block diagram of the calibrator, and one possible application are shown in Figs 2 and 3. The unit is shown schemati-

(A)

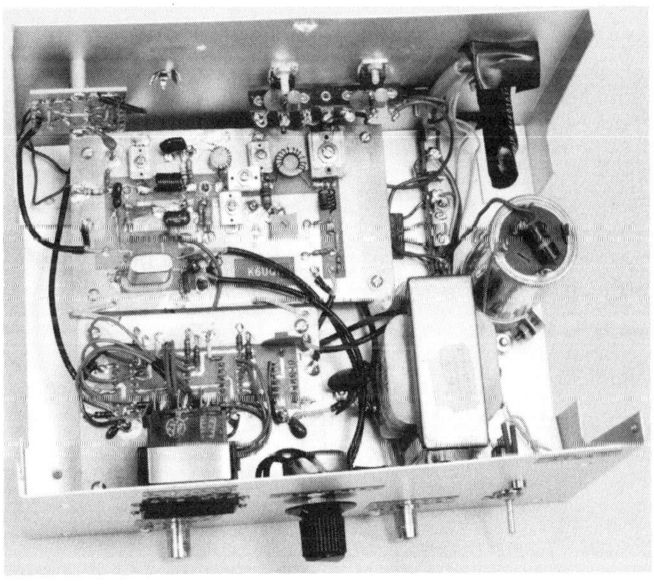

(B)

Fig 1—At A, a front-panel view of the Single-Knob, Single-Crystal Calibrator. At B, an interior view shows the placement of the three circuit boards and the power supply.

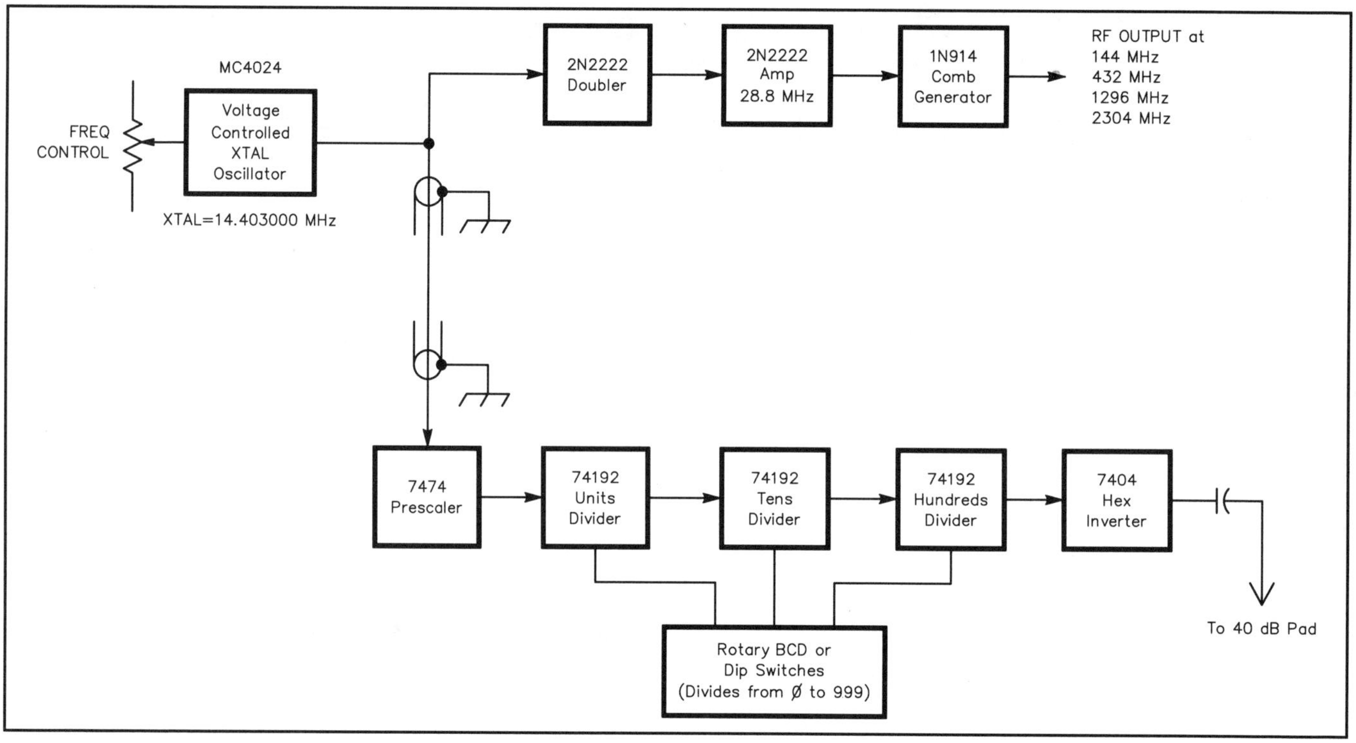

Fig 2—Block diagram of the calibrator.

cally in Fig 4, and a suitable power supply is shown in Fig 5. Fig 6 shows a 40-dB attenuator pad, depicted in the block diagram.

The heart of the calibrator is the Motorola MC4024 TTL chip, an astable, dual voltage-controlled multivibrator, operating as a crystal-controlled oscillator. The crystal is a Croven fundamental-mode, AT-cut, high-drive crystal cut for 14.403000 MHz. It provides a rock-solid output at 14.400000 MHz when tuned by the calibration potentiometer. This circuit is designed to oscillate far enough below the crystal frequency to provide mid-range tuning for the calibration potentiometer.

Two outputs are taken from the oscillator at 14.4 MHz. The first is doubled to 28.8 MHz and amplified by the 2N2222 stages. This signal is applied to a comb generator that provides the VHF/microwave output (at frequencies spaced 28.8 MHz apart). A 1N914 diode is used in the comb generator. A more-efficient step-recovery diode is discussed in the references, but the 1N914 produces useful outputs well above 2304 MHz.

The second output is prescaled by a factor of four by a 7474 dual type-D edge-triggered flip-flop. This stage is needed to reduce the divider's input frequency to compensate for the time delays incurred in the 74192 up/down counters, and provide an accurate reading on the BCD (binary-coded decimal) frequency division switches. A 7404 hex inverter buffers the output.

By selecting the proper division ratio on the BCD switches (described later), many VLF signals and their harmonics can be selected for calibrating your HF/VHF receiver, and for referencing the WWV signal available in your area. For example: 14.4 / 4 = 3.6 MHz. This 3.6-MHz signal, divided by 144 provides 25-kHz reference markers, plus the usual WWV markers at 2.5, 5, 10, 15 and 20 MHz. Other useful divider ratios are 36 (100 kHz), 72 (50 kHz) and 360 (10 kHz). To determine the harmonic interval of the marker generator, use the formula

3.6 MHz / 3-digit BCD display × 1000 = marker frequency (kHz)

The output of the 7404 hex inverter is quite strong, so a 40-dB attenuator attenuates the signal level to make its harmonics approximately equal in strength to the WWV signal.

Construction

Point-to-point wiring or a PC board are equally suitable for this circuit. I built mine on a PC board. PC board templates and part-placement diagrams are given in Figs 7 through 12. Perhaps the most tedious part of the job is wiring the BCD switches. Remember that the first divider is for units, the second for tens and the third for hundreds. When mounting the BCD switches on the front panel, be sure they read hundreds-tens-units. The BCD switch is wired as a 5-V common system, and *not* as a BCD complement (where the common is ground).

Construct the comb generator circuit carefully in terms of layout and lead dress. Remember, the RF output frequency can be as high as 10,368 MHz if you are careful.

Alignment

There is no question but that the best way to tune up a comb generator is with a spectrum analyzer. By using your receivers as signal-strength indicators, however, you can do an acceptable job. I assume you can find your frequency to within 14 MHz. Since the basic oscillator frequency is 14.4 MHz, there are several unwanted frequency spikes spaced 14.4 MHz apart in the comb, particularly at the lower frequen-

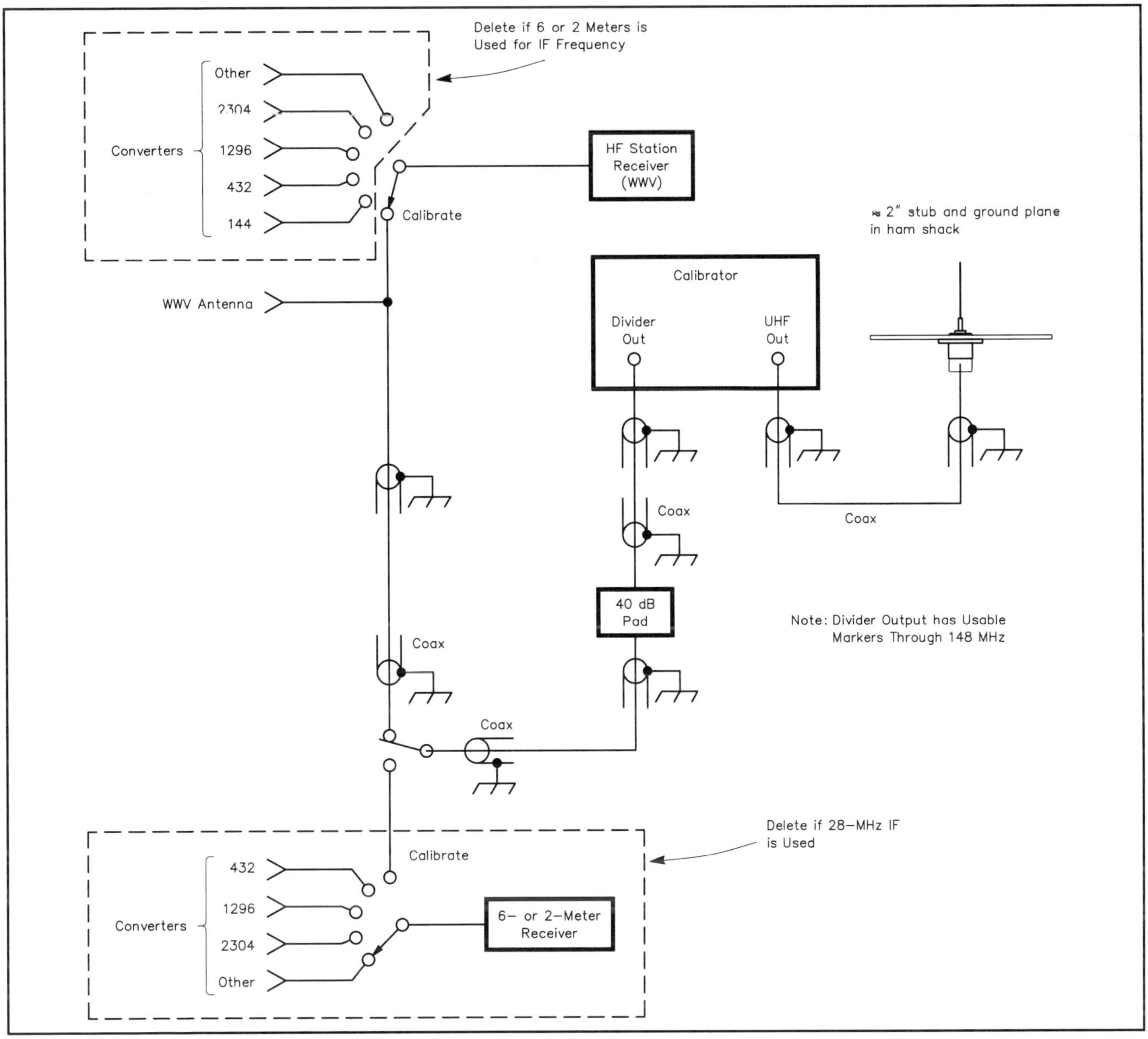

Fig 3—Here's one application of the calibrator. See text for more information.

cies. You can minimize their effects considerably by tuning with the aid of a spectrum analyzer, but the spurs shouldn't be a problem if you use high-quality microwave receivers as tune-up indicators.

The first step in alignment without a spectrum analyzer is to set the 1N914 bias pot to its maximum value. Monitor the voltage across the pot with a sensitive voltmeter, and adjust the doubler and amplifier stages for maximum voltage—approximately 2 to 3 V. Turn the bias pot to approximately zero resistance. At this point, connect a 2-in. stub antenna to the VHF/microwave output. You should be able to hear the calibration signal on your receivers. Now, use your station receivers, preferably the highest frequency receiver available, and adjust the other controls for maximum signal strength—and you're done.

If your wattmeter has a 10-mW range, an alternative tune-up procedure is useful. Set the diode bias pot as above, and tune the entire unit for a maximum power indication. Output should be between 3 and 5 mW, which represents the summation of all harmonics from that port. Set the bias pot to zero resistance. The power output should drop to about 2 mW, which indicates proper operation.

Using the Calibrator

To calibrate an HF receiver, first tune in WWV at any frequency on which reception is good. Then couple the output of the decade dividers to the receiver antenna input, through the 40-dB attenuator pad. I use a coaxial tee connector at the receiver antenna jack. Set the BCD switches to provide the desired output markers. Now, carefully tune the calibration potentiometer while watching the receiver S meter. When the needle starts to flop around, you know you're getting close to

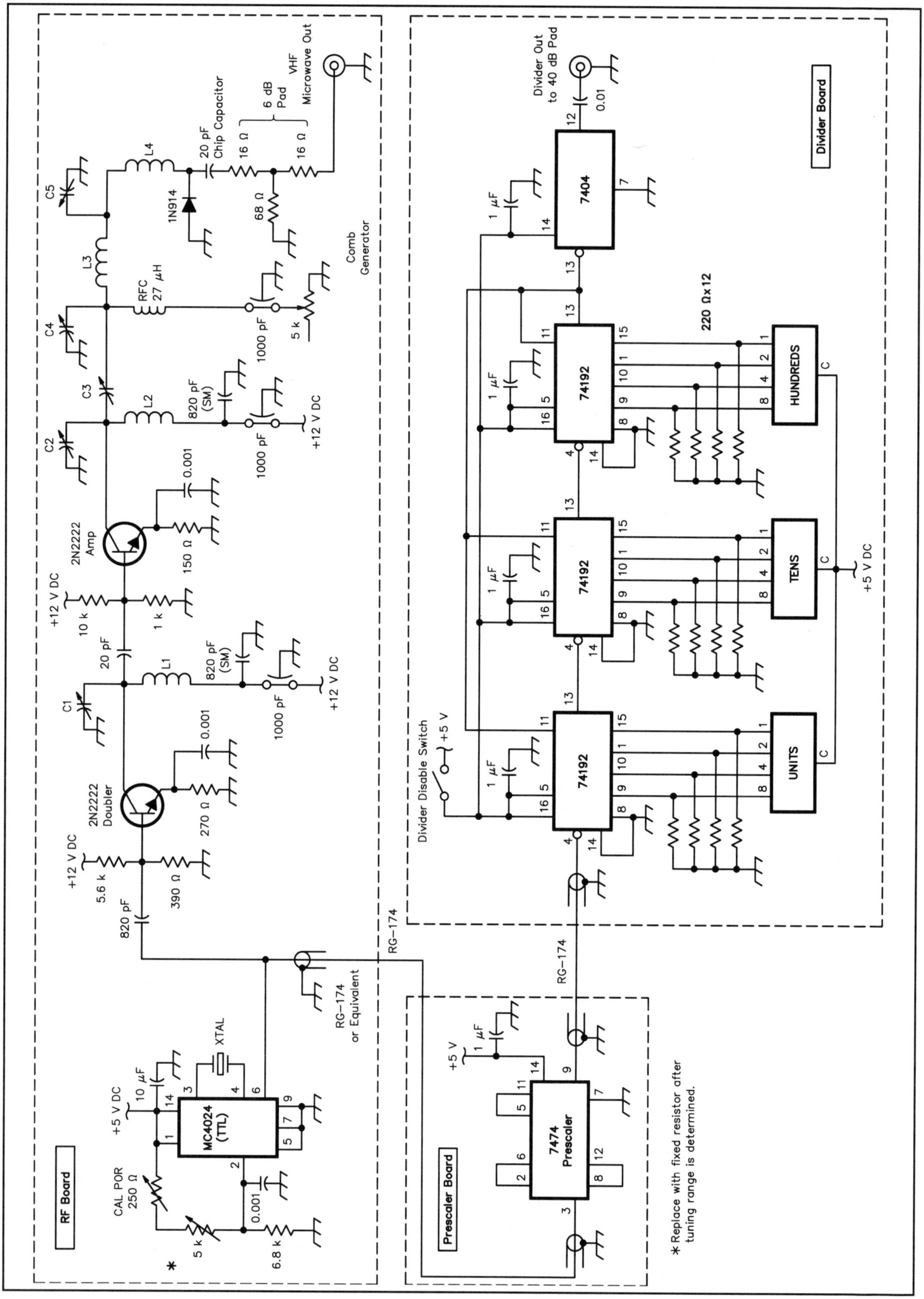
RF Board
Prescaler Board
Divider Board
MC4024 (TTL)
XTAL
CAL POR 250 Ω
5 k
6.8 k
0.001
10 µF
+5 V DC
RG–174 or Equivalent
RG–174
820 pF
5.6 k
390 Ω
+12 V DC
2N2222 Doubler
270 Ω
0.001
C1
L1
20 pF
820 pF (SM)
1000 pF
10 k
1 k
2N2222 Amp
150 Ω
C2
L2
C3
C4
RFC 27 µH
L3
C5
L4
1N914
20 pF Chip Capacitor
16 Ω
16 Ω
68 Ω
6 dB Pad
VHF Microwave Out
Comb Generator
7474 Prescaler
+5 V
1 µF
Divider Disable Switch
74192
UNITS
TENS
HUNDREDS
+5 V DC
220 Ω x12
7404
Divider Out to 40 dB Pad
0.01
* Replace with fixed resistor after tuning range is determined.

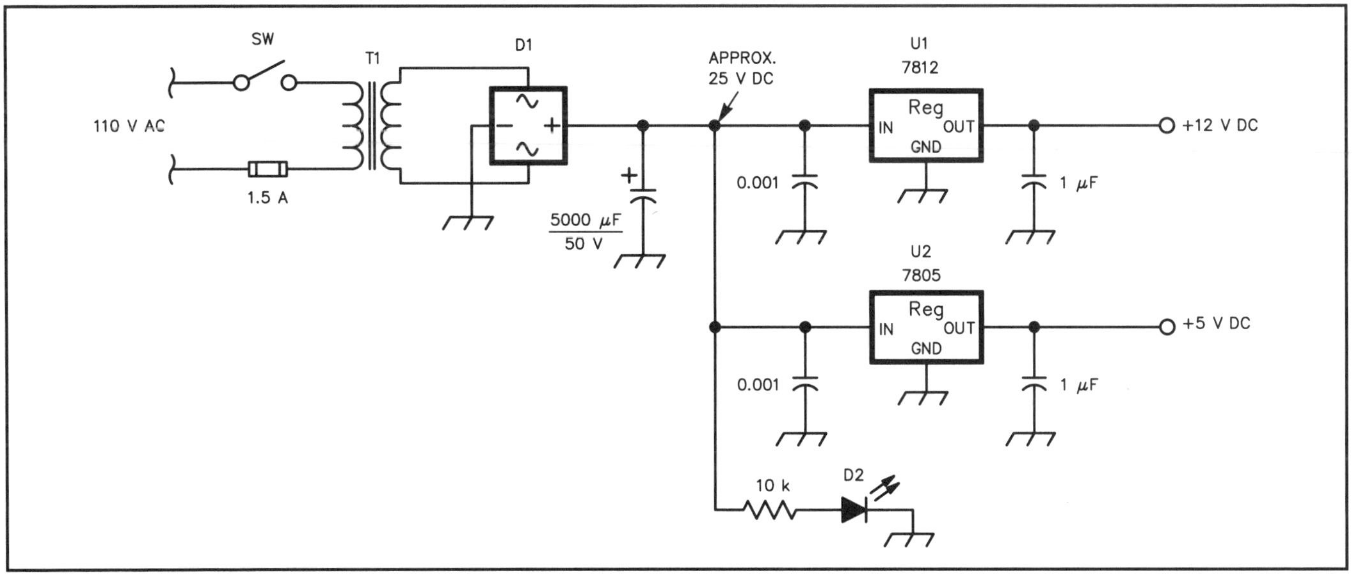

Fig 5—Schematic diagram of a power supply suitable for use with the calibrator.

D1—Bridge rectifier 100 PIV, 4 A.
D2—Red LED.
T1—Power transformer, 18-VAC, 2-A.
U1—Regulator IC, +12 V, 1 A, 7812 or equiv.
U2—Regulator IC, +5 V, 1 A, 7805 or equiv.

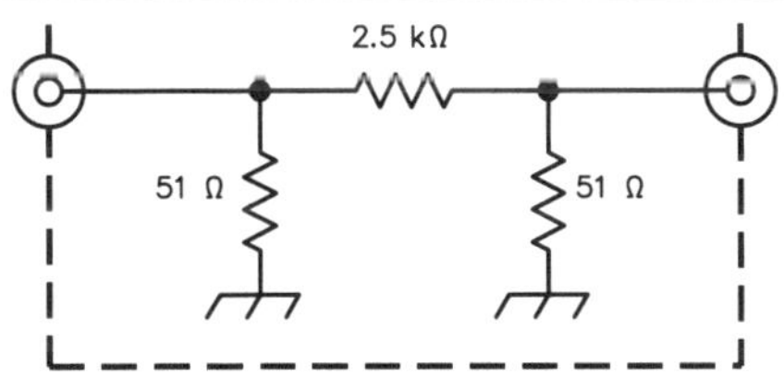

Fig 6—Schematic diagram of the 40-dB attenuator, used to make the calibrator signal strength equal to the received strength of WWV.

←

Fig 4—Schematic diagram of the calibrator.

C1, C2, C4—Trimmer capacitor, 4 to 40 pF.
C3—Trimmer capacitor, 1.5 to 20 pF.
C5—Trimmer capacitor, 30 to 450 pF.
L1—12 t no. 20, ¼-in. ID, air-wound, close-spaced.
L2—15 t no. 26 on T37-6 core.
L3—13 t no. 26 on T50-6 core.
L4—5 t no. 20, ¼-in. ID, air-wound, 0.35-in. long.

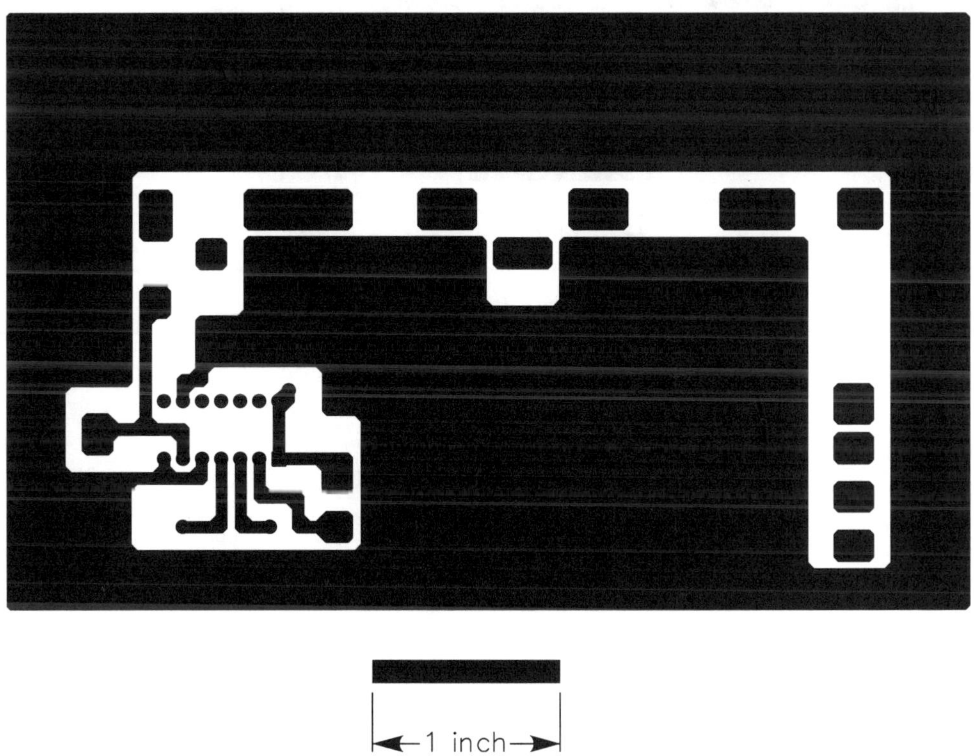

Fig 7—RF board template.

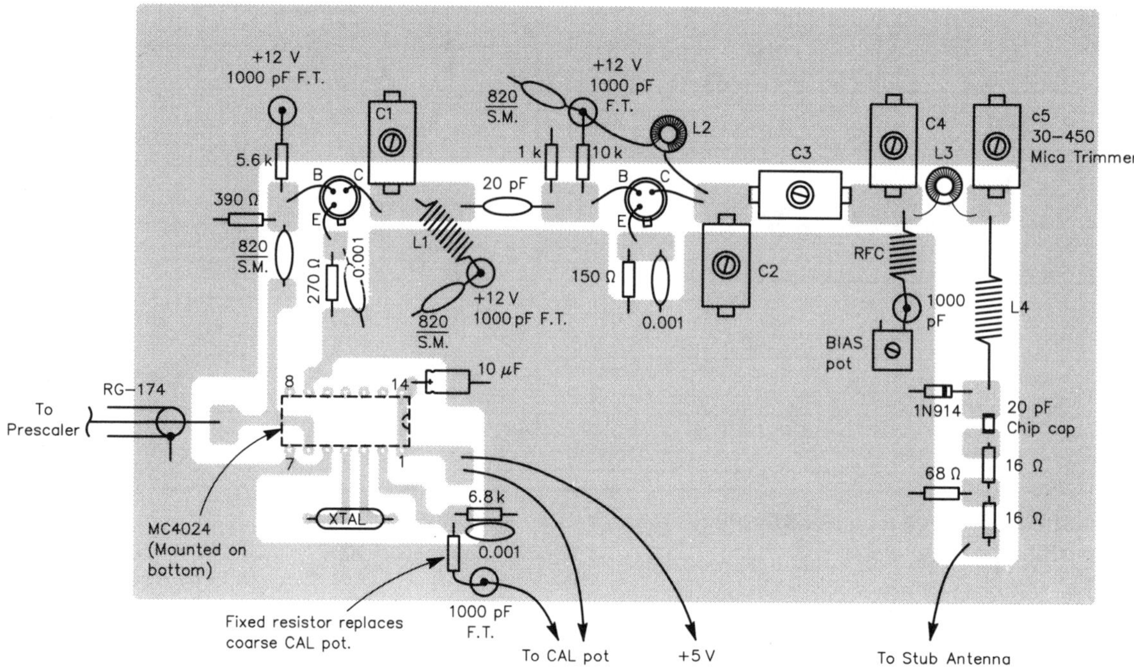

Fig 8—RF board part placement. All components *except* the IC are mounted on the foil side.

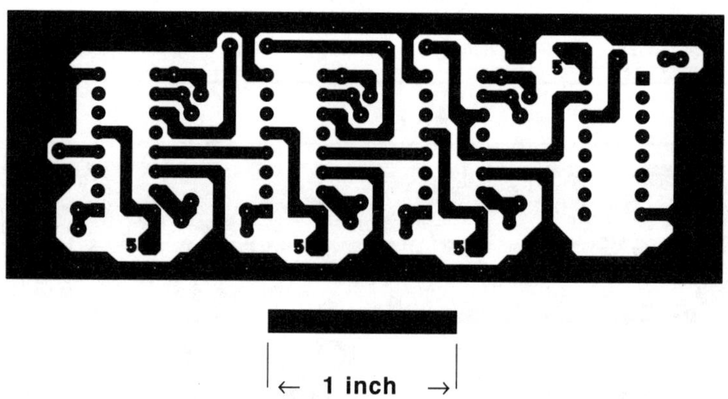

Fig 9—Decade-divider board template.

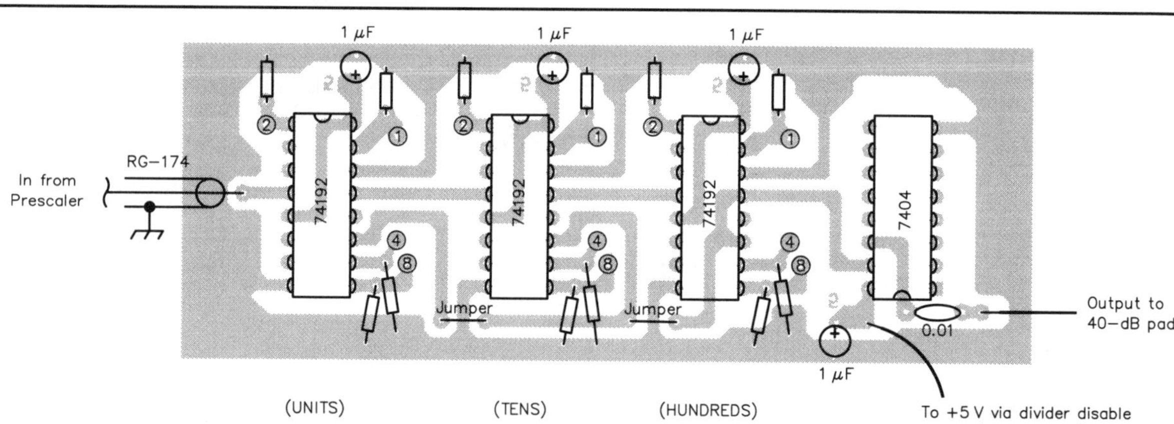

Fig 10—Decade-divider board part placement. All points labeled "5" are connected by jumpers.

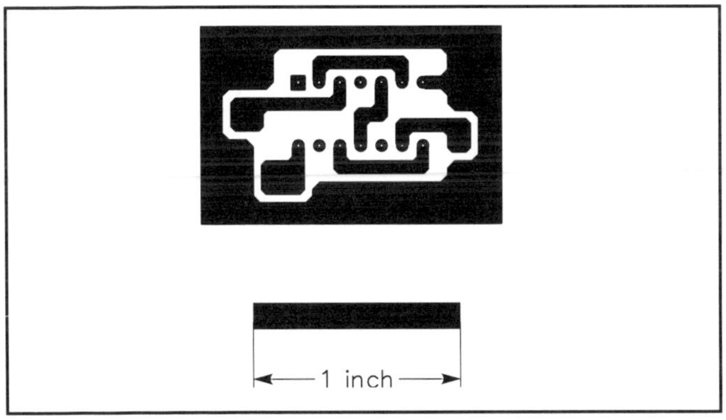

Fig 11—Prescaler board template.

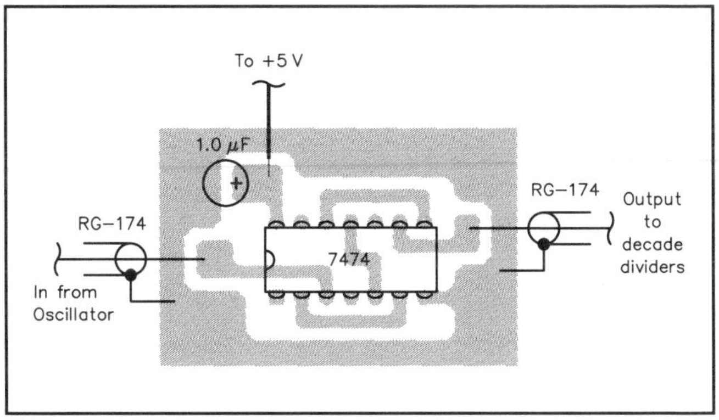

Fig 12—Prescaler board part placement.

zero beat. When the S meter fluctuates at 1 Hz or less, you've achieved true zero beat with WWV. You'll be able to hear the beating effects on your receiver as you adjust the calibration pot.

Once the oscillator has been calibrated against WWV, the output from its microwave port is also extremely accurate, and can be used as a frequency standard well into the microwave region. If you use the 20-MHz WWV signal and adjust the calibrator so the beat note is less than 1 Hz, the error at 1296 MHz is only about 65 Hz. You can get greater accuracy by adjusting the calibration pot until the S-meter pointer is nearly stationary for a long period, say 15 seconds. The divider on-off switch removes the 5-V supply from the dividers to eliminate the VLF output and its harmonics when only VHF/UHF output is desired.

If you use 6 or 2 meters as a tunable IF, first calibrate the unit against WWV, then use the divider output to check the calibration and tracking of your VHF receiver. Harmonics of the divider will show up as markers, up to at least 148 MHz.

Conclusion

Scheduled long-haul microwave tropo DX is made easier when you know you're on the right frequency. With GaAsFET amplifiers at the antenna, I receive the calibrator signal weakly on 1296 and 2304 MHz. For the higher frequencies you may need more direct coupling, unless you have terrific preamps. You'll find that the calibrator signal is remarkably stable—perhaps more so than some signals you are used to receiving!

Other crystal frequencies, prescaling integrals and more amplification prior to the comb generator will improve or extend the unit's capability to other bands, such as 50 or 222 MHz.

About the
American Radio Relay League

The seed for Amateur Radio was planted in the 1890s, when Guglielmo Marconi began his experiments in wireless telegraphy. Soon he was joined by dozens, then hundreds, of others who were enthusiastic about sending and receiving messages through the air—some with a commercial interest, but others solely out of a love for this new communications medium. The United States government began licensing Amateur Radio operators in 1912.

By 1914, there were thousands of Amateur Radio operators—hams—in the United States. Hiram Percy Maxim, a leading Hartford, Connecticut, inventor and industrialist saw the need for an organization to band together this fledgling group of radio experimenters. In May 1914 he founded the American Radio Relay League (ARRL) to meet that need.

Today ARRL, with more than 170,000 members, is the largest organization of radio amateurs in the United States. The League is a not-for-profit organization that:

- promotes interest in Amateur Radio communications and experimentation
- represents US radio amateurs in legislative matters, and
- maintains fraternalism and a high standard of conduct among Amateur Radio operators.

At League headquarters in the Hartford suburb of Newington, the staff helps serve the needs of members. ARRL is also International Secretariat for the International Amateur Radio Union, which is made up of similar societies in more than 100 countries around the world.

ARRL publishes the monthly journal *QST*, as well as newsletters and many publications covering all aspects of Amateur Radio. Its headquarters station, W1AW, transmits bulletins of interest to radio amateurs and Morse code practice sessions. The League also coordinates an extensive field organization, which includes volunteers who provide technical information for radio amateurs and public-service activities. ARRL also represents US amateurs with the Federal Communications Commission and other government agencies in the US and abroad

Membership in ARRL means much more than receiving *QST* each month. In addition to the services already described, ARRL offers membership services on a personal level, such as the ARRL Volunteer Examiner Coordinator Program and a QSL bureau.

Full ARRL membership (available only to licensed radio amateurs) gives you a voice in how the affairs of the organization are governed. League policy is set by a Board of Directors (one from each of 15 Divisions). Each year, half of the ARRL Board of Directors stands for election by the full members they represent. The day-to-day operation of ARRL HQ is managed by an Executive Vice President and a Chief Financial Officer.

No matter what aspect of Amateur Radio attracts you, ARRL membership is relevant and important. There would be no Amateur Radio as we know it today were it not for the ARRL. We would be happy to welcome you as a member! (An Amateur Radio license is not required for Associate Membership.) For more information about ARRL and answers to any questions you may have about Amateur Radio, write or call:

ARRL Educational Activities Dept
225 Main Street
Newington CT 06111-1494
(860) 594-0200

Prospective new amateurs call:
800-32-NEW HAM (800-326-3942)

E-mail: ead@arrl.org
Internet: http://www.arrl.org/

Notes

Notes

Notes

Notes

Notes

MICROWAVE
PROJECTS
MANUAL

PROOF OF
PURCHASE

FEEDBACK

Please use this form to give us your comments on this book and what you'd like to see in future editions, or e-mail us at **pubsfdbk@arrl.org** (publications feedback).

Where did you purchase this book?

☐ From ARRL directly ☐ From an ARRL dealer

Is there a dealer who carries ARRL publications within:

☐ 5 miles ☐ 15 miles ☐ 30 miles of your location? ☐ Not sure.

License class:

☐ Novice ☐ Technician ☐ Technician Plus ☐ General ☐ Advanced ☐ Amateur Extra

---

Name ______ ARRL member? ☐ Yes ☐ No

______ Call Sign ______

Daytime Phone ( ) ______ Age ______

Address ______

City, State/Province, ZIP/Postal Code ______

If licensed, how long? ______

Other hobbies ______

______

Occupation ______

For ARRL use only	MWPROJ
Edition	1 2 3 4 5 6 7 8 9 10 11 12
Printing	2 3 4 5 6 7 8 9 10 11 12

From ______________________

____

____

Please affix postage. Post Office will not deliver without postage.

EDITOR, MICROWAVE PROJECTS MANUAL
AMERICAN RADIO RELAY LEAGUE
225 MAIN STREET
NEWINGTON CT 06111-1494

please fold and tape